# Information Technologies for Remote Monitoring of the Environment

Vladimir F. Krapivin and Anatolij M. Shutko

# Information Technologies for Remote Monitoring of the Environment

Springer

Published in association with
Praxis Publishing
Chichester, UK

PRAXIS

Professor Dr. Vladimir F. Krapivin
Institute of Radioengineering and Electronics
Russian Academy of Sciences
Moscow
Russian Federation

Professor Dr. Anatolij M. Shutko
Institute of Radioengineering and Electronics
Russian Academy of Sciences
Moscow
Russian Federation

SPRINGER–PRAXIS BOOKS IN ENVIRONMENTAL SCIENCES

ISBN 978-3-642-20566-8 ISBN 978-3-642-20567-5 (eBook)
DOI 10.1007/978-3-642-20567-5
Springer Heidelberg New York Dordrecht London

Library of Congress Control Number: 2011944828



Cover design: Jim Wilkie
Project management: OPS Ltd., Gt. Yarmouth, Norfolk, U.K.

Printed on acid-free paper

Springer is part of Springer Science+Business Media (www.springer.com)

# Contents

# Preface

The growing amount of published work dedicated to global ecological studies indicates just how urgent nature conservation has become. The question of finding the principles that underlie the co-evolution of humankind and nature is being posed with ever-increasing persistence. Scientists across the globe are attempting to find ways of formulating the laws that govern the processes that are going on in the environment. Many national and international programs have been set up to study the biosphere and climate in the quest to find a means of resolving the conflict between human society and nature. However, attempts to find efficient ways of regulating human activity globally have come up against major difficulties. Arguably one of the most difficult is the absence of an adequate knowledge base pertaining to climatic and biospheric processes as well as the incompleteness of databases concerning global processes going on in the atmosphere, in the ocean, and on land. Another difficulty is the inability of modern science to formulate the requirements that global databases need to meet for reliable evaluation of the state of the environment and forecasting its development over the long term.

A dramatic aspect of anthropogenic activity is its influence on the biospheric water cycle. This influence is global and is composed of a hierarchy of regional changes, especially in arid areas. This is the reason being able to reconstruct biospheric water systems is such a major element of climate system monitoring.

Many scientists are trying to find answers to these questions. Most suggest that an efficient way of resolving the conflicts between nature and human society would be creation of a unified planetary-scale adaptive geoinformation monitoring system (GIMS). Such a system should be based on knowledge bases and global datasets that are constantly updated. The adaptive nature of such a system should be provided by continuously correcting the data acquisition mode and by varying the parameters and structure of the global model.

The attempts of many scientists to create a global model—the major component of the GIMS—have not been implemented so far as a result of a limited knowledge base. The creation of a model that adequately represents the real processes going on in the environment is clearly not possible: on the one hand, complex models including every last environmental parameter are sure to lead to boundless multidimensionality and, on the other hand, simplified models including restricted numbers of parameters are clearly not fit for purpose. Moreover, creation of an adequate global model cannot be done simply because of the practical impossibility of providing the exhaustive information required. The uncertainty attached to simulation of socioeconomic processes is inevitable. In attempting to surmount these difficulties many researchers have resorted to the methods of game theory, evolution modeling, and describing the behavior of systems as sets of scenarios. Such an approach takes expert opinions into account. However, without a concept of global changes all such attempts are doomed to failure. There needs to be a mechanism for singling out information-carrying elements and determining their interrelation so that global model redundancy is kept to a minimum.

The first steps toward understanding the key problems that need to be resolved in global ecology and climate were made by Kondratyev (1990, 1998a, b, 1999, 2000a, b). The discussions in Krapivin (1993, 1995, 1996, 2000a, b) and Shutko (1986) have made synthesizing a global model of natural processes by considering sets of spatial scales possible.

The present book proposes ideas that could help solve these problems and describes a simulation system based on sets of computer algorithms that process data from global and regional monitoring. A global simulation model that describes spatial interactions in the nature–society system is synthesized. The model comprises blocks describing the biogeochemical cycles of carbon, nitrogen, sulfur, phosphorus, oxygen, and ozone; the global hydrologic balance in liquid, gaseous, and solid phases; the productivity of soil–plant formations and their definition; photosynthesis in ocean ecosystems with depth and surface inhomogeneity taken into account; demographic processes and anthropogenic changes. The model is designed to be connected to a global climate model.

Creating an information interface between the global model, algorithms, and experiments makes it possible to use global sets of vegetation-related parameters, datasets about meteorological fields and hydrologic processes, and other global environmental and anthropogenic data to forecast global change. The information interface supports the relationships between different components of a global model.

The theoretical part of the book has chapters that describe various algorithms and models. The applied part of the book considers specific problems of environmental dynamics. Development of a universal information technology to estimate the state of environmental subsystems under various climatic and anthropogenic conditions is the purpose of having theoretical and applied parts. Combination of geographical information system techniques with modeling technology to estimate the functioning of the nature–society system is the basic idea behind the approach proposed in this book. This idea can be implemented by using new methods to spatiotemporally reconstruct incomplete data.

Algorithms, models, methods, criteria, and software are created with the objective of synthesizing a GIS with modeling functions capable of complex estimation of the state of nature–society subsystems. The newly developed Geoinformation Monitoring System (GIMS = GIS + Model) is focused on systematic observations and evaluation of the environment related to changes attributable to the human impact on environmental subsystems. An important functional aspect of the integrated system is the possibility of early warning of undesirable changes in the environment.

Various applications of GIMS technology are described. There are chapters describing how GIMS technology can be applied to the study of the dynamics of radionuclear pollutants, heavy metals, and oil hydrocarbons in the Angara/Yenisey river system and in the Arctic Basin; to estimation of the Peruvian Current and Okhotsk Sea ecosystems; to estimation of excess $CO_2$ distribution in the biosphere; and to the study of the Aral/Caspian aquageosystem water regime.

The book also deals with microwave radiometric methods. These are the traditional methods of remote sensing of the Earth's surface from aircraft and satellites. Combined use of microwave remote sensing, mathematical modeling of the environment, data processing, and decision-making procedures is proposed. This book aims to focus the attention of the reader on microwave radiometric technology as one of the most powerful technologies in radiophysics for the remote sensing of lakes, seas, oceans, rivers, agricultural land, irrigated land, desert areas, forested areas, wetlands, snow-covered ground, and ice in the wavelength range from 0.5–2 cm to 21–30 cm.

This book brings together the wide spectrum of theoretical and applied techniques used to estimate global change. It presents the theoretical and practical notions of modeling calculations associated with environmental and human systems in a systematic way. Applied mathematicians, hydrologists, geophysicists, ecologists, socioeconomists, and other researchers of global change will find this book useful.

# Figures

# Tables

# Abbreviations and acronyms

| | |
|---|---|
| AHRS | Altitude and Heading Reference System |
| AAMU | Alabama Agricultural and Mechanical University |
| AARS | Asian Association on Remote Sensing |
| ACMR | Airborne C-band Microwave Radiometer |
| ACRS | Asian Conference on Remote Sensing |
| ACS | Aral–Caspian System |
| ACWS | Aral Caspian Water System |
| ADEOS | ADvanced Earth Observation Satellite |
| AIC | Aerial Industrial Camera |
| AIDS | Acquired Immune Deficiency Syndrome |
| AIP | American Institute of Physics |
| AIRS | Atmospheric InfraRed Sounder |
| AMAP | Arctic Monitoring and Assessment Program |
| AMSR2 | Advanced Microwave Scanning Radiometer-2 |
| ANWAP | Arctic Nuclear Waste Assessment Program |
| APS | Atmosphere–Plant–Soil |
| ARCSS | ARCtic System Science program |
| ARCUS | Arctic Research Consortium of the United States |
| AVHRR | Advanced Very High Resolution Radiometer |
| AVSS | Atmosphere–Vegetation–Soil System |
| AYRS | Angara–Yenisey River System |
| AYRSSM | AYRS Simulation Model |
| CALRS | Cluster Analysis focused on the account of Local Reading of Sensors |
| CASRS | Cluster Analysis focused on signal Space of Remote Sensors |
| CBS | Climate–Biosphere–Society |
| CCSS | Carbon–Climate–Society System |
| CIESIN | Center for International Earth Science Information Network |

| | |
|---|---|
| CM | Climate Model |
| CNES | Centre National d'Etudes Spatiales (Space Agency of France) |
| CR | Club of Rome |
| CRC | Chemical Rubber Company |
| CSI | Canopy Structure Index |
| DAS | Data Application System |
| DC | Direct Current |
| DEM | Digital Elevation Model |
| DMS | Decision Making System |
| DMSP | Decision Making Sequential Procedure |
| DMSP-I | DMSP Input |
| DOE | Department Of Energy |
| EDB | Extendable Data Base |
| EMW | ElectroMagnetic Waves |
| ENSO | El Niño–Southern Oscillation |
| EOS | Earth Observing System |
| EOS-PM1 | Earth Observing System–Progress Mission-1 |
| EOSDIS | EOS Data and Information System |
| EP | Evaporation/Precipitation |
| EPA | Environmental Protection Agency |
| ESA | European Space Agency |
| ESAIEP | Expert System for Adaptive Identification of Environmental Parameters |
| ESECEZ | Expert System for Ecological Control of Estuary Zones |
| ESTAR | Electronically Scanned Thin Array Radiometer |
| ETM+ | Enhanced Thematic Mapper Plus |
| EUP | Enterprise Unified Process |
| EUSAR | EUropean Synthetic Aperture Radar |
| EWT | Equivalent Water Thickness |
| FAA | Federal Aviation Administration |
| FAO | Food and Agriculture Organization |
| FASIR | Fourier-Adjusted, Solar zenith angle corrected, Interpolated and Reconstructed data |
| FMC | Fuel Moisture Content |
| FPAR | Fraction of PAR |
| FWHM | Full-Width-Half-Maximum |
| GC | Gas Chromatograph |
| GCC | Global Carbon Cycle |
| GCOM | Global Change Observation Mission |
| GCOS | Global Climate Observation System |
| GCP | Global Carbon Project |
| GCS | Ground Control Station |
| GDP | Gross Domestic Product |
| GEO | Group on Earth Observations |
| GEOSS | Global Earth Observation System of Systems |

| | |
|---|---|
| GHG | GreenHouse Gas |
| GIGAS | GEOS, INSPIRE, and GMES an Action in Support |
| GIMS | Geo-Information Monitoring System |
| GIMSAF | GeoInformation Monitoring System of Agricultural Functions |
| GIS | Geographical Information System |
| GMES | Global Monitoring for Environment and Security |
| GMNSS | Global Model of the NSS |
| GMS | Geostationary Meteorological Satellite |
| GNSS | Global Navigation Satellite System |
| GOOS | Global Ocean Observing System |
| GPS | Global Positioning System |
| GSD | Grand Space Distance |
| GSM | Global Simulation Model |
| GTOS | Global Terrestrial Observation System |
| GWP | Global Warming Potential |
| HFC | HydroFluoroCarbon |
| HSCaRS | Hydrology, Soil Climatology, and Remote Sensing |
| IAHS | International Association of Hydrological Sciences |
| IASI | Infrared Atmospheric Sounder Interferometer |
| ICID | International Commission on Irrigation and Drainage |
| ICLIPS | Integrated assessment of CLImate Protection Strategies |
| ICP | Inductively Coupled argon Plasma |
| IEEE | Institute of Electrical and Electronics Engineers |
| IFSAR | InterFerometric Synthetic Aperture Radar |
| IGARSS | International Geoscience And Remote Sensing Symposium |
| IGBP | International Geosphere–Biosphere Program |
| IHDP | International Human Dimensions Program |
| IIASA | International Institute for Applied Systems Analysis |
| IIC | Information Interface with Computer |
| IMAGE | Integrated Model to Assess the Greenhouse Effect |
| IMARC | Intelligent Multi-frequency Airborne polarimetric Radar Complex |
| INEL | Idaho National Engineering Laboratory |
| INSPIRE | INfrastructure for SPatial InfoRmation in the European community |
| IPCC | Intergovernmental Panel on Climate Change |
| IREE | Institute of Radio-Engineering and Electronics |
| ISGGM | Integrated System for Global Geoinformation Monitoring |
| ISTC | International Science and Technology Center |
| ITHEA | Institute of information THEories and Applications |
| JERS | Japanese Earth Resources Satellite |
| KIT | Karlsruhe Institut für Technologie |
| LAI | Leaf Area Index |
| LOSAC | L-band Ocean Salinity Airborne Campaign |
| LPJ-DGVM | Lund–Potsdam–Jena Dynamic Global Vegetation Model |

| | |
|---|---|
| LUT | LookUp Table |
| LW | LongWave |
| MACS | Microwave Autonomous Copter System |
| MBB | Marine Biota Block |
| MCBS | Magnetosphere–Climate–Biosphere–Society |
| MCRM | Markov Chain Reflectance Model |
| MDA | Model-Driven Architecture |
| MEAP | Millennium Ecosystems Assessment Program |
| MEM | Microwave Emission Model |
| METOP | METeorological Operational Polar |
| MFIMS | Multi-Functional Information-Modeling System |
| MGBC | Model of the Global Biogeochemical Cycle |
| MGOC | Model of the Global Oxygen Cycle |
| MGSC | Model of Global Sulfur Cycle |
| MISR | Multi-angle Imaging SpectroRadiometer |
| MODIS | MODerate resolution Imaging Spectroradiometer |
| MOS | Marine Observation Satellite |
| MR | Microwave Radiometer |
| MRSD | Microwave Remote Sensing Division |
| MSP | Multi-channel SpectroPolarimeter |
| MUE | Model of the Upwelling Ecosystem |
| NASA | National Aeronautical and Space Administration |
| NASDA | NAtional Space Development Agency (Japan) |
| NCAR | National Center for Atmospheric Research |
| NDVI | Normalized Differential Vegetation Index |
| NHC | National Hurricane Center |
| NIST | National Institute of Standards and Technology |
| NN | Neural Network |
| NOAA | National Oceanic and Atmospheric Administration |
| NRI | Neural-Robotics, Inc. |
| NSS | Nature–Society System |
| OAS | Ocean–Atmosphere System |
| OECD | Organization for Economic Co-operation and Development |
| OSE | Okhotsk Sea Environment |
| PALS | Passive and Active L and S band system |
| PAR | Photosynthetically Active Radiation |
| PCE | Peruvian Current ecosystem |
| PIRATA | PIlot Research moored Array in the Tropical Atlantic |
| POLDER | POLarization and Directionality of Earth's Reflectances |
| PSM | Pollution Simulation Model |
| PSR/C | Polarimetric Scanning Radiometer/C-band |
| PWI | Plant Water Indexes |
| RAMA | Research moored Array for Monsoon Analysis |
| RAS | Russian Academy of Sciences |
| RCA | Russian Custom Academy |

| | |
|---|---|
| RFI | Radio Frequency Interference |
| RT | Radiative Transfer |
| RWC | Relative Water Content |
| SAR | Synthetic Aperture Radar |
| SCAN | Soil Climate Analysis Network |
| SCIAMACHY | SCanning Imaging Absorption SpectroMeter for Atmospheric CartograpHY |
| SCOPE | Scientific Committee On Problems of the Environment |
| SGLI | Second-generation GLobal Imager |
| SGM | Spatial Global Model |
| SGPE | Southern Great Plains Experiment |
| SHF | Super High Frequency |
| SIBA | Simple Index for Burned Areas |
| SIL | Scattering Index over Land |
| SMEX | Soil Moisture EXperiments |
| SMMR | Scanning Multichannel Microwave Radiometer |
| SMOS | Soil Moisture and Ocean Salinity |
| SMOSE | Simulation Model of the Okhotsk Sea Environment |
| SPF | Soil–Plant Formation |
| SPOT | Système Probatoire d'Observation de la Terre |
| SRVI | Simple Ratio Vegetation Index |
| SSAPP | Simulation System for Atmosphere Pollution Physics |
| SSM/I | Spatial Sensor Microwave/Imager |
| SSMAE | Spatial Simulation Model of the Arctic Ecosystem |
| SST | Sea Surface Temperature |
| SSURGO | Soil SURvey GeOgraphic |
| STARRS | Salinity, Temperature, And Roughness Remote Scanner |
| STW | Computer SofTWare |
| SVI | Spectral Vegetation Index |
| SR | Simple Ratio |
| TAO | Tropical Atmosphere Ocean |
| TIM | Theory–Information Model |
| TIR | Thermal Infrared Radiometer |
| TIROS-N | Television InfraRed Observational Satellite N |
| TM | Thematic Mapper |
| TOA | Top-Of-Atmosphere |
| TORRO | TORnado and storm Research Organization |
| TRITON | TRIangle Trans-Ocean buoy Network |
| TRMM | Tropical Rainfall Measuring Mission |
| TSFP | Turbulence and Shear Flow Phenomena |
| TTP | Technology Transfer and Promotion |
| UAV | Unmanned Aerial Vehicle |
| UML | Unified Modeling Language |
| UNEP | United Nations Environment Program |
| USDA | United States Department of Agriculture |

| | |
|---|---|
| USEPA | United States Environmental Protection Agency |
| USSR | Union of Soviet Socialist Republics |
| VA | Volt-Amper |
| VDC | Volts Direct Current |
| VHF | Very High Frequency |
| VI | Vegetation Indexes |
| VWC | Vegetation Water Content |
| WAAS | Wide Area Augmentation System |
| WCRP | World Climate Research Program |
| WISE | Web-based Inquiry Science Environment |
| WIT | Wessex Institute Transactions |
| WMO | World Meteorological Organization |
| WTARS | Winfred Thomas Agricultural Research Station |
| XRF | X-Ray Fluorescence |

# About the authors

**Vladimir F. Krapivin** was educated at the Moscow State University as mathematician. His Ph.D. in radiophysics he received from Kotelnikov Institute of Radioengineering and Electronics of Russian Academy of Sciences (KIRE RAS) in 1966. The Doctoral Dissertation in geophysics was defended him at the Moscow Institute of Oceanology of RAS in 1973. He became Professor of Radiophysics in 1987 and Head of the Applied Mathematics Department at the KIRE RAS in 1972. Vladimir Krapivin is a full member of the Russian Academy of Natural Sciences. He has specialized in investigating global environmental change by the application of modeling technology and remote sensing. During last years, Vladimir Krapivin deals with study of ocean-atmosphere system instability in connection with the tropical cyclones. He has published more 350 papers and 23 books in the fields of ecoinformatics, theory of games, remote sensing and global modeling.

**Anatolij M. Shutko** was educated as an engineer in radio communication at the Popov Electrotechnical Institute of Communication. He received his PhD in radiophysics from the Kotelnikov Institute of Radioengineering and Electronics of the Russian Academy of Sciences (KIRE RAS) in 1967. His Doctoral Dissertation in radiophysics was defended by him at KIRE RAS in 1987. He became Professor of Radiophysics in 1997 and Head of Scientific Laboratory "Multi-channel Investigation of the Environment" at KIRE RAS in 1990. Anatolij was awarded the State Prize of the U.S.S.R. in 1983 and received an award from Pope John Paul II in 1986. He is a full member of the Prokhorov Academy of Engineering Sciences and the U.S. Academy for Electromagnetism. Anatolij is the Chairman of the Russian Chapter of IEEE Geoscience and Remote Sensing Society. His scientific interests lie in microwave radiometry of water objects, soils, and vegetation. He has published more than 200 papers and 5 books in the fields of ecoinformatics and remote sensing.

# Summary

The growth of needs in practically all spheres of remote environmental monitoring has resulted in numerous problems arising from the many unsolved tasks including the assessment of important characteristics of the soil-plant formations and oceans. It has become a priority to develop new concepts and approaches for assessing and dealing with natural and man-made system dynamics. A major priority is global problems connected with creation of effective information technologies to data processing within the environment studies.

The problems of microwave remote sensing technology and also substantiation of environmental monitoring systems receiving, storing, and process the necessary information for the solution of relevant problems has been analyzed. Main objective of the book lies in the working information technology for the combined use of modeling technology and microwave remote sensing measurements in the assessment of environmental system states as well as to illustrate this technology with computer calculations for various environmental problems. Different tasks related to the assessment and prognosis of natural systems dynamics basing on the microwave remote sensing measurements using mobile platforms are solved. New information technology is proposed to be used for the optimization of remote sensing monitoring systems. This technology is based on sets of computer algorithms for comprehensive analysis of data from global and regional monitoring systems. Chapters in the theoretical part of the book contain description of rigorous algorithms and environmental models. The applied part considers specific problems of environmental dynamics in areas of different countries where on-site experiments were realized to test new information technology.

This book is the result of a detailed study of the environment by means of computer algorithms and simulation models. It describes the sets of algorithms and models for comprehensive analysis of data from global and regional monitoring systems. Chapters in the theoretical part of the book contain descriptions of rigorous algorithms and environmental models. The applied part considers specific problems of environmental dynamics and the assessment of environmental parameters.

Main purpose of this book is to develop a universal information technology to estimate the characteristics of environmental subsystems both global and regional scales. Applied mathematicians, hydrologsts, geophysicists, radiophysicists and other researchers of environment will find a wealth of information in this book.

# 1

# New information technology for environmental monitoring

## 1.1 PRINCIPAL CONCEPT OF GEOINFORMATION-MONITORING TECHNOLOGY

Solving the majority of applied problems of ecodynamics as it stands today is difficult because effective methods of control of soil–plant formations (SPFs) and ocean ecosystems are insufficiently developed. The need for the creation of new effective information technology to process and interpret remote data is dictated by different areas of human activity. Many global environmental problems such as the greenhouse effect or natural disasters have major restrictions when attempts are made to forecast them. International scientific programs focus their investigations on understanding the processes that affect the physical, biological, and chemical environment for life. Unfortunately, the objective of many of these programs has not been realized. It is evident that new paradigms for studying the Earth and its environs have not been developed. However, an effective information technology that has been developed in recent years is the Geographic Information System (GIS). It engenders massive interest as it is a geographically oriented computer technology with a commercial orientation. Nevertheless, GIS suffers numerous restrictions in attempts to predict environmental dynamics. The difficulties arising here are connected with the complexity of the Earth surface and the absence of detailed data that reflect environmental dynamics. This is corroborated, for example, by the problems that were experienced during the Global Carbon Project (GCP).

Land surface properties and processes play an important role in the formation of global ecodynamics including climate change. The land surface is characterized by many parameters such as the type of soil–plant formation, leaf area index (LAI), roughness length, and albedo. These and other parameters determine the processes taking place in the atmosphere–land system: evaporation, precipitation, and photosynthesis. The Global Climate Observation System (GCOS) and Global

V.F. Krapivin and A.M. Shutko, *Information Technologies for Remote Monitoring of the Environment*, Springer Praxis Books, DOI 10.1007/978-3-642-20567-5_1, © Springer-Verlag Berlin Heidelberg 2012

Terrestrial Observation System (GTOS) supply the LAI to an accuracy of ±0.2 to 1.0 over large areas. For example, MODIS gives a 1 km global data product updated once each 8-day period throughout the year. The Multi-angle Imaging Spectroradiometer (MISR) supplies the LAI at a spatial resolution of 1.1 km every 8 days. There exist several methods to estimate the LAI (Fang and Liang, 2003):

- using the empirical relationship between the LAI and vegetation indexes (VIs);
- through inversion of a radiative transfer (RT) model;
- lookup table (LUT) method;
- neural network (NN) algorithms.

Each of these approaches has specific parameters as input information for the method. The VI approach is based on the correlation between the LAI and VI. The RT model inversion method describes the physical processes of radiative transfer in the soil–vegetation system. The LAI can be determined remotely relatively cheaply and easily using imagery from various satellites. The most commonly used satellites for determining the LAI are SPOT, NOAA AVHRR, and Landsat TM, all with differing spatial and spectral resolutions. Multi-channel image data (such as TM) and empirical relationships (such as NDVI–LAI or SR–LAI) are used to estimate the LAI, while multi-angular remote-sensing data provide more information for canopy structure. Methods employing multi-angular data and model inversion are preferentially used to estimate the LAI.

Much information about environmental subsystems is provided by the Earth Observing System (EOS). EOS technologies provide the global perspective needed for integrated, long-term, scientific, integration of information about our home planet. Changes in the Earth system are inevitable. Numerous environmental problems connected with global climate change are the subject of ongoing debate between scientists. Unfortunately, the extent of future global change remains poorly assessed. Kondratyev *et al.* (1990, 2004a) proposed a methodology for the solution of this problem based on EOS data.

The numerous instruments and platform hardware of EOS really do supply the requisite information about all the subsystems of the Earth system:

- clouds, radiation, water vapor, precipitation;
- oceans (circulation, productivity, air–sea exchange, temperature);
- greenhouse gases (GHGs) and tropospheric chemistry;
- land surface (ecosystems and hydrology);
- ice sheets, polar and alpine glaciers, and seasonal snow;
- ozone and stratospheric chemistry;
- volcanoes, dust storms, and climate change.

The EOS Program includes scientific and technical support of environmental investigations (Asrar and Dozier, 1994; Tianhong *et al.*, 2003). EOS missions and the EOS Data and Information System (EOSDIS) provide data and the infrastructure to facilitate interdisciplinary research about the Earth system. EOSDIS, which is NASA's Earth science data system, facilitates the collection of Earth science data, command and control, scheduling, data processing, and data archiving and distribution services for EOS missions. EOSDIS science operations are carried out by means of a widespread system of many interconnected centers with specific responsibilities for production, archiving, and distribution of Earth science data products. These data centers supply science data and data products to many science data users. EOSDIS is managed by the Earth Science Data and Information System (ESDIS), which is part of the Earth Science Projects Division under the Flight Projects Directorate at Goddard Space Flight Center. EOSDIS is responsible for

- processing, archiving, and distributing Earth science satellite data (e.g., land, ocean, and atmosphere data products);
- preparation of tools to facilitate the processing, archiving, and distribution of Earth science data;
- collecting metrics and user satisfaction data to continue improving services provided to users;
- ensuring scientists and the public have access to data to facilitate the study of Earth from space to advance scientific understanding and meet societal needs.

Unfortunately, there is no effective technology that can adapt this infrastructure to basic environmental problems. However, by coupling GIS and a modeling technique such a technology becomes available. GIS provides efficient analytical tools for generation of prospective maps by combining a number of processes in a knowledge-driven approach including Boolean logic combination, algebraic combination, index overlay combination, fuzzy logic and vector fuzzy logic combinations, and so on (Givant and Halmos, 2009). There are a number of model techniques that synthesize environmental problems including standard modeling languages such as the Unified Modeling Language (UML), the Model-Driven Architecture (MDA), and the Enterprise Unified Process (EUP). However, these languages do not cover all existing models of environmental systems and processes. Many models are created based on others approaches (Cracknell *et al.*, 2009). Combination of such models with GIS can be achieved by means of GIMS technology (Armand *et al.*, 1987; Krapivin and Potapov, 2002; Krapivin and Shutko, 1989; Krapivin *et al.*, 2006).

In the last few years, the global carbon cycle problem has acquired special significance because of the greenhouse effect. Knowledge of the state of SPFs and ocean ecosystems allows one to get a real picture of the spatial distribution of the carbon sinks and sources on the Earth's surface. As is well known, among the types of remote-sensing techniques, microwave radiometry proves particularly effective for the observation of SPF environmental parameters. However, these observations are a function of different environmental conditions that mainly depend on the type of SPF. That is why it is necessary to develop data-processing methods for microwave monitoring that allow the reconstruction of SPF characteristics by considerating vegetation types and that provide the possibility of synthesizing their spatial distribution.

As noted by Chukhlantsev and Shutko (1988), the problem of microwave remote sensing of the vegetation cover requires study of the attenuation of electromagnetic waves (EMWs) within the vegetation layer. The problems arising here can be solved by combining experimental and theoretical studies. The vegetation cover is commonly characterized by its varied geometry and additional parameters. Therefore, knowledge of the radiative characteristics of SPFs as functions of temporal and spatial coordinates can be acquired by means of a combination of on-site measurements and models. The general aspects of such an approach have been considered by many authors (Del Frate *et al.*, 2003; DeWitt and Nutter, 1988; Dong *et al.*, 2003; Friedi *et al.*, 2002). But these investigations were mainly restricted to investigating models describing the dependence of the vegetation medium on environmental properties as well as the correlation between the morphological and biometrical properties of the vegetation and its radiative characteristics.

A prospective approach to solving these problems is provided by GIMS technology (GIMS = GIS + Model). This approach was proposed by Krapivin and Shutko (1989, 2002). A combination of an environmental acquisition system, a model of the functioning of the typical geo-ecosystem, a computer cartography system, and a means of artificial intelligence will result in the creation of a geo-information monitoring system of typical natural elements that is capable of solving the many problems arising in the remote monitoring of the global vegetation cover. The GIMS-based approach, in the framework of EMW attenuation by the vegetation canopies, allows synthesizing a knowledge base that establishes the relationships between experiments, algorithms, and models. The links between these areas have an adaptive character giving an optimal strategy for experimental design and model structure. The goal of this chapter is to explain and assess the application of the GIMS method to the tasks of reconstructing the spatial and temporal distribution of SPF radiative characteristics.

## 1.2 THE GIMS AS A KEY INSTRUMENT FOR THE SOLUTION OF ENVIRONMENTAL PROBLEMS

### 1.2.1 Principal aspects of GIMS technology

The accumulation of knowledge, major scientific and engineering progress, and unprecedented growth of human influence on the environment have all played their part in bringing about the problem of global evaluation of the state and possible long-term forecast of the environment since the 1970s. Scientific research in this field has led to scientists throughout the world reaching the conclusion that solution of the problem of environmental quality control is only possible by creating a unified international monitoring system based on a global magnetosphere–climate–biosphere–society (MCBS) system. Many international and national programs on the environment are dedicated to realization of this system. It is within the framework of these programs that well-stocked databases of environmental parameters are created, information dynamics about natural and anthropogenic processes on various scales are accumulated, and model sets of biogeochemical, biogeocenotic, climatic, and demographic processes are prepared. The technical base of global geoinformation monitoring is well stocked by means of efficient data acquisition methods that include recording, accumulation, and processing of measurement data obtained from on board spacecraft, ground-based laboratories, and floating laboratories.

However, despite significant progress in many fields of monitoring nature the main problems remain unresolved: designing the optimum combination between all technical means, the creation of an efficient and economical monitoring structure, and the creation of ways of reliably forecasting and evaluating environmental dynamics as a result of anthropogenic activity. The experience of recent years shows that it is possible to create a global model capable of being used adaptively to make recommendations on the monitoring structure and requirements needed to build databases. A solution to this model creation problem has not been found and has led to unjustified expense to conduct new expeditions and build new observation systems.

In the last decade or so, many investigators (Kondratyev *et al.*, 2002; Krapivin and Shutko, 1989; Sellers *et al.*, 1997) have posed the problem of synthesizing a complex system to collect environmental information by coupling GIS, remote, and contact measurements with models. Such systems are called Geo-Information Monitoring Systems (GIMS) and they are aimed at systematic observation and evaluation of the environment and its changes under the effects of anthropogenic economic activity. An important aspect of the functioning of these systems is the possibility of forecasting the state of the environment and warning about signs of undesirable changes. Realization of this monitoring function is possible by applying mathematical modeling, of methods that simulate the functioning of natural complexes (Armand *et al.*, 1987; Krapivin and Varotsos, 2008; Krapivin *et al.*, 1982, 1997a; Phillips *et al.*, 1997).

The development of models of biogeochemical, biocenotic, demographic, socioeconomic, and other biospheric and climatic processes on the whole brings about the need for model requirements to meet the GIMS structure and its database. According to the proposed GIMS structure in Kelley *et al.* (1999b), simulation of the dynamics of the biosphere is an important function of the GIMS. Consequently, there is a need for a new approach to estimating the state of the biosphere. After all, the basic aim of all investigations into GIMS technology development can be broken down to the following tasks:

- finding the quickest and best way of reconstructing environmental survey systems;
- creating the conditions for optimum planning of the organizational structure of human society;

- ensuring global processes follow a purposeful direction so that they are good for humankind and do not cause damage to nature.

As was shown by the above investigators, there are balanced criteria of information selection that cover the hierarchy of causal–investigatory constraints in the biosphere. They include coordination of tolerances, depth of spatial quantization in the course of description of the atmosphere, land, and oceans, the degree to which biomes are detailed, etc. At an empirical level, expressed as expert evaluations as a result of computing experiments, these criteria make it possible to select the informational structure of the geo-informational monitoring system such that it indicates the hierarchical subordination of models at various levels.

At the same time, creation of an effective global system for control of the environmental state comes up against the requirements imposed by the regional socioeconomical structure of society. This is expressed by non-uniform development of industrial businesses whose production is ecologically suspect, by the degree to which they are concentrated by region, by differences in regional services for the control and collection of information, by the technical equipment they use, etc. Such differences inevitably have an influence on the choice of GIMS structure and its informational technical base.

Thus, the hierarchical structure of the combination of mathematical models that are used in GIMS is determined by natural, climatic, and socioeconomical factors, as well as by technical potentialities. The degree of detail of models depends on the level of their entry into the common structure and especially on the spatiotemporal characteristics simulated at the given level of natural processes.

Global ecoinformatics suggests developing banks of models for various processes in the biosphere on the basis of their spatial non-homogeneity and combining them with existing global databases comprising functioning systems for environmental observation. It is further suggested to cooperate with specialists who have developed climatic, biospheric, and socioeconomic models with the aim of creating a global model of the climate–biosphere–society (CBS) system. As a subsequent improvement of the model we may study the interaction of the CBS system with processes in near-Earth space (mainly the magnetosphere) and extend it to creation of the MCBS model. As a result a system will be created that is capable of forecasting the development of natural processes and evaluating long-term consequences of large-scale action on the environment. Application of this system will encompass the problems facing environmental protection on global, continental, regional, and local scales by means of expert examination of changes in the structure of topsoil, hydrological regimes, and atmospheric air composition.

Realization of the MCBS model would permit all international and national means of environment monitoring to be integrated into a complex structure and would provide a tool to evaluate objectively the quality of the environment under all states. By filling the system with up-to-date efficient processing techniques of data monitoring, the wide spectrum of problems of identification of pollution sources can be solved, thus eliminating conflicts because of trans-border flows of contaminated material.

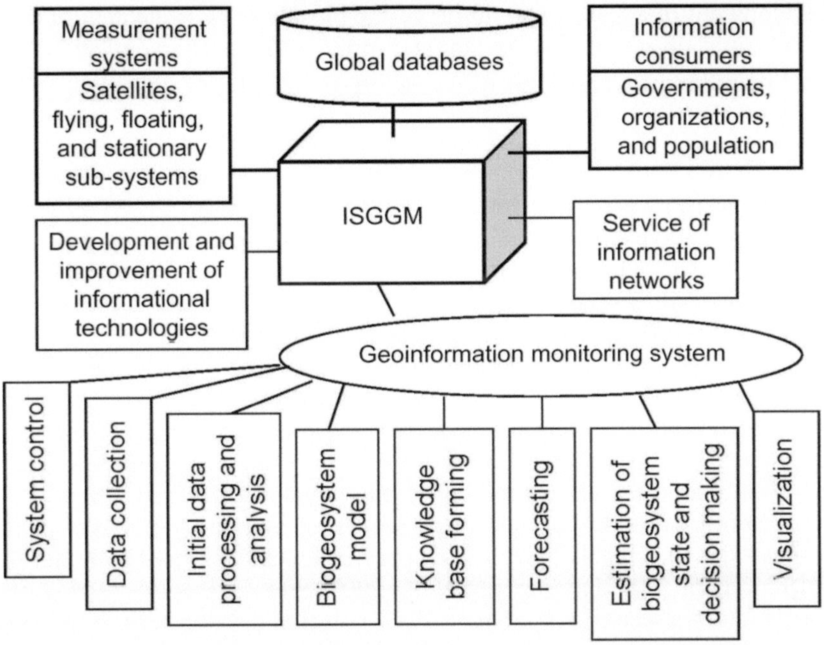


**Figure 1.1.** Integrated System for Global Geoinformation Monitoring (ISGGM) and its structure.

New information technologies in global modeling within the framework of the Integrated System for Global Geoinformation Monitoring (ISGGM) (Figure 1.1) will be used to create a principally new structure of monitoring that depends on data of various quality and many mathematical and physical models of various types. Evolution technology will solve the many contradictions that arise from non-completeness and undetermined information bases, fragmentary knowledge about the laws of nature, the absence and underdevelopment of instrumental systems in the field of experimental simulation (Bukatova, 1992; Hushon, 1990; Krapivin *et al.*, 1997a, b).

The ISGGM functional structure is based on the idea of bringing about evolutionary neurocomputer technology. Such a realization will take on the architecture of a modern computer system with access to a network of evolution-type tutorial servers. This will lead to an ISGGM structure that will simulate the MCBS system with a lookahead capability of evaluating nature protection measures and other structural decisions in how human society can best interact with nature.

The practical embodiment of MCBS model creation requires carrying out purposeful complex studies into the following:

- Systematization of global change and understanding of biospheric processes and the structure of biospheric levels.

- Development of a conceptual model of the biosphere as an element of the global geoinformation-monitoring system (Krapivin and Kondratyev, 2002).
- Inventory and analysis of existing ecological databases and choice of global database structure (Bréon, 2004).
- Classification of biospheric space–time characteristics and cause–effect connections with the aim at working out a space–time scale of coordination of ecological processes (Rastetter *et al.*, 2003).
- Creation of typical model databases of ecological systems comprising biogeochemical, biogeocenotical, hydrological, and climatic processes (Cracknell *et al.*, 2009).
- Study of the processes involved in the interaction between the biosphere and climate. Discovering the regularity with which the Sun influences biospheric systems (Kondratyev, 1990; Krapivin and Varotsos, 2007).
- Systematization of information about oceanic ecosystems. Description of geophysical and trophic structures and how they can be broken down regionally and coordinated on space–time scales (Shapovalov, 2010).
- Construction of models of biocenographic fields and development of algorithms to synthesize these fields based on the oceanic block of the global model (Burkov and Krapivin, 2009; Kondratyev *et al.*, 2003b, c).
- Formation of a scenario database for co-evolution development of the biosphere and human society. Creation of demographic models. Parameterization of scientific and technical processes in the utilization of land resources (Norris and McCulloch, 2003; Pollak, 1990).
- Search for new information technologies regarding global modeling such that requirements on databases and knowledge bases can be reduced. Development of architectural and algorithmic principles such that computer systems can function much like neurons, process evolutionary information at high speed and efficiency, and integrate these systems into the ISGGM structure (Bukatova and Makrusev, 2003).
- Development of an ecological monitoring concept and creation of a theoretical ecoinformatics database. Development of methods and criteria to evaluate the stability of global natural processes. Analysis of the stability of biospheric and climatic structures (Savinikh *et al.*, 2007).
- Synthesis of the MCBS model and development of a computer means of carrying out experiments to evaluate the consequences of various anthropogenic activity scenarios (IPCC, 2005; Singer, 2008).

Analysis of investigations in recent years in these fields shows that for global modeling to be successful it is necessary to develop new methods of system analysis of complicated natural processes and new data-processing methods directed at the synthesizing balanced criteria of information selection and considering the hierarchy of cause–effect connections in the MCBS system. There exists a series of global models that can be used as input to the MCBS. For example, the multi-disciplinary IMAGE model simulates the dynamics of the global society–biosphere–climate system and makes it possible to investigate linkages and feedbacks in this system. The model consists of three fully linked subsystems: energy–industry, terrestrial environment, and ocean–atmosphere (Alcamo *et al.*, 1994). Krapivin and Kelley (2009) proposed a multi-functional model of global change in the nature–society system based on GIMS technology. Degermendzhy *et al.* (2009) developed a principally new approach to biosphere dynamics modeling by considering the biosphere–climate system as a hierarchy of interacting subsystems each of which can be described by a small-scale model.

The accumulated knowledge and experience of global monitoring makes it possible to break down the main blocks of the MCBS model: magnetosphere (Korgenevsky *et al.*, 1989), climate (Randall *et al.*, 2007), biogeochemical cycles (Shultze *et al.*, 2001), biogeocenotic processes (Cocknell *et al.*, 2006; Wang *et al.*, 2011), socioeconomic structure (George, 2002; Garsey and McGlade, 2006), scientific and engineering progress (Brucan, 1984). Development of methods to parameterize these blocks has reached the level where synthesis of the MCBS model is possible based on the principle of system coordination of individual block inputs and outputs. This depends on the principal tasks of coordinating the space–time scales of natural processes and choice of algorithms for MCBS model connection with databases being resolved. As a result of national and state borders two upper spatial levels in MCBS can be differentiated: global and continental. The national and state level encompasses three space scales: national, regional, and local. There are also intermediate levels.

Creation of the MCBS model will require systematization of models and databases at national and state levels and their connection to global models and databases. A consequence of such a systematization will be the creation of typical models of natural processes in the form of MCBS model base elements that bring together national, regional, and local levels. Bringing about a hierarchical structure for the mathematical model complex as part of the MCBS model is additional to the requirement for modeling algorithms to be used. As a result of space–time fragmentary, noisy, and incomplete data measurements of environment parameters, new modeling methods need to be found to facilitate the parameterization processes of phenomena in models under these conditions. One of these new approaches is the simulation–evolution modeling technology developed by Bukatova *et al.* (1991). This technology allows isolating in the MCBS model (besides traditional natural phenomenon models) the new types of models that describe basic parameterization processes. Hence, blocks may well appear in the MCBS model describing such processes as scientific and engineering progress, agricultural production, extraction and expenditure of mineral resources, demography, etc. (Krapivin and Varotsos, 2007, 2008).

For GIMS technology to work there need to be many algorithms and models to parameterize and process observational data. Hence, the need for systematic description of key environmental processes and subsystems. At present there exist global datasets whose use within GIMS technology is connected with their classification and systematization. However, any balance between environmental datasets of different origins is lacking and, therefore, one of the actual problems is the coordination of observations that come from airborne, satellite, and on-site measurements. Moreover, it is evident that satellite monitoring has gaps concerning the selection of sensors; optical and infrared bands prevail.

Many natural phenomena such as the hydrologic cycle, energy balance, and atmospheric $CO_2$ concentration are related to forest ecosystems. Of the many ways to measure forest characteristics, microwave remote sensing is a method that is independent of both weather conditions and the time of day measurements can be acquired. However, the use of microwave sensors requires additional investigations (see Chapter 7).

Large-scale tests of GIMS technology were carried out between 1990 and 2009 in the following countries where international testing areas were set up (Shutko *et al.*, 2010):

1. *Bulgaria*. Irrigation problems have been solved by means of GIMS technology. A multispectral remote-sensing and modeling product was created in Bulgaria during 2009 within the framework of the "Development in Bulgaria of Geo-Information Monitoring System (GIMS)" project. The final GIMS product for Bulgaria took into account the following parameters and conditions:
   - surface and subsurface soil moisture;
   - depth to the shallow water table (down to 3–5 m in dry areas);
   - biomass of vegetation above water surfaces or wet soil;
   - contours of water seepage through levees, walls, and other hydroengineering constructions;
   - contours of flooding;
   - contours of areas where drainage has failed;
   - contours of forest fire;
   - increase in temperature in land, forested, and volcano areas;
   - changes in salinity/mineralization and temperature of water surfaces;
   - rainfall;
   - melting–freezing conditions;
   - contours of water pollution in outflow zones, river deltas, lakes, harbors;
   - on-ground snow melting;
   - ice on water surfaces, roads, and runways;
   - risk assessment and emergency monitoring of situations based on the above phenomena.

2. *The Netherlands*. GIMS technology has been used to assess water leakage through dams and to understand what instruments are needed to remedy this. A flying laboratory was set up on the Twin-Commander airplane of the Dutch company Miramap. The laboratory was equipped with a number of sensors, principal among which was the Radius radiometric system supplied by the Russian company Vega.

3. *U.S.A*. Remote-sensing platforms have been set up to study soil moisture in agriculture. The following scientific and applied tasks were undertaken:
   - study of super high frequency (SHF) radiation spectra under conditions of subsurface moistening;

- study of SHF radiation spectra under conditions of reservoir obliteration;
- simultaneous determination of soil moisture, depth of underground waters, and vegetation biomass in waterlogged regions;
- study of statistical and dynamic radiation–moisture dependencies;
- application of SHF radiometry methods in geoecology and epidemiology;
- synthesizing a rover-type land-based radiometric platform equipped with radiometers with ranges of 6, 18, and 21 cm;
- instrumentation of pilotless radiometric measuring complex on board a MACS helicopter with a radiometer range of 6 cm;
- determination of above-water vegetation biomass;
- operational reception of maps showing soil moisture, underground water, and reservoir obliteration;
- detection of water leakages from irrigation systems.

4. *Vietnam*. A number of water quality and air pollution tasks have been completed using GIMS technology instruments. The main task was setting up a satellite to monitor paddy fields and rice production. Preliminary tasks consist in modeling how microwave frequencies disperse in rice vegetation and studying the radiation of rice vegetation in the microwave range.
5. *Middle Asia*. The hydrology of the Kara-Kum Channel, water leakage, waterlogging, transport of Siberian rivers to Middle Asia, and shallowing of the Aral Sea have all been studied using GIMS technology methods.
6. *Ukraine*. The following problems have been investigated:
   - ascertaining pollution levels in the Black Sea and Sea of Azov (by means of microwave radiometric determination of polluted areas in Odessa Harbor and the possibility of an aircraft platform equipped with microwave radiometers to measure water surface salinity);
   - leakage of water from beneath banks;
   - exposure of areas to disturbed drainage.

As a result of these long-term investigations, several systems have been set up and tested (Shutko *et al*., 2010). Some results are given below.

### 1.2.2 The Alabama Mesonet Soil Moisture Field Experiment

Between November 9 and November 22, 2005 an experiment on passive microwave sensing of soil moisture was carried out by a team of U.S. and Russian scientists at the Winfred Thomas Agricultural Research Station at the Alabama A&M University in cooperation with the NASA Center for Hydrology, Soil Climatology, & Remote Sensing (Archer *et al*., 2006; Shutko *et al*., 2010). The 2-week campaign took place in an area about 1 km$^2$ in size. During this period of time the weather changed from very dry to very wet and rainy. These weather conditions permitted changes in brightness temperature $T_b$ as a result of changes in soil moisture to be seen. Microwave radiometric measurements were obtained both from a manned rover-type mobile platform at a height of 2 m providing data with a ground resolution of 1.4 m and from a fully autonomous unmanned aerial vehicle (UAV) helicopter platform at an altitude of 30 m providing data with a ground resolution of 20 m, for the purpose of daily mapping of soil moisture at field scale. Results have confirmed the sensitivity of microwave radiation to soil moisture. Brightness temperature data obtained with the 21 cm sensor were well correlated ($R^2 = 0.93$) with the soil moisture values of a rather deep soil layer. As expected, 6 cm brightness temperature measurements were sensitive to surface soil moisture values. An algorithm for the retrieval of soil moisture, based on the measurement $T_b$ at both 6 cm and 21 cm, has given satisfactory results. A two-dimensional (2-D) soil moisture map of the study area was obtained from 6 cm and 21 cm measurements using both platforms.

This experiment marked a new era in Earth remote sensing of soil moisture using two low-cost platform systems, both of which were successfully demonstrated. To our knowledge, this was the first time a microwave radiometer was flown on board an unmanned helicopter without problems from radio frequency interference (RFI), vibrations, or gusts of wind. The conclusion reached was that microwave radiometry is a reliable remote-sensing technology for soil moisture determination and, thus, can serve as a tool for non-stop monitoring of the dryness index and reliably ascertaining the risk of water shortages during the whole vegetation cycle. A large part of this progress can be attributed to the annual large-scale Soil Moisture Experiments (SMEX) conducted mainly by the National Aeronautics and Space Administration (NASA) and the United States Department of Agriculture (USDA) Hydrology Laboratory.

Figures 1.2–1.4 show maps of brightness temperature data and retrieved soil moisture data obtained at different wavelengths over the test area. These maps have been reconstructed by means of the data-processing algorithms that are intrinsic to GIMS technology and that enable formation of the spatial distribution of soil moisture from trace measurements. The precision of these data is determined by areas of spatial resolution (i.e., 1.96 $m^2$ in this case). Figure 1.5 shows the correlation between data received *in situ* and remote measurements. Clearly, a model of the vertical distribution of soil moisture is necessary to improve the precision of monitoring data.

Figures 1.6 and 1.7 show the results of soil moisture measurements made by the unmanned (MACS) helicopter platform equipped with a 6 cm radiometer. The precision of these data is defined by the spatial resolution pixel size that depends on the flight altitude. In this case, the flight altitude was 30 m and the pixel area was 400 $m^2$. These microwave experiments were controlled by the Alabama Web of Ground Based Stations (Alabama Mesonet). Each of the stations provided measurements of soil moisture and soil temperature (hotness) below ground at five depths within the range from 5 to 102 cm. Also, such parameters as wind speed, air temperature, rain, and solar radiation were measured. Hence, the radiometers were calibrated to take account of the different environmental targets (Figure 1.8).

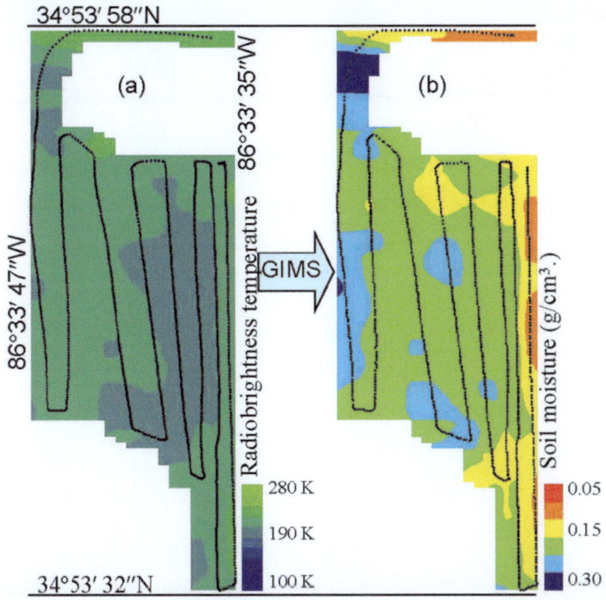


**Figure 1.2.** GPS-positioned rover traces overlaid on maps of (a) brightness temperature and (b) retrieved soil moisture data at the 21 cm wavelength obtained on November 9, 2005 during the Alabama A&M University Experiment (Shutko *et al.*, 2010).

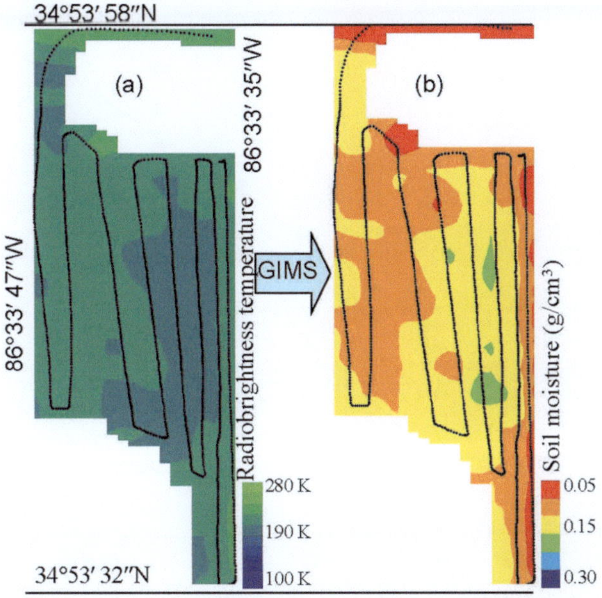


**Figure 1.3.** GPS-positioned rover traces overlaid on maps of (a) brightness temperature and (b) retrieved soil moisture data at the 21 cm wavelength obtained on November 10, 2005 during the Alabama A&M University Experiment (Shutko *et al.*, 2010).

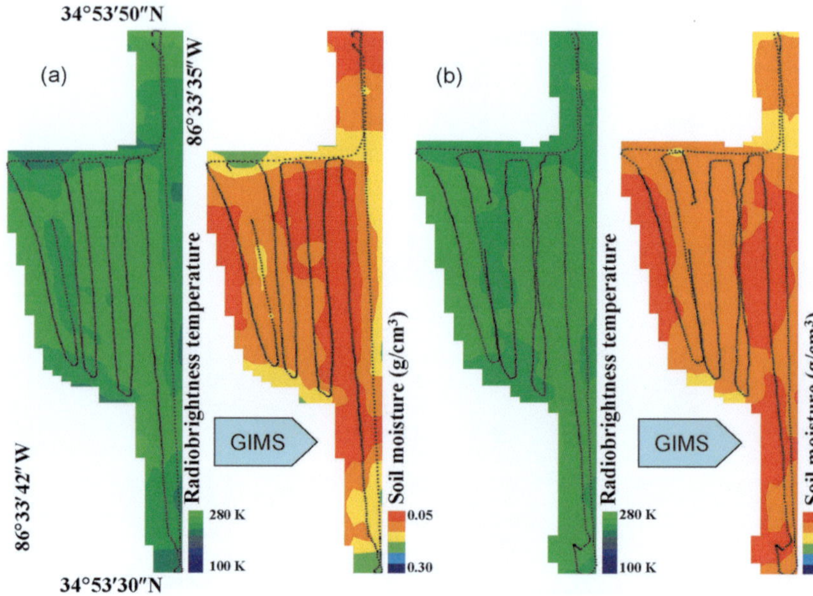


**Figure 1.4.** GPS-positioned rover traces overlaid on maps of (a) brightness temperature and (b) retrieved soil moisture data at the 21 cm wavelength obtained on November 11 (a) and 12 (b), 2005 during the Alabama A&M University Experiment (Shutko *et al.*, 2010).

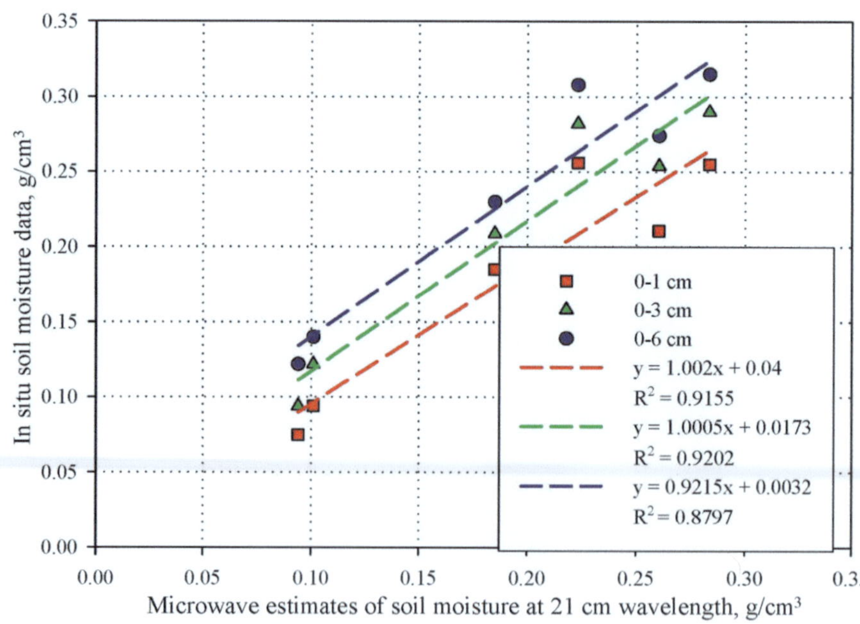


**Figure 1.5.** Comparison of microwave estimates of soil moisture at the 21 cm wavelength with *in situ* soil moisture in the 0–1 cm, 0–3 cm, and 0–6 cm layers assessed by means of the rover platform during the Alabama A&M University Experiment (Shutko *et al.*, 2010).

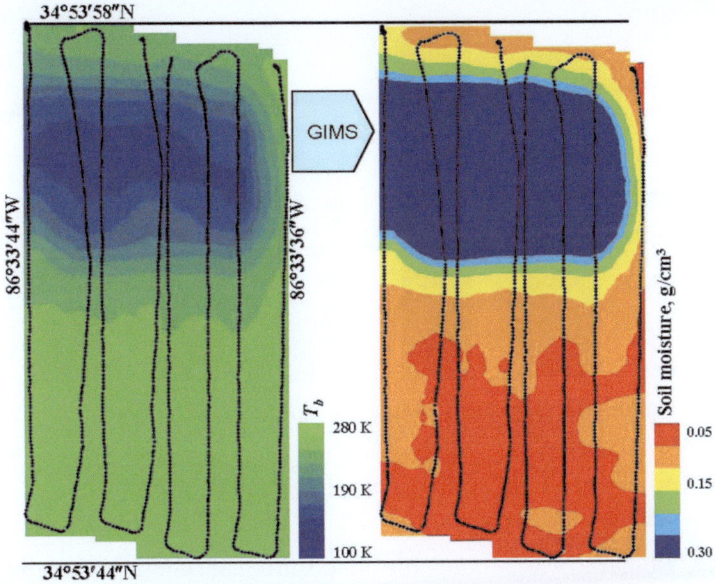


**Figure 1.6.** GPS-positioned unmanned helicopter flight line traces overlaid on maps of (a) brightness temperature and (b) retrieved soil moisture data at the 6 cm wavelength obtained on November 9, 2005 during the Alabama A&M University Experiment (Shutko *et al.*, 2010; Shutko and Krapivin, 2011).

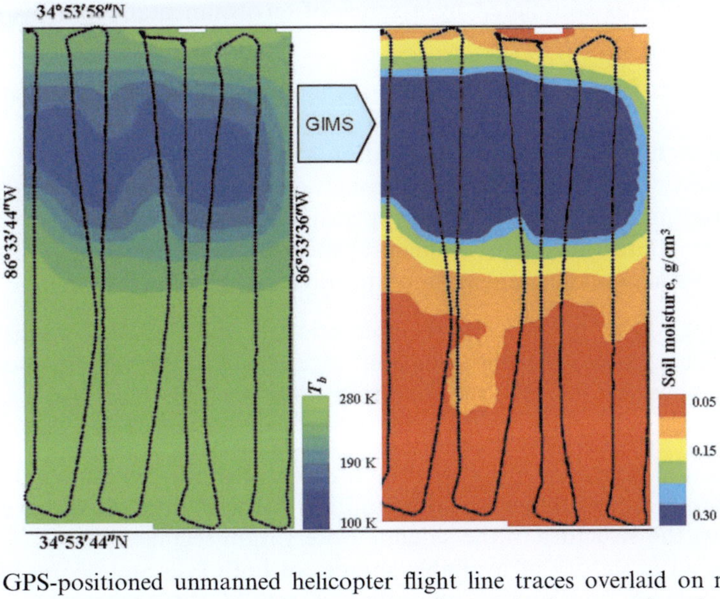


**Figure 1.7.** GPS-positioned unmanned helicopter flight line traces overlaid on maps of (a) brightness temperature and (b) retrieved soil moisture data at the 6 cm wavelength obtained on November 10, 2005 during the Alabama A&M University Experiment (Shutko *et al.*, 2010; Shutko and Krapivin, 2011).

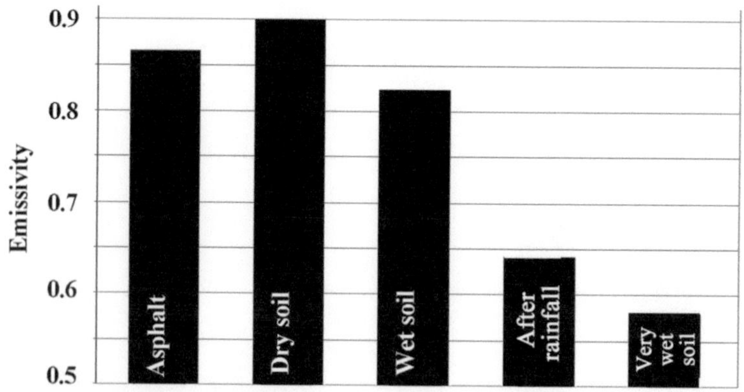


**Figure 1.8.** Six-centimeter radiometer emissivity versus qualitative soil condition (different targets) estimates.

### 1.2.3 The GIMS for Bulgaria

Remarkably, Bulgaria was among the first European countries in the last 10–15 years to engage in multi-purpose environmental monitoring work using modern hi-tech means. This was done—not at the level of private companies, nor for certain regions, nor for the solution of a specific problem—but in the widest context and over the long term.

The creation of the GIMS for Bulgaria was supported by the Bulgarian Parliament and was stimulated by a series of environmental problems (and other issues) such as:

- an increase of agricultural efficiency due to improved knowledge of soil moisture levels;
- risk assessment of floods and saturation of agricultural areas;
- risk assessment of water passing through dams;
- risk assessment of forest fires and operational decision making;
- contouring hazardous environmental processes; and
- risk assessment and emergency monitoring of negative environmental situations.

To solve these problems, an international team of top specialists from Bulgaria, Russia, and Holland was put together. The GIMS for Bulgaria was undertaken phase by phase in the following way:

(1) making an inventory of available sources of both aircraft and spacecraft information about Bulgaria, incorporating them into the GIMS structure, and adapting the structure to the satellite observation regime;
(2) making an inventory of available sources of both GIS and *in situ* information about Bulgaria and incorporating them into the GIMS structure;
(3) developing basic spatiotemporal models of Bulgarian environments, incorporating these sources of information into the GIMS structure, and building a

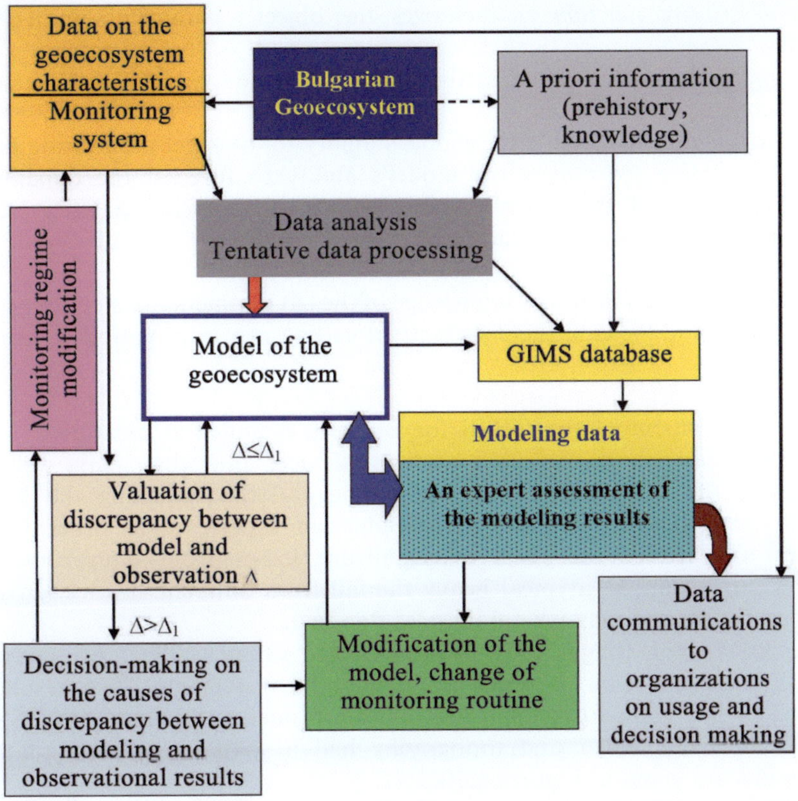


**Figure 1.9.** The structure and functional assignment of the GIMS for Bulgaria (Shutko and Krapivin, 2011).

GIMS knowledge base to encompass biospheric, climatic, and socioeconomic connections/links and their parameterization;

(4) organizing the GIMS network for data collection and processing.

Figure 1.9 explains the structure and functional assignment of the GIMS for Bulgaria. This system combines remotely sensed microwave radiometric and optical data, *in situ* measurements, and mathematical models of spatiotemporal variations of environmental parameters and enables emergency mapping and risk assessment of areas from water seepage through levees, of zones from dangerously high groundwater levels, and other hydrological, ecological, and agricultural phenomena (Shutko *et al.*, 2010). The main functions of this system are:

- the *measuring function* uses similarities between the model and the studied environmental object or process;

- the *descriptive function* characterizes the object's parameters in its various states;
- the *interpretating function* describes limits to the adaptability of the model and the scope of received solutions;
- the *explanatory function* is the model's ability to interpret monitoring data;
- the *forecasting function* is the model's ability to forecast the dynamics and characteristics of the environmental system at a given reliability based on fragmentary monitoring data and under conditions not observed by the system;
- the *criterion function* allows knowledge of the environmental system or process to be verified and looks at the possibility of getting new information by simulation experiments.

The development of the GIMS for Bulgaria is aimed at creating a powerful centralized information and management service, one that may well be of relevance to the objectives of the Integrating and Strengthening the European Research Area project. Spreading the Bulgarian experience to other European countries will reduce the risks of significant losses from dangerous natural processes. For example, atypical heavy rainfall from different storms across parts of Europe has caused massive amounts of flooding.

The GIMS for Bulgaria was tested during summer 2007. Miramap Inc.'s Flying Laboratory took a series of remote measurements. The results were processed by the Multi-Functional Information-Modeling System (MFIMS). Input data on vegetation cover and topography had a spatial resolution of $9 \times 9$ m. Some results are given in Figures 1.10–1.12.

In Figure 1.12a a dam near Nikolovo (a village southwest of Ruse) is presented. The dam was constructed in the late 1950s–early 1960s without a clay basis. There is a bypass channel on the right orthographical side, but its efficiency is poor. As a result of residential and administrative buildings being situated behind the dam in company with other reasons, much attention has been paid to a number of questions: are the hydroprotective properties of the dam of a sufficiently high standard, does it hold water well, and how safe is it. The last question was of particular concern because many residential buildings behind the dam (at a distance of about 250 m) had water in their basements and cellars. The labels (i.e., numbers) on the photos characterize humidity at the surface of the soil measured by radiometric equipment on the first day after a lengthy period of rain; humidity was about 5% in the territory in front of the dam and more than 20% in that behind the dam. The situation on the first day is presented on Figure 1.12a.

Remote sounding lasted four days and some calculations have shown that integrated humidity decreased by three to five times (Figure 1.12 b, c). All of this information was received by means of model generalization of data on the control area (a few hundred square meters in size), and by taking into account surface and underground water from depths of $1\frac{1}{2}$ to 2 meters. Figure 1.12d shows the dam's effectiveness.

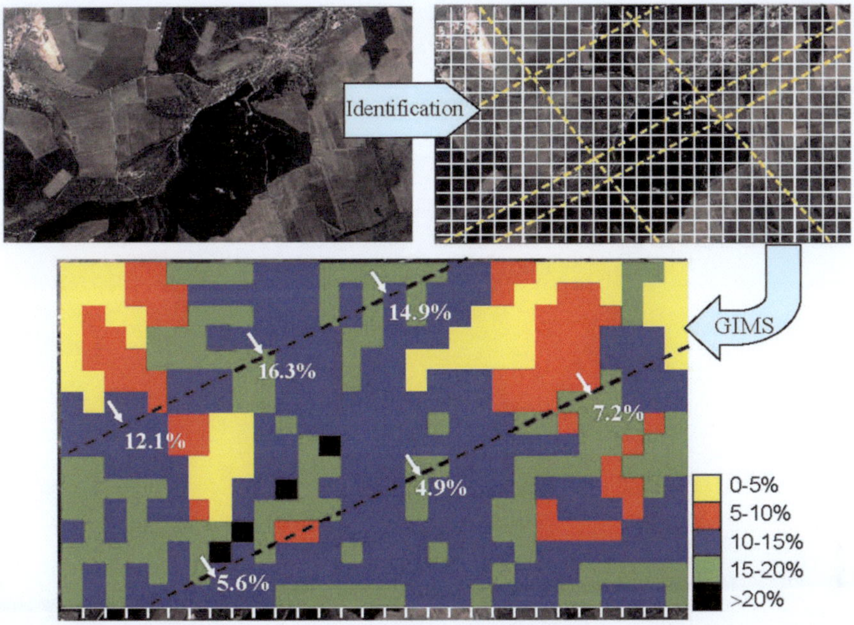


**Figure 1.10.** Stages in bringing about a monitoring procedure to assess the soil moisture distribution in and around the village of Nikolovo using land, remote, and computer calculations during the international GIMS Bulgaria 2007 experiment (Shutko *et al.*, 2010; Shutko and Krapivin, 2011). Spatial resolution is 250 m. Arrows show places where soil moisture was measured *in situ*. Dashed lines correspond to tracks taken by the flight laboratory (Soldatov, 2011).

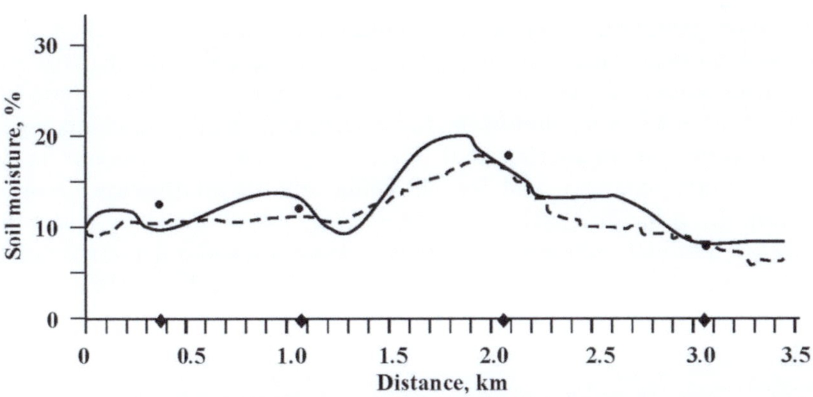


**Figure 1.11.** Comparative analysis of the results of soil moisture reconstruction by means of the MFIMS (solid curve), data of microwave monitoring (dashed line), and *in situ* measurements (◆) during the international GIMS Bulgaria 2007 experiment (Soldatov, 2011). Measurements were taken on August 2, 2007 near Nikolovo (a village in Bulgaria) by the Miramap Flying Laboratory one day after heavy rainfall (Shutko *et al.*, 2010).

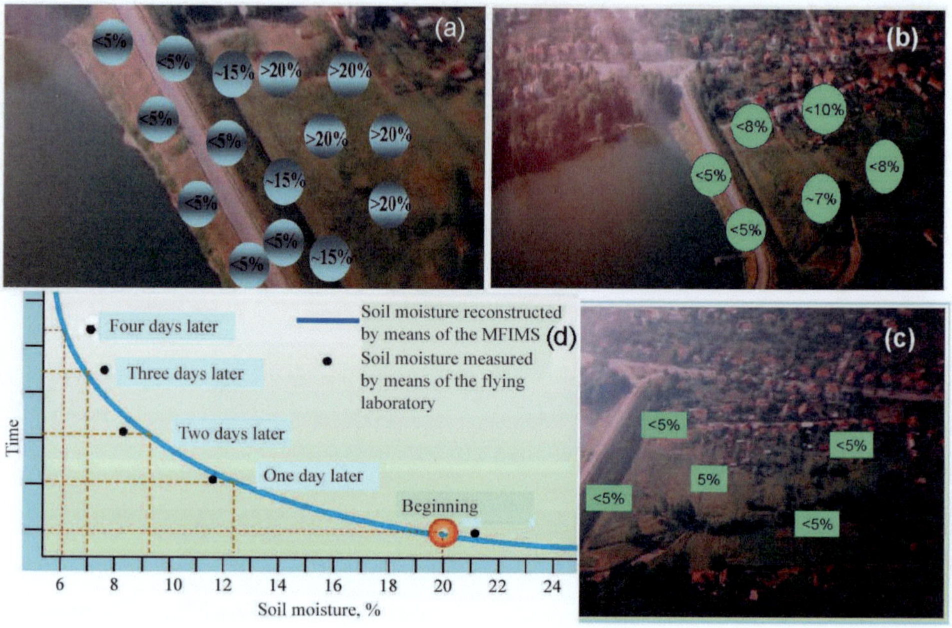


**Figure 1.12.** Results of soil moisture level retrieval after rain along the Rovenski Lom river (Bulgaria) based on microwave monitoring data in the dynamics of change over 1 day (a), 3 days (b), and 4 days (c). Panel (d) shows the tendency in soil moisture change received by means of the MFIMS (Soldatov, 2011).

## 1.3 THE GIMS STRUCTURE

There are many parameters that describe environmental conditions on Earth. For example, soil moisture and moisture-related parameters like depth to a shallow water table, contours of wetlands, marshy areas. Knowledge about these parameters and conditions is very important for agricultural needs, water management, land reclamation, for measuring and forecasting trends in regional to global hydrological regime changes, and for obtaining reliable information about water conservation. In principle, such information may be obtained by using on-site measurements, remotely sensed data, or by getting access to *a priori* knowledge-based data, formally accumulated and stored in GIS databases (Maguire *et al.*, 1991). However, the problem that arises here consists in answering the following questions:

- What instruments should be used to conduct so-called ground truth and remote measurements?
- What is the cost of contact and remote information?
- What should the balance be between the information content of contact and remote observations and the cost of these types of observations?

- What mathematical models should be used both to interpolate the data and extrapolate them in terms of time and space when the goals are to reduce the frequency and thus the cost of observations and to increase the reliability of forecasting the environmental behavior of observed objects?

These and other problems are solved by using a monitoring system that joins up the functions of environment data acquisition, the formation of control archives of these data, their analysis, and forecasting the characteristics of the most important processes in the environment. Such a unification is representative of the new information technology called "GIMS technology". The term "GeoInformational Monitoring System (GIMS)" describes the formula GIMS = GIS + Model that shows the relationship between the GIMS and the GIS. Clearly, the GIMS is a superset of the systems shown in Figure 1.1. There are two ways of looking at the GIMS. As far as the first is concerned the term "GIMS" is little more than a synonym of "GIS". The second is that the GIMS is an expansion of GIS. Let us keep the second view in mind and consider the constituent blocks of the GIMS. This consideration does not depend on either way of looking at the GIMS.

Let us consider the basic constituent of the GIMS as a natural subsystem interacting through biospheric, climatic, and socioeconomic connections with the global MCBS system. A model is created that describes this interaction and the functioning of various levels of the space–time hierarchy of the whole combination of processes in the subsystem. The model encompasses the characteristic features of typical elements of natural and anthropogenic processes, and model development commences with the existing information base. The model structure is oriented such that its use can be adapted to any regime (Figure 1.13).

The combination of an environment information acquisition system, a model of how a typical geo-ecosystem functions, a computer cartography system, and artificial intelligence will make it possible to create a geoinformation monitoring system of a typical natural element capable of undertaking the following tasks:

- evaluation of global change effects on the environment of typical elements of the MCBS system;
- evaluation of the role played by environmental change on typical elements undergoing climatic and biospheric changes on the Earth as a whole and in individual territories;
- evaluation of the ecological state of the atmosphere, hydrosphere, and soil–plant formations;
- formation and renewal of information structures on ecological, climatic, demographic, and economical parameters;
- operationally mapping the landscape situation;
- forecasting the ecological consequences of a variety of anthropogenic scenarios being realized;
- creating a database of land cover, natural phenomena, populated points, surface contaminations of landscapes, hydrological systems, forests;
- evaluation of natural threats to population security.

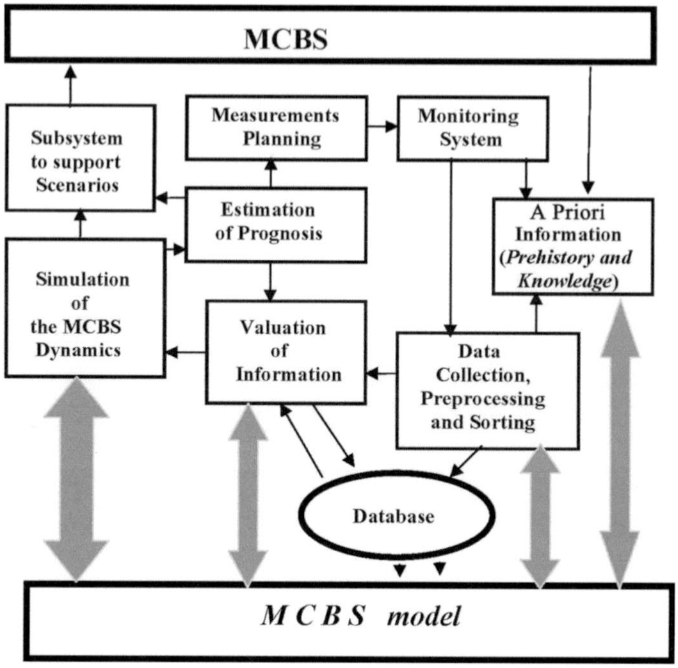


**Figure 1.13.** Conceptual block diagram of using the MCBS model in the adaptive regime of geoecoinformation monitoring.

Constructing a GIMS involves separating components of the biosphere, climate, and social medium according to a given level of spatial hierarchy. This is the reason their structure reflects these components (Table 1.1).

An important task undertaken by the GIMS is the parameterization of land cover, oceans, and atmosphere. This task is correlated with the search for effective procedures to evaluate model parameters using various sensors. For example, Ferrazzoli and Guerriero (1996) discussed using passive microwave radiometers to resolve this when the coefficients of the forest model are evaluated. It is well known that effective monitoring of the forest ecosystem is possible using microwave sensors. The range covered by microwave sensors has an advantage over visible and infrared sensors. However, the potential of microwave sensors is not revealed in their work. The use of microwave sensors to address many environmental problems—such as desertification, climate change, and the greenhouse effect—can help to make prognostic estimations of global processes more precise. Forests play significant roles in the regulation of the global biogeochemical cycles of greenhouse gases. It is known that optical systems are highly effective at sensing leaf parameters but they are unable to sense a woody biomass. As Ferrazzoli and Guerriero (1996) noted, at the P (0.45 GHz) and L (1.2 GHz) bands the backscatter coefficient of dense forest is much higher than that of agricultural fields, particularly at HV polarization. On the other hand, a "saturation" effect is observed, so that increase in the backscatter coefficient is small or absent when a dry biomass is larger than ~50–100 t/ha.

Of course, spaceborne microwave radiometers suffer from poor spatial resolution. This is the reason direct use of data received by means of spaceborne or airborne microwave radiometers to evaluate forest parameters has little value. However, GIMS technology guarantees to extract useful information from these data in a highly effective way. For this to happen it is necessary to use a model describing the studied phenomenon. Let us consider a model of the forest medium, which can be represented as having a three-level crown–trunks–soil structure. The crown is filled with scatters representing leaves, needles, twigs, and branches. There exist various schemes to approximate this structure. The most widespread approximation consists in the assumption that scatters take the forms of cylinders, disks, and ellipses uniformly located within the crown. Usually, the leaves are represented by disks of different radius, whereas branches and trunks are represented by cylinders of different categories. These widespread models are mainly theoretical. Their drawback lies in the fact that many parameters need to be evaluated, which is of course a difficult task. Nevertheless, the backscatter coefficient calculated by means of such a model can give a representation of the electromagnetic behavior of the observed medium and thereby add useful information.

The GIMS structure includes blocks serving the following main functions:

- data collection (current information about the soil–canopy system: soil moisture, depth to a shallow water table, soil salinity, biomass of vegetation, rainfall rate, etc.);
- data preprocessing, sorting, and storing in the data bank;
- modeling (simulation) of different kinds of ecological, hydrological, agricultural, climatological processes in different geophysical and environmental systems (these blocks contain a variety of models of crop productivity, the functioning of irrigation systems, geo-ecology, and the epidemiology of certain vector-borne diseases, etc.);
- estimation of the current state of a specific geophysical system;
- forecasting the state of this system at a future time;
- feedback support.

The data collection block consists of two main sub-blocks, namely:

- a sub-block for collecting prior knowledge-based information;
- a sub-block for collecting current data of *in situ* measurements and remote observations from mobile, aircraft, and satellite platforms.

The first of these two sub-blocks plays a very important role in the GIMS by making a model of the observed geo-system. It includes the relationships

**Table 1.1.** Functional structure of the GIMS.

| *GIMS item* | *Item functions* |
|---|---|
| Acquisition and Express Analysis Data Subsystem | Experimental planning of the structure of an environmental data acquisition system using satellites, flying laboratories, and movable and stationary ground observation points. Equipping laboratories with software and hardware tools allowing determination of the degree of environment contamination, the ecological situation, mapping of characteristic formations, detection of soil sub-surface centers suffering ecological damage, performing all-weather typification of land covers and detection of permafrost disturbances, oil spills, forest states, and pollution of bodies of water. |
| Initial Processing and Data Acquisition Subsystem | Methods and algorithms for synchronous analysis of aerospace information and ground measurements using space–time interpolation methods. Retrieving data and their reduction to a unique time moment. Determination of model parameters. Data-thematic classification and space–time combination images in the optical, IR, and microwave ranges. Trace measurements from diverse devices. |
| Computer Mapping Subsystem | Computer map formation algorithms with labels evaluating the ecological situation. Multilevel scaling and fragmentation of a given territory. Overlaying of output maps with the information needed by the user by means of a user interface. |
| Atmosphere State Evaluation Subsystem | Models of the spread of atmospheric contamination due to evaporation and burning of the products of oil, natural gas, and industrial enterprises. Atmosphere dust content evaluation. Gas and aerosol composition of the near-Earth atmosphere layer and maps forecasting their distribution over the Earth surface. |
| Soil–Plant Cover State Evaluation Subsystem | This sub-system's software<br>• typifies background flora in view of micro-relief, soil type and salinity, humidification, and degree of soil brine mineralization<br>• reveals micro/macro-relief peculiarities and sub-surface anomalies<br>• determines the structural topology of the land cover<br>• indicates the state of forests, swamps, agricultural crops, and pastures. |
| Water Medium State Evaluation Subsystem | A complex model of the territory accounting for seasonal change of surface and river runoff, influence of snow cover and permafrost, of the precipitation regime and evapotranspiration. A model of water quality dynamics for the hydrological network of the territory. |
| Ecological Safety and Population Health Risk Evaluation Subsystem | Algorithms that evaluate damage to nature, economic stability, and population health depending on environmental changes connected with natural trends of meteorological, biogeochemical, biogeocenotic, microbiologic, radiologic, and other natural processes, as well as with increases in stress of anthropogenic origin. |

| *GIMS item* | *Item functions* |
|---|---|
| Ecological and Sanitary Disturbance Cause Identification Subsystem | Environmental contamination sources. Source coordinates, power, and possible time of non-planned contamination. Dynamic characteristics of pollution sources. *A priori* unknown pollution sources and directions of possible trans-border pollution transfer. |
| Intelligent Support Subsystem | Mathematical algorithms providing user-intelligent support in complex analysis of objective information formed within the framework of a simulation experiment. Object dialogue with MCBS models places the necessary information in a convenient form for users. Data-processing corrections. Knowledge base on anthropogenic, demographic, and socioeconomic processes in the territory under study. |

between different geo-systems and the way in which they function as well as the relationships between the geo-system itself and the surrounding media.

The second sub-block consists of three boxes providing the following operations:

- *In situ* measurements including taking samples (probes) and utilizing current information accessible from meteorological stations and other accessible sources.
- Remote observations from aircraft (piloted and unmanned). Different types of sensors are used for remote observations. These include sensors operating in the optical, infrared, and microwave bands of electromagnetic wavelengths. The data they measure can potentially be used in the GIMS. In the majority of cases we have been looking at in this box, data were collected by passive microwave radiometers and to a lesser extent by infrared radiometers. These sensors provide measurements of the following parameters:
  (a) soil moisture;
  (b) depth to a shallow water table;
  (c) biomass of vegetation above water surfaces (rice crop, wetland vegetation) or above wet ground;
  (d) temperature assessment of water surfaces, land cover, and densely vegetated canopies.
- Satellite observations of the Earth–atmosphere–ocean–biosphere system. Optical sensors provide a crop type and land use classification with a ground resolution of about 1 to 50 m. Infrared sensors provide measurements of the surface temperature of land, water, and the upper boundary of vegetative canopies with a ground resolution of 10 to 100 m. Microwave radiometers provide estimates of soil moisture, the general condition of vegetation, temperature variations at the ocean surface and of the surface of dry land with a spatial resolution of around 20 to 50 km.

Ongoing algorithmic and programming maintenance of GIMS results in the following functions being available:

- parameterizing current and prognostic weather and climatic conditions based on *in situ* and remote observations as a result of running the required models;
- providing cluster analysis;
- modeling the productivity of a number of agricultural crops;
- modeling the functioning of typical irrigation systems;
- collecting *in situ*, aircraft, and spacecraft observations (with sorting and storing capabilities);
- estimating trends in the development of different geo-systems and evaluating the difference (discrepancy) between measured data and model predictions;
- current service in data flow control and parameter assessments;
- data processing (interpolation, extrapolation) using different methods;
- estimating soil moisture profiles and total water content in the 1 m soil layer using microwave data from mobile and aircraft observations and *a priori* knowledge-based information about the soil;
- modeling variations in soil concentration;
- modeling variations in humus characteristics;
- modeling the soil–canopy system growth of 30 plant types;
- mapping the results of modeling, measuring, interpolating and extrapolating;
- visualizing data as figures, graphics, tables, etc.

There are other programs and models used in global change studies such as the carbon circulation model and the oxygen circulation model (Krapivin and Varotsos, 2007, 2008).

## 1.4 EVOLUTIONARY TECHNOLOGY FOR ENVIRONMENTAL MODELING

Synthesis of an intelligent support subsystem of the GIMS is based on methods that simulate evolution (Bukatova, 1992; Bukatova and Makrusev, 2003; Elinson, 1983; Gulyaev *et al.*, 1987, 1989). These methods were the foundation for a new science called "ecoinformatics" (Krapivin and Potapov, 2002). This science lies at the boundary of such sciences as neurocybernetics, cognitive psychology, artificial intelligence, theory of systems, theory of survivability, and systemology.

Recent advances in informatics and information technology have enabled mathematical modeling and computer technology applications to enter such domains as ecology, biophysics, and medicine. The notion of model experiment has long lost its novelty only to be replaced by the term "computing experiment", which is used in many studies covering a vast range of subjects including biospheric experiments (Kondratyev *et al.*, 2003b, 2004b). All such works imply *a priori* availability of a more or less adequate model implemented as an array of tools in an algorithmic language. To manipulate a model to carry out a series of specific computing experiments one needs a general purpose computer. It is at this point that the researcher may face insurmountable difficulties caused by constraints on the computer's memory and speed of operation.

Despite many such endeavors indicating that modern hardware is capable of handling relatively complex models, at the same time there is the need for constant improvement of modeling techniques and the researcher has to face a conflict between his/her desire to enhance the accuracy of a model and limited computer capabilities. Building a model that is completely representative of a real world entity is clearly not feasible: on the one hand, taking account of all the parameters of the entity leads to the "evil of multi-dimensionality" and, on the other hand, simple models that can only cope with a small number of parameters are simply not up to the task of simulating the complex entities under consideration. Moreover, in projects that involve ocean physics, geophysics, global ecology, socioeconomics, etc. the building of an adequate model is in principle impossible because of the unattainability of information completeness. Such systems can only deal with applied problems from the domains of global ecology, biophysics, and medicine. Furthermore, difficulties in these subject areas arise at the early stages of research (e.g., when attempting to formulate a model).

What is to be done when currently available knowledge does not allow a mathematical model of an entity or a process to be synthesized? Evolutionary-type learning computers may well be the answer to this question. We shall retain the term "model" although it is used here in a somewhat different sense. What it implies here is the description of entities changing over time in an unpredictable manner and, by virtue of this, ensures the irremovability of information uncertainty at any moment. Such natural systems are studied in global ecology, geophysics, biophysics, and medicine. Consequently, a model treated in a broad sense must be able to continuously adapt to the changing behavior and structure of the observed entity. It is clear that universal models can be built only by synthesizing particular models. Models of this kind are implemented when dealing with recognition and prediction problems.

Thus, let a real world object $A$ have some unknown operational algorithm and have a previous operational history of known finite length. We need to efficiently simulate the functioning of $A$ by using models built for some system operating in real time. It is necessary to build a sequence of improving models $\{A_k\}$, where $k = 1, 2, \ldots$. Cybernetics began to ponder how this could be attained virtually from the moment the first, primitive by today's standard, machines emerged. It was then that the idea of creating artificial intelligence and designing "thinking machines" appeared, giving rise to the development of robotics. Being aware that a computer is an obedient executor of some program, cybernetics doggedly worked on the possibility of imparting a measure of unpredictable behavior to the machine when innovative facets were noticed. Treating intelligence as the ability to correctly respond to a novel situation, scientists came to the conclusion that machines capable of adapting to the level of individual components and their structural organization are in fact feasible.

The concept of model-free learning by computers has been discussed in recent years. The main problem is how to teach a computer to change the structure of

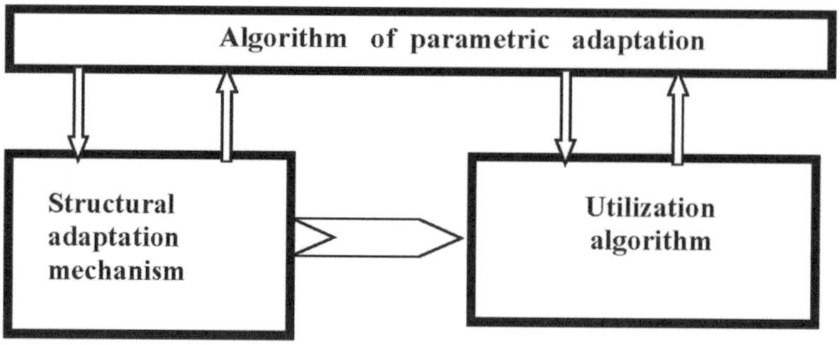


**Figure 1.14.** Schematic representation of the concept of evolutionary modeling (Kondratyev *et al.*, 2002).

relationships between its elements. An analogy with the neural operation of the brain as established in neurophysiology was a breakthrough and dramatically advanced the scale of artificial intelligence capabilities. Fogel *et al.* (1966) gave impetus to an entirely new cybernetic trend.

Evolutionary modeling on the whole can be represented by a hierarchical two-tier procedure (Figure 1.14). On the first tier there are two constantly alternating processes conventionally termed "structural adaptation" and "utilization" processes. At the $k$th stage of adaptation at the time of operation of the structural adaptation algorithm, models of the sequence $\{A_{s,i}\}$, $i = 1, \ldots, M_s$ are synthesized. Special rules form a "memory" of the most effective models $\langle A_s^1, \ldots, A_s^k \rangle$. At the utilization stage following the stage of adaptation, the system selects the most efficient model. A schematic diagram of the $i$th step of the $k$th stage of adaptation of models is represented in Figure 1.15. The "object" block here denotes that the real world object is defined as having some previous history.

The evolutionary selection of models procedure provides for virtually time-unlimited operation of the system under irremovable information uncertainty. Apart from previous history, which as a rule does not meet the requirements of traditional statistical analysis, the researcher has no other available information. Under such conditions one obviously has to use as much available information as possible (in particular, that on operation of the adaptation and utilization stages). In the parametric adaptation algorithm, this information is used as soon as the first-tier parameters are adapted (i.e., the characteristic number for each mode of changes; utilization of the multiplicity of change modes; length of the change mode list; the distribution of change modes in the list; the distribution of $k$-models of $A_s^i$; memory volume at the stage of adaptation; prehistory length; etc.).

The endeavors to search for a building block base to synthesize a network with a variable structure in an attempt to create special flexible hardware based on the new principles of information processing have offered up some results. For example, recent advances in microelectronics have helped solve the problem of selecting components for structures with variable fields of relations (Bukatova, 1992).

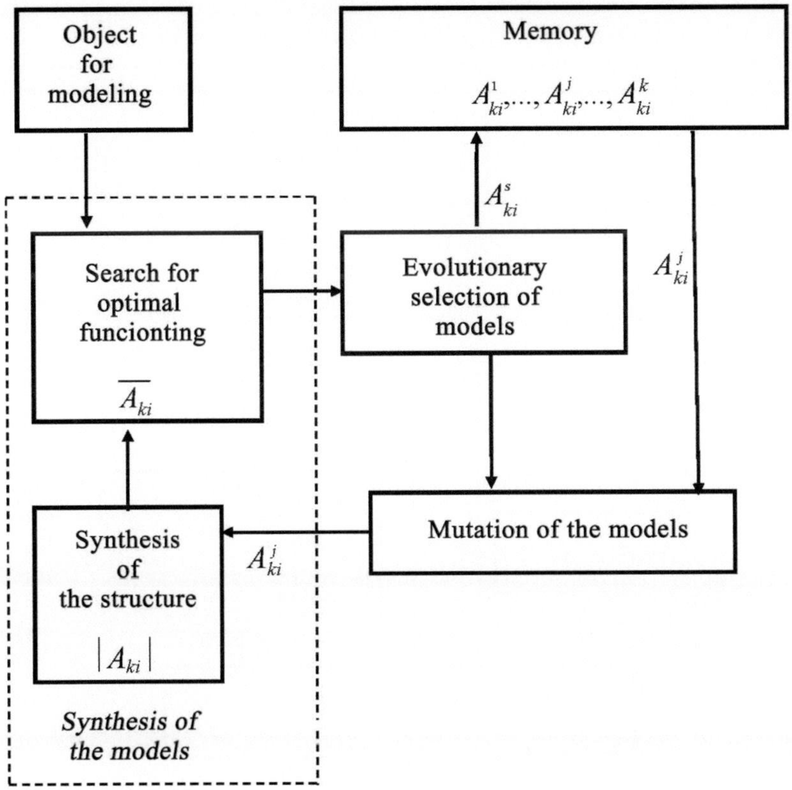


**Figure 1.15.** Conceptual flow diagram of the $i$th step at the $k$th stage of the adaptation process.

Unification and specialization are characteristic features of evolutionary software as a consequence of minimum *a priori* information, effective mechanisms for adaptation, and the modular principle of realization. With today's orientation to up-to-date personal computer engineering and diverse range of application problems, these specific features have made it possible to work out evolutionary computation technologies, in which a set of software modules has been brought about so that an active dialogue can take place with a user, in addition to adjusting the evolutionary facility to the specificity of the problem being solved (Bukatova *et al.*, 1991). Regardless of the field of application, evolutionary technology with software support is characterized by adaptability, flexibility, dynamism, self-correction, while the main distinction consists in its high effectiveness and adjustability under conditions of maximum information uncertainty, including that which is irremovable. This approach permits the adaptive procedure in Figure 1.13 to be realized. Evolutionary modeling technology really does give a new structure to the GIMS (Figure 1.16).

Evolutionary modeling is effective at bringing about an adaptation process within GIMS technology to correct the functional and parametric structure of the

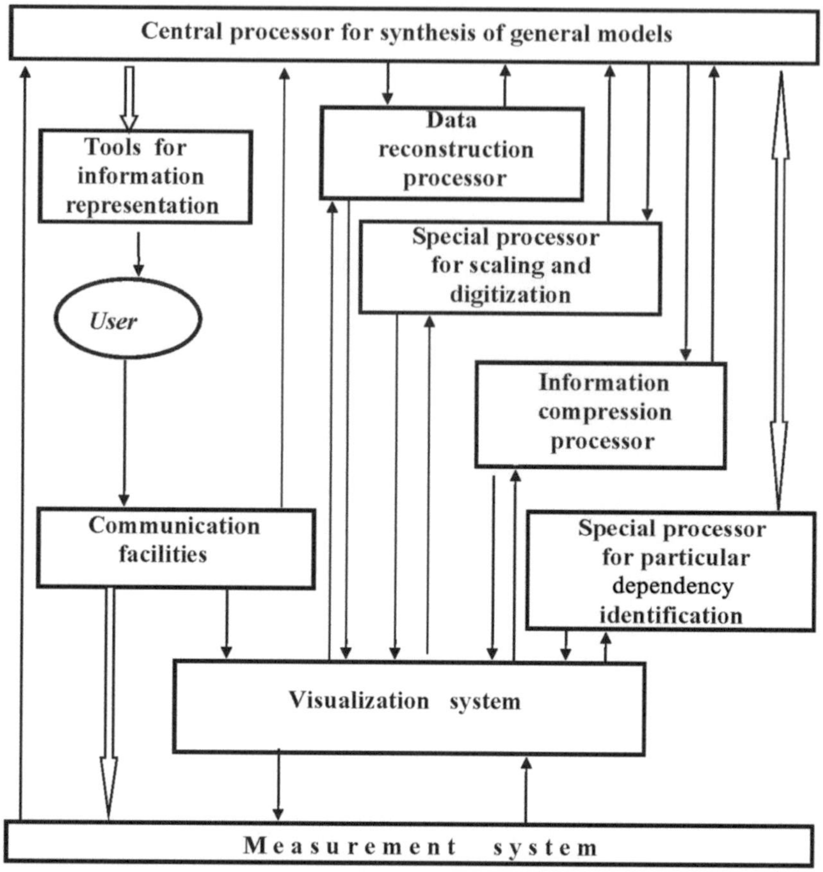


**Figure 1.16.** Structure of the geoinformation monitoring system and its evolutionary devices.

basic model. There has been a search for correlations to describe the local function within the framework of common adaptation procedures that other methods can use. For example, Fang and Liang (2003) studied two LAI retrieval schemes using a neural network algorithm:

- extraction of LAI from atmospherically corrected surface reflectance;
- extraction of LAI from top-of-atmosphere (TOA) raw radiances detected by the Enhanced Thematic Mapper Plus (ETM+) sensor.

LAI can be estimated by inverting a radiative transfer (RT) model based on Landsat ETM+ reflectance and radiance data. The first step in this procedure is the creation of an appropriate lookup table (LUT) method. Then the Markov Chain Reflectance Model (MCRM) is used.

## 1.5 A GLOBAL MODEL AS A UNIT OF THE GIMS

Effective solution of real problems of global ecodynamics is clearly impossible using observation data in isolation. Kondratyev *et al.* (2004a) developed a model of global ecodynamics that considered correlations between the basic components of the nature–society system (NSS). However, traditional approaches to building a global model imply a number of difficulties for algorithmic description of many socioeconomic, ecological, and climatic processes, resulting in researchers having to deal with information uncertainty. These approaches to global modeling simply ignore such uncertainty and, consequently, the structure of the resultant models does not adequately cover the real processes. Evolutionary modeling makes it possible to remove this drawback by synthesizing a combined model whose structure is subject to adaptation against the background of the history of a system of the biosphere and climate components. Implementation of such a model can also be combined in various classes of models using conventional software and hardware and special purpose evolutionary-type processors. The way in which they are combined is diverse depending on the spatiotemporal completeness of databases.

Global modeling abounds with examples of unsolvable problems that are encountered when looking for ways to describe scientific and technological advances as well as human activity which manifests itself in many ways. No lesser difficulties arise when modeling climate, which is described by superimposing processes with different temporal variability rates. As to the completeness of description in the global model, it is impossible to clearly delineate the amount of information available and the extent of spatial and structural detail required. Therefore, without going into the natural–philosophical analysis of global problems and skirting the issue of the ultimate solution to global modeling, let us confine ourselves to discussing only one of the possible approaches in which evolutionary modeling implemented on special processors can help overcome the computing, algorithmic, and other difficulties of global modeling. What this does imply is that a search for effective traditional-type models should be excluded or rejected. At present, building global biogeocenotic models is not seen as difficult. Many such models have been created and information is currently being gathered to support them. The history of the interaction of the biosphere with the climate system and human society is simply not known well enough, which is one of the hurdles to describing, for example, climatic cycles in the history of this interaction. To build a global model that takes into account the interaction of the biosphere, climate system, magnetosphere, etc. it is necessary to apply the evolutionary approach, which helps overcome uncertainties in the description of this interaction. By adjusting such a model to the history of the prescribed cycle we will obtain a model that implicitly traces various regularities of biospheric dynamics in the past and allows for forecast assessments in the same temporal cycle. A special processor version of this model completely removes all the existing algorithmic and computing hurdles that arise as a result of the great dimensionality of the global model and removes the conditions of irreducible nonparametric uncertainty.

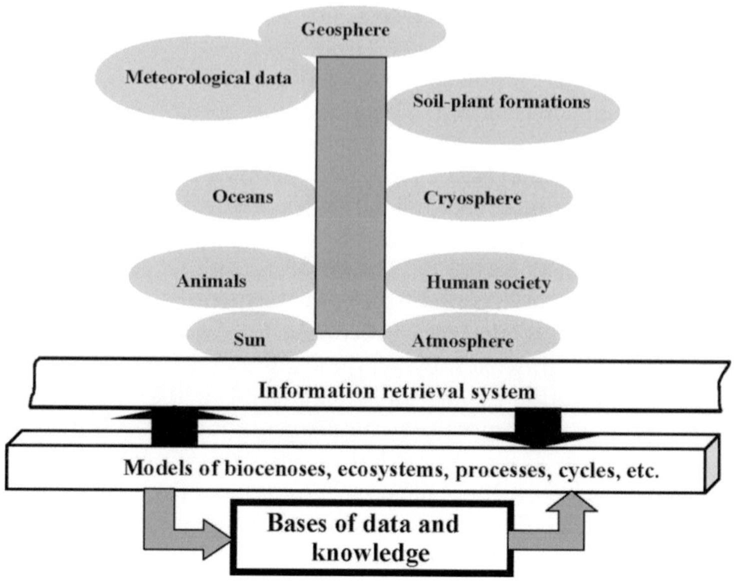


**Figure 1.17.** Organization of the global model of NSS functioning.

Figures 1.17 and 1.18 explain the conceptual diagrams of the new global model. The data archive is formed here as two structures. Data for the "machine models of biosphere processes" type are stored as climatic maps and as tables of coefficients of model equations. It is necessary to fill in all the cells of schematic maps. Data of the second type are represented as fragments recorded disparately (possibly irregularly) in time and space (i.e., $CO_2$ concentration, temperature, precipitation, pressure, population numbers, availability of resources, etc.). Data of this type are used to adjust the evolutionary processor to the given class of graph (e.g., finite automata, models). As a result of this procedure the model is adapted to the history of the prescribed time cycle. As shown by Bukatova (1992) a stable forecast of 0.75–0.95 reliability covering several temporal steps is produced. Just how extensive forecasts are depends on the depth of history versus the background of the saturation effect. Given the need for a forecast under changing conditions in the trends of human economic activities, an evolutionary processor is adjusted to the assigned scenario, thus automatically providing for simulation of the corresponding response of the biosphere to this change.

The suggested structure of the global model thus ensures a flexible combination of models of traditional and evolutionary types. The proposed approach helps escape the need to model nonstationary (i.e., climatic, socioeconomic, demographic, etc.) processes and provides for uncertainty to be overcome. A model of the new type makes it possible to go from learning experiments to assessment of biospheric viability with regard to actual trends of anthropogenic stresses in all regions of the globe.

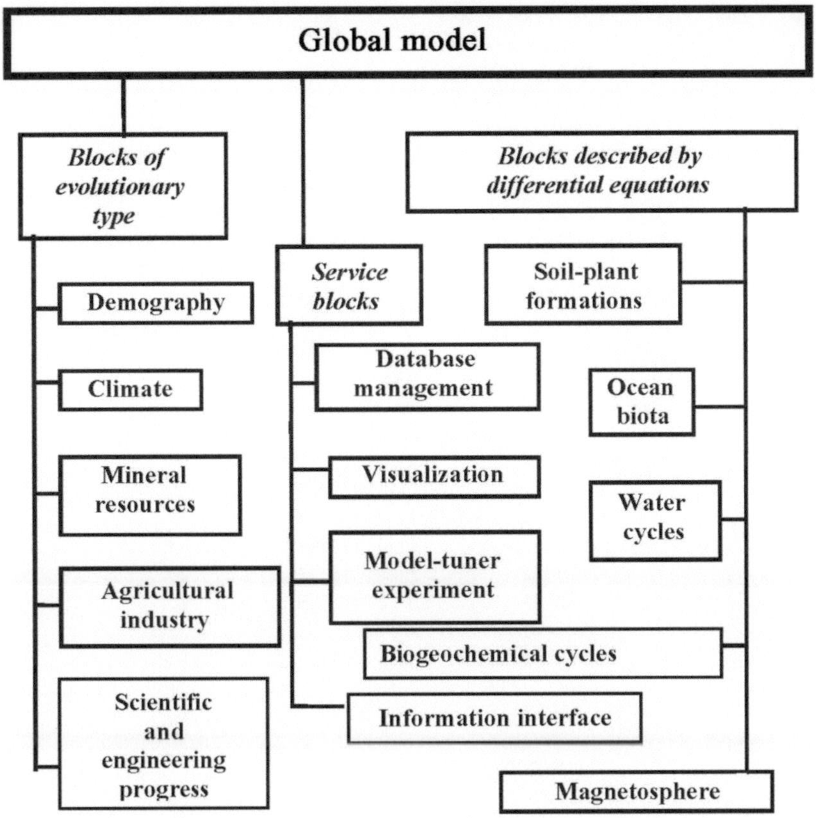


**Figure 1.18.** Structure of global model based on using evolutionary blocks.

By departing from established global modeling techniques based on new information technology, it is possible to create a global system of monitoring with the global model as part of the system support. The structure of such a system is represented in Figure 1.19 and Table 1.2. Application of evolutionary computer technology allows categorizing the whole system as a class of subsystems whose structure is variable and makes it adaptable to changes in the natural process or entity under observation. Furthermore, it becomes possible to heterogeneously detail the natural systems under study as phase variables and to select the geographical grid during sampling analysis of the planetary surface (i.e., arbitrary insertion of significant regional level regularities becomes possible).

The automatic system for processing global information is aimed at acquiring combined models that reflect the real-time scale of climatic and anthropogenic changes in the biosphere based on known history (or, rather, its simulation). The system relies on a bank of biosphere process models and, using the software of other blocks and a scenario of anthropogenic behavior formulated at input, provides for prompt assessment of the environmental state and for forecast assessments within the framework of the scenario. Another advantage of such an

```
    W    150   120    90    60    30    0     30    60    90   120   150    E
N ....................****. **********........................................ N
  ....................AAA. . **********. . . . . **. . . . . . . . . . . . . . . A. . . . . . . . . . . . . . .
  ............AAA. AAAA. . . . **********. . . . . . . . . . . . . . C. . . . . AAAAA. . . . . A. . . . . . . .
  . . . . CCCC. . . . . CCA. A. AAAC. . . ******. . . . . . . . . LLL. . . . . C. CCCCCMMMMCAAAAAAAAA. . . . .
  MC. FFFMFMFFFFCCCCCAA. . MM. . ***. . . L. . . . . . DDDFF. FFFFCLLLFFFMFFFFFMAAAMMMMMMM
  . . . . LLFFFFCCCCCCCC. . . C. . . . . L. . . . . . . . . . RG. DDDDDDDDDFFFFDDDDDDDDDMMFFMCCM. .
  . . . . . L. . . FFFFGGGFF. . . CFC. . . . . . . . . . . +. . . R. RGGGGGGGDGGGGGGGGDDDDDDM. . . L. . . .
  . . . L. . . . . . +RGGGGGDDF. FFFF. . . . . . . . . . ++. ++RRRRXXXXGXXXXGGGGGGDDGDDD. . . L. . . .
50. . . . . . . . . . . RRXWXXGGDDDDD. D. . . . . . . . . . . . +++++XXWWVVVVVVVGGGGGGGGGDDD. . . . . . . . 50
  . . . . . . . . . . . . RXWWX++. +++G. . . . . . . . . . . . +++++XXWWV. S. @@QQSGVVVVGXRR. . . . . . . . . .
  . . . . . . . . . . . . BBBXXX++++. . . . . . . . . . . . . ++. . P++. . WQ. KS&QQQSSSSSSXX++. . +. . . . . . .
  . . . . . . . . . . . . RBUWXP+PP. . . . . . . . . . . . . PP. . . . P. WWWW&K&&QQSSSSSQXX. +. . P. . . . . . . .
  . . . . . . . . . . . . . @UUUPPPP. . . . . . . . . . . . . . &&&&. . . . . &&WUUUUUQBBQQ+++. . . P. . . . . . . . .
30. . . . . . . . . . . . . @@@UP. . P. . . . . . . . . . . . . HHHHHHHHHHHHHHTHHHJPQQQQPPPP. . . . . . . . . . . . 30
  . . . . . . . . . . . . . . @@@. . . . . . . . . . . . . . . . . HHHHHHHHH. HH. TTTHNNYZZPPPP. . . . . . . . . . . .
  . . . . . . . . . . . . . . . U@. . . J. . . . . . . . . . . TTTTTTTTTTTTTTT. . . NNI. YYYI. . . . . . . . . . . . .
  . . . . . . . . . . . . . . . . YYZ. . . J. . . . . . . . . . TTTTTTTTTTT. TT. . . . JN. . YYY. . . . . . . . . . . . . .
  . . . . . . . . . . . . . . . . . . YZ. . . . . . . . . . . . . JJJJJTJJJNTT. . . . . NN. . . YY. . Y. . . . . . . . . . .
10. . . . . . . . . . . . . . . . . . . . Z. JJ. . . . . . . . . . JJJJJJJINNTT. . . . . N. . . . Y. . . . . . . . . . . . . . . 10
  . . . . . . . . . . . . . . . . . . . . ZJJZ. . . . . . . . . . YJZYJJJJJJT. . . . . . . . . . Z. . . . . . . . . . . . . . .
  . . . . . . . . . . . . . . . . . . . . ZZZZJ. . . . . . . . . . . . ZYYYYJNN. . . . . . . . . . Z. . Z. . . . . . . . . . . .
  . . . . . . . . . . . . . . . . . . . QZZZZJZZ. . . . . . . . . . YYZYJJJ. . . . . . . . . . . . Z. Z. . . . . . . . . . . .
  . . . . . . . . . . . . . . . . . . . WZZZZJJJJJ. . . . . . . . . JYYNNJ. . . . . . . . . . . . . . . . . . ZZ. . . . . . .
10. . . . . . . . . . . . . . . . . . . . ZZZZJJJJ. . . . . . . . . . JNNNN. . . . . . . . . . . . . . . . . . . . . Z. . . . . . 10
  . . . . . . . . . . . . . . . . . . . . WWZZZJJY. . . . . . . . . . JNNNN. . Z. . . . . . . . . . . . . . . NN. . . . . . . . .
  . . . . . . . . . . . . . . . . . . . . . WWNZYJY. . . . . . . . . . JJJNJ. J. . . . . . . . . . . . . . . NNNNN. . . . . . .
  . . . . . . . . . . . . . . . . . . . . . TNNYYY. . . . . . . . . . . TJJJJ. N. . . . . . . . . . . . . JJTTJJJ. . . . . . .
  . . . . . . . . . . . . . . . . . . . . . TNNYP. . . . . . . . . . . . HHJE. . . . . . . . . . . . . . . . JJTTTTNN. . . . . .
30. . . . . . . . . . . . . . . . . . . . . UUEY. . . . . . . . . . . . . . HEE. . . . . . . . . . . . . . . . . HHTTTJJ. . . . . . 30
  . . . . . . . . . . . . . . . . . . . . . UUEI. . . . . . . . . . . . . . E. . . . . . . . . . . . . . . . . . . U. . . UUP. . . . . .
  . . . . . . . . . . . . . . . . . . . . . UEE. . . . . . . . . . . . . . . . . . . . . . . . . . . . . . . . . . . . . . . . . . . P.
  . . . . . . . . . . . . . . . . . . . . +VV. . . . . . . . . . . . . . . . . . . . . . . . . . . . . . . . . . . . . . P. . . . +. .
  . . . . . . . . . . . . . . . . . . . . +V. . . . . . . . . . . . . . . . . . . . . . . . . . . . . . . . . . . . . . . . . . . . . .
50. . . . . . . . . . . . . . . . . . . . +E. . . . . . . . . . . . . . . . . . . . . . . . . . . . . . . . . . . . . . . . . . . . . . 50
  . . . . . . . . . . . . . . . . . . . . . +. . . . . . . . . . . . . . . . . . . . . . . . . . . . . . . . . . . . . . . . . . . . . .
S . . . . . . . . . . . . . . . . . . . . . . . . . . . . . . . . . . . . . . . . . . . . . . . . . . . . . . . . . . . . . . . . . . . . S
    W    150   120    90    60    30    0     30    60    90   120   150    E
```

**Figure 1.19.** The spatial distribution of soil–plant formation over a 4° × 5° geographical grid. Notations are given in Table 1.2.

automatic system to process global information consists in it formulating the entry and bound conditions in the study of regional systems for particular built-in models and virtually substituting the field measurements of those conditions.

The first version of a global model (Krapivin *et al.*, 1982) was oriented towards rigid spatiotemporal detail and, therefore, required a large amount of information. The subsequent development of an automatic system to process global information as a result of evolutionary technology has made it possible to discard the generally accepted regular geophysical grid in archive development

**Table 1.2.** Identifiers adopted in the GSM for designation of soil–plant formations. Designations: $\mathfrak{R}$ is the SPF identifier, $\sigma$ is the area of soil–plant formation ($10^6$ km$^2$), $P$ is the production of soil–plant formation (kgC $\cdot$ m$^{-2}$ yr$^{-1}$), $\Phi$ is phytomass (kg m$^{-2}$), $H$ is dead organic matter (kg m$^{-2}$).

| $\mathfrak{R}$ | *Soil-plant formation (SPF)* | $\Phi$ | $H$ | $\sigma$ | $P$ |
|---|---|---|---|---|---|
| A | Arctic deserts and tundras | 0.4 | 1.3 | 2.55 | 0.17 |
| B | Alpine deserts | 0.8 | 16.9 | 1.15 | 0.47 |
| C | Tundra | 1.9 | 5.6 | 2.93 | 0.36 |
| D | Mid-taiga forests | 22.5 | 10.8 | 5.73 | 0.63 |
| E | Pampas and grass savannas | 3.8 | 21.0 | 1.66 | 1.11 |
| F | North taiga forests | 10.0 | 8.1 | 5.45 | 0.54 |
| G | South taiga forests | 23.5 | 14.5 | 6.60 | 0.65 |
| H | Sub-tropical deserts | 0.1 | 1.4 | 7.16 | 0.12 |
| I | Sub-tropical and tropical thickets of grass–tugai type | 45.0 | 21.6 | 0.90 | 1.96 |
| J | Tropical savannas | 0.1 | 2.0 | 17.10 | 1.35 |
| K | Solonchaks | 45.0 | 21.6 | 0.38 | 0.18 |
| L | Forest tundra | 3.8 | 9.0 | 1.55 | 0.65 |
| M | Mountain tundra | 1.9 | 5.5 | 2.23 | 0.38 |
| N | Tropical xerophytic open woodlands | 10.0 | 15.1 | 9.18 | 1.42 |
| P | Sub-tropical broad-leaved and coniferous forests | 43.0 | 22.2 | 5.75 | 1.72 |
| Q | Alpine and sub-alpine meadows | 1.9 | 24.0 | 3.54 | 0.76 |
| R | Broad-leaved coniferous forests | 25.0 | 25.1 | 2.12 | 0.87 |
| S | Sub-boreal and saltwort deserts | 0.2 | 8.1 | 2.69 | 0.25 |
| @ | Sub-boreal wormwood deserts | 0.8 | 8.8 | 1.99 | 0.35 |
| T | Tropical deserts | 0.4 | 4.9 | 11.50 | 0.18 |
| U | Xerophytic open woodland and shrubs | 3.8 | 15.0 | 3.91 | 0.56 |
| V | Dry steppes | 0.8 | 12.6 | 2.66 | 0.38 |
| W | Moderately arid and arid steppes | 1.9 | 33.0 | 4.29 | 0.79 |
| X | Forest steppe (meadow steppe) | 1.9 | 38.0 | 3.72 | 0.74 |
| Y | Variable humid deciduous tropical forests | 60.0 | 20.5 | 7.81 | 2.46 |
| Z | Humid evergreen tropical forests | 60.0 | 21.6 | 10.40 | 3.17 |
| + | Broad-leaved forests | 45.0 | 24.8 | 7.21 | 1.25 |
| & | Subtropical semi-deserts | 0.4 | 12.1 | 2.08 | 0.45 |
| * | Absence of vegetation | 0.0 | 0.0 | 14.6 | 0.00 |

(Wilson and Henderson-Sellers, 1985) and solve this problem using algorithms for spatiotemporal information recovery (Armand *et al.*, 1987).

The climatic component of the NSS presents the greatest difficulty when synthesizing a global model, because it is characterized by many feedbacks, most of which are unstable. Among them are ice–albedo, water vapor–radiation, cloudiness–radiation, aerosol–radiation; there are of course many others. The functioning of the Earth climatic system is determined by the state of the atmosphere, oceans, cryosphere, and the surface of continents that have land biota, lakes, rivers, ground waters, and various anthropogenic structures. To construct a model of the climate requires consideration of numerous factors, whose role in the formation of climate has been studied inadequately. So far, attempts to derive a numerical model of the Earth's climatic system have not been able to offer up results that could be used in a global simulation model (GSM). Neither is it safe to say that realization of Global Carbon Project (GCP) plans will clarify the problem. Questions concerning the problem of the greenhouse effect will remain unanswered until climate can be parameterized.

There are two approaches to synthesizing the climatic system. One is based on the inclusion of biospheric components in existing climate models or in those under development. The other approach consists in developing a unit within the numerical model of the biosphere to simulate the dependences of biospheric components on climate parameters. In the former case, there are unstable solutions to differential equations concerning the respective systems, which makes it difficult to predict global environmental changes. In the latter case there is a possibility to obtain steady prognostic estimates of environmental changes, but their reliability depends on the accuracy of parameterization of correlations between the elements of climate and the biosphere. The second approach is preferable because it permits combining climate models that can be described at the scenario level with numerical models of the biosphere. Detailed analysis of the problems of climate modeling and assessment of the present state of affairs can be found in Marchuk and Kondratyev (1992), Kondratyev (1999), Kondratyev and Johannessen (1993). Some models of individual components of the NSS, which correspond to the second approach, are discussed here. Among them are models of atmospheric general circulation, atmosphere–ocean interaction, sensitivity of climate parameters to boundary conditions on the Earth's surface, and the interaction of biogeochemical and climatic processes (Chen *et al.*, 2000; Lin *et al.*, 2000).

The climatic system is a physico-chemical biological system with unlimited degrees of freedom. Therefore, any attempts to model this complex system encounter serious difficulties. This explains the diversity of parametric descriptions of some processes in this system. For a global model with a digitization time step up to one year, the use of two versions can be acceptable. The first version consists in combined use of correlations between particular processes taking place during formation of the climatic situation over a given area and climate scenarios. The second version is based on the use of the global-monitoring data, which are the basis for data series about climatic parameters spatiotemporally and used to retrieve a complete pattern of their spatial distribution. A widespread correlation function is the dependence of variations in the average temperature $\Delta T_g$ of the atmosphere on its $CO_2$ content:

$$\Delta T_g = \begin{cases} 2.5[1 - \exp\{-0.82(\xi - 1)\}], & \xi \geq 1; \\ -5.25\xi^2 + 12.55\xi - 7.3, & \xi < 1 \end{cases} \tag{1.1}$$

where $\xi$ is the ratio between the present content of $CO_2$ in the atmosphere $C_a(t)$ and its pre-industrial level $C_a(1850)$.

As can be seen from (1.1), $T_g$ is an increasing function of the amount of atmospheric $CO_2$. A 20% increase of $CO_2$ in the atmosphere raises the temperature by 0.3°C. A doubling of atmospheric $CO_2$ causes an increase of $T_g$ by 1.3°C. Detailed analysis of the function (1.1) and comparison of the observed combined variations of $\Delta T_g$ and $\xi$ show that the use of model (1.1) makes it possible to simplify the climate unit of the NSS model. In particular, if $(\Delta T_g)_{2[CO_2]}$ is calculated by (1.1) with a doubled concentration of atmospheric $CO_2$, then to assess present trends in changes of $T_g$ the following formula can be used:

$$\Delta T_g = (\Delta T_g)_{2[CO_2]} \ln \xi / \ln 2, \tag{1.2}$$

where, from the assumed estimates, the pre-industrial value of $C_a(1850) = 270$ ppm. The formula (1.2) satisfactorily approximates known data with an error of about 50%. It follows from (1.2) with $C_a(1980) = 338$ ppm that $\Delta T_g = 1.3$ K, whereas the real warming is estimated by many authors at 0.6 K.

Of course, the results of recent discussions about the greenhouse effect in connection with the growth of $CO_2$ partial pressure in the terrestrial atmosphere should be reflected in a GSM. Formula (1.1) takes into account the impact of $CO_2$. According to Mintzer (1987), there is a possibility to broaden this by considering the temperature effect of other greenhouse gases:

$$\Delta T_\Sigma = \Delta T_{CO_2} + \Delta T_{N_2O} + \Delta T_{CH_4} + \Delta T_{O_3} + \Delta T_{CFC\text{-}11} + \Delta T_{CFC\text{-}12},$$

where

$$\begin{aligned}
\Delta T_{CO_2} &= -0.677 + 3.019 \ln[C_a(t)/C_a(t_0)], \\
\Delta T_{N_2O} &= 0.057\{[N_2O(t)]^{1/2} - [N_2O(t_0)]^{1/2}\}, \\
\Delta T_{CH_4} &= 0.19\{[CH_4(t)]^{1/2} - [CH_4(t_0)]^{1/2}\}, \\
\Delta T_{O_3} &= 0.7[O_3(t) - O_3(t_0)]/15, \\
\Delta T_{CFC\text{-}11} &= 0.14[CFC\text{-}11(t) - CFC\text{-}11(t_0)\}, \\
\Delta T_{CFC\text{-}12} &= 0.16[CFC\text{-}12(t) - CFC\text{-}12(t_0)].
\end{aligned}$$

The value of $t_0$ is identified with the year 1980, when GHG concentrations were known.

A simple formula of the latitudinal distribution of average temperature over the globe can be found in the schemes proposed by Sergin (1974):

$$T(\varphi) = T_g + \gamma(\sin^2 \varphi_T - \sin^2 \varphi), \tag{1.3}$$

where $\varphi$ is the latitude in radians; $\gamma$ is the pole–equator temperature difference; and $\varphi_T$ is the latitude at which $T(\varphi) = T_g$.

Latitudinal temperature variations can be described for the Northern Hemisphere:

$$T_a(\varphi, t) = T_e - 2\varphi(T_e - T_N)/\pi$$

and for the Southern Hemisphere

$$T_a(\varphi, t) = T_e - 2\varphi(T_e - T_S)/\pi,$$

where

$$T_N = \begin{cases} T_{N,\min} + 2t(T_{N,\max} - T_{N,\min})/t_\Delta, & t \in [0, 0, 5t_\Delta]; \\ T_{N,\min} + 2(t_\Delta - t)(T_{N,\max} - T_{N,\min})/t_\Delta, & t \in (0, 5t_\Delta, t_\Delta]; \end{cases}$$

$$T_S = \begin{cases} T_{S,\max} + 2t(T_{S,\min} - T_{S,\max})/t_\Delta, & t \in [0, 0, 5t_\Delta]; \\ T_{S,\max} + 2(t_\Delta - t)(T_{S,\min} - T_{S,\max})/t_\Delta, & t \in (0, 5t_\Delta, t_\Delta]; \end{cases}$$

where $T_{N,\min}(T_{S,\min})$ and $T_{N,\max}(T_{S,\max})$ are the minimum and maximum temperatures at the North (South) Pole, respectively, in degrees Celsius; $t_\Delta$ is the duration of the year in units of measurements $\Delta$; $T_e$ is the equatorial temperature of the atmosphere in degrees Celsius. Many authors use these estimates: $T_{N,\min} = -30°\text{C}$, $T_{N,\max} = 0°\text{C}$, $T_{S,\min} = -50°\text{C}$, $T_{S,\max} = -10°\text{C}$, $T_e = 28°\text{C}$.

Of course, these zonally averaged temperatures are widely scattered, which leads to considerable errors. To reflect the role of various factors more accurately in changes of the basic climatic parameter, temperature, it is necessary to calculate the contribution of each factor separately. This can be done by adding the role of feedbacks: $\Delta T_{a,\text{final}} = \Delta T_a + \Delta T_{a,\text{feedback}}$. In some models authors introduce a feedback coefficient $\beta$: $\Delta T_{a,\text{final}} = \beta \Delta T_a$. The $\beta$ parameter is expressed by amplifying index $g$: $\beta = 1/(1-g)$. The value of the $g$ index is equivalent to albedo $\alpha$, which globally is a function of $T_a$. This dependence can be roughly approximated by the formula

$$\alpha(T_a) = \begin{cases} \alpha_{\text{ice}} & \text{for } T_a \le T_{\text{ice}}, \\ \alpha_{\text{free}} & \text{for } T_a \ge T_{\text{free}}, \\ \alpha_{\text{free}} + b(T_{\text{free}} - T) & \text{for } T_{\text{ice}} < T_a < T_{\text{free}}. \end{cases}$$

where $T_{\text{ice}}$ and $T_{\text{free}}$ are average planetary temperatures at which the whole surface of the Earth is ice covered or ice free, respectively; and $b$ is the coefficient of transition between critical conditions of the Earth's albedo. It is usually assumed that $T_{\text{ice}} \in [263, 283]$ K.

Simple rough climate models can be specified by considering the characteristic times at which feedbacks function. The lag time of reactions within the NSS varies widely, and its consideration is necessary when assessing the consequences of changes within one or several climate subsystems. In particular, the supplies of cold in the Antarctic ice sheet are so huge that, to raise its temperature to 0°C, the average temperature of the World Ocean should be lowered by 2°C (i.e., to get $T_0 = 3.7°\text{C}$ instead of $T_0 = 5.7°\text{C}$). According to data that characterize the flow of particles emitted to the atmosphere (Kondratyev *et al.*, 2003a), it will take hundreds of years for this to happen. The observed rate of the anthropogenic climate warming is not capable of such energy supplies.

Anthropogenic impact on the climate system is manifested through GHG emissions, albedo change due to variations in Earth's covers, interference with the water cycle, and atmospheric pollution. Aerosol particles of radius $10^{-7} \pm 10^{-2}$ cm are observed at almost all altitudes of the atmosphere. Particles of non-anthropogenic origin get to the atmosphere from land or ocean surfaces and result from *in situ* chemical reactions. Particles of anthropogenic origin are mainly the result of fossil fuel burning.

The effect particles have on atmospheric temperature is explained by solar radiation reaching the Earth (mainly within the range 0.4–4 μm) being partially reflected and absorbed by them. The global albedo of the surface–atmosphere system varies too. Moreover, particles affect the processes of moisture condensation in the atmosphere, since they participate in the formation of clouds, rain, and snow. Take the heat balance equation of the surface–atmosphere system:

$$(1-\alpha)E_0^* + E_a - \sigma T_s^4 = 0, \tag{1.4}$$

where $T_s$ is the average efficient temperature of the emission from the system close to the average energy level near the surface 400 mb; $E_0^* = 0.487\,\text{cal/cm}^2$ min is the mean hemispherical intensity of incoming solar radiation; $\alpha$ is albedo; $\sigma = 8.14 \cdot 10^{-11}\,\text{cal/cm}^2$ min is the Stefan–Boltzmann constant; and $E_a$ is the total intensity of anthropogenic sources of energy per unit surface.

Let albedo $\alpha = \alpha_0 - \Delta\alpha$, where $\alpha_0 = 0.35$ is the albedo value under today's (2012) conditions, and $\Delta\alpha$ is that fraction of albedo determined by the impact of anthropogenic aerosols. From equation (1.4) we can obtain an expression for temperature:

$$T_s = [E_0^*(1-\alpha)/\sigma]^{1/4}[1 + \Delta\alpha(1-\alpha_0)^{-1} + (E_a/E_0^*)(1-\alpha_0)^{-1}]^{1/4}. \tag{1.5}$$

Assuming $\Delta\alpha \ll 1$ and $E_a/E_0^* \ll 1$, we can expand the function on the right-hand side of equation (1.5) into a Taylor series by degrees $\Delta\alpha$ and $E_a/E_0^*$ and write the first terms of the series:

$$T_s \approx [E_0^*(1-\alpha_0)/\sigma]^{1/4}\{1 + 0.25\Delta\alpha(1-\alpha_0)^{-1}[1 + E_a/E_0^*]\}. \tag{1.6}$$

It follows from (1.6) that temperature under weak anthropogenic impacts is the sum of terms describing bonds in the surface–atmosphere system, ignoring anthropogenic factors, and terms $T_1$ and $T_2$, which represent the contribution of heat and aerosol emissions, respectively:

$$T_1 = 0.25(1-\alpha_0)^{-1}[E_0^*(1-\alpha_0)/\sigma]^{1/4} E_a/E_0^* \approx 96.046 E_a/E_0^*,$$

$$T_2 = 0.25(1-\alpha_0)^{-1}[E_0^*(1-\alpha_0)/\sigma]^{1/4}\Delta\alpha \approx 96.046\Delta\alpha.$$

Note that the contribution made by $T_1$ is currently very small. Assuming that $E_a = 4 \cdot 10^{-5}\,\text{cal/cm}^2$ min and, hence, $E_a/E_0^* = 8.21 \cdot 10^{-4}$, then $T_1 = 0.0079$°C. Thus, the direct impact of global energy on atmospheric average temperature is now negligible. It follows from the expression for $T_1$ that to raise the atmospheric temperature by 0.5°C by thermal emissions, the condition $E_a/E_0^* = 0.0052$ has to be satisfied, and this means a factor-of-63.4 increase of anthropogenic heat fluxes to the environment. This is equivalent to the energy release of annually burning $570 \cdot 10^9$ t of additional fuel.

If we assume that energy production is proportional to population size, then $T_1 = 96.046 k_{TG} G \sigma_S / E_0^*$, where $G$ is the population density in people per square kilometer; $\sigma_S$ is the land area in square kilometers; and $k_{TG}$ is the per capita amount of produced energy in calories per minute.

If we ignore the impact of aerosols on the thermal regime of the atmosphere, then direct radiation $E$, its change $dE$, and change of atmospheric turbidity $dB$ will be related as $dE/E = -k_B\, dB$, where $k_B = 0.1154\,\mathrm{km}^2/\mathrm{t}$ is the proportion coefficient, $B$ is the amount of anthropogenic aerosols in tonnes per kilometer. After integrating this equation, we obtain $E = E_0^*(1 - \alpha_0)\exp(-k_B B)$. On the other hand, according to the definition of albedo, $E = E_0^*(1 - \alpha) = E_0^*(1 - \alpha_0 + \Delta\alpha)$. Equating these expressions for $E$, we obtain $\Delta\alpha = -(1 - \alpha_0)[1 - \exp(-k_B B)]$. Hence, temperature change as a result of pollution of the atmosphere by anthropogenic aerosols is equal to:

$$T_2 = -0.25[E_0^*(1 - \alpha_0)/\sigma]^{1/4}[1 - \exp(-k_B B)] = -62.43[1 - \exp(-k_B B)].$$

Since the average emission of anthropogenic aerosols is estimated by many authors at $300 \cdot 10^6$ t/yr and the average lifetime of aerosols in the atmosphere is estimated at three weeks, $17.262 \cdot 10^6$ tonnes of particles, on average, reside in the atmosphere. In this case it follows from the formula for $T_2$ that atmospheric temperature should decrease by 0.84°C/yr.

Some experts consider the factor of atmospheric turbidity $B_T$, instead of index $\beta$, defining $B_T$ as the ratio of the coefficient of attenuation by solar energy in the real atmosphere $\alpha_r$ to the coefficient of attenuation in an ideal atmosphere $\alpha_I$: $B_T = \alpha_r/\alpha_I = (\alpha_I + \alpha_W - \alpha_A)/\alpha_I$, where $\alpha_W$ and $\alpha_A$ are the coefficients of attenuation by water vapor and by aerosols, respectively. The following estimates are assumed in a GSM:

$$B_T = \begin{cases} 3 & \text{in mid-latitudes,} \\ 3.5 & \text{in tropical latitudes,} \\ 2 & \text{at reduced contents of dust and water vapor.} \end{cases}$$

Many experts want global climate modeling to accurately and completely take into account all possible feedbacks and components of the climate system. However, this leads to complicated mathematical problems whose solution necessitates a huge amount of data and, in most cases, the solutions of respective equations turn out to be unstable. Therefore, the use of such complicated models as subunits of the global model of the NSS leads to a negative result (i.e., it is not possible to construct an efficient model). The best approach is to combine climate models with data from global monitoring. The scheme of this combination is very simple. The existing surface and satellite systems used to monitor climate-forming processes cover some of the cells $\{\Omega_{ij}\}$ of the Earth surface. Over these cells, measurements are made of temperature, cloudiness, water vapor content, aerosols and gases, albedo, and the various parameters of energy fluxes. By employing simple climate models and spatiotemporal interpolation methods it is possible to reconstruct, based on these measurements, the complete distribution pattern of climatic parameters over the whole territory $\Omega$.

## 1.6 STANDARDIZATION AND UNIVERSALIZATION OF GIMS FUNCTIONS

The GIMS includes a set of items whose functions are determined by a number of standard tasks. These items form the informational architecture of the monitoring system, which includes the basis of models that describe the environmental subsystems that brings about the possibility of environmental diagnostics. A GIMS-oriented system shell is formed to formalize input information. An additional database level is synthesized to have multiple assignments of semantic structures to real environmental subsystems with variable dimensions. Let us designate these structures by the matrix symbol $A_i = \|a_{i_1,\ldots,i_s}\|$, where element $a_{i_1,\ldots,i_s}$ matches the object, process, phenomenon, event, or other environmental bifurcation. Actually, matrix $A_i$ is the parametric image of a real environmental subsystem with its specific features. Parameter $s$ reflects the informational-type dimension of the subsystem section. The structures $\{A_i\}$ identify both the spatial distribution of subsystem components and their types and parameters. The basic structures of $\{A_i\}$ have four dimensions: $i_1$ by latitude, $i_2$ by longitude, $i_3$ by height, and $i_4$ by time. Other structures of $\{A_i\}$ determine model coefficients, types of SPFs, precipitation, temperature, radiation, etc.

The semantic structures $\{A_i\}$ called identifiers are used by basic models for the formation of initial fields, validation of model output, and preparation of final or intermediate reports. Land cover classification is the main function of the GIMS. The identifier of basic land cover classes enables correspondence between different types of SPFs and their parameters and spatial structure. An example of such an identifier is given in Figure 1.20. Each of the identifier elements can take on a vector structure when describing the various classifications of land cover and allows formation of global land cover classification maps by means of interpolation and extrapolation of satellite data.

The GIMS database together with the $\{A_i\}$ structures consists of information about model coefficients and a set of scenario fragments. The structures $\{A_i\}$ link the knowledge base to the database. Each symbol of $A_i$ is decoded in conformity with spatial ranking and reflects just how reliable the description of the environmental subsystem is both qualitatively and quantitatively.

## 1.7 A GIMS-BASED SIMULATION EXPERIMENT

The estimation and forecast of the state of the environmental system can be given by a simulation experiment, the conceptual and functional scheme of which is

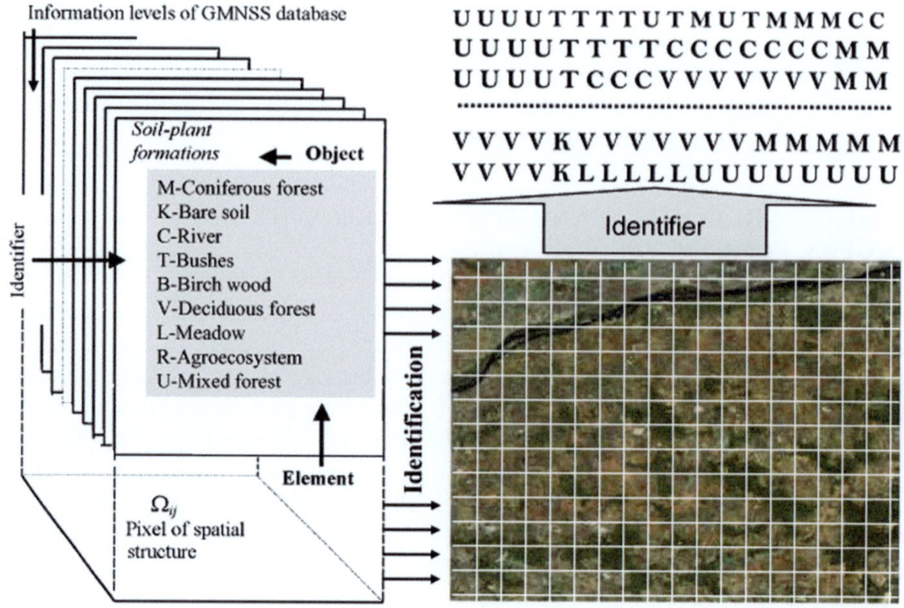


**Figure 1.20.** Cartographic identification and formation scheme of the GMNSS database.

shown in Figure 1.13. This scheme assumes various approaches to synthesizing a set of models to describe all aspects of the interactions of environmental bodies with their physical, biological, and chemical properties. The main idea of a GIMS-based simulation procedure is to estimate any deviation between measured and predicted trends of the environmental subsystem, which can be used to make a decision on planning a monitoring regime or for correcting model characteristics. Using this process to set up dynamical monitoring of a specific natural object requires the adaptation and possible addition of new data and model items. This is the reason the basic model structure must be oriented to the adaptive regime of its use. A specialized identifier controls this regime making it possible to select between various regimes that are both manually and automatically relocatable. Moreover, one of the GIMS subsystems has an intelligent support function. Mathematical algorithms compiled as part of the software provide the user with intelligent support when undertaking complex analysis of objective information gathered within the framework of the simulation experiment. The necessary information for objective dialogue with the basic model is provided in a convenient form for the user. The introduction of data-processing corrections is also catered for.

The orientation of the GIMS—for example, on problems related to attenuation of electromagnetic waves by vegetation canopies or on the classification of terrestrial carbon sinks—requires setting up a specific knowledge base and synthesizing a subsystem with functions to model and evaluate the soil–plant cover within the area studied.

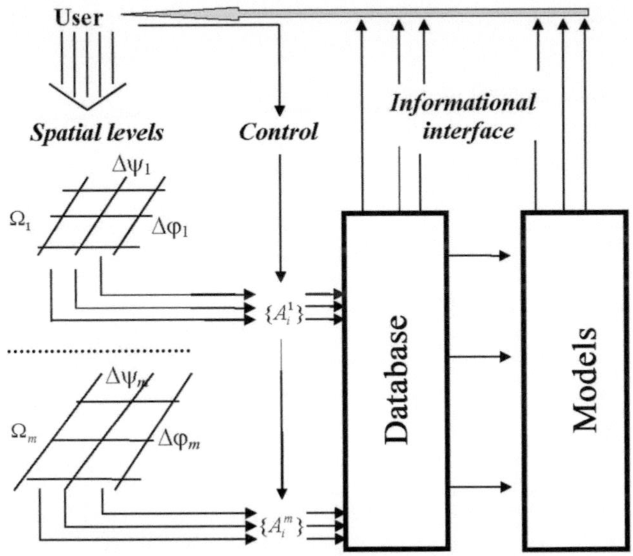


**Figure 1.21.** Block diagram showing how simulation experiments are controlled during environmental investigations using a GSM.

GIMS input data are usually formed in conformity with spatial discretization of the Earth's surface. The basic type is a geographic grid with latitude and longitude steps of $\Delta\varphi$ and $\Delta\psi$, respectively. Actual application of the existing database within the framework of the GIMS leads to nonuniform structures for different units. Depending on the specific features of the task under consideration, the real grid structure may be linked to regional features. The interactive mode of the GIMS allows the regime to be adapted to the real natural system through a set of identifiers. This is demonstrated in Figure 1.21. The GIMS database has identifier sets for the spatial distribution of SPFs (Figures 1.19 and 1.22). The user can form such identifiers according to concrete discretization of the space $(\Delta\varphi, \Delta\psi)$ using his/her own datasets. An example of the structure of the identifier level of the GIMS database is given in Figure 1.19.

Thus, the GIMS problem shell is capable of formalizing an image of the environmental subsystem based on the standard set and allows the use of informational channels to simulate experimental investigations. The principal scheme of GIMS application to retrieve the status of land covers using the Internet is given in Figure 1.23.

## 1.8 GIMS-BASED REMOTE-SENSING RESEARCH PLATFORMS

The solution of specific environmental tasks requires optimizing both instrumental and algorithmic means. This can be brought about by means of GIMS technology

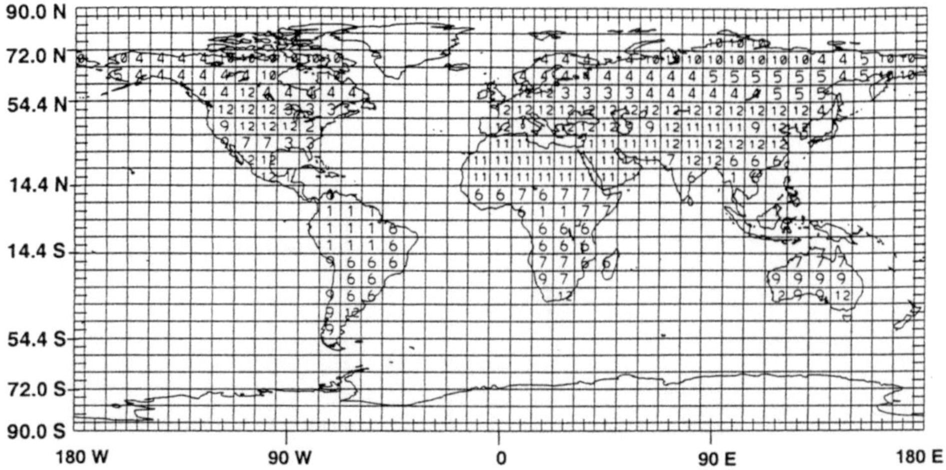


**Figure 1.22.** Spatial distribution of land biomes over a 7.2° × 9.0° geographical grid (Sellers *et al.*, 1997). Designations: 1—deciduous evergreen trees (tropical forest), 2—broad-leaved deciduous trees, 3—mixed broad-leaved and coniferous trees, 4—coniferous evergreen trees, 5—coniferous trees for felling, 6—broad-leaved trees with subforest, 7—meadows, 8—forest-covered meadows, 9—shrubs with bare soil, 10—dwarf trees and shrubs (tundra), 11—soil without vegetation cover, 12—agricultural plants.

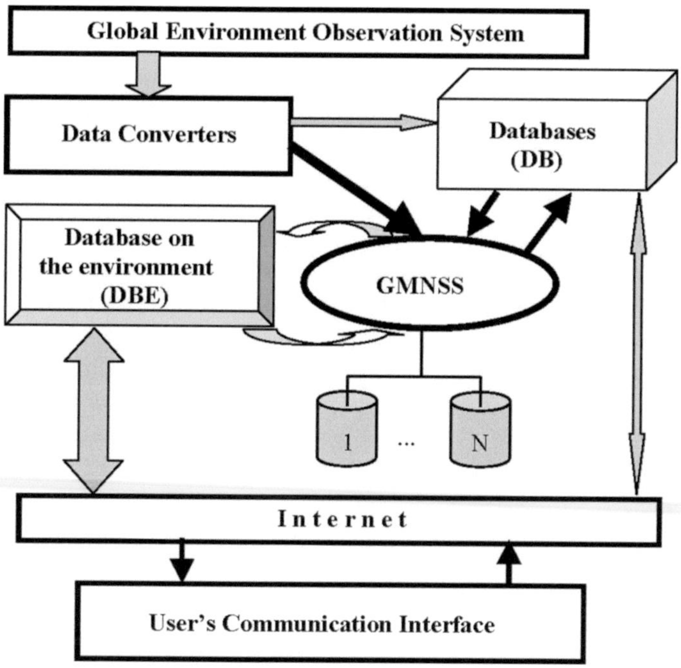


**Figure 1.23.** Block diagram of information flows within the GIMS structure set up in such a way to solve the tasks involved in the control of soil–plant formation.

**Figure 1.24.** Airborne SAR complex IMARC showing how its antennas are deployed (Shutko *et al.*, 2010).

methods. The ratio between experimental and theoretical methods in this case is defined by the complexity of the environmental system that is to be studied. Cumulative experience of the GIMS application is based on resolving national problems in many countries where different remote-sensing platforms have been set up. Historically, this started in 1980 by setting up flying laboratories on an AN-2 bi-plane, a twin-engine IL-14 aircraft, a four-engine IL-18 aircraft, and on three types of helicopter: MI-2, MI-8, and Ka-26. In the ensuing years, this set of carriers was widened.

An effective multi-frequency polarimetric synthetic aperture radar system was created in the 1990s by the Russian corporation Vega for surface and subsurface sensing. This is the airborne Intelligent Multi-frequency Airborne Polarimetric Radar Complex (IMARC) synthetic aperture radar (SAR) system (see Figure 1.24). This corporation created the Radius radiometer system, which operates at wavelengths of 0.8 cm, 2 cm, 5.5 cm, 21 cm, and 43 cm (Figure 1.25, Table 1.3). The main features of the IMARC SAR complex are given in Table 1.4.

The IMARC is a four-wavelength polarimetric airborne SAR system designed at Vega (Kutuza *et al.*, 2000). The basics of this radar are given in Table 1.4. The radar system operates at wavelengths: X (3.9 cm), L (23 cm), P (68 cm), and VHF (2.54 m); polarizations in all bands: VV, HH, VH, and HV; spatial resolution is around $12 \pm 8$ m; maximum swath is 24 km. The carrier of this system was a TU-134A twin-turbine jet airplane (Figure 1.24), but other aircraft were also used. The positioning of IMARC antennas on the airplane is shown in Figure 1.24. The main IMARC SAR mission goals were to map the characteristics of Earth's covers (including the hydrological regimes of soils), to map ground terrains covered by vegetation while eliminating the noise of vegetation, and to produce elevation

**Figure 1.25.** General view of the scanning SHF Radius radiometric system (2 cm and 5 cm) (Shutko *et al.*, 2010).

models to detect areas with on-ground and underground irregularities, etc. (Shutko *et al.*, 2010).

IMARC can be used to carry out the following tasks:

- surface sensing of ocean, ice, and vegetation;
- subsurface sensing of objects that scatter and deep layers.

Multi-frequency polarimetric SAR gives new possibilities for operational remote sensing of the sea, soil, vegetation, ice cover, and other Earth surfaces. The algorithms used for deep-layer remote sensing, which is undertaken layer by layer, make it possible to solve a narrow set of environmental problems.

**Table 1.3.** Characteristics of the Radius scanning radiometer. Power consumption is 200–300 W, power supply is 27 VDC; weight is 100–120 kg; $H$ is height above ground.

| *Frequency* | *Wavelength* | *Band* | *Pixels/scan* | *Resolution* | *Mode* |
|---|---|---|---|---|---|
| 37 GHz | 0.8 cm | Ka | 32 | $0.04H$ | Scanning |
| 15.2 GHz | 2 cm | X | 16 | $0.08H$ | Scanning |
| 5.5 GHz | 5.5 cm | C | 6 | $0.13H$ | Scanning |
| 1.4 GHz | 21 cm | L | 2 | $0.5H$ | Scanning |
| 0.7 GHz | 43 cm | S | 1 | $H$ | Non-scanning |

**Table 1.4.** The IMARC SAR complex parameters.

| *Parameter* | *Value* | | | |
|---|---|---|---|---|
| Frequency | X | L | P | VHF |
| Wavelength (cm) | 3.9 | 23 | 68 | 254 |
| Polarization | VV, HH, VH, HV | | | |
| Resolution (m) | 4–6 | 8–10 | 10–15 | 15–20 |
| Antenna gain (dB) | 30 | 14–17 | 14–17 | 9–11 |
| Width in azimuth (deg) | 18 | 24 | 24 | 40 |
| Width in elevation (deg) | 24 | 24 | 24 | 60 |

The larger the wavelength, the higher the influence of deeper soil layers—this fact allows the development of methods of thick-layer deep sensing using multi-frequency radar systems. For subsurface sensing long-wave P and VHF bands are required.

Information about the soil property (soil moisture) profile can be received by analyzing scattering measurements at different wavelengths. To calculate the influence of the soil moisture profile in the backscattering cross section it is necessary to develop models of reflection from layers at different depths. Solution of the inverse problem can be obtained by setting the measurement of backscatter at several wavelengths and at different polarization modes. For complete information about the soil moisture profile it is necessary to solve the problem of image interpretation in a broad band of wavelengths including the meter band where attenuation by soil and vegetation is comparably low.

The results of a multi-band radar survey obtained with the help of the four-band airborne SAR IMARC (Vega) illustrate the possibility of measuring hydrological soil regimes and water lens allocation in the Kara-Kum Desert. Lenses of underground water at depths of 50–70 m were detected. The results were validated by boring a control well. Radar images reveal

(1) the dry riverbed of the Uzboy;
(2) sand dunes 6–15 m in height;
(3) underground water lenses;
(4) transmission facilities.

Application of GIMS technology depends on just how complex the solution to the environmental tasks under consideration is in a given time period at a given precision. An effective instrument using GIMS technology was synthesized by Miramap (a private-sector Dutch company) in response to a European Space Incubator initiative from the ESA Technology Transfer & Promotion (TTP) Office. The TTP Office contributes to funding space-based technology and know-how for the benefit of Europe's economy and science. The innovative microwave radiometer (MR) produced by Miramap was nominated for the Dutch Innovation Prize in 2005 and was featured in several newspapers and magazines such as the *Dutch Financial Times*.

The Miramap instrument consists of three microwave sensors in the X-band, C-band, and L-band, all of which are Global Navigation Satellite System (GNSS) integrated. The X-band and C-band sensor makes a conical scan at a constant incidence angle over a wide swath, while the L-band sensor makes a twin-beam oscillating scan (Figure 1.26 and Tables 1.5 and 1.6). The small instrument sizes

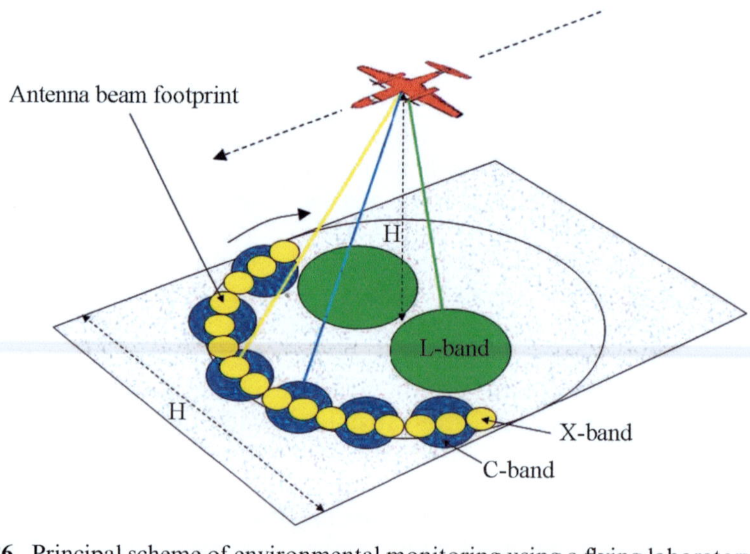


**Figure 1.26.** Principal scheme of environmental monitoring using a flying laboratory equipped with the Radius scanning microwave radiometric system.

**Table 1.5.** Miramap microwave sensor specifications.

| *Parameter/Band* | *X-band* | *C-band* | *L-band* |
|---|---|---|---|
| Frequency (GHz) | 15.2 | 5.5 | 1.4 |
| Wavelength (cm) | 2.0 | 5.5 | 21 |
| Pixels/Scan | 16 | 6 | 2 |
| Incidence angle (deg) | 30 | 30 | 15 |
| Beamwidth (deg) | 3.5 | 5 | 25 |
| Polarization | H | H | H |
| Sensitivity (K/s) | 0.15 | 0.2 | 1 |
| Absolute accuracy (K) | ±5 | ±5 | ±5 |

**Table 1.6.** Microwave Radiometer Mapping Company (Miramap, Noordwijk, The Netherlands): sensor specification.

| *Sensor* | *Type* | *Wavelength* | *Project specs* | *Use* |
|---|---|---|---|---|
| Digital photo camera | Rollei AIC 50 mm lens | Visible 0.4–0.7 μm | 10 cm GSD sub-pixel precision | Detailed visible interpretation |
| Lidar scanner | Optech Altimeter | SW infrared 1,064 nm | 2 cm GSD 0.1 m precision | Elevation model |
| Passive microwave scanner | Radius (IREE–Vega design) | Microwave 2, 5, 21 cm | 5 m GSD 0.15 K | (Sub) surface detection of wet and dry areas |
| Thermal camera | Flir Systems | LW infrared 7.5–13 μm | 3 m GSD 0.1°C | Surface temperature |

and weights enable use of a low-cost light aircraft as the observing platform, providing decision makers with a new affordable tool. The platform on which these instruments are flown is a reliable twin-engine Aero Commander (Figures 1.27 and 1.28). The aircraft is specially modified to simultaneously carry a range of other instruments, such as (digital) photogrammetric cameras, lidar scanners, and thermal infrared and multi-spectral sensors. The capability to measure such a comprehensive range of remotely sensed parameters from a single low-cost airborne platform is a first worldwide.

**Figure 1.27.** Miramap aircraft and its on-board sensors.

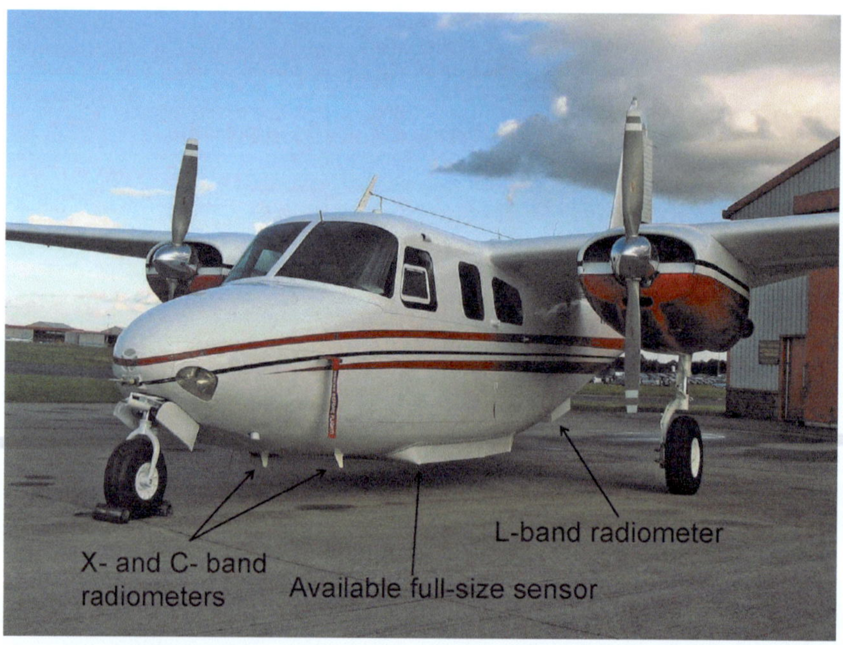


**Figure 1.28.** Miramap sensor aircraft (Shutko *et al.*, 2010).

**Table 1.7.** Miramap's platforms and remote-sensing instrumentation.

| *Platforms/Instrumentation* | *Aircraft laboratory* | *Unmanned plane/ Helicopter* | *Specific car/ Rover* | In situ *data-collecting instruments* |
|---|---|---|---|---|
| Microwave radiometers | Exist | Exist | Exist | Exist |
| Infrared radiometers | Exist | Exist | Exist | Exist |
| Optical color digital cameras | Exist | Exist | Exist | Exist |
| Lidar (3-D) land surface relief measurer | Exists | Doesn't exist | Doesn't exist | Doesn't exist |
| Georadar | Doesn't exist | Doesn't exist | Exists | Exists |

Armed with the set of sensors shown in Tables 1.5–1.7, Miramap provides customers with the parameters and environmental conditions given in Table 1.8, which include

- surface soil moisture;
- underground moistening;
- depth to a shallow water table (down to 2 m in humid areas and 3–5 m in arid/dry areas);
- surface-located and shallowly buried metal objects of reasonable size under dry ground conditions;
- contours of water seepage through hydrotechnical constructions (levees, dams, drainage systems that have fallen into disrepair, different kinds of leaks);
- biomass of vegetation above a water surface or wet ground;
- increase in temperature in land, forested, and volcano areas;
- changes in salinity/mineralization and temperature of a water surface;
- water surface pollution, oil slicks on a water surface;
- on-ground snow melting;
- ice on a water surface and on roads, runways.

**Table 1.8.** Miramap product specifications.

| *Parameters* | *Operating range* | *Maximum absolute error* |
|---|---|---|
| Soil moisture (g/cm$^3$) | 0.02–0.5 | 0.07 |
| Depth to water table (m) | 0.05–5 | 0.3–0.6 |
| Plant biomass (kg/m) | 0–3 | 0.2 |
| Pollutant concentrations (ppt) | 1–30 | 1–5 |

Some indexes of the effectiveness of GIMS technology as a result of being used by Miramap's Flying Laboratory are the following:

(1) Soil moisture content
- operating range is 0.02–0.5 $g/cm^3$
- maximum absolute error is 0.05 $g/cm^3$ (when vegetation biomass is less than 2 $kg/m^2$) and 0.07 $g/cm^3$ (when vegetation biomass is greater than 2 $kg/m^2$).

(2) Depth to a shallow water table
- operating range is 0.2–2 m (for humid or swampy areas) and 0.2–5 m (for dry arid areas, deserts)
- maximum absolute error is 0.3–0.6 m.

(3) Plant biomass (above wet soil or water surface)
- operating range is 0–3 kg/m;
- maximum absolute error is 0.2 kg/m.

(4) Salt and pollutant concentration of water areas (offshore zones, lakes):
- operating range is 1–300 ppt
- maximum absolute error is 1–5 ppt
- relative error is 0.5 ppt.

Large-scale investigations into the creation of microwave research carrying platforms were carried out by the Microwave Remote Sensing Division (MRSD) between 2002 and 2005. MRSD was part of the NASA Center for Hydrology, Soil Climatology & Remote Sensing (HSCaRS) at Alabama Agricultural & Mechanical University (AAMU). MRSD was capable of performing microwave radiometric data interpretation and conducting studies under field conditions, from mobile platforms to unmanned helicopters.

Antennas, radiometers, a data collection system, and an embedded Global Positioning System (GPS) receiver were mounted on a manned rover-type mobile platform and an unmanned MACS helicopter platform (MACS stands for Microwave Autonomous Copter System) to measure radiation from the soil–plant system. All radiometers were mounted on a foldable mounting panel to observe horizontally polarized radiation when folding the panel between the nadir through zenith looking angle. GPS information was used to register microwave readings to a common coordinate system of the study area. All data were stored on a 256 MB memory card. The data capture rate was set to one measurement per second in each of the radiometric channels and GPS readings.

The manned rover-type mobile platform was a modified Gator utility vehicle. This two-seater vehicle had a 286 cc, air-cooled, four-cycle gasoline engine. Its towing capacity was 500 lb (226 kg) and it had a top speed of 20 mph (32 km/h). The instrument platform (or mounting frame) for the radiometers and other instruments was assembled at the AAMU Research Station. The aluminum folding panel (1.5 m × 1.5 m) connecting all system components was designed so that the incidence angle from 0° (nadir) to 180° (sky) could be easily obtainable (Figures 1.29 and 1.30 and Table 1.9). The radiometers were mounted with the antennas viewing off to the right-hand side of the platform at an incidence angle

**Figure 1.29.** General view of the manned rover-type mobile platform equipped with three portable microwave radiometers, operating at wavelengths 6 cm, 18 cm, and 21 cm. Also equipped with a folding panel of 1.5 m × 1.5 m for instrumentation installation along with a GPS receiver, data acquisition system, and power supply battery (Shutko *et al.*, 2010).

of 10°. The rover radiometer shuttled back and forth in a north–south direction at a speed of 2–5 mph. Using the specially developed remote-sensing system, data were obtained from a height of 2 m and provided a spatial resolution of 1.4 m of the land area.

The unmanned MACS helicopter platform was equipped with a 6 cm radiometer (incidence angle 5°) mounted on the nose of an AutoCopter™ onto a stabilized gimbal with a pan/tilt interface that attenuates vibrations (Figures 1.31 and 1.32 and Table 1.10). MACS is a modified AutoCopter™, a small unmanned helicopter platform that can fly autonomously (fly a pre-programmed flight path) or semi-autonomously (with an operator directing the maneuvers). This is a product of Neural-Robotics, Inc. (NRI) of Huntsville, AL.

The unmanned helicopter's advantage lies in its patented flight control system consisting of multiple neural network modules working together. The result is an autonomous helicopter that adapts to changing conditions and provides an extremely stable platform for hundreds of applications. The AutoCopter™ is 2.18 m in length (from tip of tail rotor to tip of main rotor) and weighs approximately 13.6 kg. It carries a payload of up to 6.8 kg. Basic avionics consist of a PC/104

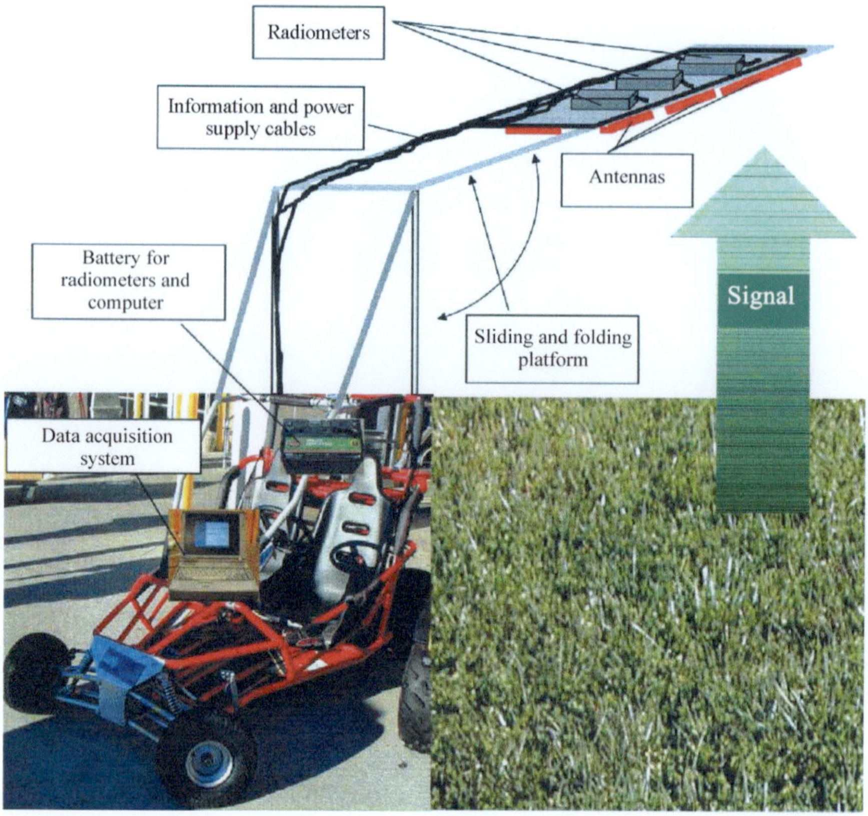


**Figure 1.30.** Subsystems and microwave radiometers on the rover's mobile platform.

**Table 1.9.** Characteristics of the rover microwave platform.

| *Parameter* | *Value* |
|---|---|
| Frequencies | 1.41, 1.67, and 5 GHz |
| Polarization | H or V |
| Antennas | 19.7″ × 19.7″ and 16.3″ × 16.3″ (slot array); 7.9″ × 6.5″ (dipole array) |
| Incidence angle | Fixed angles 0–30° off nadir |
| Sensitivity | 0.5 K |
| Beamwidth (3 dB) | 30° |
| Weight | 12 kg |
| Power required | 30 VA |

**Figure 1.31.** The Microwave Autonomous Copter System (MACS) unmanned aerial vehicle (UAV), a helicopter equipped with a 6 cm radiometer, data acquisition system, GPS receiver, and power supply battery (Shutko *et al.*, 2010).

**Figure 1.32.** Installation procedure for the unmanned helicopter's on-board radiometer.

**Table 1.10.** Characteristics of the microwave autonomous copter system.

| *Parameter* | *Value* |
|---|---|
| ***Physical characteristics*** | |
| Length | 3.58 m |
| Weight | 22.68 kg |
| Main blades | 1.03 m |
| Engine | 120 cc |
| Range | 161 km |
| Payload capacity | 20.41 kg |
| Fuel tank | 32 oz |
| ***Passive sensor packages*** | |
| Wavelengths | 3, 6, or 21 cm |
| Sensitivity | 0.3–0.5 K |
| 3 db Beamwidth | 30° |
| Power supply | 27 VDC |
| Power consumption | 15 VA |
| Weight | 2–6 kg |
| ***Others*** | |
| Frequencies | 1.4 and/or 5 GHz |
| Polarization | H or V |
| Antennas | Flat panels w/30° 3 dB |
| Incidence angle range | Fixed angles 0–30° |
| Sensitivity | 0.5 K |
| Flight time | 2 hours |

computer, Altitude and Heading Reference System (AHRS), GPS receiver (WAAS compatible), downward-pointing range finder (ultrasonic sensor), barometric pressure sensor, and heading hold gyro. The standard transmitter is used as the "ground station".

A ground control station (GCS) employing the WayPlanner flight-planning program was used with the AutoCopter™. This is a self-contained Windows-based application that unlocks the power of fully autonomous flight. The program enables mission planning in two dimensions using stored satellite images. Flight plans were uploaded to the AutoCopter™ via a data link enabling the aircraft to take off, climb, fly its programmed route, and land fully autonomously. The programmed flight consisted of an autonomous launch with 16 waypoints, a climb to 30 meters, followed by a transition to forward flight at a velocity of 2 m/s and auto-landing.

During flight the aircraft had the ability to state data (aircraft altitude, speed, and other parameters) in real time on the GCS screen in both two and three dimensions. While maintaining airspeed, altitude, and heading, eight flight lines were flown at a distance of ~500 m at 30 m intervals. The time it took the helicopter to fly from waypoint to waypoint (north–south direction) was ~4.5 minutes totaling ~40 minutes of flight time. Because the aircraft flew below 500 ft it was exempt from FAA regulations. Using a specially developed remote-sensing system, data obtained from an altitude of 30 m provided a spatial resolution of 20 m of the land area.

## 1.9 SOME RESULTS OF THE ALABAMA SOIL MOISTURE FIELD EXPERIMENT

### 1.9.1 Microwave radiometers, radar systems, and thermal infrared radiometers

There are several types of instrumentation that can be used for environmental investigations. For examination of soil type, or surface moisture, the ones that spring to mind are thermal infrared radiometers, active radar systems, and passive microwave radiometers. A system of non-scanning single-beam microwave radiometers was designed in Kotelnikov's Institute of Radioengineering & Electronics (Shutko *et al.*, 2010). The basic technical parameters of this system are given in Tables 1.11 and 1.12. A common view of portable radiometers is given in Figures 1.33 and 1.34.

Radar systems, including Interferometric Synthetic Aperture Radar (IFSAR), share the same frequency band as microwave radiometers. Therefore, they have similar wave propagation characteristics, although radar systems are actively transmitting and measuring return values. The radar/IFSAR systems rely on surface texture to deduce surface boundary conditions.

**Table 1.11.** Set of portable microwave radiometers.

| *Parameter* | *Characteristic* |
|---|---|
| Frequency (GHz) | 1.41; 1.67; 5.0 |
| Polarization | V or H |
| Antennas<br>Slot array<br>Dipole array | <br>19.7″ × 19.7″; 16.3″ × 16.3″<br>7.9″ × 6.5″ |
| Incidence angle | Fixed angles 0–30° off nadir |
| Sensitivity (K) | 0.5 |
| Beamwidth (3 dB) | 30° |
| Platform | Boom truck, land rover, small aircraft |
| Weight (kg) | 12 |
| Power required (VA) | 30 |

**Table 1.12.** Characteristics of microwave radiometers designed at the Kotelnikov Institute of Radioengineering and Electronics, Russian Academy of Sciences (Shutko and Krapivin, 2010).

| *Frequency* (GHz) | *Wavelength* (cm) | *Range* | *Visual field* (deg) | *Spatial resolution* (*H*) | *Type* |
|---|---|---|---|---|---|
| 13.3 | 2.25 | X | 30 | 0.65 | Single beam |
| 5 | 6 | C | 30 | 0.65 | Single beam |
| 1.4 | 18/21 | L | 30 | 0.65 | Single beam |

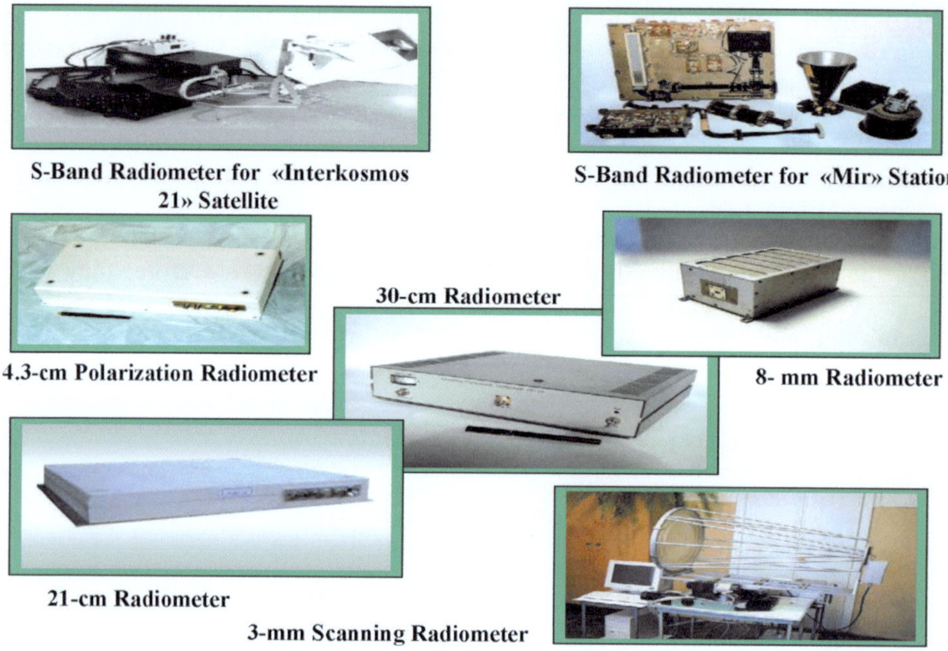


**Figure 1.33.** Set of portable radiometers produced by the Special Design Office of the Kotelnikov Institute of Radioengineering and Electronics (part of the Russian Academy of Sciences).

Passive microwave systems only record naturally emitted radiation, which is a function of soil moisture content within the top few centimeters/decimeters and is less influenced by surface roughness. From a given height above the ground surface radar systems have a much smaller ground sample distance, or pixel size, than passive systems. To compensate, passive systems can be flown on light aircraft at very low altitudes, as low as 500 feet above the mean terrain.

Thermal infrared radiometers (TIRs) and microwave radiometers both measure natural radiation from a thermodynamic process. TIR sensors are further

**Figure 1.34.** Twenty-one centimeter microwave radiometer produced by the Special Design Office of the Kotelnikov Institute of Radioengineering and Electronics (part of the Russian Academy of Sciences).

influenced by cloud, fog, smoke, solar radiation, and other similar atmospheric conditions. However, TIRs operating in the 10–12.5 μm wavelength range are mainly sensitive to temperature variations that take place within a thin surface layer of the object being imaged. This temperature variation could be the result of many things, such as warm water, steam leaking from underground pipes, poor insulation properties of buildings, effluent from a manufacturing process, etc.

Microwave radiometers, on the other hand, measure natural radiation at millimeter, centimeter, and decimeter wavelengths, which is primarily a function of soil moisture content in the first couple of centimeters/decimeters of the Earth's surface.

### 1.9.2 The field experiment

As mentioned earlier, from November 9 to November 22, 2005 an experiment on passive microwave remote sensing of soil moisture was carried out by a team of U.S. and Russian scientists at the Winfred Thomas Agricultural Research Station (WTARS) at the Alabama A&M University. The 2-week campaign took place in an area about 1 $km^2$ in size. Figure 1.35 shows the outline of the main area using a digital orthophoto image from 2005 of the WTARS. This area was selected because of its relatively smooth topography, availability of historical data, and an 8-acre irrigation pond. The historical data included map layers at a scale of 1:24,000 such as digital elevation models (DEMs), topographic, land use, and soil surveys. The Soil Survey Geographic (SSURGO) database was used to extract several parameters regarding the study site soil system such as soil type, surface

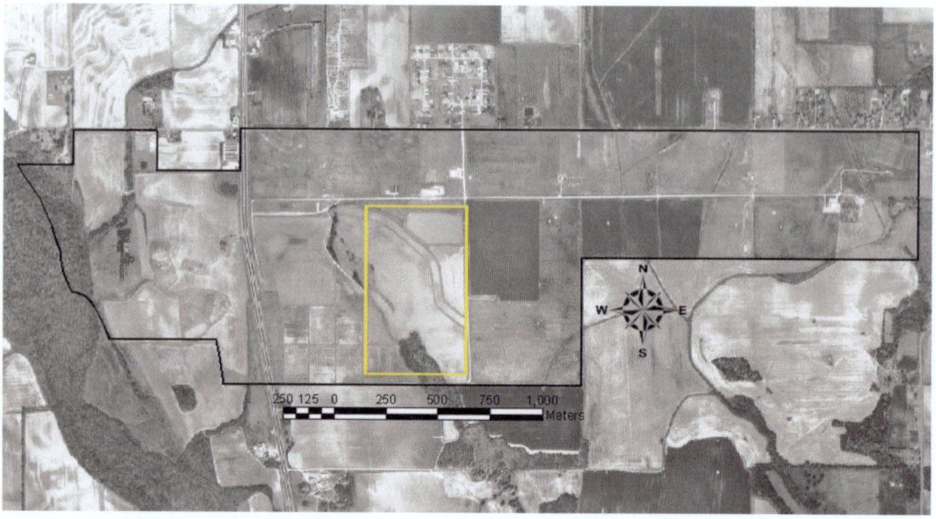


**Figure 1.35.** An outline of the study area is shown over an orthophoto image from June 2004 of the Winfred Thomas Agricultural Research Station, Alabama A&M University. An 8-acre irrigation pond is featured in the top center of the image.

texture, and permeability. The site is made up of two soil types, Decatur and Abernathy (Archer *et al.*, 2006). The surface texture ranged from a silt loam to silty clay. About 90% of the area is covered by soils with moderate infiltration rates. The organic matter content in the top 6 cm was less than 3%.

Another important resource in the area is the Alabama Mesonet. This is a network of 23 meteorological and soil moisture stations located in northern Alabama and southern Tennessee. One station located in close proximity to the study area was used to collect soil moisture and temperature, air temperature, relative humidity, wind speed and direction, and rainfall measurements. These stations are part of the USDA Soil Climate Analysis Network (SCAN). They are fully automated and provide near real-time observations at 5-minute intervals. More information on the Alabama Mesonet can be found in Tsegaye *et al.* (2005).

### 1.9.3 Materials and methods

#### *1.9.3.1 Ground truth measurements*

As part of this investigation, gravimetric soil moisture, surface temperature, and deep soil temperature were collected at random over some 10 locations of bare soil. These measurements were taken daily in the late morning, around 11:00 AM (local time), to coincide with ground overpasses of the rover and unmanned helicopter flight overpasses. Gravimetric soil samples were collected using 0–1 cm, 0–3 cm, and 0–6 cm scoop samplers. The samples were sealed in aluminum cans and their moisture content determined gravimetrically by weighing a moist sample, oven-drying it at 105°C, reweighing, and calculating the mass of water lost as a percentage of the mass of the dried soil. Gravimetric values were later converted to volumetric soil moisture using soil bulk density estimates. Soil bulk density estimates were determined using undisturbed soil cores taken at known depth increments of known volumes. These measurements (made on November 15) ranged from 1.16 to 1.30 g/cm$^3$ and increased with depth. Soil temperature measurements were measured daily by inserting a soil thermometer probe vertically to depths of 0–1, 0–3, and 0–6 cm. The probe was allowed to stabilize (for at least 2 minutes) before readings were recorded in degrees Celsius. Additionally, a small portable thermal infrared radiometer was used to estimate surface temperature by measuring thermal emission in the 8–14 μm wavelength range.

### *1.9.3.2 Microwave sensor systems*

Three microwave radiometers operating at the C-band (5.0 GHz) and two L-band (1.66 GHz and 1.42 GHz) wavelengths of 6, 18 and 21 cm, respectively, were used to investigate soil moisture retrieval. These unique remote-sensing radiometers were developed by the Institute of Radio-Engineering & Electronics (IREE) but managed and maintained in cooperation with Alabama A&M University. The characteristic antenna sizes were $20 \times 20 \times 3$ cm for the 6 cm radiometer and $60 \times 60 \times 3$ cm for the 18 and 21 cm radiometers (Figure 1.36). The radiometers

**Figure 1.36.** The microwave radiometers, which operate at wavelengths 6 cm, 18 cm, and 21 cm. These radiometers were used for remote-sensing measurements by means of the Microwave Autonomous Copter System (MACS), as shown in Figure 1.37.

**Figure 1.37.** Another view of the MACS (Shutko *et al.*, 2010).

weighed between 2–6 kg and were of Dicke type having a sensitivity of 0.2 K for a one-second integration time. The radiometers worked in non-scanning mode with a field of view of 30°, yielding a ground resolution 0.7 times the altitude. The system could operate with a supply voltage of 27 V.

Calibration of the radiometers mounted on the manned rover platform was based upon measurements of two external calibration targets of known brightness temperature, a cold target (sky) and a hot target (absorber material). The brightness temperature of the sky was assumed to be about 5 K. The emissivity of the absorbers was assumed to be about 0.99. The temperature (in degrees Celsius) of the absorbers was measured with a hand-held infrared probe before each measurement. Data from the sky and absorber measurements were used to produce a calibration regression equation (voltages versus brightness temperatures) from which brightness temperatures were derived from radiometric readings taken on different days. Based on the calibration results, the accuracy of the radiometric measurements was estimated at ±3 K. For the radiometer installed on board the unmanned helicopter platform, two natural environmental objects were chosen as external calibration targets (a pond and a forest).

### 1.9.4 Results and discussion

#### *1.9.4.1 Meteorological measurements*

Temporal variations in rainfall, air temperature, relative humidity, and wind speed are the main forces influencing fluctuations of soil moisture and soil temperature.

These forces were assumed to be constant for the entire study area because of its small size. Initially dry at the beginning of the experiment, four rainfall events of 3.8 mm (23:00–24:00), 45.7 mm (08:00–24:00), 22.6 mm (09:00–24:00), and 36.1 mm (06:00–21:00) occurred on days 10, 14, 15, and 21, respectively. The averaged air temperature range from 0.59°C to 21.16°C was slightly higher than the climatological norm for the month of November, typically ranging from 4.83°C to 16.44°C. Daytime sky conditions were usually clear to partly cloudy. Relative humidity ranged from 52% to 97%. Daytime relative humidity was lower during the first week of the experiment with an average of 80.59% compared with 83.09% the second week. The wind speed averaged 1.95 m $s^{-1}$ throughout the experiment. The highest wind speed of 4.30 m $s^{-1}$ was obtained on November 15 and the lowest occurred on November 11. Wind speeds were higher in the first week (3.20 m $s^{-1}$) compared with the second week (1.63 m $s^{-1}$).

#### *1.9.4.2 Soil moisture measurements*

Meteorological conditions during the experiment resulted in ground moisture values covering the full range of brightness temperature observations. At the beginning of the study (November 9, 2005) the soil surface was very dry. Over the night of November 10 there was a trace amount of rainfall (3.8 mm), which resulted in a slight increase of soil moisture estimates in the late morning of November 11. On November 14 and 15 there were rainfall totals of 50 mm, which saturated the surface and marked the starting point of the dry-down sequence. After November 15, there was a gradual downward trend in soil moisture. A time series of observed mean volumetric soil moisture content for the 0–1 cm, 0–3 cm, and 0–6 cm soil layers from gravimetric samples collected at random locations within the study area are shown in Figure 1.2. Also included are the rainfall amounts. The mean surface moisture content responded predictably to rainfall, increasing after rain events and decreasing thereafter. Surface soil moisture ranged from 0.07 $cm^3/cm^3$ (0–1 cm) and 0.12 $cm^3/cm^3$ (0–6 cm) under dry conditions to 0.25 $cm^3/cm^3$ (0–1 cm) and 0.32 $cm^3/cm^3$ (0–6 cm) under wet.

#### *1.9.4.3 Surface temperature measurements*

The temperature of the soil surface is dependent on several meteorological and physical parameters, which include soil moisture. The maximum surface temperature was obtained at about 13:00 hours on November 9, 2005, with an average value of 29.6°C. After the rainfall event on November 21, surface temperature dropped to its lowest value of 8.6°C recorded during the study period. Surface temperature measurements exhibited the largest change in values between the day prior to the rainfall event and the day after the rainfall event, creating the steepest slope between temperature values during the study period. Between November 9 and November 22 surface temperature changed to around 21°C. Mean surface temperature was close to air temperature with surface temperature averaging

14.57°C andair temperature averaging 10.47°C throughout the experiment. These results indicate that meteorological data of air temperature could serve as a reasonable proxy for surface temperature. Water surface temperature (pond) ranged from 3.13°C to 19.46°C.

#### *1.9.4.4 Soil temperature measurements*

The soil temperature profile varied from layer to layer at a given site and time. This fluctuation was likely mainly due to the intensity and distribution of precipitation, daily fluctuations in air temperature, vegetation cover, and duration of moisture status. Soil temperature measurements at depths of 0–1 and 0–6 cm were very close, within 3°C of one another, and exhibited the same pattern in both increases and decreases in temperature over time. However, over the 2-week study, no significant change in temperature at any depth was observed. Soil temperature ranged from 7.20°C to 30.40°C in the top 1 cm and from 5.90°C to 25.30°C at the 6 cm depth. Soil temperature peaked on November 9 and reached a minimum on November 15. The values decreased with increase in depth.

#### *1.9.4.5 Rover-based brightness temperature analysis*

Two-dimensional (2-D) brightness temperature ($T_b$) data collected during the experiment are discussed in the following section. These discussions include the establishment of relationships between the response of 6 cm and 21 cm data over bare soil. Results relative to the different wavelengths are examined separately. Brightness temperature measurements inside the study area averaged 226 K at the 21 cm wavelength throughout the experiment. Figures 1.4 and 1.38 show 2 days of 21 cm (L-band) $T_b$ measurements and retrieved soil moisture distribution for the study area covered by rover ground overpasses. The results from the measurements on November 9 and 10 are shown as a 2-D image. About eight north–south ground overpasses at a height of 2 m (ground resolution 1.4 m) were needed to cover the area. The black dotted line is the GPS indicated position. The soils were very dry to begin with; there was little rainfall over the area in the month prior to the first data obtained on November 9. On November 9, 21 cm (L-band) values over the land surface averaged ~245 K and displayed a variation in the range 230–280 K. Subsequent measurements on November 10 showed an increase in surface brightness (247 K), implying a general decrease of surface soil moisture with time in the area (Figure 1.4b). These results show interesting patterns that are obviously associated with drying of the soil surface. Over the night of November 10 there was a trace amount of rainfall, which lowered the $T_b$ values in late morning of November 11 to 237 K. On November 16 (the wettest day of the experiment) $T_b$ averaged 205 K. Figures 1.2 and 1.3 show the brightness temperature measurements and retrieved soil moisture estimates at 21 cm and 6 cm after rainfall on November 21, 2005. In general, it caused a distinctive change in $T_b$ values measured after this date. The 2-D $T_b$ map derived from that day displayed

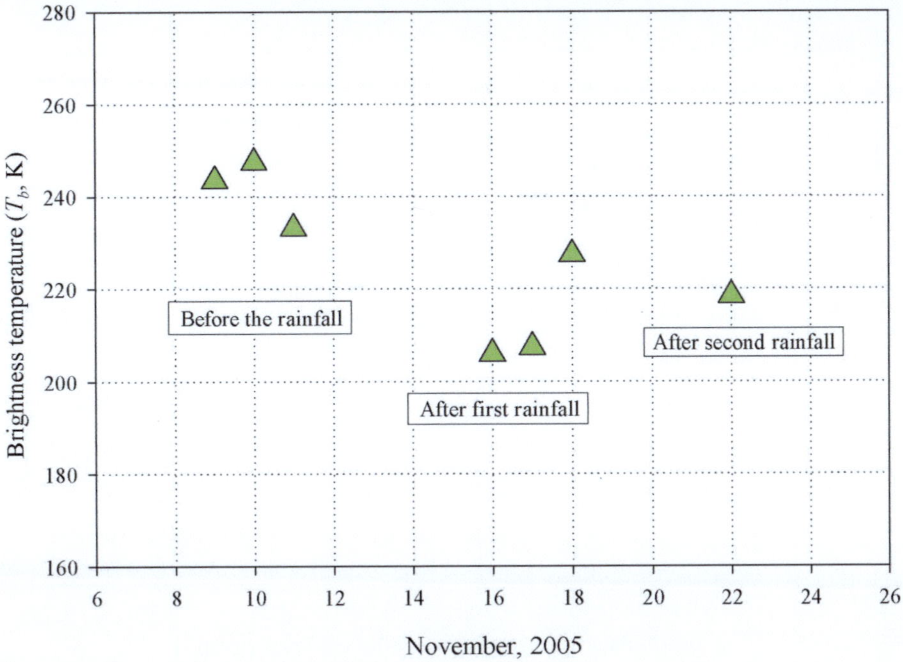


**Figure 1.38.** Time series of mean brightness temperature ($T_B$) measurements at 21 cm, obtained inside the study area from the rover platform.

a variation of values from below 200 K to more than 250 K. The overall results of 6 cm $T_b$ measurements tended to produce higher $T_b$ values than the 21 cm $T_b$. Variations between the two wavelengths may indicate that the 21 cm $T_b$ is lower as a result of deeper sensing of microwave radiation at this wavelength. Based on these measurements, the dependence of aggregated brightness temperature at 6 and 21 cm wavelengths versus volumetric soil moisture data was obtained (Figure 1.39). It indicated good agreement with theoretical predictions for uniformly moistened soil. In particular, the experimental data of radiation sensitivity to changes in volumetric data of soil moisture (i.e., $-207.83\ \text{K/cm}^3$), lay within the range of theoretical estimates (i.e., $-200$ to $-300\ \text{K/cm}^3$), as presented below

$$S_{21} = \frac{\Delta T_{b_{21}}}{\Delta m} = \frac{T_{b_{\max 21}} - T_{b_{\min 21}}}{m_{\min} - m_{\max}} = \frac{247.4 - 206}{0.12 - 0.32} = -207 \frac{\text{K}}{\text{g/cc}}.$$

Theoretical assessments were $S_{21} = -200\ \text{Kg}^{-1}\text{cm}^{-3}$ or higher.

### *1.9.4.6 Rover-based retrieved soil moisture at the 6 and 21 cm wavelengths*

Comparisons of model-estimated soil moisture ($\text{cm}^3/\text{cm}^3$) at 21 cm and observed field-measured soil moisture ($\text{cm}^3/\text{cm}^3$) for three surface layers gathered by the manned rover platform are shown in Figure 1.40. The model performed well in

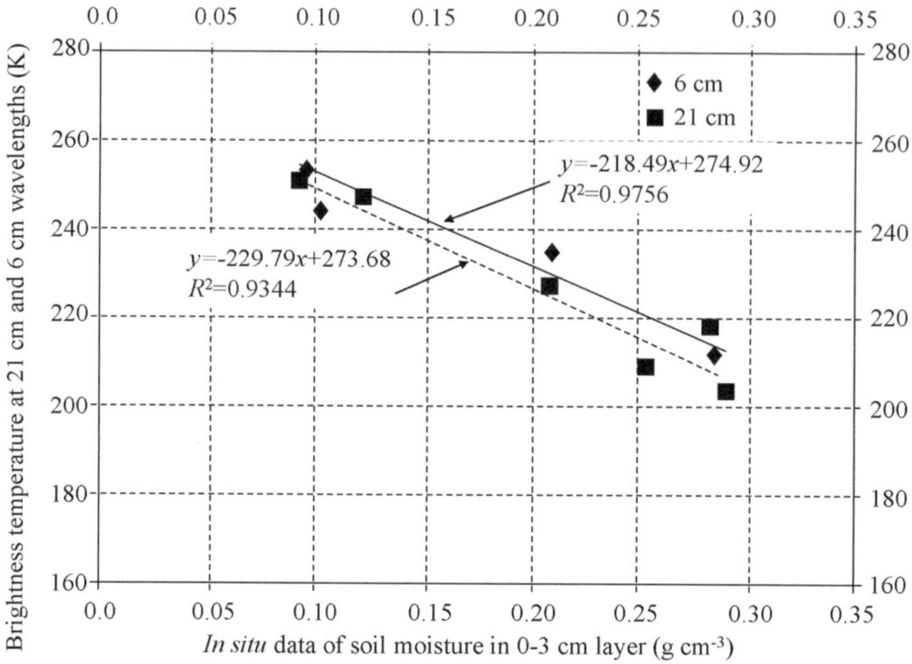


**Figure 1.39.** Comparisons of brightness temperature measurements at 21 cm and 6 cm wavelengths with *in situ* data of soil moisture in the 0–3 cm layer (rover platform).

estimating surface soil moisture at both 21 and 6 cm compared with gravimetrically determined data. Highly significant correlations $R^2$ were obtained between remotely sensed surface moisture at 21 cm and 6 cm and validating data obtained through gravimetric sampling. In general, linear regressions of soil moisture for the 0–1 cm ($R^2 = 0.91$), 0–3 cm ($R^2 = 0.92$), and 0–6 cm ($R^2 = 0.87$) fit the measured *in situ* data very well (average 0.89). During these investigations, it was confirmed that the extent to which soil depth is affected by microwave radiation depends on the observation wavelengths. This leads to the conclusion that multi-wavelength observations are capable of retrieving the soil moisture profile.

#### *1.9.4.7 Unmanned helicopter-based brightness temperature analysis*

The 6 cm (20 m ground resolution) data collected from overpasses made by the unmanned helicopter were analyzed to provide brightness temperatures for comparison with *in situ* data. The data analyzed corresponded to November 9 and 10, 2005. The purpose of the tests was to evaluate the feasibility of using an unmanned helicopter platform to acquire passive microwave remote-sensing datasets. The two flights provided data over a wide range of brightness temperatures. The full flight trajectory is shown in Figures 1.41a and 1.42a. The black dotted line is the GPS indicated position overlaid on a 2-D map of $T_B$ measurements.

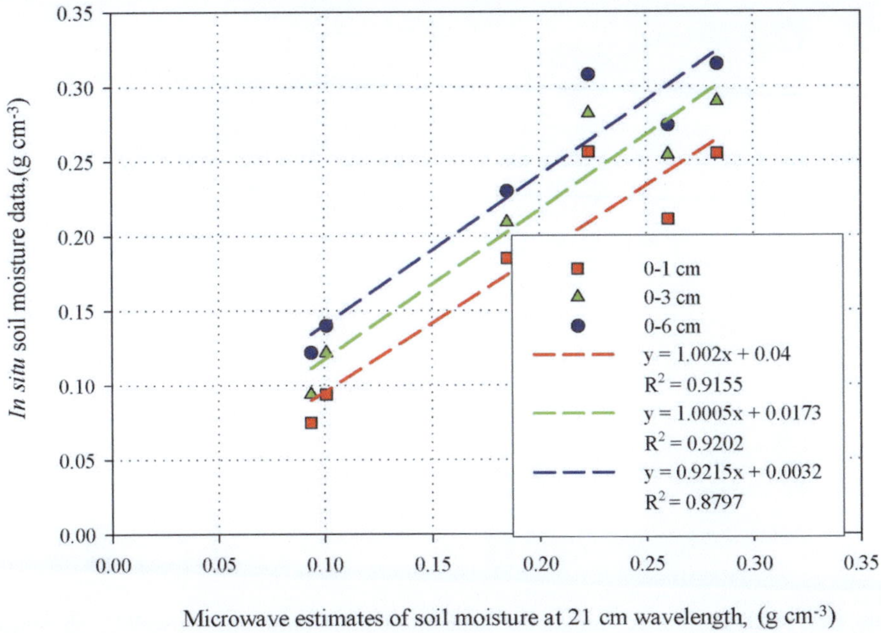


**Figure 1.40.** Comparison of microwave estimates of soil moisture at the 21 cm wavelength with *in situ* soil moisture in the 0–1 cm, 0–3 cm, and 0–6 cm layers (rover platform).

Unfortunately, during the experiment only 2 days of flying time were used, due to technical problems with the helicopter. Many hours targeted for science flights were lost to the science team while work was being done on the aircraft. On the days of flight, the aircraft flew flawlessly with no problems from radio frequency interference (RFI), vibrations, and/or gusts of wind. Data obtained at the 6 cm wavelength were used for comparison with soil moisture measurements. Good results confirmed the sensitivity of the wavelength to soil moisture. The range of $T_b$ was between 100 and 280 K showing a significant range of variations. This range of variation can be attributed to an 8-acre irrigation open-water pond. The area with high $T_b$ values (>230 K) warmed up markedly on November 10, suggesting a decrease in surface soil moisture over the period.

### *1.9.4.8 Unmanned helicopter-based retrieval of soil moisture at 6 cm*

Figures 1.41b and 1.42b show the soil moisture values (g/cm$^3$) retrieved between November 9 and 10. The results show variation in soil moisture over bare soil and alongside the open-water pond. Comparisons between estimated and ground-measured soil moisture are shown along with rover C-band estimates over the same area and time frame in Figure 1.43. They show good correlation at the 0–1 cm depth. These findings are in agreement with those of Njoki and O'Neil (1982).

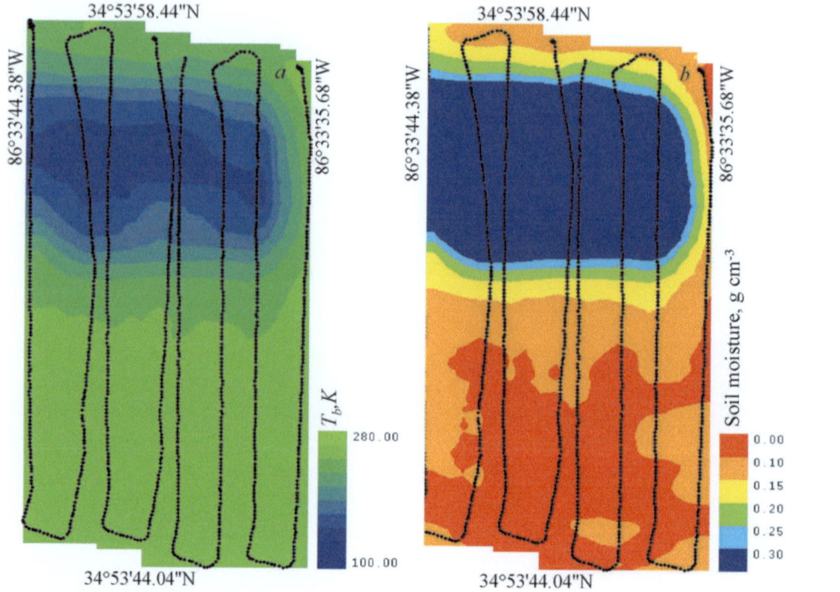


**Figure 1.41.** GPS-positioned unmanned helicopter flight lines traces overlaid on maps of (a) brightness temperature and (b) retrieved soil moisture data at the 6 cm wavelength obtained on November 9, 2005.

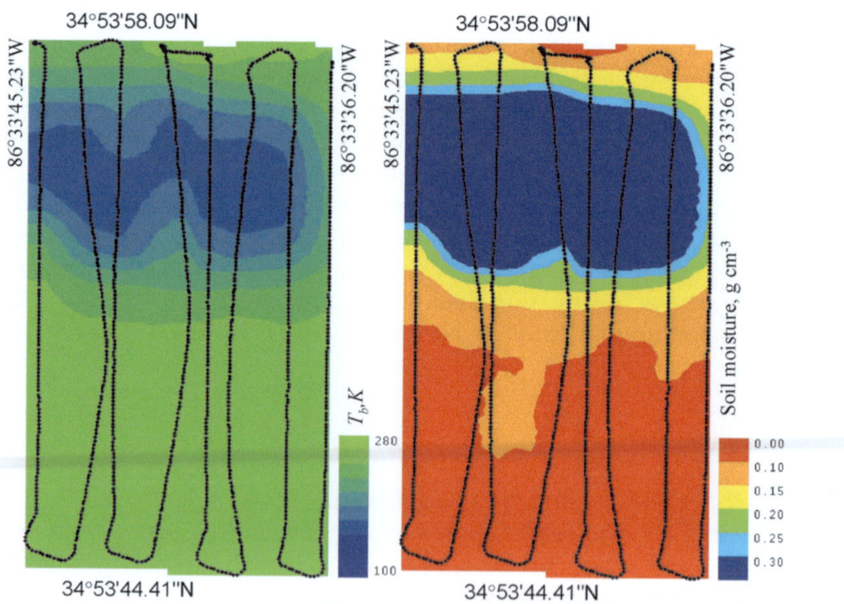


**Figure 1.42.** GPS-positioned unmanned helicopter flight lines traces overlaid on maps of (a) brightness temperature and (b) retrieved soil moisture data at the 6 cm wavelength obtained on November 10, 2005.

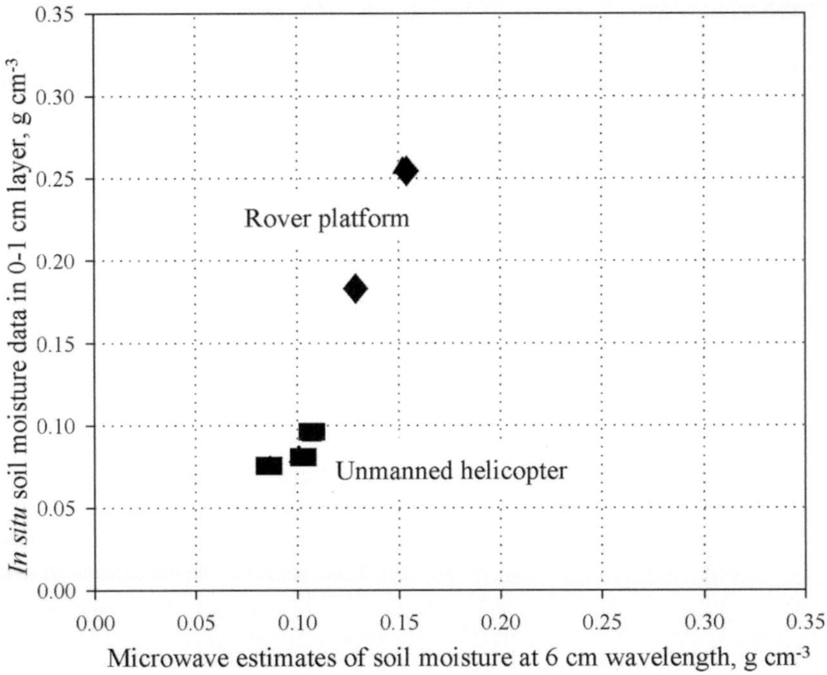


**Figure 1.43.** Comparison of microwave estimates of soil moisture at the 6 cm wavelength with *in situ* soil moisture data in the 0–1 cm layer (unmanned helicopter).

### 1.9.5 Summary and conclusion

Microwave radiometric measurements were obtained from a manned rover-type mobile platform at a height of 2 m (which provided data at a ground resolution of 1.4 m) and a fully autonomous unmanned aerial vehicle (UAV) helicopter platform at an altitude of 30 m (which provided data at a ground resolution of 20 m) for the purpose of investigating daily mapping of soil moisture at the field scale. Results confirmed the sensitivity of microwave radiation to the moisture of the soil. Data obtained with the 21 cm sensor were found be to well correlated ($R^2 = 0.93$) with the soil moisture values of a rather deep soil layer. As expected, the 6 cm microwave brightness temperature was sensitive to surface soil moisture values. An algorithm for the retrieval of soil moisture, based on $T_b$ measurement at both 6 cm and 21 cm, gave satisfactory results. A two-dimensional soil moisture map of the study area was obtained from the 6 cm and 21 cm measurements using both platforms. In this chapter, the authors embarked on a new era in Earth remote sensing of soil moisture using two low-cost platform systems that have been successfully demonstrated. To our knowledge, this was the first time ever that a microwave radiometer was flown on board an unmanned helicopter with no problems from RFI, vibrations, and gusts of wind. Future work includes the development of a three-dimensional mapping system by integrating image data with land use/land cover information of a given location.

## 1.10 CONCLUSION

The GIMS has been successfully tested and applied in many countries of the world including the former U.S.S.R. (Russia, Moldova, Uzbekistan, Turkmenistan), Bulgaria, Poland, Hungary, Vietnam, Cuba, the U.S.A., and Holland. The advantages of applying the GIMS has been proven to consist principally in the following capabilities:

- to plan the *in situ* and remote-sensing parts of experiment by optimizing customer requirements for quality and cost of data;
- to simulate data in the form of maps having actual along-flight line measurements by taking *in situ* measurements and remote observations without using scanning technologies and, therefore, making the GIMS extremely cost-effective;
- to forecast/predict the behavior of those environmental/geophysical systems examined;
- to expand the scale of examined areas from regional to global by combining the volume/amount and meaning of *in situ* and remote observations in the best possible way and by using such operational data in the modeling so that the requisite algorithmic and programming tasks are fulfilled.

Thus, the GIMS provides remote sensing and on-ground (*in situ*) sampling by means of GIS information and mathematical modeling of physico-chemical processes in selected areas. GIMS technology can effectively be applied to solving many agricultural, hydrological, environmental, and other Earth-related problems. The GIMS technology solves numerous environmental tasks that GIS technology struggles with, such as:

- the study of environmental processes and systems that change quickly and are unstable;
- to assess ecosystems in regions where monitoring systems do not exist;
- to optimizae environmental monitoring systems;
- to reconstruct spatial images of systems based on restricted information;
- to reveal the characteristic features involved in the way environmental systems evolve.

Currently, methods set up for local diagnostics of the environment clearly do not permit complex assessment of the state of a natural object or process, especially when this element of the environment covers a vast area. In geoinformation monitoring, problems occur when it is necessary to make a decision in real time and applied means of data collection and processing are in short supply or nonexistent. The scheme in Figure 1.44 addresses this shortfall by proposing a monitoring procedure that formalizes decision making about operational diagnosis

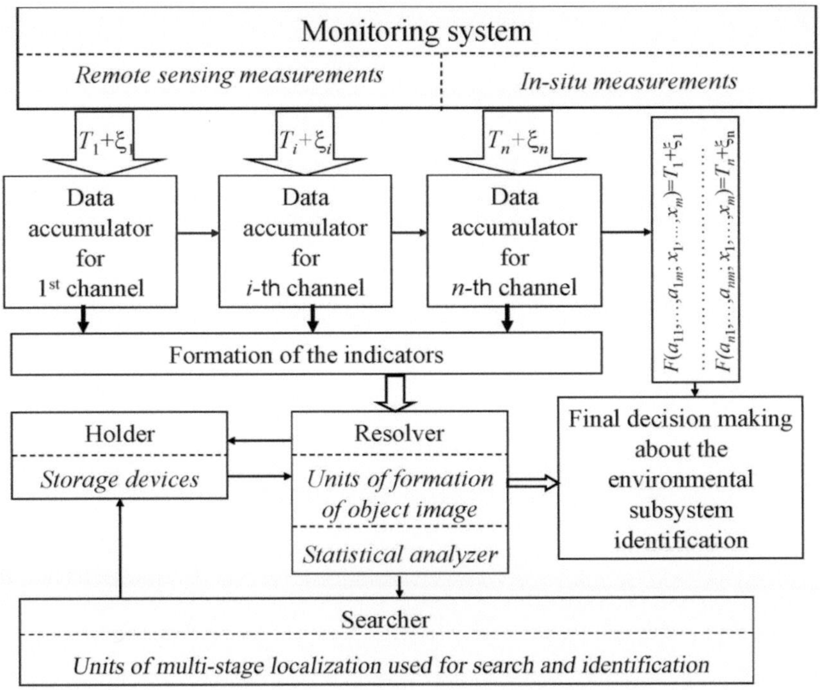


**Figure 1.44.** Block diagram of the monitoring system designed to detect and identify natural subsystems in the environment in real time by means of measuring devices and data-processing algorithms.

of the environmental subsystem under study. According to this scheme, the regular control of environmental elements can be brought about by efficient use of technical measuring means and algorithms for observational data processing (Kondratyev and Krapivin, 2004).

# 2

# Survivability and biocomplexity

## 2.1 INTRODUCTION

One of the main problems facing modern science is the estimation of biosphere survivability under conditions of increasing anthropogenic impact. Such estimation can be brought about by using a biosphere model. Well-known attempts at synthesizing a global model have given unsatisfactory results. The global models that have been created are educational in nature and are not fit for purpose for real estimations. In the last couple of decades many investigators proposed creating reliable and effective systems that are capable of considering the environmental state globally. Generally, this proposal included developing technical means for the collection, storage, and transfer of the data on nature's state, on the one hand, and the development of methods to process these data, on the other hand. Current means for collecting information about natural objects and processes make it possible to form a dataset covering large territories—even the whole biosphere. Remote means of environmental monitoring have become especially effective. The aim of this chapter is to formulate a basic model of biosphere survivability and to propose a new view on global modeling. The behavior of any system is determined by the value it can place on the different terms characterizing the state of the system. By interacting with an external medium and, in particular, with other systems, the values of these terms can vary in one way or another. For any technological or biological system it must always be possible to show changes in the field of characteristic parameters in which the system can be considered to be functioning. Outside this field the system does not exist.

Thus, we can substitute the complex behavior of a system by describing the behavior depicting this system as a point in phase space of the characteristic parameters. If a change in any coordinate leads to the disappearance of the depicting point from the allowable field, the system collapses (i.e., the organism as a whole perishes).

V.F. Krapivin and A.M. Shutko, *Information Technologies for Remote Monitoring of the Environment*, Springer Praxis Books, DOI 10.1007/978-3-642-20567-5_2, © Springer-Verlag Berlin Heidelberg 2012

Significant variables are not identical regarding the degree of threat they pose to the system. Such variables as the oxygen content of blood or the structural integrity of the medulla oblongata cannot tolerate any significant changes, since such changes would almost invariably lead to immediate death. However, there are changes, such as in the temperature of individual areas of the skin where sharp fluctuations do not necessarily lead to such an eventuality. Separating all variables characterizing the state of the system into significant variables makes it possible to simplify the behavioral strategy of a system as it interacts with an external environment or other systems.

A system is defined by its structure and behavior. The behavior of such a system is aimed at providing uninterrupted functioning by means of a correspondingly organized structure and behavior. The characteristic of a complex system to actively withstand the hostile action of an external medium is referred to in this book as "survivability" (Abrahamson, 1989; Corcoran, 2005; EPA, 2005; Fleishman, 1965; Kondratyev *et al.*, 2003d; Krapivin, 1978; Krapivin and Nazaryan, 1995; Starke, 2004; Svirezhev, 1987; Tait, 1987).

In the present chapter an analysis is made of a system whose elements are subdivided into working, defending, and active external agents of the system, which suppress or neutralize the hostile actions of an external medium. By taking into account the possibility of suppressing the hostile action of an external medium and the vulnerability of all the elements of the system, it becomes possible to reach a theoretical game definition of the problems. The use of game theory to investigate the survivability of complex systems makes it possible to classify the most unfavorable action of an external medium on the system and to work out the best strategy behavior for the system. This study of antagonistic situations between systems enables the mechanisms involved in the adaptability of living systems to the varying conditions of an external medium to be understood.

## 2.2 PRINCIPAL DEFINITIONS

The biosphere is a complex unique system. Looked at historically, humankind was but an element of the biosphere. However, at the present time the problem of co-evolution between human society ($H$) and nature ($N$) has arisen. The influence of human activity on natural systems has reached global scales, but it is still possible to divide anthropogenic and natural processes conditionally. The use of system analysis permits carrying out a more formal description of this division. There are commonly two interacting systems:

- $H$ deals with technologies, sciences, economics, sociology, agriculture, industry, etc.;
- $N$ deals with climatic, biogeocenotic, biogeochemical, hydrological, geophysical, and other natural processes.

Practical problems when investigating complex systems involve evaluating their effectiveness and, in particular, their stability under the indeterminate conditions in which they function. The theory of the potential effectiveness of complex systems is used to resolve these problems (Fleishman, 1970). A constructive mathematical apparatus making it possible to solve the different problems that arise when optimizing the structure and behavior of $H$ and $N$ functioning in certain situations has been developed within the framework of this theory.

The systems $H$ and $N$ are determined by their structure (i.e., the number of elements and relations among them) and behavior (responses to impacts). The internal behavior of such a system is aimed at maintaining its uninterrupted functioning. The external behavior of the system is aimed at achieving a certain outside goal. The temporal stability of a complex system is a necessary property without which all its other properties become meaningless. This is connected with the structural stability of the material composition and energy balance of the complex system as well as with the regularity of its responses to the same external impacts.

A breach in the stability of a system may result from internal causes (the aging of its elements) or external causes associated with the unfavorable influence of the environment (an ill-intentioned enemy, in particular). The survivability of biological systems is determined by the environmental conditions that apply to them, humankind's interference with nature being an important factor. Keeping this in mind when trying to construct artificial biological systems leads to the problem of finding the best way to do this, increase in the productivity of a biological system being the main optimality criterion.

On the global scale, the problem of interacting system survivability is complicated by a hierarchy of interaction levels. For a complete explanation of $H$ and $N$ systems their openness has to be taken into account. It is normal to consider the interaction of two open complex systems $H$ and $N$ as defined by their goals $H_G$ and $N_G$, structures $H_S$ and $N_S$, and behavior $H_B$ and $N_B$, respectively. Fleishman (1970) suggested that the functioning of such systems should be described by the equations involved in $(V, W)$ exchange. In other words, the interaction of an open system with the environment (or other system) is represented as a process whereby the system exchanges a certain quantity $V$ of resources spent in exchange for a certain quantity $W$ of resources consumed. The aim of the systems is the most advantageous $(V, W)$ exchange (i.e., it tries to get maximum $W$ in exchange for minimum $V$). $V$ is a complex function of the structure and behavior of both systems:

$$V = V(W, H_S, N_S, H_G, N_G) = V(W, H, N). \tag{2.1}$$

As a result of interaction, the systems $H$ and $N$ get the following $(V, W)$ exchanges:

$$\left.\begin{aligned} V_{H,\max} &= V_{H,\max}(W_H, H^*, N^*) = \max_{\{H_B,H_S\}} \min_{\{N_B,N_S\}} V_H(W_H, H, N), \\ V_{N,\max} &= V_{N,\max}(W_N, H^*, N^*) = \max_{\{N_B,N_S\}} \min_{\{H_B,H_S\}} V_N(W_N, H, N), \end{aligned}\right\} \tag{2.2}$$

where $H^*$ and $N^*$ are the optimal $H$ and $N$ systems, respectively.

From equations (2.1) and (2.2) we can see that the value of $(V, W)$ exchange depends on the goal of the system and may vary within certain limits: $V_{1,\min} \leq V_H \leq V_{1,\max}$, $V_{2,\min} \leq V_N \leq V_{2,\max}$, where $V_{i,\min}$ $(i = 1, 2)$ corresponds to the case when both systems are most aggressive, and $V_{i,\max}$ $(i = 1, 2)$ to the case when they are most cautious. In a word there is a *spectrum* of interactions between $H$ and $N$. For a formal description of these interactions we shall divide all the elements of both systems into three classes: the working (functioning), protective (defensive), and active elements, the latter designed to act on the environment. In short, we shall refer to the working elements of systems $H$ and $N$ as $a$ and $b$ elements, to the protective elements as $R_a$ and $R_b$ elements, and to the active elements as $C_a$ and $C_b$ elements, respectively.

Let us assume that before interaction the systems $H$ and $N$ have certain limited energy resources (i.e., vital "substrates") $V_a$ and $V_b$, where $V_a = \{V_{aj}, j = 1, \ldots, m_a\}$, $V_b = \{V_{bi}, i = 1, \ldots, m_b\}$. These substrates generate working elements in such a way that the substrate $V_{aj}$ $(V_{bi})$ can generate $H_j$ $(N_i)$ $a$ $(b)$ elements of the $j$th ($i$th) type of values $a_j$ $(b_i)$.

The protective and active elements of each system are generated by the working elements. First of all, the protective $E^a_{Rm}$ $(E^b_{Rm})$ and active $E^a_{Cm}$ $(E^b_{Cm})$ substrates are created which, in their turn, generate $R$ and $C$ elements of the $m$th type. These processes are described by the following dependences:

$$E^a_{Rm} = E^a_{Rm}(V_a, H_1, \ldots, H_{m_a}) = \sum_{j=1}^{m_a} w^a_{mj} f^a_{jR}(V_{aj}, H_j),$$

$$E^b_{Rm} = E^b_{Rm}(V_b, N_1, \ldots, N_{m_b}) = \sum_{j=1}^{m_b} w^b_{mj} f^b_{jR}(V_{bj}, N_j),$$

$$E^a_{Cm} = E^a_{Cm}(V_a, H_1, \ldots, H_{m_a}) = \sum_{j=1}^{m_a} w'^a_{mj} f^a_{jC}(V_{aj}, H_j),$$

$$E^b_{Cm} = E^b_{Cm}(V_b, N_1, \ldots, N_{m_b}) = \sum_{j=1}^{m_b} w^b_{mj} f^b_{jC}(V_{bj}, N_j),$$

where $w^{a(b)}_{mj}$, $w'^{a(b)}_{mj}$, and $f^{a(b)}$ are current weights and functions, respectively.

Let us assume that, as a result of such hierarchical synthesis, elements in the systems $H$ and $N$ have at the beginning of the interaction (i.e., $t = 0$):

(1) $m_j$ and $n_j$ working elements of the $j$th type with values $a_j$ and $b_j$, respectively, where

$$\sum_{j=1}^{m_a} a_j H_j = M_a(0), \qquad \sum_{j=1}^{m_b} b_j N_j = M_b(0); \tag{2.3}$$

(2) $r_a$ and $r_b$ types of protective elements, the $m$th type having $\alpha_m$ and $\beta_m$ elements, and

$$\sum_{m=1}^{r_a} \alpha_m = M_{R_a}(0), \qquad \sum_{m=1}^{r_b} \beta_m = M_{R_b}(0); \tag{2.4}$$

(3) $s_a$ and $s_b$ types of active elements, the $m$th type having $\nu_m^a$ and $\nu_m^b$ elements, and

$$\sum_{m=1}^{s_a} \nu_m^a = D_a(0), \qquad \sum_{m=1}^{s_b} \nu_m^b = D_b(0); \tag{2.5}$$

respectively.

In the discrete case, change in the average number of system elements that have survived until moment $t_{i+1}$ will be described by the following relations:

$$H_s(t_{i+1}) = \max\{0, H_s(t_i) - \sigma_{hs}^n(t_i)p_{hs}^n(t_i)\}, \qquad s = 1, \ldots, m_h \tag{2.6}$$

$$\alpha_j(t_{i+1}) = \max\{0, \alpha_j(t_i) - \sigma_{Rj}^n(t_i)p_{Rj}^n(t_i)\}, \qquad j = 1, \ldots, r_h \tag{2.7}$$

$$\nu_m^h(t_{i+1}) = \max\{0, \nu_m^h(t_i) - \sigma_{Cm}^n(t_i)p_{Cm}^n(t_i)\}, \quad m = 1, \ldots, s_h \tag{2.8}$$

$$N_l(t_{i+1}) = \max\{0, N_l(t_i) - \sigma_{nl}^h(t_i)p_{nl}^h(t_i)\}, \qquad l = 1, \ldots, m_n \tag{2.9}$$

$$\beta_s(t_{i+1}) = \max\{0, \beta_s(t_i) - \sigma_{Rs}^h(t_i)p_{Rs}^h(t_i)\}, \qquad s = 1, \ldots, r_n \tag{2.10}$$

$$\nu_m^n(t_{i+1}) = \max\{0, \nu_m^n(t_i) - \sigma_{Cm}^h(t_i)p_{Cm}^h(t_i)\}, \quad m = 1, \ldots, s_h, \tag{2.11}$$

where the $\sigma_{\omega i}^{a(b)}(t)$ values characterize the external behavior of both these systems:

$$\bar{H}_e^{(i)} = \{\|\sigma_{bl}^a\|, \|\sigma_{Rs}^a\|\}, \qquad \bar{N}_e^{(i)} = \{\|\sigma_{as}^b\|, \|\sigma_{Rj}^b\|\};$$

and $p_{\omega i}^{a(b)}(t)$ are the respective probabilities of death of the elements as a result of their interaction.

The following limiting conditions should be taken into account here:

$$\sum_{i=0}^{T} \left\{ \sigma_{Cs}^b(t_i)p_{Cs}^b(t_i) + \sum_{j=1}^{m_b} \sigma_{bj}^a(t_i) + \sum_{j=1}^{r_b} \sigma_{Rj}^a(t_i) + \sum_{j=1}^{s_b} \sigma_{Cj}^a(t_i) \right\} = \nu_l^h(0), \tag{2.12}$$

$$\sum_{i=0}^{T} \left\{ \sigma_{Cs}^a(t_i)p_{Cs}^a(t_i) + \sum_{j=1}^{m_a} \sigma_{aj}^b(t_i) + \sum_{j=1}^{r_a} \sigma_{Rj}^b(t_i) + \sum_{j=1}^{s_a} \sigma_{Cj}^b(t_i) \right\} = \nu_l^n(0). \tag{2.13}$$

The stochastic solution of equations (2.1)–(2.13) can be in the unrealized form in practice. There are many real situations when the realization of $H^*$ or $N^*$ system is impossible. Some tasks and algorithms were described by Krapivin and Klimov (1995, 1997).

## 2.3 SURVIVABILITY MODEL

Let us consider the interaction of the two systems within the framework of the diagram shown in Figure 2.1. At the start, systems $H$ and $N$ have, respectively, $N_a(0)$ and $N_b(0)$ working elements, $N_{Ra}(0)$ and $N_{Rb}(0)$ protective elements, and $M_a(0)$ and $M_b(0)$ active agents for undertaking action against an external medium. In this case we shall assume that the initial structures $H_S$ and $N_S$ of the systems are uniformly filled with elements. This means that at time $t = 0$ in any sphere with a fixed radius $\varepsilon$, which is completely confined within system $H$, there are constant numbers of elements.

We shall consider that all elements of systems $H$ and $N$, independent of their spatial location, are accessible to the same degree to active agents of the external medium. The interaction of the systems consists in the situation in which each system in a fixed interval of time $[0, T]$ at discrete moments $t_i = ih$, $i = 0, 1, 2, \ldots, k$ $(k = [T/h])$ can determine its behavior by a set of numbers:

$$H_B = \{m_b(t_i), m_{Rb}(t_i), \rho_i\}, \qquad N_B = \{m_a(t_i), m_{Ra}(t_i), r_i\},$$

where $m_a$ and $m_b$ are parts of those $C_b$ and $C_a$ elements aimed at destroying $a$ and $b$ elements, respectively; analogously, parts $(1 - r_i)m_{Ra}$ and $(1 - \rho_i)m_{Rb}$ of $C_b$ and $C_a$ elements are directed at destroying the corresponding protective elements. In the course of time portions of $C_a$ and $C_b$ elements uniformly fill the opposite system, and in this way the elements of the systems become with time *uniformly depleted*.

It is assumed that system $H(N)$ is put out of operation if at time $t_i \le T$, $N_a(t_i) \le \theta_a N_a(0)$ $(N_b(t_i) \le \theta_b N_b(0))$; that is, if more than the $(1 - \theta_a)$th $((1 - \theta_b)$th$)$

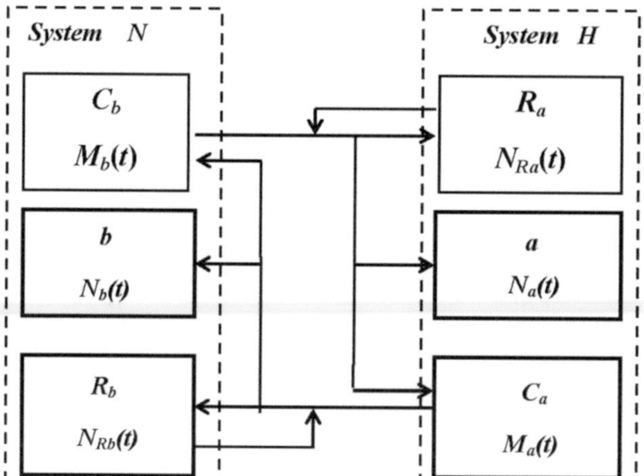


**Figure 2.1.** Schematic diagram of the interaction between two systems in the survivability problem (Kondratyev *et al.*, 2002).

portion of the initial number of its working elements is out of operation, where $0 \leq \theta_a, \theta_b \leq 1$. On the other hand, when $N_a(T) > \theta_a N_a(0)$ $(N_b(T) > \theta_b N_b(0))$, then system $H(N)$ is considered to have survived. The assignment of values $\theta_a$ and $\theta_b$ is determined by the peculiarities of the system under consideration. It is clear that the smaller the number $\theta$, the more *survivable* the system.

Further, we shall assume that one of the elements of system $H$ or $N$ is put of operation by the action of one of the elements $C_a$ or $C_b$ with a probability $p_1[N_{Ra}(t)]$ or $p_2[N_{Rb}(t)]$, respectively. Consequently, in the interval of time $[t_i, t_i + h]$ with step length $h$ there occur on average the following changes in the structure of the system:

$$\left.\begin{aligned} N_a(t_i) - N_a(t_i + h) &= m_a(t_i)p_1[N_{Ra}(t_i)], \\ M_a(t_i) - M_a(t_i + h) &= r_i m_{Ra}(t_i)[N_{Ra}(t_i) + m_b(t_i) + m_{Rb}(t_i), \\ N_{Ra}(t_i) - N_{Ra}(t_i + h) &= (1 - r_i)m_{Ra}(t_i)p_1[N_{Ra}(t_i)], \\ N_b(t_i) - N_b(t_i + h) &= m_b(t_i)p_2[N_{Rb}(t_i)], \\ M_b(t_i) - M_b(t_i + h) &= \rho_i m_{Rb}(t_i)p_2[N_{Rb}(t_i)] + m_a(t_i) + m_{Ra}(t_i), \\ N_{Rb}(t_i) - N_{Rb}(t_i + h) &= (1 - \rho_i)m_{Rb}(t_i)p_2[N_{Rb}(t_i)]. \end{aligned}\right\} \tag{2.14}$$

The above-discussed interaction scheme of the two systems can be readily implemented using game theory methods. Indeed, from equations (2.14) it follows that in the interval of time $[0, T]$ $(T = kh$, we can take $h = 1$ here) both systems lose working $a$ and $b$ elements in the following amounts:

$$Q_1 = \sum_{n=0}^{k-1} m_a(n)p_1[N_{Ra}(n)], \qquad Q_2 = \sum_{n=0}^{k-1} m_b(n)p_2[N_{Rb}(n)]. \tag{2.15}$$

In this case, system $H$ with its behavior of $H_B(t) = \{m_b(t), m_{Rb}(t), \rho(t)\}$ tends to minimize the function $Q_1$ (which characterizes its losses) and to maximize $Q_2$ (the losses of system $N$). On the other hand, system $N$ with its behavior $N_B(t) = \{m_a(t), m_{Ra}(t), r(t)\}$ tends to maximize function $Q_1$ and minimize function $Q_2$.

In practical problems, the win function is taken as that characteristic of the antagonistic situation that describes a given conflict most fully. In this case such a function is $Q = Q_1 - Q_2$. The maximizing participant in this case will be system $N$, the minimizing opponent system $H$. In this manner, solution of the set problem on the optimal behavior of the two systems in an antagonistic situation is reduced to solution of a game with a win function:

$$Q(H_B, N_B) = \sum_{n=0}^{i-1} \{m_a(n)p_1[N_{Ra}(n)] - m_b(n)p_2 N_{Rb}(n)]\}, \tag{2.16}$$

where, according to equation (2.14),

$$\left.\begin{aligned} N_{Ra}(n) &= N_{Ra}(0) - \sum_{i=0}^{n-1}(1 - r_i)m_{Ra}(i)p_1[N_{Ra}(i)], \\ N_{Rb}(n) &= N_{Rb}(0) - \sum_{i=0}^{n-1}(1 - \rho_i)m_{Rb}(i)p_{21}[N_{Rb}(i)], \end{aligned}\right\} \tag{2.17}$$

and conditions are imposed on the behavior of the system related to limitations of the $C_a$ and $C_b$ elements:

$$\left.\begin{aligned} &\sum_{i=0}^{k-1}\{r_i m_{Ra}(i)p_1[N_{Ra}(i)] + m_b(i) + m_{Rb}(i)\} = M_a(0), \\ &\sum_{i=0}^{k-1}\{\rho_i m_{Rb}(i)p_2[N_{Rb}(i)] + m_a(i) + m_{Ra}(i)\} = M_b(0). \end{aligned}\right\} \tag{2.18}$$

Thus, we have a $k$-step survival game. At the start of each step, system $H$ provides a certain amount of its resources $u = \{m_b, m_{Rb}, \rho\}$ and $N$ provides a certain amount of its resources $v = \{m_a, m_{Ra}, r\}$, so that limitations (2.18) are maintained. As a result of this distribution of resources, system $N$ gets the advantage:

$$R(u, v, M_a, M_b) = m_a p_1[N_{Ra}]. \tag{2.19}$$

However, $N$'s win and $H$'s losses are not counted on the basis of their initial resources and cannot be added to the remaining amounts of $C$ elements. After each step of the game (equation 2.14), a change in the resources available to the participants takes place, and as a result of the $k$-step process the total win of system $N$ can be described by the equation:

$$\begin{aligned} Q_k &= Q_k[u_0, u_1, \ldots, u_{k-1}; v_0, v_1, \ldots, v_{k-1}; M_a(0), M_b(0)] \\ &= \sum_{n=0}^{k-1}\{m_a(n)p_1[N_{Ra}(n)] - m_b(n)p_2[N_{Rb}(n)]\}. \end{aligned} \tag{2.20}$$

There are several methods that can help analyze this $k$-step process. One can consider this $k$-step game as a one-step game, in which case system $H$ must select simultaneously a plurality of the vector $\{u_0, u_1, \ldots, u_{k-1}\}$, and system $N$ a plurality of the vector $\{v_0, v_1, \ldots, v_{k-1}\}$, where the selection of $u_k$ and $v_k$ depends on the previous values obtained according to equation (2.14).

To solve this complex problem, we suggest substituting it by the similar problem of multi-step optimization. The values of this game can be expressed as:

$$\begin{aligned} V_k &= \max_F \min_G \left\{ \int Q_k \, dF(v_0, v_1, \ldots, v_{k-1}) \, dG(u_0, u_1, \ldots, u_{k-1}) \right\} \\ &= \min_G \max_F \{\cdots\} \end{aligned} \tag{2.21}$$

where the distribution functions $F$ and $G$ are determined on the boundaries of the complex form:

$$\begin{aligned} &0 \le v_0 \le M_b(0) && 0 \le u_0 \le M_a(0) \\ &0 \le v_1 \le M_b(1) && 0 \le u_1 \le M_a(1) \\ &\vdots && \vdots \\ &0 \le v_{k-1} \le M_b(k-1) && 0 \le u_{k-1} \le M_a(k-1). \end{aligned} \tag{2.22}$$

By utilizing the optimality principle and taking into account the dependence $V_k = V_k[M_a(0), M_b(0)]$, we obtain the following functional equation:

$$\begin{aligned} &V_{n+1}[M_a(0), M_b(0)] \\ &\quad = \max_F \min_G \left[ \iint_{\substack{0\le u\le M_a(n-1) \\ 0\le v\le M_b(n-1)}} \{R(u,v) + V_n[M_a(n-1), M_b(n-1)]\}\, dF(v)\, dG(u) \right] \\ &\quad = \min_G \max_F [\cdots] \end{aligned} \tag{2.23}$$

where

$$\begin{aligned} V_1[M_a(0), M_b(0)] &= \max_F \min_G \left[ \iint_{\substack{0\le u\le M_a(0) \\ 0\le v\le M_b(0)}} R(u,v)\, dF(v)\, dG(u) \right] \\ &= \min_G \max_F [\cdots]. \end{aligned} \tag{2.24}$$

Finding a solution to these functional equations is difficult. This was the reason that the so-called *curse of dimensionality* was introduced. In this chapter we shall present the solution to some specific cases. However, this is carried out in more detail by Nitu *et al.* (2000a). To solve the proposed problem, we shall begin with a case where the $C$ and $R$ elements are indistinguishable in both systems. A diagram of the interaction between systems $H$ and $N$ is shown in Figure 2.2. By taking into account the designations used in this diagram, we obtain the number of $C_a$ and $C_b$ elements that have reached the $C_b$ and $C_a$ elements of the opposite system in a one-step operation. They are, respectively, $\max\{0, a_C - N_{Rb}\}$ and $\max\{0, b_C - N_{Ra}\}$. Therefore, the numbers of $C_a$ and $C_b$ elements that are put out of operation on average are $\max\{0, b_C - N_{Ra}\}p_1[N_{Ra}]$ and $\max\{0, a_C - N_{Rb}\}p_2[N_{Rb}]$. Consequently, after one step of the game there remain in the system the following numbers of $C_a$ and $C_b$ elements that are not put out of operation: $\max[0, M_a - \max(0, b_C - N_{Ra})p_1[N_{Ra}]]p_1[N_{Ra}]$ and $\max[0, M_b - \max(0, a_C - N_{Rb})p_2[N_{Rb}]]p_2[N_{Rb}]$. Turning to the multi-step situation, we introduce the win function (2.16). The aim of system $N$ is to destroy system $H$ or, more precisely, to put all of the latter's $a$ elements out of operation. For this purpose system $N$ cannot use all its $C_b$ elements, since it will then be left defenseless against the $C_a$ elements. An analogous situation holds true for system $H$. After each successive step of the game each system is compelled to release a

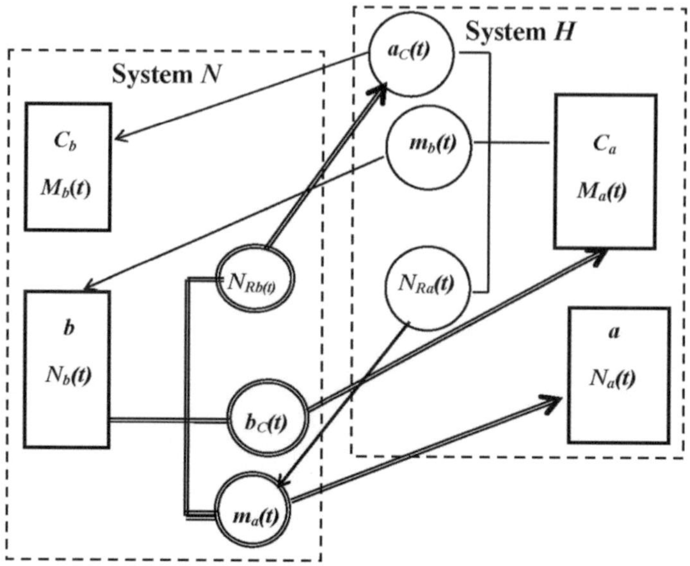


**Figure 2.2.** Schematic diagram of mutual interchangeability.

certain number of working $a$ and $b$ elements. The magnitude of this payoff is proportional to the difference $M_b - b_C - N_{Rb} - (M_a - a_C - N_{Ra})$. Now the problems on both sides become clear. System $N$ must strive to provide such a number of $C_b$ elements in all directions so as to maintain the maximum value of this difference. That is, it must provide the largest possible number of elements $m_a$ for the destruction of $a$ elements and thus increase its winnings. However, we shall also consider the presence of an analogous distribution of $C_a$ elements and provide sufficient protection for system $N$ through the maximum distribution of $C_a$ elements. An analogous situation exists for system $H$. For the $(n-1)$ moves that remain before the end of the game, we have:

$$\left.\begin{aligned} M_a(n-1) &= \max\{0, M_a(n) - \max[0, b_C(n) - N_{Ra}(n)p_1[N_{Ra}(n)]\}p_1[N_{Ra}(n)], \\ M_b(n-1) &= \max\{0, M_b(n) - \max[0, a_C(n) - N_{Rb}(n)]p_2[N_{Rb}(n)]\}p_2[N_{Rb}(n)]. \end{aligned}\right\} \quad (2.25)$$

The payoff for the entire game according to equation (2.20) will be:

$$Q_k = \sum_{n=1}^{k} \{M_b(n) - b_C(n) - N_{Rb}(n) - [M_a(n) - a_C(n) - N_{Ra}(n)]\}. \quad (2.26)$$

The functional equation (2.23) will acquire the following form:

$$\begin{aligned} V_{n+1} &= \max_{H_B} \min_{N_B} \{M_b(n+1) - b_C(n+1) - N_{Rb}(n+1) - [M_a(n+1) - a_C(n+1) \\ &\quad - N_{Ra}(n+1)] + Q_k[M_a(n), M_b(n)]\} \\ &= \min_{N_B} \max_{H_B} \{\cdots\}. \end{aligned} \quad (2.27)$$

Since at the end of the game $Q_0 = 0$, we obtain from equation (2.26) for $k = 1$:

$$Q_1 = M_b(1) - b_C(1) - N_{Rb}(1) - M_a(1) + a_C(1) + N_{Ra}(1). \tag{2.28}$$

From equation (2.28), one step before the end of the game, we obtain the following optimal strategies:

$$N_{Rb}(1) = N_{Ra}(1) = 0, \qquad a_C(1) = b_C(1) = 0 \tag{2.29}$$

and the prize of the game is $V_1 = M_b(1) - M_a(1)$. This means that in the last step of the game both systems direct all their $C_a$ and $C_b$ elements that have remained from the previous steps towards the destruction of $b$ and $a$ elements, respectively. Analogously, two steps before the end of the game, we have:

$$Q_2 = M_b(2) - M_a(2) + a_C(2) - b_C(2) + N_{Ra}(2) - N_{Rb}(2) + V_1[M_a(1), M_b(1)] \tag{2.30}$$

where $V_1 = M_b(1) - M_a(1)$; and

$$\left.\begin{aligned} M_a(1) &= \max\{0, M_a(2) - \max[0, b_C(2) - N_{Ra}(2)]p_1[N_{Ra}(2)]\}p_1[N_{Ra}(2)], \\ M_b(1) &= \max\{0, M_b(2) - \max[0, a_C(2) - N_{Rb}(2)]p_2[N_{Rb}(2)]p_2[N_{Rb}(2)]. \end{aligned}\right\} \tag{2.31}$$

It is obvious at this step of the game that the participants have no pure strategies. Therefore, solution of this game is impossible analytically, and it can only be obtained in a concrete case by the numerical method. Modeling of the game provides some understanding of the nature of its solution. Indeed, with the aid of a computer it is possible either to construct a model of the game and to gather statistics or to solve the functional equation (2.23) numerically and, with the aid of heuristic concepts, to investigate the dependence of strategies on the initial conditions. Of course, such an approach for a short interval of modeling cannot give any significant information concerning the solution. Nevertheless, this is the only possible approach at the present time. The feeling of hopelessness in a specific situation should not deter us from seeking a solution by analytical methods. The importance of obtaining analytical solutions is obvious, since they have an advantage over numerical solutions in that they make possible the detection of general regularities of the optimal behavior of complex systems in antagonistic situations. The importance of analytical solutions was pointed out by Krapivin (1978), who showed that a single numerical solution cannot replace an analytical solution in which the quantitative description of the phenomenon is most concentrated. In the case examined here, when the participants have no information concerning the action of the opponent in the process of the entire game, the solution of particular cases with the aid of a computer enables us to obtain the following quantitative description of optimal strategies.

Let $M_b(n)/M_a(n) = \delta_n$. Then if $\delta_n \gg 1$, system $N$ in the initial stage of the game has a pure strategy, but it is more advantageous for system $H$ to adopt a mixed strategy, using the tactics of deception. During the first steps, system $N$ destroys only $C_a$ elements and it is only during the last steps that it destroys $a$ elements. By using the corresponding probability mechanism, system $H$ must direct all its $C_a$ elements with probability $p_1$ towards the destruction of $C_b$ elements, with probability $p_2$ towards the defense of its own elements, and with probability $1 - p_1 - p_2$ towards the attack of $b$ elements. In the case when $\delta_n \cong 1$, then at this stage of the game $p_1 + p_2 = 1$, $m_b(n) = 0$, and the behavior of the system becomes symmetrical. During the last steps of the game systems $H$ and $N$, independent of the magnitude of $\delta_n$, change over to the strategies $m_a(t) \neq 0$, $m_b(t) \neq 0$, and $b_C = a_C = N_{Ra} = N_{Rb} = 0$ (i.e., to the destruction of working elements).

Let us consider a particular case where the systems have no protective elements (i.e., $N_{Ra} = N_{Rb} = 0$) and, therefore, $p_1 = p_2 = 1$. Then we obtain:

$$M_a(n-1) = \max\{0, M_a(n) - b_C(n)\}, \quad M_b(n-1) = \max\{0, M_b(n) - a_C(n)\}. \quad (2.32)$$

One step before the end of the game, the payoff according to equation (2.31) is $V_1 = M_b(1) - M_a(1)$ and the optimal strategies $a_C^*(1) = b_C^*(1) = 0$; that is, the systems release all their $C$ elements in an attempt to destroy the $a$ and $b$ elements, respectively.

Next, by assuming $n = 2$, according to equations (2.30) and (2.31) we obtain:

$$Q_2[a_C(2), b_C(2)] = \begin{cases} Q_{21} & \text{when } b_C(2) \geq M_a(2), a_C(2) \geq M_b(2); \\ Q_{22} & \text{when } b_C(2) < M_a(2), a_C(2) \geq M_b(2); \\ Q_{23} & \text{when } b_C(2) \geq M_a(2), a_C(2) < M_b(2); \\ Q_{24} & \text{when } b_C(2) < M_a(2), a_C(2) < M_b(2), \end{cases} \quad (2.33)$$

where $Q_{21} = M_b(2) - b_C(2) - M_a(2) + a_C(2)$; $Q_{22} = M_b(2) - 2M_a(2) + a_C(2)$; $Q_{23} = 2M_b(2) - b_C(2) - M_a(2)$; and $Q_{24} = 2M_b(2) - 2M_a(2)$.

The solution to the game with the win function (2.33) has the following form:

$$V_2 = 2M_b(2) - 2M_a(2), \qquad b_C^*(2) = M_a(2); a_C^*(2) = M_b(2). \quad (2.34)$$

Actually, if account is made of the real situation, the optimal strategies of both systems two steps before the end of the game will be as follows:

$$a_C^*(2) = \min\{M_a(2), M_b(2)\}, \qquad b_C^*(2) = \min\{M_a(2), M_b(2)\}. \quad (2.35)$$

Therefore at the penultimate step when $\delta_2 > 1$, system $N$ releases a portion of its force and system $H$ releases all its forces against the $c$ elements of the other system.

Through analogous reasoning, we can see that $n$ steps before the end of the game the strategies of systems $H$ and $N$ will be:

$$a_C^*(n) = \min\{M_a(n), M_b(n)\}; \qquad b_C^*(n) = \min\{M_a(n), M_b(n)\}. \quad (2.36)$$

If in the process of the game $C$ elements are clearly not replaced in systems $H$ and $N$, then it follows from equations (2.35) and (2.36) that it makes sense to conduct the game in two steps; and the two following cases are distinguishable:

(1) when $M_a(0) = M_b(0)$, both systems release all their $C$ elements into battle against the $C$ elements of the other system, but in so doing the systems themselves survive;

(2) when $M_a(0) > M_b(0)$, system $N$ releases all the elements into battle against $C_a$ elements, and system $H$ releases $M_b(0)$ of $C_a$ elements into battle against $C_b$ elements, and $M_a(0) - M_b(0)$ of $C_a$ elements into battle against $b$ elements. Consequently, system $H$ survives in every case, while system $N$ survives only when $M_a(0) - M_b(0) < \theta_b N_b(0)$.

Now, let us consider the case in which $C_a$ and $C_b$ elements are mutually indifferent. In this case, by supposing $r = \rho = 0$ in equations (2.17) and (2.18), we obtain:

$$\left.\begin{aligned} N_{Ra}(n) &= N_{Ra}(0) - \sum_{i=0}^{n-1} m_{Ra}(i) p_1[N_{Ra}(i)], \\ N_{Rb}(n) &= N_{Rb}(0) - \sum_{i=0}^{n-1} m_{Rb}(i) p_2[N_{Rb}(i)] \end{aligned}\right\} \tag{2.37}$$

and

$$\sum_{n=0}^{k-1} [m_a(n) + m_{Ra}(n)] = M_b(0), \qquad \sum_{n=0}^{k-1} [m_b(n) + m_{Rb}(n)] = M_a(0). \tag{2.38}$$

The solution to the game of the two systems, $H$ and $N$, with the win function as shown in equation (2.16) and under conditions as expressed in equations (2.37) and (2.38), is reduced to the problem of maximizing two functions:

$$Q_1 = \sum_{n=0}^{k-1} m_a(n) p_1[N_{Ra}(n)] = \max_{Ra} \tag{2.39}$$

$$Q_2 = \sum_{n=0}^{k-1} m_b(n) p_2[N_{Rb}(n)] = \max_{Rb} \tag{2.40}$$

From equations (2.39) and (2.40) it is evident that the optimal strategies of both sides consist in destroying $a$ and $b$ elements when $p_1[N_{Ra}(n)] = p_2[N_{Rb}(n)] = 1$. If the $C$ elements act independently of one another, and the probability of putting the $C$ elements out of operation by a single $R$ element is equal to a constant value $D_a$ and $D_b$ for the $H$ and $N$ systems, respectively, then

$$p_1[N_{Ra}(n)] = \exp[-d_a N_{Ra}(n)], \qquad p_2[N_{Rb}(n)] = \exp[-d_b N_{Rb}(n)], \tag{2.41}$$

where, $d_a = -\theta_1 \ln(1 - D_a)$, $d_b = -\theta_2 \ln(1 - D_b)$ are the *effectiveness* coefficients of the $R_a$ and $R_b$ elements; and $\theta_1$ and $\theta_2$ are the *density* of the location of elements in the $H$ and $N$ system, respectively.

When $k = 1$, the optimal strategies of the behavior of the systems will be $m_a^*(0) = M_b(0)$ and $m_b^*(0) = M_a(0)$. This is natural since in the last step it makes no sense to destroy protective elements. When $k = 2$, from equations (2.37)–(2.40) we obtain the optimal strategies in the form:

$$\left.\begin{aligned} m_a^*(0) = m_b^*(0) = m_{Ra}^*(1) = m_{Rb}^*(1) = 0, \quad m_a^*(1) = d_a^{-1}\exp[d_a N_{Ra}(0)], \\ m_b^*(1) = d_b^{-1}\exp[d_b N_{Rb}(0)], \quad m_{Ra}(0) = M_b(0) - d_a^{-1}\exp[d_a N_{Ra}(0)], \\ m_{Rb}(0) = M_a(0) - d_b^{-1}\exp[b_b N_{Rb}(0)]. \end{aligned}\right\} \tag{2.42}$$

The cost of the game is:

$$V_2 = d_a^{-1}\exp\{-1 + d_a M_b(0)\exp[-d_a N_{Ra}(0)]\} \\ -d_b^{-1}\exp\{-1 + d_b M_a(0)\exp[-d_b N_{Rb}(0)]\}. \tag{2.43}$$

From equation (2.43) it is evident that for system $N$ to destroy system $H$ it is necessary that the following inequality holds:

$$d_a M_b(0)\exp[-d_a N_{Ra}(0)] \geq l_n[d_a e(1-\theta_a)N_a(0)] \tag{2.44}$$

from which we have:

$$M_b(0) \geq d_a^{-1}\exp[d_a N_{Ra}(0)]\{1 + l_n[d_a(1-\theta_a)N_a(0)]\}. \tag{2.45}$$

Similarly, we obtain the condition for system $H$:

$$M_a(0) \geq d_b^{-1}\exp[d_b N_{Rb}(0)]\{1 + l_n[d_b(1-\theta_b)N_b(0)]\}. \tag{2.46}$$

Thus, for system $N$ to destroy system $H$ it is necessary that inequality (2.45) holds. Analogously, for system $H$ it is necessary that inequality (2.46) holds. The number of $a$ and $b$ elements, as a rule, is fixed, as follows from considerations related to the work of the system. The number of $R$ elements, which perform protective functions, can be best selected when certain physical parameters in the problem are fixed and when *a priori* information concerning the number of $C$ elements of the opposite system is available. For example, if the effectiveness of all $R$ elements is constant independent of their number, then in order for system $H$ to survive, the necessary number of protective elements must satisfy the inequality:

$$N_{Ra}(0) \geq d_a M_b(0) l_n / \{d_a(1 + l_n[d_a(1-\theta_a)N_a(0)])\}. \tag{2.47}$$

When the number of $C_b$ elements is constant and the survivability of system $H$ decreases, the necessary number of protective elements clearly increases rapidly. In this case, the greater their effectiveness, the smaller the number of protective elements required for carrying out one and the same task. With an increase in survivability the necessary number of protective elements can be decreased.

Now, let systems $H$ and $N$ have a fixed amount $E_a$ and $E_b$ of a certain substratum (e.g., energy) and be able to distribute it evenly between their protective elements, so that for each fraction of $R_a$ and $R_b$ elements there is $E_{1a} = E_a/N_{Ra}(0)$ and $E_{1b} = E_b/N_{Rb}(0)$, respectively. Then the efficiency coefficients of the protective elements must increase as the $E_{1a}$ and $E_{1b}$ portions increase, since in this case the probabilities of $D_a$ and $D_b$ increase. Therefore, let $D_a = 1 - \exp\{-F_a F_{1a}^{\alpha}\}$, $D_b = 1 - \exp\{-F_b F_{1b}^{\beta}\}$, where $F_a$, $F_b$, $\alpha$, and $\beta$ are independent of the number of protective elements. This then gives the following expressions for the coefficients of effectiveness of the protective elements:

$$\left.\begin{aligned} d_a &= -\theta_a \ln(1 - D_a) = -\theta_a F_a E_{1a}^{\alpha} = G_a N_{Ra}^{-\alpha}(0), \\ d_b &= -\theta_b \ln(1 - D_b) = -\theta_b F_b E_{1b}^{\beta} = G_b N_{Rb}^{-\beta}(0), \end{aligned}\right\} \tag{2.48}$$

where $G_a = \theta_a F_a E_a^{\alpha}$, $G_b = \theta_b F_b E_b^{\beta}$. From equations (2.47) and (2.48) we obtain the following transcendental equations for the number of necessary protective elements in systems $H$ and $N$:

$$N_{Ra}(0) = N_{Ra}^{\alpha}(0) G_a^{-1}\{\ln[G_a M_b(0)/f_1] - \alpha \ln N_{Ra}(0)\}, \tag{2.49}$$

$$N_{Rb}(0) = N_{Rb}^{\beta}(0) G_b^{-1}\{\ln[G_b M_a(0)/f_2] - \beta \ln N_{Rb}(0)\}, \tag{2.50}$$

where

$$f_1 = \ln[eG_a(1 - \theta_a) N_a(0) N_{Ra}^{-\alpha}], \qquad f_2 = \ln[eG_b(1 - \theta_b) N_b(0) N_{Rb}^{-\beta}(0)].$$

Close study of equations (2.49) and (2.50) reveals that in this case the number $N_{Ra}(0)$ is very sensitive with respect to changes in the quantity $\ln[(1 - \theta_a) N_a(0)]$. What is more, this is natural, since with the increase in the number of $R_a$ elements their effectiveness sharply decreases. It is clear that there exists a certain optimal level for the number of protective elements. This level is defined by the assigned survivability of the system. It is better to have a small number of $R_a$ elements of high effectiveness than a large number of $R_a$ elements of low effectiveness. When $k = 3$, from equations (2.37)–(2.39) we have:

$$Q_1 = \sum_{i=1}^{3} m_a(i-1) \exp[-\alpha N_{Ra}(i-1)], \tag{2.51}$$

where

$$\left.\begin{aligned} N_{Ra}(1) &= N_{Ra}(0) - m_{Ra}(0) \exp[-d_a N_{Ra}(0)], \\ N_{Ra}(2) &= N_{Ra}(0) - m_{Ra}(0) \exp[-d_a N_{Ra}(0)] - m_{Ra}(1) \exp[-d_a N_{Ra}(1)] \\ &\quad \times m_a(1) + m_a(0) + m_a(2) + m_{Ra}(1) + m_{Ra}(0) + m_{Ra}(2) \\ &= M_b(0). \end{aligned}\right\} \tag{2.52}$$

Equations (2.51) and (2.52) show that the function $Q_1$ reaches its maximum value when $m_a^*(0) = m_{Ra}^*(2) = m_{Ra}^*(1) = 0$ with this maximum value being independent of the distribution of $C_a$ elements during the last two steps. In this case, $Q_1^*(3) = Q_1^*(2)$. An analogous result is also obtained for system $N$.

Thus, for an identical game, say system $N$ can for instance destroy at the most $Q_1^*(2) = M_b(0) \exp[-d_a N_{Ra}(0)]$ of $a$ elements; when the number of steps is greater than 2, system $N$ can destroy no more than $Q_1^*(2) = (ed_a)^{-1} \exp[d_a Q_1^*(1)]$ of $a$ elements. Therefore, by comparing $Q_1^*(1)$ with $Q_1^*(2)$ we can see that in the analyzed antagonistic situation the best strategy for both systems is to carry out all allowable operations in two steps. During the first step the action of a portion of each system's $C$ elements is to be set against the protective elements of the opposite system, and during the second step the action of the remaining force is to be set against the working elements of the opposite system. This conclusion completely agrees with the conclusions reached by Krapivin (1978), which were obtained using different methods.

In the models of the interaction of the two systems that have just been examined, it was assumed that the effectiveness of protective elements does not change with respect to time. However, this assumption in many real situations must be withdrawn. Let us consider a case where both systems can vary the effectiveness of protective elements from step to step, so that at each step the effectiveness is independent of the number of protective elements. Let the effectiveness of $R_a$ and $R_b$ elements be equal to $d_{1a}$ and $d_{1b}$, respectively, during the first step in the two-step case. Similarly, during the second step the effectiveness acquires values $d_{2a}$ and $d_{2b}$ so that $d_{1a} + d_{2a} = 2d_a, d_{1b} + d_{2b} = 2d_b$; that is, the summed value of the effectiveness does not exceed a constant value. As a result, we obtain the following matrix game:

$$\begin{array}{c|ccc|}
 & d_{1a} = d_{2a} & d_{1a} < d_{2a} & d_{1a} > d_{2a} \\
d_{1b} = d_{2b} & Q_{11} & Q_{12} & Q_{13} \\
d_{1b} < d_{2b} & Q_{21} & Q_{22} & Q_{23} \\
d_{1b} > d_{2b} & Q_{31} & Q_{32} & Q_{33}
\end{array},$$

where

$$\left.\begin{aligned}
Q_{11} &= (ed_a)^{-1}\exp\{M_b(0)d_a\exp[-d_aN_{Ra}(0)]\} - (ed_b)^{-1}\exp\{M_a(0)d_b\exp[-d_bN_{Rb}(0)]\}, \\
Q_{22} &= Q_{33} = Q_{23} = Q_{32} = (ed_{2a})^{-1}\exp\{(d_{1a} - d_{2a})N_{Ra}(0) + d_{2a}M_b(0)\exp[-d_{1a}N_{Ra}(0)]\} \\
&\quad - (ed_{2b})^{-1}\exp\{(d_{1b} - d_{2b})N_{Rb}(0) + d_{2b}M_a(0)\exp[-d_{1b}N_{Rb}(0)]\}, \\
Q_{12} &= Q_{13} = (ed_{2a})^{-1}\exp\{(d_{1a} - d_{2a})N_{Ra}(0) + d_{2a}M_b(0)\exp[-d_{1a}N_{Ra}(0)]\} \\
&\quad - (ed_b)^{-1}\exp\{M_b(0)d_b\exp[-d_bN_{Rb}(0)]\}, \\
Q_{21} &= Q_{31} = (ed_a)^{-1}\exp\{M_a(0)d_a\exp[-d_aN_{Ra}(0)]\} \\
&\quad - (ed_{2b})^{-1}\exp\{(d_{1b} - d_{2b})N_{Rb}(0) + d_{2b}M_b(0)\exp[-d_{1b}N_{Rb}(0)]\}.
\end{aligned}\right\} \tag{2.53}$$

The matrix of this game has a saddle-shaped point under the following conditions $M_a(0) \le N_{Rb}(0)\exp[d_bN_{Rb}(0)]$, $M_b(0) \le N_{Ra}(0)\exp[d_aN_{Ra}(0)]$ and that it takes place in the real system.

Thus, for both systems it is advantageous during the first step to provide protective elements with a small amount of effectiveness and to increase their effectiveness during the second step. This is natural, since it is better to lose a greater number of protective elements during the first step, thus securing a reliable defense of the working elements during the second step of the game.

From equation (2.53) it follows that the losses of $a$ and $b$ elements by systems $H$ and $N$, respectively, will amount to:

$$\left.\begin{aligned} Q_{1^*} &= \exp\{-1 + (d_{1a} - d_{2a})N_{Ra}(0) + d_{2a}M_b(0)\exp[-d_{1a}N_{Ra}(0)]\}/d_{2a}, \\ Q_{2^*} &= \exp\{-1 + (d_{1b} - d_{2b})N_{Rb}(0) + d_{2b}M_a(0)\exp[-d_{1b}N_{Rb}]\}/d_{2b}. \end{aligned}\right\} \quad (2.54)$$

In order to determine the optimum value of the effectiveness of the $R_a$ and $R_b$ elements for each step of the game, we must find:

$$\min_{(d_{1a},d_{2a})} Q_{1^*}(d_{1a}, d_{2a}) \quad \text{and} \quad \min_{(d_{1b},d_{2b})} Q_{2^*}(d_{1b}, d_{2b}).$$

From equation (2.54) we obtain:

$$\frac{\partial Q_{1^*}}{\partial d_{1a}} = 1 + (2d_a - d_{1a})\{2N_{Ra}(0) - M_b(0)[1 + N_{Ra}(2d_a - d_{1a})] \times \exp[-d_{1a}N_{Ra}(0)]\} = 0, \quad (2.55)$$

$$\frac{\partial Q_{2^*}}{\partial d_{1b}} = 1 + (2d_b - d_{1b})\{2N_{Rb}(0) - M_a(0)[1 + N_{Rb}(0)(2d_b - d_{1b})] \times \exp[-d_{1b}N_{Rb}(0)]\}, \quad (2.56)$$

From equation (2.55) it is evident that if the equation

$$\{(2d_a - d_{1a})/(d_a - d_{1a})\}\exp[-d_{1a}N_{Ra}(0)] = 0.5N_{Ra}(0)/M_b(0) \quad (2.57)$$

has a real root $0 \leq d^*_{1a} \leq 2d_a$ then system $H$, by utilizing this root for its own optimum strategy, can guarantee the average losses of elements not exceeding $Q^*_1 = [e(2d_a - d^*_{1a})]^{-1}$. If equation (2.57) does not have a root in the interval $[0, 2d_a]$, the optimum strategy is then determined either by the root of equation (2.55) or by $d^*_{1a} = 0$. In particular, when $M_b(0) = N_{Ra}(0)$, then $d^*_{1a} = 0$.

Similar calculations are carried out for equation (2.56).

## 2.4 STABLE STRATEGIES WITHIN THE SURVIVABILITY MODEL

The task of equations (2.11)–(2.13) is solved by means of game theory algorithms. There are many models that simulate the above situations describing the system interaction. Below the two-player antagonistic game is considered.

In game $\Gamma(M, [0, 1])$ with gain function $M(x, y)$ $(0 \leq x, y \leq 1)$ player I can receive no fewer than

$$v_1 = \max_F \min_G \int_0^1 E_1(F)\, dG(y) = \max_F \min_y E_1[F(y)],$$

where $F$ and $G$ are strategies of the first and second players, respectively:

$$E_1[F] = \int_0^1 M(x, y)\, dF(x).$$

Similarly, player II can receive no more than

$$v_2 = \min_G \max_F \int_0^1 E_2(G)\, dF(x) = \min_G \max_x E_2[G(x)],$$

where

$$E_2[G] = \int_0^1 M(x,y)\, dG(y).$$

Obviously if there exist $F^*$ and $G^*$ such that

$$E_1[F^*] = E_2[G^*] = v_1 = v_2 = v \tag{2.58}$$

then the functions $F^*$ and $G^*$ are the optimal strategies of the players.

Let us consider the game with the gain function:

$$\left.\begin{aligned}
&M(x-y) = \alpha_{k+1} && \text{for } x-y \le -\sum_{j=1}^{k} \varepsilon_j, \\
&M(x-y) = \alpha_i && \text{for } -\sum_{j=1}^{i} \varepsilon_j < x-y \le -\sum_{j=1}^{i-1} \varepsilon_j \quad (i=\overline{1,k}); \\
&M(x-y) = \beta_j && \text{for } \sum_{i=1}^{j-1} \delta_i < x-y \le \sum_{i=1}^{j} \delta_i \quad (j=\overline{1,s}); \\
&M(x-y) = \beta_{s+1} && \text{for } x-y > \sum_{i=1}^{s} \delta_i,
\end{aligned}\right\} \tag{2.59}$$

where $\alpha_i$ and $\beta_j$ are arbitrary real numbers; $\varepsilon_j$ and $\delta_i$ are positive quantities representing the step lengths of function $M$; and $k$ and $s$ are natural positive numbers. The number of steps for function $M$ is $k+s+2$.

Let players I and II choose arbitrary values $x$ and $y$ from $[0,d]$, respectively. The next theorem is valid.

**Theorem 2.1** *If equation*

$$\sum_{i=1}^{k} (\alpha_{i+1} - \alpha_i)\lambda^{-\psi_i} + \sum_{j=1}^{s} (\beta_j - \beta_{j+1})\lambda^{\gamma_j} = \beta_1 - \alpha_1 \tag{2.60}$$

*where*

$$\psi_i = \sum_{j=1}^{i} \delta_j, \qquad \gamma_j = \sum_{i=1}^{j} \delta_i$$

*has at least one root $\lambda^* = \rho \exp(i\omega)$ such that $\omega \le \pi/(l+r)$, then stable strategies exist in the game with gain function (2.59) and they are optimal. Under this the* *game solution has the form:*

$$F^*(x) = (\rho_0)^x[C_1 \cos(\omega_0 x) + C_2 \sin(\omega_0 x)] + B, \quad -r \le x \le l;$$
$$G^*(y) = D[1 - (\rho_0)^y \cos(\omega_0 y)], \quad 0 \le y \le n;$$
$$v = [\beta_{s+1}(\rho_0)^n - \alpha_{k+1} \cos(\omega_0 n)]/[(\rho_0)^n - \cos(\omega_0 n)],$$

*where the root of equation (2.60)* $\lambda^* = \rho_0 \exp(i\omega_0)$ *has minimal argument and maximal module; r, l, and n are minimal natural numbers that are greater than* $\psi_k$, $d + \gamma_s$, *and d, respectively:*

$$D = (\beta_0)^n/[(\rho_0)^n - \cos(\omega_0 n)];$$
$$B = \cos(\omega_0 n)[\cos(\omega_0 n) - (\rho_0)^n]^{-1};$$
$$C_1 = [(\rho_0)^{n-1} \sin(\omega_0 r) + (\rho_0)^{n-r} \cos(\omega_0 n) \sin(\omega_0 l)]/\{\sin[(r+l)\omega_0][(\rho_0)^n - \cos(\omega_0 n)]\};$$
$$C_2 = [(\rho_0)^{n-r}\cos(\omega_0 r) - (\rho_0)^l \cos(\omega_0 n)\cos(\omega_0 l)]/\{(\rho_0)^{l-r}\sin[(r+l)\omega_0][(\rho_0)^n - \cos(\omega_0 n)].$$

The proof of the theorem is based on solving equation (2.58) under function (2.59).

Let us now consider the game with gain function $M(x-y)$. In analogy with (2.58), we have:

$$\int_0^1 M(x-y)p_1(x)\,dx = \int_0^1 M(x-y)p_2(y)\,dy = v \tag{2.61}$$

where $dF(x) = p_1(x)\,dx$; and $dG(y) = p_2(y)\,dy$.

Equations (2.61) are solved by means of the Fourier transform:

$$R_2(-p^2)v = R_1(-p^2)p_1(y) \tag{2.62}$$
$$R_2(-p^2)v = R_1(-p^2)p_2(x) \tag{2.63}$$

where $R_1(\omega^2)$ and $R_2(\omega^2)$ are polynomials with real coefficients defined from equations (2.61).

**Theorem 2.2** *If the solution* $\bar{p}_1(x) \ge 0$ *of equation (2.62) exists, the game with gain function* $M(x-y)$ *has optimal stable strategies* $p_1^*(x)$ *and* $p_2^*(y)$ *such that*

$$p_1^*(x) = p_2^*(x) = K\bar{p}_1(x) + \sum_{i=1}^{\infty}[A_i\delta^{(i-1)}(x) + B_i\delta^{(i-1)}(x-1)],$$

*where* $\delta(x)$ *is the delta function; and the constants* $K$, $A_i$, *and* $B_i$ *are determined from (2.61) and the conditions:*

$$\int_0^1 p_1^*\,dx = \int_0^1 p_2^*\,dy = 1.$$

**Example 2.1.** Let us pay attention to the game $\Gamma(M, [0,1])$ with gain function $M(x-y) = a[\delta(x-y) + d\exp\{-b|x-y|\}]$, where $a$, $b$, and $d$ are arbitrary constants.

It is easy to see that Theorem 2.2 gives the following solution to this game:

$$p_1(x) = p_2(x) = vb^2(1 + A\exp\{\gamma x\} + B\exp\{-\gamma x\})[a\gamma^2]^{-1},$$

$$v = a\gamma^2 b^{-2}[1 + A(e^{\gamma} - 1)/\gamma - B(e^{-\gamma} - 1)/\gamma]^{-1},$$

where $\gamma^2 = b^2 + 2db$,

$$A = (\gamma - b)b^{-1}[(\gamma + b)^2 - (\gamma^2 + b^2)e^{-\gamma}][(\gamma + b)^2 e^{\gamma} - (\gamma - b)^2 e^{-\gamma}]^{-1},$$

$$B = -(\gamma + b)b^{-1}[(\gamma^2 + b^2)e^{\gamma} - (\gamma - b)^2)][(\gamma + b)^2 e^{\gamma} - (\gamma - b)^2 e^{-\gamma}]^{-1}.$$

**Example 2.2.** It is easy to find the solution to game $\Gamma(M, [0,1])$ with gain function

$$M(x-y) = \sum_{i=1}^{n} a_i \exp(-b_i|x-y|).$$

From Theorem 2.2 it follows that

$$p_1(x) = p_2(x) = v\left\{a[\delta(x) + \delta(x-1)] + \sum_{j=1}^{n-1} c_j[\exp(-b_j x) + \exp\{-b_j(1-x)\}] + \theta_{2n}/\rho_{2n}\right\}$$

$$v = \left\{2a + \sum_{j=1}^{n-1} 2a_j[1 - \exp(-b_j)]/b_j + \theta_{2n}/\rho_{2n}\right\}^{-1},$$

where $a$ is an arbitrary root of $R_2(\omega^2)$; and $c_j$ ($j = 1, \ldots, n$) are roots of $R_1(\omega^2)$.

**Example 2.3.** Let us consider the game with gain function:

$$M(x,y) = \begin{cases} 1 & \text{for } (x,y) \in G_1, \\ a & \text{for } (x,y) \in G_a, \\ b & \text{for } (x,y) \in G_b, \end{cases}$$

where $x \in X$, $y \in Y$, $G_1 \cup G_a \cup G_b = X * Y$, $G_1 = \{(x,y) : \rho(x,y) < \varepsilon, x \in K_m\}$, $G_a = \{(x,y) : \rho(x,y) \geq \varepsilon\}$, $G_b = \{(x,y) : \rho(x,y) < \varepsilon, x \in X \setminus K_m\}$; $\varepsilon$ is an arbitrary value; $X$ and $Y$ are $m$-dimensional bit cubes; $K_m$ is an arbitrary set having volume 0.5; $\rho(x,y)$ is the distance between $x$ and $y$; and $a = k\varepsilon^m$.

It is obvious that

$$p_1^*(x) = p_2^*(x) = \begin{cases} 2/(1+\lambda) & \text{for } x \in K_m, \\ 2\lambda/(1+\lambda) & \text{for } x \in X \setminus K_m, \end{cases}$$

$$v = \varepsilon^m(2\lambda b\pi^{0.5m}\Gamma(m/2)(1+\lambda)^{-1} - kb) + O(\varepsilon^m),$$

where $\lambda = (1-a)/(b-a)$ and $\Gamma(m/2)$ is a gamma function.

## 2.5 BIOCOMPLEXITY RELATED TO ECOSYSTEM SURVIVAVILITY

The interaction of various elements and processes in the global nature–society system (NSS) has recently attracted the attention of many scientists. Attempts to estimate and predict the dynamics of this interaction have been made in different scientific spheres. One of these attempts is the Biocomplexity Program set up in the U.S.A. by the National Scientific Foundation, within which plans to study and understand relationships between the dynamics of complexity of biological, physical, and social systems and trends in changes of the present habitat. Within the framework of this program, the complexity of the system interacting in some way or another with the environment is connected with phenomena appearing as a result of global-scale contact of a living system with the environment.

Biocomplexity is a derivative of the biological, physical, chemical, social, and behavioral interactions of environmental subsystems, including living organisms and global population. As a matter of fact, the notion of biocomplexity in the environment is closely connected with the rules of biosphere functioning as a combination of its forming ecosystems and natural-economic systems of different scales, from local to global. Therefore, to determine biocomplexity and to estimate it, a combined formalized description is needed of the biological, geochemical, geophysical, and anthropogenic factors and processes taking place at a given level of the spatiotemporal hierarchy of units and scales.

Biocomplexity is a characteristic feature of all systems of the environment connected with life. Elements of this manifestation are studied within the framework of the theory of stability and ecosystem survivability. Note should be taken here that biocomplexity includes indicators of the degree of mutual modification of interacting systems, and this means that biocomplexity should be studied considering both the spatial and biological levels of organization. The difficulty lies in the complicated behavior of the object under study, especially if the human factor is considered, as a result of which the number of stress situations in the environment is constantly growing.

Humankind has accumulated a great deal of knowledge about environmental systems. Use of this knowledge to study biocomplexity is possible within the framework of synthesizing a global model that reflects the laws of interactions between environmental elements and permits assessing just how efficienct it is at constructing different scenarios in the development of human society, based on the actual data of ground and satellite measurements. It it this problem that serves the basis of all questions set forth by the Biocomplexity Program.

Studies of the interaction process are aimed as a rule at understanding and assessing the consequences of a given interaction. The reliability and accuracy of these assessments depend on criteria that serve as the basis for expert examination and recommendations. At present, there is no agreed method to select such criteria for lack of a single scientifically substantiated approach to ecological normalization of economic forcings on the environment. The choice of such criteria determines the accuracy of the ecological expertise made available to those making decisions about existing and planned human activities and the representativeness of global geoinformation monitoring data.

The processes taking place in the environment can be represented as the totality of interactions between its subsystems. Since a human is one of its elements, it is impossible to definitely divide the environment, for instance, into the biosphere and society: everything on the Earth is correlated and interconnected. The point is to find mechanisms to describe such correlations and interdependences that would reliably reflect the environmental dynamics and answer the questions formulated in the Biocomplexity Program:

1. How does the complexity of biological, physical, and social systems in the environment manifest themselves and change?
2. What mechanisms lie behind the spontaneous development of numerous phenomena in the environment?
3. How do systems of the environment with living components, including those created by humans, react and adjust themselves to stress situations?
4. In what ways do information, energy, and matter move within the systems of the environment and though their levels of organization?
5. Is it possible to predict the system's adaptability and to give prognostic estimates of its changes?
6. How does humankind affect and respond to biocomplexity in natural systems?

One can add many other, no less important questions. For instance, up to what level of complexity should spaceborne observation systems be improved so that their information was of a sufficiently high standard to reliably estimate the state of the environment, if only at the moment of receiving this information? No less important is the question about optimal allocation of the means of geoinformation monitoring at different levels of its organization. Finally, one of the main questions in modern environmental science is estimation of biosphere survivability under conditions of increasing anthropogenic impact. In our opinion such an estimation can be brought about by using the global NSS model and by applying a biocomplexity index. This chapter proposes an approach to resolving this. The general idea is in the combined use of biological complexity and survivability as indicators of the NSS state. Using state-of-the-art simulations, the results presented in this chapter provide an evaluation of the NSS capability to survive under different scenario realizations.

### 2.5.1 Biocomplexity and survivability indicators

Processes that have their origin in the environment can be presented as the combination of interactions between its subsystems. The human subsystem is a part of the environment and it is impossible to divide the environment into separate subsystems such as biosphere and society. The problem is to search for methodologies to describe existing feedbacks between nature and humanity and to simulate dynamic tendencies in the NSS reliably. Unfortunately, the part of the

NSS that is responsible for the quality of modeling climatic processes introduces instability in the modeling results. This is the reason that we suppose below that the NSS climatic component can be replaced by a scenario describing stable climatic trends during the time interval of investigation. What is actually studied is the NSS.

Let us introduce the scale symbol $\Xi$ of biocomplexity ranging from the state where all interactions between environmental subsystems are broken to the state where they correspond to natural evolution. In this case, we have an integrated indicator of the environmental state including bioavailability, biodiversity, and survivability. It reflects the level of all types of interactions among environmental subsystems. In reality, specific conditions exist where these interactions are changed and transformed. For example, under the biological interaction of consumer/producer type or competition-for-energy-resource type there exists some minimal level of food concentration where contacts between interacting components cease. Physical, chemical, and other types of interactions in the environment commonly depend on specific critical parameters. Environmental dynamics is regulated by these parameters and the main task is its parametrical description. Biocomplexity reflects these dynamics.

All of this corroborates the fact that biocomplexity is related to categories that are difficult to measure empirically and to express quantitatively. However, we will try to transfer truly verbal tautological reasoning to formalized quantitative definitions. For the transition to gradations of the scale $\Xi$ with quantitative positions it is necessary to postulate that relationships between two values of $\Xi$ are of the type $\Xi_1 < \Xi_2$, $\Xi_1 > \Xi_2$, or $\Xi_1 = \Xi_2$. In other words, there always exists a value of the scale $\rho$ that defines a biocomplexity level $\Xi \to \rho = f(\Xi)$, where $f$ is a certain transformation of the biocomplexity concept to a number. Let us attempt to search for a satisfactory model to simulate the verbal biocomplexity image in constructive terms, subordinating it to formal description and transformation. With this purpose in mind $m$ subsystems of the NSS are selected. The correlations between these subsystems are defined by the binary matrix function $X = \|x_{ij}\|$, where $x_{ij} = 0$, if subsystems $B_i$ and $B_j$ do not interact, and $x_{ij} = 1$, if subsystems $B_i$ and $B_j$ are interacting. Then, any one point $\xi \in \Xi$ is defined as the sum $\xi = \sum_{i=1}^{m} \sum_{j>i}^{m} x_{ij}$. Of course, there arises the need to overcome uncertainty for which it is necessary to complicate the scale $\Xi$ (e.g., by introducing weight coefficients for all NSS subsystems). The origin of these coefficients depends on the type of subsystem. This is the reason three basic subsystem types are selected: living, nonliving, and vegetation. Living subsystems are characterized by their density, determined by the number of their elements or by biomass value per unit area or volume. Vegetation is characterized by the type and portion of occupied territory. Nonliving subsystems are measured by their concentration per unit area or volume of the environment. In the common case, certain characteristics $\{k_i\}$, corresponding to the significance of subsystems $\{B_i\}$, are assigned to every subsystem $B_i$ $(i = 1, \ldots, m)$. As a result we obtain a better definition of the formula to move from the biocomplexity concept to the scale $\Xi$ of its indicator:

$$\xi = \sum_{i=1}^{m} \sum_{j>i}^{m} k_j x_{ij}. \tag{2.64}$$

It is clear that $\xi = \xi(\varphi, \lambda, t)$, where $\varphi$ and $\lambda$ are the geographical latitude and longitude, respectively, and $t$ is the current time. For the territory $\Omega$ the biocomplexity indicator is defined as the mean value:

$$\xi_\Omega(t) = (1/\sigma) \int_{(\varphi,\lambda) \in \Omega} \xi(\varphi, \lambda, t)\, d\varphi\, d\lambda$$

where $\sigma$ is the area of $\Omega$.

Thus, the indicator $\xi_\Omega(t)$ is the characteristic of integrated NSS complexity that reflects the individuality of its structure and behavior at each time $t$ in space $\Omega$. According to the laws of natural evolution a decrease (increase) in $\xi_\Omega(t)$ will correspond to an increase (decrease) of biocomplexity and the survivability of the nature–anthropogenic systems. Since a decrease in biocomplexity disturbs biogeochemical cycles and leads to a decrease in stress on the nonrenewal of resources, then the binary structure of the matrix $X$ changes direction to intensify resource impoverishment technologies. The vector of energy exchange between NSS subsystems is moved to a position where the survivability level of the NSS is reduced.

The global simulation model is constructed with the spatial resolution of the Earth's surface in which $\Delta\varphi$ represents latitude and $\Delta\lambda$ longitude. In other words, the NSS space $\Omega$ is divided into a set of pixels $\Omega_{ij}$ ($\Omega = \bigcup \Omega_{ij}$; $\Omega_{ij} = \{((\varphi, \lambda)$; $\varphi_i \le \varphi < \varphi_{i+1}$; $\lambda_j \le \lambda < \lambda_{j+1}$; $i = 1, \ldots, N$; $j = 1, \ldots, M$; $N = [180/\Delta\varphi]$; $M = [360/\Delta\lambda]\}$). Each cell $\Omega_{ij}$ has its own biocomplexity indicator value:

$$\xi_\Omega(i,j,t) = (1/\sigma_{ij}) \int_{(\varphi,\lambda) \in \Omega_{ij}} \xi(\varphi, \lambda, t)\, d\varphi\, d\lambda. \tag{2.65}$$

The value $\xi_\Omega(i,j,t)$ calculated by formula (2.65) reflects the topological structure of the matrix $X(i,j,t)$. Consequently, there exist $n = N \cdot M$ matrixes and biocomplexity indicators to characterize NSS biocomplexity. As part of the computer experiment there arises a set of numerical characteristics of NSS biocomplexity distributed in space and time. Integrated NSS biocomplexity indicators can be calculated for any arbitrary area $\omega \in \Omega$:

$$\xi_\omega(t) = (1/\sigma_\omega) \sum_{(\varphi_i,\lambda_j) \in \omega} \xi_\Omega(i,j,t). \tag{2.66}$$

This can be average NSS biocomplexity by longitude or latitude zone, by ocean or sea aquatory, by country or state territory, etc.

NSS survivability is closely connected with biocomplexity. In the common case the behavior of any system is determined by the values characterizing the state of the system, which can take different terms. Upon interaction with an external medium and, in particular, with other systems, the values of these terms can vary in one way or another. For any technological or biological system, it is always possible to show the field of change of the characteristic parameters, wherein the system can be considered to be functioning. Outside this field the system does not exist. Thus, one can substitute the complex behavior of a system by depicting this system as a point in the phase space of the characteristic parameters. If the change of any coordinate leads to the disappearance of the depicting point from the allowable field, the system collapses (the organism as a whole perishes).

Let us restate what we said in the introduction to this chapter (Section 2.1) about significant variables not being identical regarding the degree of threat they pose to the system. Such variables as the oxygen content of blood or the structural integrity of the medulla oblongata cannot tolerate any significant changes, since such changes would almost invariably lead to immediate death. However, there are changes, such as in the temperature of individual areas of the skin where sharp fluctuations do not necessarily lead to such an eventuality. Separating all variables characterizing the state of the system into significant variables makes it possible to simplify the behavioral strategy of a system as it interacts with an external environment or other systems. A most constructive approach to describing the global environment was proposed by Gorshkov *et al.* (2000). They consider that a different path of development compatible with long-term environmental safety lies through conservation and restoration of a substantial part of the Earth's biosphere to its natural nonperturbed state, bearing in mind the stabilizing potential of the natural biota of Earth with respect to the global environment. The problem of biosphere survivability really does correlate with the mechanisms of biotic regulation, physical and biological stability, the sensitivity of biota, ecological limitations, and other basic principles of biology. In general, each living organism clearly plays a role in global change. The problem lies in describing this role to estimate the significance of interactions between the hierarchy of biospheric elements having various spatial scales and different influences on the levels of biological organization.

Following Krapivin (1978, 1996) a survivability indicator can be taken from trophic relations between ecosystem components and represented by the equation:

$$\nu(t) = \frac{\sum_{i=1}^{m} \iint_{(\varphi,\lambda)\in\Omega} \int_0^{z_0} B_i(\varphi,\lambda,z,t)\, d\varphi\, d\lambda\, dz}{\sum_{i=1}^{m} \iint_{(\varphi,\lambda)\in\Omega} \int_0^{z_0} B_i(\varphi,\lambda,z,t_0)\, d\varphi\, d\lambda\, dz},$$

where $B_i$ is the $i$th element of the NSS.

Indicators such as these and others help to determine the state of an environmental subsystem based on restricted information. Remote sensing of the environment is characterized by a series of such indicators as NDVI, LAI, and SIL, which are used widely in many studies. Of the important problems that are the subject of many international environmental programs the study of forest ecosystems is arguably the most pressing.

Anthropogenic and natural biomass burning has become an ordinary event in the world. The development of an effective technology for atmosphere pollution control by means of satellite system is long overdue. However, there exist difficulties connected with spatial resolution, the temporal frequency of satellite overpasses, and cloudiness. The temporal dynamics of fire cannot be correlated with the interval between two consequent satellite overpasses over the area that is ablaze. Boschetti *et al.* (2003) proposed a methodology of using data acquired by the European Meteosat and the Japanese GMS to detect burned areas in different tropical environments. The methodology is based on a multiple threshold approach applied to thermal radiance and to a spectral index specific to burned surfaces.

Of the informational indexes available, the Simple Index for Burned Areas (SIBA) is the best for adaptation to observational conditions. It allows various features of the burned area to be enhanced:

- low albedo;
- high temperature;
- temperature higher than the surrounding pixels in a window large enough to encompass the burned areas.

SIBA can be described as a normalized function:

$$\mathrm{SIBA}_{ij} = I_{ij}/(I_{ij}+1),$$

where

$$I_{ij} = \left[\frac{(T_{ij}-a)(T_{ij}-b)}{c\rho_{\mathrm{TOA}_{ij}}(\hat{T}_{ij}-b)}\right]^2;$$

$T_{ij}$ is the surface temperature (K) of pixel $\Omega_{ij}$; $\hat{T}_{ij}$ is the mean surface temperature (K) in a window of $60 \times 60$ pixels around pixel $\Omega_{ij}$; $a = 248$, $b = 273$, $c = 300$; and $\rho_{\mathrm{TOA}_{ij}}$ is the top-of-atmosphere reflectance of pixel $\Omega_{ij}$.

Numerous calculations made by Boschetti *et al.* (2003) show that SIBA behaves differently in some significant cases: water bodies, clouds, and burnt surfaces. SIBA maintains the capability to detect burned areas in different areas and under different conditions. This capability depends on the sensor type.

The introduction of integral characteristics for use as indicators of the state of the environmental subsystem (as demonstrated by many authors) allows satellite monitoring of soil–plant formations. The most appropriate index here is the Scattering Index over Land (SIL), which permits different surfaces to be distinguished:

$$\mathrm{SIL} = 451.9 - 0.44T_b(19\ \mathrm{GHz}) - 1.755T_b(22\ \mathrm{GHz}) + 0.00575T_b^2(22\ \mathrm{GHz}) - T_b(85\ \mathrm{GHz});$$

$$\mathrm{SIL} = \begin{cases} 10\ \mathrm{K} & \text{precipitation area;} \\ 13\ \mathrm{K} & \text{snow cover;} \\ 15\ \mathrm{K} & \text{desert or semidesert territory.} \end{cases}$$

SIL facilitates solution of a precipitation problem:

$$RR(\text{mm/h}) = 0.00513 \cdot \text{SIL}^{1.9468}.$$

There exist other correlations as functions of integral indexes. For example, low atmospheric temperature can be estimated by means of the following formula:

$$T(\text{K}) = 58.08 - 0.39T_{bv}(19\text{ GHz}) + 1.21T_{bv}(22\text{ GHz}) - 0.37T_{bv}(37\text{ GHz}) + 0.36T_{bv}(85\text{ GHz}).$$

According to this, when $T > T^*$ precipitation is possible, where

$$T^* = \begin{cases} 242.5 + 5\cos\theta & \text{for } T_b(53.6\text{ GHz}) \le 248\text{ K;} \\ 0.667[T_b(53.6\text{ GHz}) - 248] + 252 + 6\cos\theta & \text{for } T_b(53.6\text{ GHz}) > 248\text{ K;} \end{cases}$$

where $\theta$ is the satellite zenith angle.

The NDVI and LAI indexes help to assess the water content of vegetation (kilograms per square meter):

$$m_v = \begin{cases} 1.9134(\text{NDVI})^2 - 0.3215(\text{NDVI}) & \text{when NDVI} \le 0.5; \\ 4.2857(\text{NDVI})^2 - 1.5429 & \text{when NDVI} > 0.5. \end{cases}$$

All these correlations help to form (as part of the GIMS) an effective algorithm to assess forest conditions that may be susceptible to fire. Maki *et al.* (2004) proposed the following procedure to predict fire outbreak and propagation. The prediction of fire outbreak, propagation, and scale in forested areas depends mainly on wind direction, vegetation water status, topography, but other factors are involved. Vegetation water status is the most important parameter determining the risk of fire. Maki *et al.* (2004) calculated the vegetation water status at ground level by means of three definitions:

(i) fuel moisture content (FMC);
(ii) equivalent water thickness (EWT);
(iii) relative water content (RWC).

FMC is defined as the ratio between the quantity of water in vegetation and either the fresh or dry weight of vegetation:

$$\text{FMC} = \frac{\text{FW} - \text{DW}}{\text{FW (or DW)}} \times 100(\%).$$

EWT is the ratio between the quantity of water and the area $\sigma$:

$$\text{EWT} = \frac{\text{FW} - \text{DW}}{\sigma}(\text{g/cm}^2).$$

RWC is calculated by means of the following formula:

$$\text{RWC} = \frac{\text{FW} - \text{DW}}{\text{TW} - \text{DW}},$$

where FW is the full weight; DW is the dry weight; and TW is the turgid weight.

### 2.5.2 The nature–society system biocomplexity model

The NSS consists of subsystems $B_i$ ($i = 1, \ldots, m$) whose interactions are built up over time as a result of the functions of many factors. NSS biocomplexity indicates the structural and dynamic complexity of its components. In other words, NSS biocomplexity is formed under the interaction of its subsystems $\{B_i\}$. Over the course of time, subsystems $B_i$ can change their state and, consequently, change the topology of the relations between them. The evolutionary mechanism of adaptation of subsystem $B_i$ to the environment allows the hypothesis that each subsystem $B_i$, independent of its type, has structure $B_{i,S}$, behavior $B_{i,B}$, and goal $B_{i,G}$ such that $B_i = \{B_{i,S}, B_{i,B}, B_{i,G}\}$. The strivings of subsystem $B_i$ to achieve certain preferable conditions are represented by its goal $B_{i,G}$. The expedience of structure $B_{i,S}$ and the purposefulness of behavior $B_{i,B}$ for subsystem $B_i$ are estimated by the effectiveness with which goal $B_{i,G}$ is achieved.

As an example, let us consider the process of fish migration. The investigations of many authors have revealed that this process is accompanied by purposeful behavior. From these investigations it follows that fish migrations are subordinated to the principle of complex maximization of effective nutritive rations, subject to favorable environmental conditions (temperature, salinity, dissolved oxygen, pollution level, depth). In other words, the travel of migrating species takes place at characteristic velocities toward the maximum gradient of effective food, subject to ecological restrictions. This is the reason we can formulate that goal $B_{i,G}$ of the fish subsystem is toward increasing its food supply and that behavior $B_{i,B}$ consists in finding the optimum route to attain goal $B_{i,G}$.

Since the interactions of subsystems $B_i$ ($i = 1, \ldots, m$) are connected with chemical and energy cycles, it is natural to suppose that each subsystem $B_i$ accomplishes the geochemical and geophysical transformation of matter and energy to remain in a stable state. The formal approach to this process consists in supposing that interactions between NSS subsystems are represented as a process whereby the systems exchange a certain quantity $V$ of resources spent in exchange for a certain quantity $W$ of resources consumed. We shall call this process "$(V, W)$ exchange".

The goal of subsystem $B_i$ is the most advantageous $(V, W)$ exchange (i.e., it tries to get maximum $W$ in exchange for minimum $V$). The quantity $W$ is a complex function of the structure and behavior of interacting subsystems, $W = W(V, B_i, \{B_k, k \in K\})$, where $K$ is the set of subsystem numbers interacting with subsystem $B_i$.

Let us designate $B_K = \{B_k, k \in K\}$. Then, the following $(V, W)$ exchange is the result of interactions between subsystem $B_i$ and its environment $B_K$:

$$W_{i,0} = \max_{B_i} \min_{B_K} W_i(V_i, B_i, B_K) = W_i(V_i, B_{i,\text{opt}}, B_{K,\text{opt}})$$

$$W_{K,0} = \max_{B_K} \min_{B_i} W_K(V_K, B_i, B_K) = W_K(V_K, B_{i,\text{opt}}, B_{K,\text{opt}}).$$

Hence, it follows that some range of the goal of subsystem $B_i$ exists which defines the levels of $V_i$ and $V_K$. Since limiting factors are defined by nature, then it is natural to suppose in this case that some level $V_{i,\min}$ exists when subsystem $B_i$ ceases to spend its energy resources on external resources (i.e., if $V_i \le V_{i,\min}$, subsystem $B_i$ concentrates on regenerating its internal resources). In other words, when $V_i \le V_{i,\min}$, any decrease in the biocomplexity indicator $\xi_\Omega(t)$ takes place at the expense of breaking off interactions between subsystem $B_i$ and other subsystems. Commonly, the structure of $V_{i,\min}$ is checkered (i.e., the changeover of $x_{ij}$ from state $x_{ij} = 1$ to state $x_{ij} = 0$ is not realized for all $j$ at the same time). Actually, in any trophic pyramid of living subsystems producer/consumer-type relationships cease when the consumer biomass concentration falls below some critical level. In other cases the interactions between subsystems $\{B_i\}$ can be stopped at the expense of various combinations of its parameters. The parametrical description of possible situations of interactions between subsystems $\{B_i\}$ can be realized in the NSS simulation model.

### 2.5.3 Simulation experiments

#### *2.5.3.1 The Okhotsk Sea ecosystem case*

The Okhotsk Sea ecosystem (OSE) is a significant element of the biosphere whose evaluation requires development of a common criterion. The OSE Biocomplexity Index helps to explain many processes regulating the interactions between biotic components, hydrodynamic effects, and energy fluxes. Traditional estimates of the contributions from different processes within the OSE deal with the study of local or special parameters. This makes understanding correlations between OSE components and forecasting their dynamics impossible. Moreover, a simple index would make it possible to evaluate the state of the OSE by means of ordinary calculations.

The OSE has a trophic graph showing interactions at many levels between biological, chemical, and physical processes. OSE biocomplexity consists of numerous sets of biotic regulations determining the fundamental properties of living objects. An important property of OSE living components is that all biological species exist in the form of populations. All processes and phenomena observed in the OSE are characterized by a certain degree of physical and biological stability, which is a function of external and internal fluxes of energy. External fluxes of energy are defined as solar irradiation, the influence of the Pacific Ocean, and anthropogenic interventions. In the absence of an external flux of energy, the OSE tends toward a state of thermodynamic equilibrium, which is characterized by the maximum degree of chaos possible in a given system (Gorshkov *et al.* 2000).

In any event, OSE dynamics is a complex function of many parameters having different chemical, physical, and biological character. A biocomplexity index has to reflect this and to characterize the biological stability of the whole aquageoecosystem.

The Okhotsk Sea is a typical highly productive sea whose ecosystem has to function under severe climatic conditions. The spatiotemporal structure of the basic hydrological and ecological characteristics of the Okhotsk Sea is heterogeneous. The chemical, physical, and biological processes occurring in seawater have been studied by many authors to assess their bioproductivity. According to the investigations made by Terziev *et al.* (1993), the following structural discretization of the Okhotsk Sea geoecosystem can be realized. Five ecological layers exist. Layer 1 is where photosynthesis is greatest. It is situated above the thermocline and lies at depths of 20–30 m. It corresponds to the wind-mixed layer. Layer 2 occupies depths from 30 to 150 m. It has a low temperature and oxygen saturation of about 80–90%. Layer 3 is characterized by low oxygen saturation (15–20%). It lies at depths of 150–750 m. Layer 4 extends from 750 m down to a depth of 1,500 m. This layer has the lowest oxygen saturation (10–15%). Lastly, layer 5 is located deeper than 1,500 m. It is characterized by oxygen saturation of 25–30%.

The Okhotsk Sea aquatory is divided into zones having specific ecological features (Suzuki, 1992). The spatial distribution of fish depends on seasonal conditions and to a great extent correlates with the layers just mentioned. The use of the sea's biological resources is a function of this distribution. Fishing intensity essentially depends on knowledge of the biomass distribution in zones that have their own specific environmental conditions. Various authors (Aota *et al.*, 1992; Nitu *et al.*, 2000b; Plotnikov, 1996) have tried to resolve this by using models that simulate ecosystem dynamics. However, the modeling results have not always turned out to be sufficiently representative and to reflect the classification of sea zones according to their productivity. The biocomplexity indicator is one such simple form capable of identifying these zones. It has been shown by many investigators that highly productive Okhotsk Sea zones are characterized by a complex multilevel trophic graph (Terziev *et al.* 1993). However, this effect is not universal. For instance, the ecosystem of the Peruvian Current is highly productive in zones where the trophic graph is short (Krapivin, 1996). These situations can be distinguished by migration processes. Hence, the biocomplexity of ecosystems can be formed in various ways.

Let us consider the following components of the Okhotsk Sea ecosystem (Table 2.1). The trophic pyramid $X = \|x_{ij}\|$, where $x_{ij}$ is a binary value equal to "1" or "0" under the existence or absence of nutritive correlations between the $i$th and $j$th components, respectively. Let us define the biocomplexity function as:

$$\xi(\varphi, \lambda, z, t) = \sum_{i=1}^{20} \sum_{j=1}^{19} x_{ij} C_{ij}, \tag{2.67}$$

where $\varphi$ and $\lambda$ are the geographical latitude and longitude; $t$ is current time; $z$ is depth; $x_{ij} = 1$ if $B_m \geq B_{m,\min}$ and 0 if $B_m < B_{m,\min}$ where $B_{m,\min}$ is the minimal biomass of the $m$th component consumed by other trophic levels; $C_{ij} = k_{ji} B_{i,*} / \sum_{j,+}$ is the nutritive pressure placed by the $j$th component on the $i$th component; $\sum_{i,+} = \sum_{m \in S_i} k_{im} B_m$ is the real store of food available to the $i$th component; $B_{m,*} = \max\{0, B_m - B_{m,\min}\}$; $k_{im} = k_{im}(t, T_W, S_W)$ $(i = 1, \ldots, 17)$ is the satisfaction index of the nutrition requirements of the $i$th component at the expense of the biomass of the $m$th component; $k_{im}$ $(i = 18, 19)$ is the trans-

**Table 2.1.** Trophic pyramid of the Okhotsk Sea ecosystem considered when calculating the biocomplexity indicator. Notation: $B_{14}$ *Reinhardti ushippoglossoi des matsuurae*; $B_{15}$ *Clupeapallasi pallasi* Val.

| *Consumers of energy and matter* | *Sources of energy and matter* | | | | | | | | | | | | | | | | | | |
|---|---|---|---|---|---|---|---|---|---|---|---|---|---|---|---|---|---|---|---|
| | $B_1$ | $B_2$ | $B_3$ | $B_4$ | $B_5$ | $B_6$ | $B_7$ | $B_8$ | $B_9$ | $B_{10}$ | $B_{11}$ | $B_{12}$ | $B_{13}$ | $B_{14}$ | $B_{15}$ | $B_{16}$ | $B_{17}$ | $B_{18}$ | $B_{19}$ |
| Phytoplankton, $B_1$ | 0 | 0 | 0 | 0 | 0 | 0 | 0 | 0 | 0 | 0 | 0 | 0 | 0 | 0 | 0 | 0 | 0 | 1 | 0 |
| Bacterioplankton, $B_2$ | 0 | 0 | 0 | 0 | 0 | 0 | 0 | 0 | 0 | 0 | 0 | 0 | 0 | 0 | 0 | 0 | 0 | 0 | 1 |
| Microzoa, $B_3$ | 1 | 1 | 0 | 0 | 0 | 0 | 0 | 0 | 0 | 0 | 0 | 0 | 0 | 0 | 0 | 0 | 0 | 0 | 1 |
| Microzoa, $B_4$ | 1 | 1 | 0 | 0 | 0 | 0 | 0 | 0 | 0 | 0 | 0 | 0 | 0 | 0 | 0 | 0 | 0 | 0 | 0 |
| Carnivores, $B_5$ | 0 | 1 | 1 | 1 | 1 | 0 | 0 | 0 | 0 | 0 | 0 | 0 | 0 | 0 | 0 | 0 | 0 | 0 | 1 |
| Zoobenthic animals, $B_6$ | 1 | 1 | 1 | 1 | 1 | 0 | 0 | 0 | 0 | 0 | 0 | 0 | 0 | 0 | 0 | 0 | 0 | 0 | 1 |
| Flat-fish, $B_7$ | 0 | 0 | 0 | 0 | 0 | 1 | 1 | 1 | 1 | 1 | 1 | 1 | 1 | 1 | 1 | 1 | 1 | 0 | 0 |
| Coffidae, $B_8$ | 0 | 0 | 0 | 0 | 1 | 1 | 1 | 1 | 1 | 1 | 1 | 1 | 1 | 1 | 1 | 1 | 1 | 0 | 0 |
| *Ammodytes hexapterus*, $B_9$ | 0 | 0 | 0 | 0 | 0 | 1 | 1 | 0 | 0 | 1 | 1 | 1 | 0 | 0 | 0 | 1 | 0 | 0 | 0 |
| *Mallotus*, $B_{10}$ | 0 | 0 | 0 | 0 | 1 | 1 | 1 | 0 | 0 | 0 | 1 | 0 | 0 | 0 | 0 | 0 | 0 | 0 | 0 |
| *Theragra chalcogramma*, $B_{11}$ | 0 | 0 | 0 | 0 | 1 | 0 | 0 | 0 | 0 | 1 | 1 | 1 | 0 | 0 | 0 | 0 | 0 | 0 | 0 |
| Salmonidae, $B_{12}$ | 0 | 0 | 0 | 0 | 0 | 1 | 1 | 0 | 0 | 1 | 1 | 1 | 1 | 0 | 0 | 1 | 0 | 0 | 0 |
| Coryphaenoides, $B_{13}$ | 0 | 0 | 0 | 0 | 1 | 1 | 1 | 1 | 1 | 1 | 1 | 1 | 1 | 1 | 1 | 1 | 1 | 0 | 0 |
| $B_{14}$ | 0 | 0 | 0 | 0 | 1 | 1 | 1 | 1 | 1 | 1 | 1 | 1 | 1 | 1 | 1 | 1 | 1 | 0 | 0 |
| $B_{15}$ | 0 | 0 | 0 | 0 | 1 | 1 | 0 | 0 | 0 | 0 | 0 | 0 | 0 | 0 | 0 | 0 | 0 | 0 | 0 |
| Crabs, $B_{16}$ | 0 | 0 | 0 | 0 | 0 | 1 | 0 | 1 | 1 | 1 | 0 | 1 | 0 | 0 | 0 | 1 | 1 | 0 | 0 |
| *Laemonema longipes*, $B_{17}$ | 0 | 0 | 0 | 0 | 1 | 1 | 0 | 0 | 0 | 0 | 0 | 0 | 0 | 0 | 0 | 0 | 1 | 0 | 0 |
| Biogenic salts, $B_{18}$ | 0 | 0 | 0 | 0 | 0 | 0 | 0 | 0 | 0 | 0 | 0 | 0 | 0 | 0 | 0 | 0 | 0 | 0 | 1 |
| Detritus, $B_{19}$ | 1 | 1 | 1 | 1 | 1 | 1 | 1 | 1 | 1 | 1 | 1 | 1 | 1 | 1 | 1 | 1 | 1 | 0 | 0 |
| People, $B_{20}$ | 1 | 1 | 1 | 1 | 1 | 1 | 1 | 1 | 1 | 1 | 1 | 1 | 1 | 1 | 1 | 1 | 1 | 1 | 1 |

formation coefficient from the $m$th component to the $i$th component; $k_{i,20}$ represents anthropogenic influence on the $i$th component; $S_i = \{i : x_{ij} = 1, j = 1, \ldots, 19\}$ is the food spectrum of the $i$th component; $T_W$ is water temperature; and $S_W$ is water salinity.

Let us designate the aquatory of the Okhotsk Sea by $\Omega = \{(\varphi, \lambda)\}$. The value of the biocomplexity indicator for any area $\omega \in \Omega$ is determined by the formula:

$$\xi_\omega(z_1, z_2, t) = (1/\sigma_\omega) \int_{(\varphi,\lambda) \in \omega z_1} \int_{z_1}^{z_2} \xi(\varphi, \lambda, z, t) \, d\varphi \, dz,$$

where $[z_1, z_2]$ is the water layer located between the depths of $z_1$ and $z_2$.

The maximum value of $\xi = \xi_{max}$ ($\approx 20$) is reached during the spring–summer when nutrition relationships in the OSE are extended, the intensity of energy exchanges is increased, and horizontal and vertical migration processes are stimulated. In winter the value of $\xi$ approaches $\xi_{min}$ ($\approx 8$). The spatial distribution of $\xi$ reflects the local variability of the food spectrum for the various components. Figure 2.3 and Table 2.2 show examples of such a distribution. Comparison of this distribution with that of zones with industrial fish accumulations (Terziev *et al.*, 1993) shows that there is a correlation between these distributions.

The indicator $\xi$ reflects the level of complexity of the OSE. A change in $\xi$ is a consequence of migration processes and the variability of nutritive interactions. In

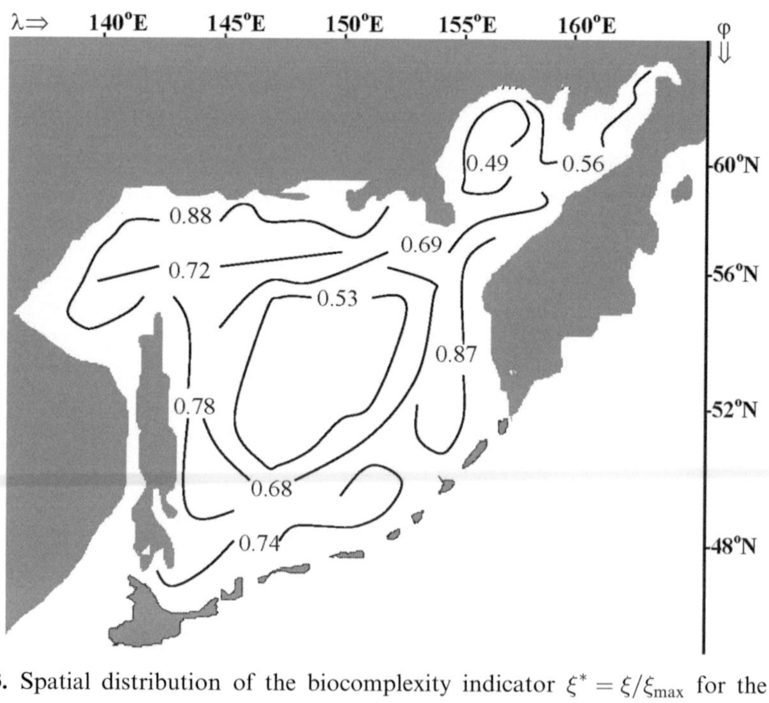


**Figure 2.3.** Spatial distribution of the biocomplexity indicator $\xi^* = \xi/\xi_{max}$ for the spring–summer.

**Table 2.2.** Estimates of the biocomplexity indicator $\xi^*$ for different layers in spring–summer and in winter.

| *Season* | *Layer* | | | | |
|---|---|---|---|---|---|
| | *1* | *2* | *3* | *4* | *5* |
| Spring–summer | 0.89 | 0.93 | 0.62 | 0.34 | 0.21 |
| Winter | 0.31 | 0.49 | 0.71 | 0.39 | 0.22 |

these processes subsystem $B_{20}$ plays the role of an external source of change in other components. These changes are interpreted in terms of fishing and impacts causing variations in component biomass.

Calculations show that basic variability in $\xi^* = \xi/\xi_{\max}$ is caused by migration processes. Under these conditions, there occurs a quick redistribution of the interior structure of matrixes $X$ and $\|C_{ij}\|$. For instance, according to Terziev *et al.* (1993) many fish during spring migrate to the shelf zone and during winter they move to the central aquatories of the sea. Therefore, the value of $\xi^* \to 1$ during spring and $\xi^* \to 0.6$ during winter for the shelf zone. This means that the biocomplexity of the Okhotsk sea ecosystem in the shelf decreases by 40% in winter in comparison with spring. For the central aquatories the value of $\xi^*$ varies near 0.7 throughout the year. Such stability of the biocomplexity indicator is explained by the balance between nutrition correlations and productivity during spring, summer, and winter.

It has been established that variability in $\xi^*$ stimulates changes in fish concentrations that are controlled by environmental conditions. Specifically, during spring time the larval Pacific hering (*Clupea pallasi*) occupies the area with $T_W < 5°\text{C}$. Other fish have an elective depth for feeding and spawning (Terziev *et al.*, 1993). All these processes influence the variability of $\xi^*$. More detailed investigation of correlations between the value of $\xi^*$ and the structural and behavioral dynamics of the Okhotsk Sea ecosystem requires additional studies.

This section has introduced a means of moving from a verbal description of biocomplexity to a numerical scale. In future studies it will be necessary to take into consideration various factors such as bottom relief, climate trends, ice field dynamics, detailed components of the trophic pyramid, bottom sediments, and current structure. Moreover, it is necessary to add members describing anthropogenic impacts on the ecosystem considered socioeconomically to formula (2.67).

#### 2.5.3.2 *Upwelling ecosystem case*

It is known that upwelling zones characterized by vertical water motions have high productivity. An upwelling zone is the result of many phenomena: water removal from the coastline by the wind, changes in ocean currents, etc. Water velocity and the stability of an upwelling zone are determined by a set of synoptic parameters. The most specific value for the vertical speed of water in an upwelling zone is $0.77 \cdot 10^{-3}$ cm · s$^{-1}$. The depths at which water flows start vary within 200 m.

Let us proceed from the concept of successive development of a community from the time of its origin in a region invaded by deep water until its climax in the oligotrophic convergence region. In between these times the system develops and moves along with the water flow. In addition, the total energy of the system and its structure (spatial, trophic, and specific) are changed. The observed general characteristics of these changes occurring in time and space are now available, and one of the major criteria determining the adequacy of the model is its agreement with the actual picture observed in the oceans.

Let us suppose ecosystem motion from the upwelling zone is homogeneous in the horizontal plane. The ecosystem state is characterized by depth $z$ with step $\Delta z$ ($\approx$10 m) and by time $t$ with interval $\Delta t$ (daily). The horizontal speed of the water current from the upwelling zone is $V = V_\varphi = V_\lambda$, so that the distance of the water volume from the upwelling zone equals $\Delta r = (\Delta\varphi^2 + \Delta\lambda^2) = V \cdot \Delta t$.

The ecosystem state at each layer $z = \text{const}$ is determined by light intensity $E(z,t)$, by nutrient salt concentration $n(z,t)$, and by the biomass of detritus $d(z,t)$, phytoplankton $p(z,t)$, bacterioplankton $b(z,t)$, protozoa $Z_1(z,t)$, microzoa $Z_2(z,t)$, small-sized herbivores $Z_3(z,t)$, large-sized herbivores $Z_4(z,t)$, small predators like *Cyclopodia* $Z_5(z,t)$, intermediate predators like *Calanoida* $Z_6(z,t)$, and large predators like *Chaetognatha* and *Polychaeta* $Z_7(z,t)$. The protozoa include infuzorii and radiolarii. The microzoa include the nauplii stages of copepods. Based on plankton-feeding studies made by many authors (Vinogradov *et al.*, 1972) the small-sized herbivores are now believed to include, apart from the protozoa and nauplii, young copepod stages of *Calanoida* and adult copepods whose size does not reach 1.0 mm such as *Clausocalanus*, *Acrocalanus*, *Paracalanus*, *Calocalanus*, etc. The group of large-sized herbivores consists of animals whose size exceeds 1.0 mm such as *Undinula*, *Eucalanus*, *Rhincalanus*, *Neocalanus*, *Lucicutia*, juveniles of *Euphausiacea*, etc. The group of omnivores includes *Centropages*, *Pleuromamma*, *Scolecithrix*, *Undeuchaeta*, *Conchoecia*, etc., while *Chaetognatha*, *Candacia*, *Euchaeta*, *Cyclopodia*, etc. are grouped with the carnivores.

It is accepted that 30% of the bacterioplankton biomass is held in natural clots of size greater than 3–5 μm, which can be consumed by the herbivores ($Z_3$ and $Z_4$). The microzoa ($Z_2$), protozoa ($Z_1$), and small-sized herbivores ($Z_3$) can consume unclotted bacterioplankton as well.

The trophic relations between the components are described by means of the energetic principle (Figure 2.4). Biomass, production, respiration, mortality, and rations are measured by the energy scale in calories per cubic meter or calories per square meter.

When estimating changes in the system over time, it was assumed that water takes more than 60 days to cover the distance from the upwelling zone to the oligotrophic zone of planetary convergence. Figure 2.5 gives a representation of the changes in biomasses as functions of time. We see that the phytoplankton

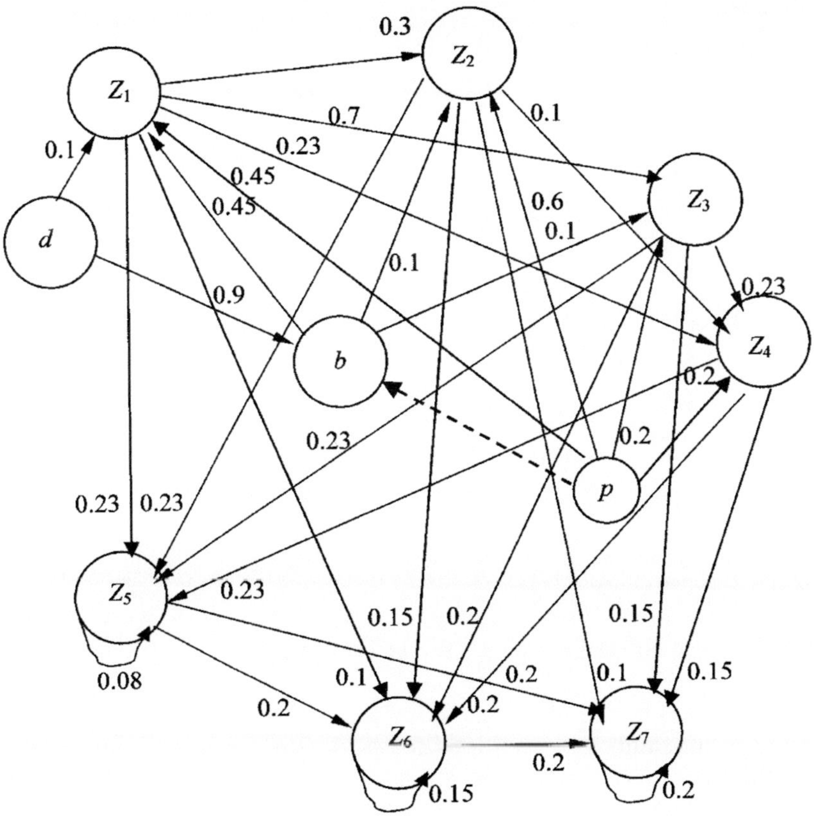


**Figure 2.4.** Scheme of the trophic interactions between components of the upwelling ecosystem. The values of coefficients $C_{ij}$ are shown on the arrows' shafts.

biomass increases most rapidly here, reaching its maximum ($\approx 4{,}500\,\text{cal}\cdot\text{m}^{-2}$) on the 5th to 10th day of the existence of the system. After this, the phytoplankton biomass decreases. The peak bacterioplankton biomass is reached on the 10th to 15th day. Small-sized herbivores lag somewhat behind phytoplankton in development, and their biomass reaches its peak only on the 30th day. Nevertheless, its influence along with that of nutrient salt decrease leads to a sharp drop in phytoplankton and bacterioplankton biomasses. Namely, inverse chains of the community give weak contributions to $R_p$ and $R_b$. After the 40th day, the phytoplankton biomass mainly functions at the expense of biogenic elements arriving in the eutrophic zone across the thermocline from deeper layers. Subsequently, the phytoplankton biomass decreases relatively slowly. This is the time that stability factors begin to influence the community at the expense of exterior energy flows.

The carnivores prove to be even more inertial than the herbivores as their biomass attains its peak only on the 35th to 50th day. This is the time (i.e., the 50th to 60th day) that the system reaches its quasistationary state characterized by low concentrations of all living components.

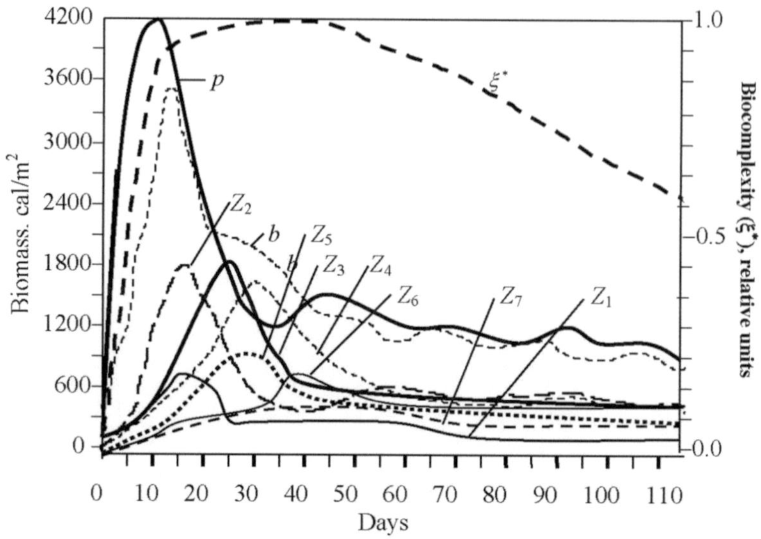


**Figure 2.5.** Time dependence of the total biomasses of living components of the upwelling ecosystem in the 0–200 m water layer.

## 2.6 COMPUTER SIMULATION OF ACID RAIN

Key to understanding the state of today's global ecodynamics and ecosystem survivability is the study of conditions for acid rain formation and their prediction. For the first time, this problem was widely discussed at the 28th General Assembly of the International Unit on Theoretical and Applied Chemistry held in Madrid in September 1975. Subsequent conferences and various international programs have made it possible to accumulate data and knowledge in this sphere. Consequently, we now know that sulfur compounds emitted to the atmosphere from natural and anthropogenic sources are an important precursor of acid rain, which seriously damages the environment. Sulfur resides in the atmosphere mainly in the form of gas-phase $SO_2$ and $H_2S$ as well as sulfate ion $SO_4^{2-}$.

Sulfur dioxide is a basic precursor of acid rain. Its concentration at the surface level is estimated at $1\,\mu g\,m^{-3}$. The participation of $SO_2$ in acid rain formation takes place in two ways: through dry deposition onto a wet surface and formation of $H_2SO_4$ directly in the atmosphere with subsequent deposition either onto land surfaces or in water basins. Sulfur dioxide reacts with water to give sulfuric acid: $SO_2 + H_2O + \frac{1}{2}O_2 \rightarrow H_2SO_4$. These processes are shown in Figure 2.6. $SO_2$ residence in the atmosphere depends strongly on the means of its removal. As a result of dry deposition, $SO_2$ is removed from the atmosphere in 7.6 days and its transformation to $SO_4^{2-}$ in 13 days. The rate of dry deposition depends on the type of surface and many other environmental parameters. This rate averages $2\,cm \cdot s^{-1}$ over land and $0.9\,cm\,s^{-1}$ over the ocean. As a result of the combination of the processes of $SO_2$ removal from the atmosphere its residence can shorten to 4.8 days.

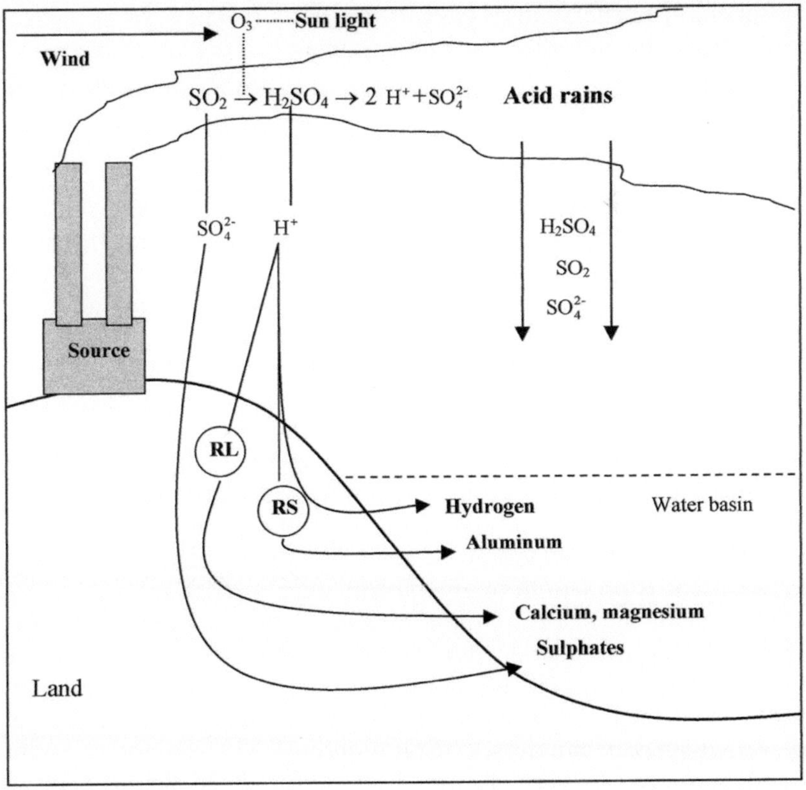


**Figure 2.6.** A conceptual scheme of the impact of anthropogenic sulfur emissions on the quality of the aquatic medium. Notation: RL, reactions with limestone minerals; RS, reactions with aluminum-containing silicate minerals.

A simplified formula for acid rain is: acid rain $= H_2O + SO_2 + NO_2$. Natural and anthropogenic emissions of $SO_2$ to the atmosphere are responsible for 60 to 70% of acid rain globally. Deposition from the atmosphere of excessive sulfate is estimated at $360\,\mathrm{Tg \cdot yr^{-1}}$ at an average emission rate to the atmosphere of $110\,\mathrm{Tg \cdot yr^{-1}}$, with 31% of excessive sulfate in rain water being anthropogenic in origin. On the whole, anthropogenic sources emit to the atmosphere >90% S. These sources are:

- coal burning (coal contains 2–3% S and its burning gives $SO_2$);
- oil burning and refining (the power of sulfur sources is four to five times lower than that of coal burning);
- ore melting to obtain metals such as copper, nickel, and zinc;
- volcanic eruptions;
- organic decomposition;
- weathering of sulfur-containing rocks ($\sim 15\,\mathrm{Tg \cdot yr^{-1}}$);

- input of sulfur to the atmosphere with sea spray ($\sim$45 $Tg \cdot yr^{-1}$);
- sulfate fertilizers and their subsequent input to the atmosphere with dust ($\sim$10 $Tg \cdot yr^{-1}$);
- aviation and car engines.

Natural sources of sulfur compounds in the form of hydrocarbons, dimethylsulfide, carbonylsulfide, and methylmercaptan include soils, marshes, forests, volcanoes, the hydrosphere, and agricultural soils. Dimethylsulfide from the surface of the World Ocean reaches the atmosphere and is rapidly oxidized to give sulfates that reside in the atmosphere for no more than 5 days. Available estimates put the sulfur emission of volcanoes annually to the atmosphere from 4 million to 16 million tonnes of sulfides (recalculated for $SO_2$). Sulfur-containing compounds also form as a result of geothermal activity and the activity of living organisms on land and in water. Rivers bring sulfur to seas and oceans at the rate of $\sim$100 $Tg \cdot yr^{-1}$. Natural sources of sulfur are rather small. For instance, in the U.S.A. and Canada the emissions of sulfur products from natural sources constitute, respectively, no more than 4 and 18% of total sulfur emissions.

The spectrum of anthropogenic sources of sulfur compounds is diverse and variable because it is present in many minerals such as coal, oil, iron, copper, and other ores. Humankind's use of these minerals leads to sulfur emissions to the atmosphere despite the use of purification devices. The main by-product of industrial processes and fossil fuel burning is sulfur dioxide. For instance, in the U.S.A., $SO_2$ emissions can be traced to electric power stations 67%, fossil fuel burning 3%, industrial enterprises 15%, transport 7%, and other sources 8%. Globally, these indicators vary strongly both in space and time. For example, in contrast to the U.S.A., electric power stations in Canada emit 20% $SO_2$, whereas non-ferrous metallurgy emits 43% $SO_2$. On the whole, in Canada industrial enterprises are the main source of $SO_2$ (74%). It should be noted that the sources of sulfur in the U.S.A. are responsible for more than 50% of acid rain in Canada, and territories bordering Quebec province get up to 75% of acid rain as a result of $SO_2$ emissions in the U.S.A. Available estimates put the trans-boundary transport of $SO_2$ from the U.S.A. to Canada at 3.5 million to 4.2 million tonnes per year.

In most countries, as a consequence of the intense use of petroleum, the problem of acid rain has caused economic problems. Petroleum in the recent past contained sulfur ranging from 150 to 600 ppm. Economic losses because of acid rain forced developed countries to find technologies to reduce the content of sulfur in petroleum. The expected level of 30–50 ppm targeted for 2005 was reached.

The structure of sulfur dioxide emissions is similar in most countries. In Germany, power stations contribute about 90% of the total emissions of sulfur dioxide, while industrial enterprises and transport contribute only 7.5% and 2.5%, respectively. Emissions of sulfur compounds to the atmosphere at high latitudes are clearly seasonal in nature. On reaching the atmosphere from different sources in a given territory, sulfur compounds can be transported by air masses for long distances and deposited in other territories. Knowledge of the spatial distribution of the concentration of sulfur compounds together with meteorological information should make it possible to predict acid rain.

An expert system that takes into account the totality of models of atmospheric transport of pollutants, and at the same time includes a model of the sulfur cycle in the environment as an independent unit, would make such predictions possible. Of course, there are difficulties here as a result of a limited global database and the absence of some functional descriptions. These difficulties can be overcome using a global simulation model within which the sulfur cycle is parameterized with due regard to the role of the many subsystems of the biosphere and anthropogenic processes. Moreover, inclusion of the sulfur unit in a GSM would broadens its functions, since it is dictated by the dependence of biotic processes on the content of sulfur in biospheric compartments. Available data on the supplies and fluxes of sulfur compounds in the atmosphere, soils, vegetation cover, and hydrosphere would enable mathematical relationships to be formulated to simulate the global sulfur cycle.

Globally, the sulfur cycle is a mosaic of local fluxes of its compounds carried by other elements as a result of water migration and atmospheric processes. The conceptual schemes of global and regional sulfur cycles have been described in detail by many authors (Kondratyev *et al.*, 2006; Krapivin and Kondratyev, 2002; Nitu *et al.*, 2004). However, the available models have been designed for autonomous functioning and application, which does not facilitate their inclusion in a GSM without substantial changes in their parametric spaces. We now offer a solution to this problem.

Sulfur in its non-metallic state is widespread in nature and is a component of global biogeochemical cycles. From the human perspective, sulfur belongs to the group of elements that can negatively affect vital media. The harmful impact of sulfur on the environment is manifested mainly through acid rain. Water basin acidity due to acid rain and subsequent transport of sulfur compounds with the runoff from adjacent lands are manifestations of this impact.

Processes causing the acidification of water basins are mainly connected with anthropogenic sources of sulfur and, of course, other chemical elements such as nitrogen. In the pre-industrial period, the acidity of inland water basins never dropped below $pH = 8$. With the growing anthropogenic impact on the environment, water basin acidity increased with many basins dropping to $pH = 5.7$ in the middle of the last century. At present, in highly industrialized areas natural water basin acidity is about $pH < 5$. The increasing trend of acidity causes serious problems for fish reserve control. Many lakes and rivers in North America and Europe have been excessively acidified, resulting in the aquatic biota in them suffering irreversible changes. For instance, in the U.S.A. (Stoddard *et al.*, 2003) about 4.2% of lakes and 2.7% of river systems are in such a state that their capability to neutralize high acidity naturally is non-existent. The chronic excess of acidification of water basins ($pH < 5.2$) leads to irreversible changes in ecosystems and to a decrease in their survivability.

As seen in Figure 2.6, many factors affect the composition of water, the most important of which being vegetation cover and soil type in the aquatic system's basin. Therefore, acidity regulation can only be brought about by considering all the factors involved as a complex, which is only possible using numerical models. Experimental technologies cannot be used for this purpose due to the unique character of natural systems. Acid rain is a serious cause of forest damage, especially coniferous forests. As a rule, forests grow in regions that have sufficiently high rainfall and, hence, can get large doses of harmful elements when rainfall turns acid damaging leaves and pine needles and changing the soil composition. At $pH \in [2, 2.6]$ vegetation productivity drops drastically and at $pH \leq 2$ young shoots wither. On the whole, at $pH \in [0.5]$ there is a danger of acid rain and changes to the parameters involved in soil–plant formation. The degree of danger depends on the climatic zone and the type of soil–plant formation. At $pH \geq 5.6$, precipitation is not a threat to the environment.

The soils of most forests have $pH \in [3.2, 5.5]$. The stability of ion exchange processes in the soil is the reason for this interval. It is this stability that preserves the living conditions for the root systems of trees. An excess of cations $H^+$ in the soil affects the leaching of nutrients while the conversion of insoluble aluminum compounds to soluble ones leads to ion-forming centers being substituted and, as a consequence, external conditions for the root system get seriously compromised.

For global assessment of the role acid rain plays in the environment, a database is needed that can characterize regional pH levels and give the structure of pH-forming processes. Partial databases have been formed in many developed countries, but they are not up to the task of parameterizing the global pattern of acid rain formation. As shown by Safai *et al.* (2004), such data have been collected for India. Using observational data of the rain composition in India for the period 1984–2002, Safai *et al.* (2004) found the content of $SO_4$ and $NO_3$ in rain water in the rainy season steadily growing over many regions of India, but the pH level remaining within the alkalinity range. Such data and knowledge of trans-boundary fluxes of sulfur make it possible to calculate pH levels with due regard to the growth of industrial production and development of transport. Another example of database accumulation for acid rain control is an analysis of trends in changes of $SO_2$ and $SO_4^{2-}$ concentrations in the atmosphere over urban territories of the western and mid-Western regions of the U.S.A. for the period 1990–1999. The important thing here is assessment of the spatial variability of these concentrations (30–42%), which makes it possible to more reliably calculate the parameters of the respective equations in the biogeochemical units of models of atmospheric aerosol transport.

The model of the global sulfur cycle (MGSC) proposed here is a unit of the GSM whose inputs and outputs are compatible with other units of the global model. In contrast to hydrogen, sulfur compounds cannot be attributed to long-lived elements of the biosphere. Therefore, the spatial digitization of the sulfur unit's natural and anthropogenic reservoirs should be planned to reflect the local distributions of sulfur in the vicinity of its sources in such a way as to enable estimation of the intensities of inter-regional fluxes of sulfur compounds. The version of the sulfur unit proposed here, in contrast to known hydrodynamic models of long-distance transport, takes into account the fluxes of sulfur com-

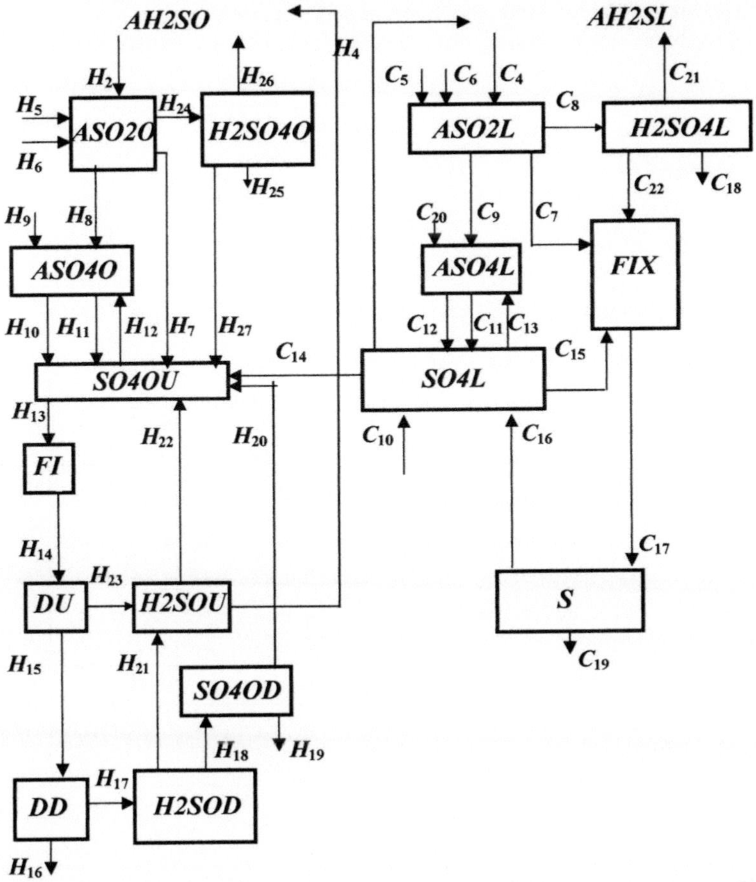


**Figure 2.7.** The scheme of sulfur fluxes in the environment considered in the MGSC. Notation is given in Table 2.3.

pounds between the hydrosphere, atmosphere, soil, and biota. The model does not consider vertical stratification of the atmosphere. The characteristics of sulfur fluxes over land and oceans averaged vertically are calculated. The spatial digitization of the biosphere and the World Ocean corresponds to a criterion inherent in the GSM. A block scheme of the model of the biogeochemical cycle of sulfur is shown in Figure 2.7, and a description of the fluxes of sulfur compounds is given in Table 2.3. This scheme can be applied to every cell $\Omega_{ij}$ of the Earth's surface and every compartment $\Omega_{ijk}$ of the World Ocean. The interaction between cells and compartments is organized through the climate unit of the GSM. Therefore, the equations of the sulfur unit lack terms reflecting the dynamic pattern of the spatial transformation of sulfur reservoirs. With due regard to notations assumed in Figure 2.7 and in Table 2.3, the equations describing the balance relationships between the reservoirs of sulfur compounds are written in the form of ordinary differential equations (Krapivin and Kondratyev, 2002).

**Table 2.3.** Characteristics of land and hydrospheric fluxes of sulfur shown in Figure 2.7. Assessments of fluxes ($mg\,m^{-3}\,day^{-1}$) obtained by averaging over the respective territories.

| *Sulfur flux* | *Land* | | *Hydrosphere* | |
|---|---|---|---|---|
| | *Identifier* | *Estimate* | *Identifier* | *Estimate* |
| Volcanic eruptions | | | | |
| $H_2S$ | $C_1$ | 0.018 | $H_3$ | 0.0068 |
| $SO_2$ | $C_5$ | 0.036 | $H_5$ | 0.0073 |
| $SO_4^{2-}$ | $C_{20}$ | 0.035 | $H_9$ | 0.0074 |
| Anthropogenic emissions | | | | |
| $H_2S$ | $C_2$ | 0.072 | $H_1$ | 0.00076 |
| $SO_2$ | $C_6$ | 0.92 | $H_6$ | 0.038 |
| $SO_4^{2-}$ | $C_{10}$ | 0.47 | | |
| Oxidation of $H_2S$ to $SO_2$ | $C_4$ | 1.13 | $H_2$ | 0.3 |
| Oxidation of $SO_2$ to $SO_4^{2-}$ | $C_9$ | 1.35 | $H_8$ | 0.16 |
| Dry sedimentation of $SO_4^{2-}$ | $C_{12}$ | 0.37 | $H_{11}$ | 0.11 |
| Fallout of $SO_4^{2-}$ with rain | $C_{11}$ | 1.26 | $H_{10}$ | 0.38 |
| Biological decomposition and emission of $H_2S$ into the atmosphere | $C_3$ | 1.03 | $H_4$ | 0.31 |
| Assimilation of $SO_4^{2-}$ by biota | $C_{15}$ | 0.41 | $H_{13}$ | 1.09 |
| Biological decomposition and formation of $SO_4^{2-}$ | $C_{16}$ | 1.13 | $H_{17}$ ($H_{23}$) | 0.43 (0.12) |
| Sedimentation and deposits | $C_{18}$<br>$C_{19}$ | 0.22<br>0.11 | $H_{15}$ ($H_{25}$)<br>$H_{16}$ ($H_{19}$) | 0.98 (0.036)<br>0.55 (0.0076) |
| Wind-driven return to the atmosphere | $C_{13}$ | 0.25 | $H_{12}$ | 0.33 |
| Replenishing sulfur supplies due to dead biomass | $C_{17}$ | 0.86 | $H_{14}$ | 1.1 |
| Assimilation of atmospheric $SO_2$ | $C_7$ | 0.46 | $H_7$ | 0.18 |
| Washing out $SO_2$ from the atmosphere | $C_8$ | 0.27 | $H_{24}$ | 0.061 |
| River runoff of $SO_4^{2-}$ to the ocean | $C_{14}$ | 1.17 | | |
| Transformation of gas-phase $H_2SO_4$ to $H_2S$ | $C_{21}$ | 0.018 | $H_{26}$ | 0.0076 |
| Assimilation of the washed-out part of atmospheric $SO_2$ by biota | $C_{22}$ | 0.036 | $H_{27}$ | 0.015 |
| Oxidation of $H_2S$ to $SO_2$ in water medium | | | $H_{18}$ ($H_{22}$) | 0.045 (0.19) |
| Advection of $SO_2$ | | | $H_{20}$ | 0.38 |
| Advection of $H_2S$ | | | $H_{21}$ | 0.37 |

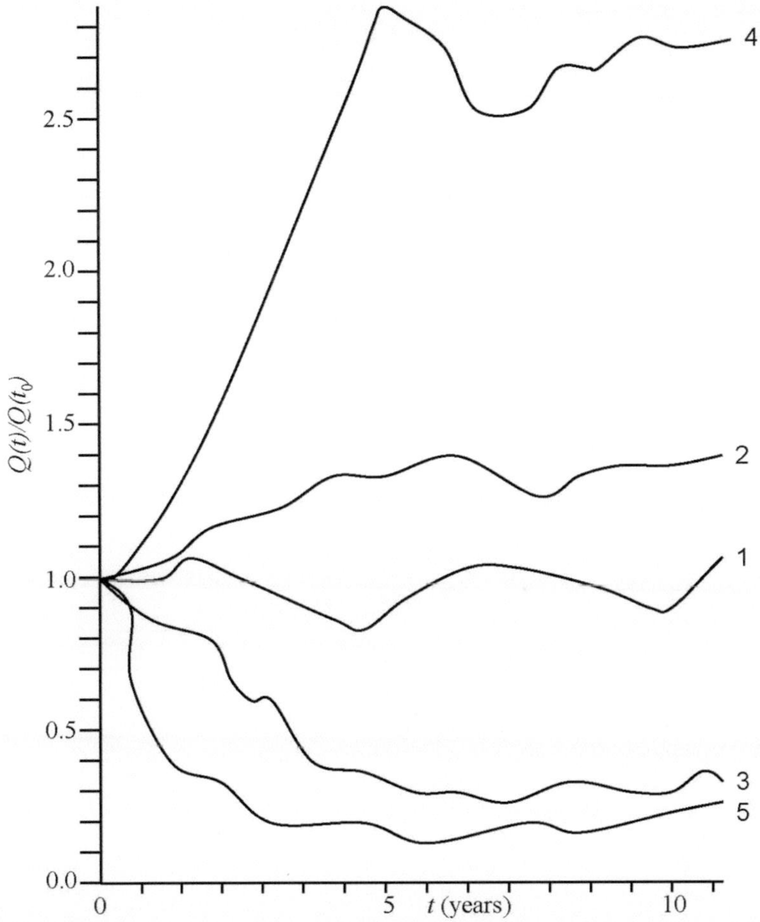


**Figure 2.8.** Dependence of the dynamics of sulfur concentrations $Q(t)/Q(t_0)$, normalized to initial conditions, averaged over $\Omega$ on initial conditions: 1—initial conditions correspond to standard data; 2—reduced by 50%; 3—increased by 50%; 4—reduced by 70%; 5—increased by 70%.

The parameterization accuracy of the GSM sulfur unit, like all units of biogeochemical cycles, is similar to that of other GSM units, and therefore there is no deregulation of the global model, and the stability of results of simulation experiments is ensured. To check this stability, we have undertaken some numerical experiments, taking the parameters of the sulfur unit from Kondratyev *et al.* (2006) and assuming $\Delta\varphi = 4°$, $\Delta\lambda = 5°$, $\Delta z_1 = 10$ m, $\Delta z_2 = 100$ m. As follows from Figure 2.8, any enhancement of sulfur supplies affects the system's dynamics during the first 2 years, whereas a decrease delays the system's return to a stationary regime for 5 years.

The curves in Figure 2.9 characterize acid rain's dependence on the level of anthropogenic activity. Calculations have shown that the pH value of precipitation

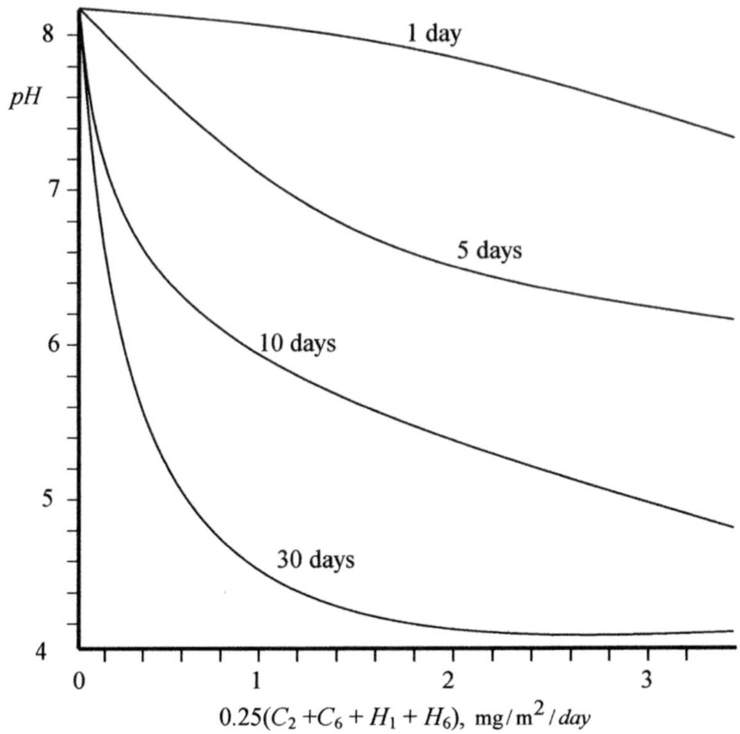


**Figure 2.9.** Dependence of the average acidity of rain on anthropogenic sulfur fluxes. Change in anthropogenic activity is assumed to be homogeneous in all territories. The curves are labeled with the time that has elapsed since the beginning of the experiment. pH is calculated with the formula $pH = \lg H^+$.

stabilizes, on average, within 30 days from the moment at which anthropogenic emissions of sulfur change. The spatial distribution of the pH of rain with an even increase in fluxes of $C_2$, $C_6$, $C_{10}$, $H_1$, and $H_6$ by 0.2% $yr^{-1}$ does not markedly change for 3 years. The ratio of acid rain rates between latitudinal bands 70°–90°N and 70°–90°S remains, on average, at a level of 2.5.

Let us now estimate the contribution of various regions to Arctic pollution. With the stable state of the mean annual concentration of gas-phase $H_2SO_4$ in the Arctic atmosphere assumed to be 100%, the contributions of countries or territories to the formation of this level are as follows: the U.S.A. 17%, Canada 21%, Europe 37%, and the eastern territory of Russia 25%. These estimates correlate with anthropogenic sulfur fluxes as a result of anthropogenic activity as determined by the relationship of all parameters of the global sulfur cycle. As seen from Figure 2.10, the intensity of biological decomposition in water is distributed non-uniformly, and this means that hydrocarbon production in the oceans is a function of the vertical structure of their ecosystems. For instance, in the Indian Ocean there are two distinct maxima of $H_2S$ production. A weak second

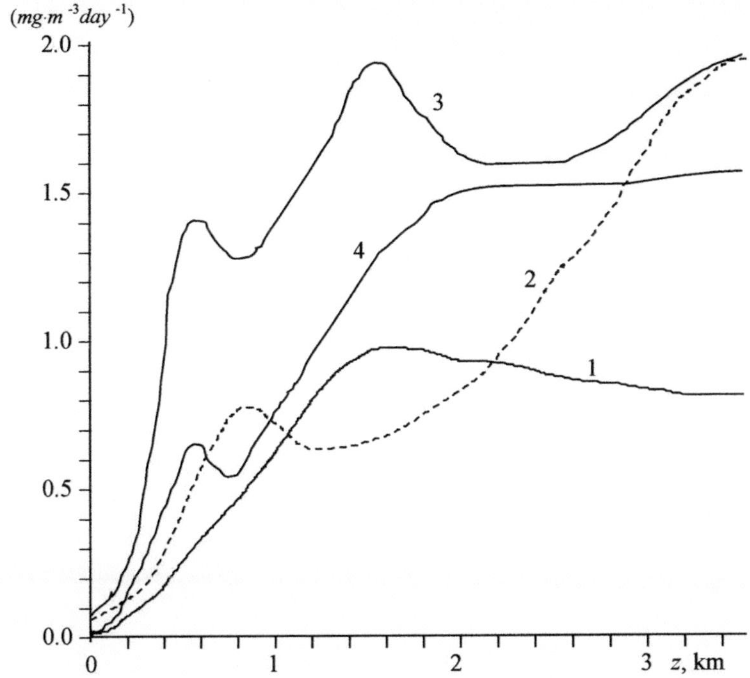


**Figure 2.10.** Average production of $H_2S$ (mg m$^{-3}$ day$^{-1}$) in the oceans: 1, Arctic Ocean; 2, Pacific Ocean; 3, Indian Ocean; 4, Atlantic Ocean.

maximum of hydrocarbon production appears at depths of ~1.5 km in the Atlantic Ocean. In other oceans, there is a single maximum of the vertical distribution of $H_2S$.

Numerical modeling as outlined above shows that reliable assessment of the spatial distributions of pH levels depends on many factors including the accuracy of parameters in equations of the MGSC unit and the form in which the GSM is referenced. To increase the reliability of acid rain forecast, it is necessary to further improve the MGSC unit by including the biogeochemical cycles of other chemicals, such as carbon bisulfide and sulfurous anhydride. By restricting consideration to sulfur dioxide, sulfates, and hydrocarbons in the models of the global sulfur cycle limits the accuracy of these models. Unfortunately, most international and national programs studying the sulfur cycle are confined to these elements.

# 3

# The NSS model as a GIMS component

## 3.1 THE PROBLEM OF GLOBAL MODELING

### 3.1.1 Some history

The problem of global environmental change is the subject of discussions in many publications (Cracknell *et al.*, 2009; Holdren, 2003; Kondratyev, 1998a, b; Kondratyev and Krapivin, 2001a–d; Krapivin, 1993; Krapivin and Kondratyev, 2002; Krapivin and Varotsos, 2007, 2008; Likens *et al.*, 1981; Pielke, 2001a, b, 2002; Purkis and Klemas, 2011; Sellers *et al.*, 1996; Svirezhev, 2002; Turekian, 1996; WB, 2009; Zhou *et al.*, 1986). Though the notion of global change can be considered sufficiently well established despite the many terminological differences that remain (especially regarding the definition of "sustainable development"), it should be made clear that here we speak mainly about the interaction between society (socioeconomic development) and nature. The most substantial features of global change consist in their multi-component character, interactivity, and non-linearity. These features complicate prognostic estimates to such an extent that the notion of prediction has recently been ousted in favor of the more vague concept of "scenarios" or "projections". The uncertainty of scenarios grows because, as a rule, there are no probabilistic estimates for some scenarios (e.g., those that relate to global climate change).

One way of overcoming these uncertainties is global ecoinformatics (a new scientific mechanism that has been intensively developed in recent years) within which information technologies have been created that facilitate combined use of various data on the past and present state of the NSS. The creation of a model that simulates how the NSS functions based on knowledge and available data, that blends with the adaptive–evolutionary concept of geoinformation monitoring, and that brings about interconnection between the NSS model and the regime of global data collection marks an important step forward in global ecoinformatics.

V.F. Krapivin and A.M. Shutko, *Information Technologies for Remote Monitoring of the Environment*, Springer Praxis Books, DOI 10.1007/978-3-642-20567-5_3, © Springer-Verlag Berlin Heidelberg 2012 

As a result, it is possible to optimize the organizational–behavioral structure of the NSS in such a way as to ensure purposeful global changes for the benefit of humans with limited damage to nature and, most importantly, to create international mechanisms for the way in which the world's population uses natural resources.

Among the major strides taken by ecoinformatics is the development of models of the various processes that take place in the NSS. Modeling refers to spheres of knowledge, the efficient expansion of which into non-traditional spheres of life during recent decades can be spoken of with great confidence. This has to do with the fact that the model has facilitated understanding of correlations between NSS fragments and made it possible to view the mosaic of processes isolated at first sight as a single whole. An especially important property of the model is its ability to reflect the present duality of the anthropogenic component. On the one hand, humans are elements of nature whose behavior is determined by nature but, on the other hand, humans can plan their behavior and transform natural environments into artificial ones.

More than three decades ago the first report of the Club of Rome (CR) (Meadows *et al.*, 1972) was published; it invoked worldwide interest. In 1972 the International Conference on the Environment was held in Stockholm; its resonance was felt worldwide. Later on, further issues relating to global change were mainly concentrated within the framework of the International Geosphere–Biosphere Program (IGBP) and the World Climate Research Program (WCRP), which then were supplemented by the International Human Dimensions Program (IHDP). These and attendant international programs of environmental studies resulted in extensive global databases on various NSS components being created and, in particular, in their trends during the last three decades being established.

Let us briefly analyze how the global ecological situation has developed in the last three decades and, most importantly, answer why, despite huge efforts and enormous expense (totaling many billions of U.S. dollars), the global ecological situation has not only failed to improve but has actually got worse. But, first of all, let us briefly discuss the models developed by Forrester and the Club of Rome.

The titles of the monographs *The Limits to Growth* (Meadows *et al.*, 1972) and *World Dynamics* (Forrester, 1971) reflected the appearance of a new fundamental concept: the development of human society (population size and growing scales of human activity) is not boundless and has already approached certain limits, mainly from the viewpoint of the levels at which irreplaceable natural resources are used. The main goals of the CR consisted in analyzing global demographic dynamics, estimating natural resources, and substantiating a model of global ecodynamics and possible scenarios of ecodynamics of the future. Meadows *et al.* (1972) formulated the principal goals as: "The goal of the project is to study a complex of problems of concern for all nations: poverty among abundance; environmental degradation; loss of confidence in public institutions; uncontrolled expansion of cities; unreliable employment; estranged young people; neglect of traditional values; inflation and other economically destructive phenomena."

Instead of an introduction, the authors of *The Limits to Growth* quoted an address by U Thant, the U.N. Secretary-General: "I do not want to overdramatize events, but I only want, based on information I have got as the U.N. Secretary-General, to make the conclusion that the U.N. participants have possibly 10 years at their disposal to regulate their old disagreements and to start a global cooperation in order to curb arms, to improve the environment, to restrain the explosive growth of population, and to stimulate efforts in the sphere of socioeconomic development. If the global cooperation does not become a first-priority goal, I am very much afraid that all these problems will reach stunning levels, beyond our capabilities to control them."

Now, three decades later, the conclusion might be drawn that U Thant (like D. Meadows) was an adherent of the concept of catastrophism: at first sight, it seems that, despite all the problems associated with the development of today's civilization, the world survived and did not reach the critical point. However, analysis of the realities of today's global ecodynamics does not permit such an optimistic conclusion. As for developments predicted by the CR and U Thant's speech, their foresightedness should be applauded. Their assessment is even more true today because further development, as a result of scientific advances, has demonstrated just how poor the success rates of large-scale events like the Second U.N. Conference on Environment and Development (Rio de Janeiro, 1992) and the U.N. Special Session (New York, 1997) to the fiasco of the World Conference on Sustainable Development (Johannesburg, 2002) turned out to be against the backdrop of a continuously worsening global ecological situation. Let us now briefly discuss the basic content of the CR report, whose authors begin with an analysis of global demographic dynamics.

In 1650, when the global population size was about 0.5 billion, the population increment constituted 0.3% annually, which corresponded to a period of population doubling every 250 years. By 1970, this had reached 3.6 billion people, and the rate of increment increased to 2.1%/year (the period of population doubling corresponded to every 36 years): the increase in population became "super-exponential". An important demographic indicator is the average lifetime of the population, which in 1650 constituted about 30 years. By 1970 it reached 53 years and has continued to increase ever since. Meadows *et al.* (1972) pointed out the lack of prospects of stabilizing global population size by the year 2000.

A second important parameter is the rate of industrial development: between 1965 and 1968 the rate of growth averaged 7% a year (the respective per capita indicator was 5%). It is important to point out that this growth was mainly concentrated in industrial countries where the rate of population growth was, however, relatively low (hence the many contradictions between developed and developing countries characteristic of today). Table 3.1 illustrates the rate of growth in gross domestic product (GDP) in different countries. The CR report correctly emphasized the low probability that the growth of both population and GDP would remain even right up to the end of the 20th century, since the many factors involved would change. However, taking the example of Nigeria, where the expectancy of the end of inter-ethnic conflicts to coincide with economic growth proved lame, reflecting a continuation and deepening of the socioeconomic

**Table 3.1.** Gross domestic product (GDP) and population of some nations in 2011.

| *Country* | *GDP* (USD bn) | *GDP per capita* (USD) | *Population* (thousands) | *Literacy rate* (%) | *Average lifetime* (years) | *Annual growth of GDP in 2010* |
|---|---|---|---|---|---|---|
| U.S.A. | 14,660 | 46,802 | 313,233 | 99 | 78 | 2.9 |
| China | 10,090 | 7,548 | 1,336,719 | 92 | 75 | 10.3 |
| Japan | 4,310 | 34,078 | 126,476 | 99 | 82 | 5.1 |
| India | 4,060 | 3,414 | 1,189,173 | 61 | 67 | 9.7 |
| Germany | 2,940 | 36,086 | 81,472 | 99 | 80 | 3.6 |
| Russia | 2,223 | 16,023 | 138,740 | 99 | 68 | 4.0 |
| U.K. | 2,173 | 34,658 | 62,699 | 99 | 80 | 1.3 |
| Brazil | 2,172 | 10,677 | 203,430 | 89 | 73 | 7.5 |
| Canada | 1,330 | 39,082 | 34,031 | 99 | 81 | 3.1 |
| Australia | 882 | 40,439 | 21,767 | 99 | 82 | 2.0 |
| Argentina | 596 | 14,269 | 41,770 | 97 | 77 | 9.2 |
| Pakistan | 465 | 2,482 | 187,343 | 50 | 66 | 4.4 |
| Ukraine | 305 | 6,762 | 45,135 | 99 | 69 | 4.2 |
| Vietnam | 277 | 3,055 | 90,550 | 94 | 72 | 6.8 |
| Qatar | 151 | 177,591 | 0.85 | 89 | 96 | 16.2 |
| Burundi | 3.4 | 333 | 11 | 59 | 59 | 3.9 |
| *World* | *74,540* | *10,759* | *6,928,199* | *82* | *67* | *4.9* |

conflicts on the African continent (as convincingly demonstrated in the growth of its foreign debt). Estimates (Meadows *et al.*, 1972) of possible levels of per capita GDP in 2000 are interesting. Though the reliability of such extrapolative estimates cannot be high, they totally disagree with reality, which testifies to the difficulty of making extremely complex political and socioeconomic predictions. A key aspect of socioeconomic dynamics turns out to be quite predictable, though: the process of economic growth left in its wake an increased contrast between rich and poor countries.

To ascertain what was necessary to provide persistent growth of both the economy and population size up to the year 2000 and beyond, Meadows *et al.*

(1972) analyzed physical (material) and social needs separately. The satisfaction of material needs required, first of all, solving provision of food, drinking water, and irreversible natural resources. Table 3.1 shows the significant discrepancy between countries concerning GDP per capita. The rank of countries in this league is headed by Qatar with a GDP per capita level of USD179,000 and Burundi brings up the rear with USD300.

Though reliable data on the amount of food available were absent, it was supposed in the CR report that about 50–60% of the population of developing countries (about one third of the global population) were underfed. The area of the arable soils in these countries constituted about 3.2 billion ha, of which about 50% (i.e., the most productive) were cultivated. According to data from the FAO (the international organization for food and health), cultivation of the remaining soils was uneconomic. From those estimates that were available, even if all the arable soils of developing nations were used and the rate of population growth continued as forecast, an acute shortage of agricultural soils would occur by 2010 (below we will see that this prediction was wrong). It was in this regard that the CR report emphasized the limited capabilities of humankind to prevent the catastrophic consequences of exponential population growth when resources were limited.

Meadows *et al.* (1972) obtained various prognostic estimates using a global model, the structural description of which caused no objections. However, subsequent publications (e.g., Krapivin and Kondratyev, 2002) demonstrated that the model did not take into account the numerous indisputable feedbacks (both direct and indirect) between society and nature and, most importantly, their spatial heterogeneity was overlooked. Nevertheless, the predictions made placed the many problems that needed resolution before specialists in the field of global modeling and prompted them to develop better technologies for environmental control. Moreover, the problems discussed in Meadows *et al.* (1972) could not be solved without using the systems involved in global monitoring of the environment.

### 3.1.2 Present state of NSS modeling

Over the last 30 years, as a result of the efforts of many scientists, the key priorities of global ecodynamics have been formulated and perspective trends to solutions of numerous new problems have been outlined (Krapivin and Varotsos, 2007, 2008; Tanaka, 2010). Clearly, to work out a global strategy of sustainable development, a constructive formalized approach to description of the NSS is needed that takes into account not only its multi-dimensional and multi-component nature but also the non-linearity and interactivity of the processes taking place in it. Many feedbacks in the NSS have strengthened in importance and prevail over other feedbacks. For example, from available estimates, about 1.2 billion people live now on less than one dollar per day; 100–150 million people suffer from asthma; 2.4 billion people need better sanitary living conditions; 150–300 million ha of cultivated soils (these represent 10–20% of all agricultural soils) have become degraded; more than 2 billion people live in conditions where drinking water, food, and normal dwellings are sub-standard. Finally, there is a trend towards increasing population mortality due to intensified terrorism and other human-caused catastrophes. All this changes the concept of the global model and requires a search for new information technologies for the control of trends in the NSS.

The CR model reflected only a narrow spectrum of feedbacks (levels) in the NSS (population, finance, pollution, food production, mineral resources). Moreover, the model failed to reflect the direct role of biospheric feedbacks or the spatial heterogeneity of these feedbacks. Therefore, it could not be objectively predictive even within the framework of successfully formulated scenarios, which introduced numerous uncertainties covering a broad spectrum of possible issues into the model. A comparison of Meadows' model with that of the global dynamics by Forrester reveals their conceptual identity both in feedbacks from level variables and ideologically. Depending on variations of initial suppositions (scenarios) on limited or unlimited irreversible resources and on stabilization of the population size, the results of prognostic estimates of the state of NSS components coincided qualitatively for both models, but differed substantially from the reality of the end of the 20th century. The basic difference consisted in estimation of population size whose rate of growth in the 1990s stabilized at about 80 million per year, passing its maximum of 87 million in the late 1980s and returning to the level of the 1970s at the beginning of the 21st century. This is explained by the time correlations considered in the CR model (and in Forrester's model) underwent unpredictable changes in connection with misunderstanding earlier poorly manifested feedbacks in the NSS. Understanding the structure and importance of heterogeneous and complex feedbacks in the present world has changed rapidly together with their dynamics, which is difficult to predict. So, as a consequence of the extension of trade relations, transport, and information networks during the last 30 years, the spatial correlation between ecological, demographic, political, and economic events has grown.

A substantial difference between the CR model and the present world concerns the use of mineral resources and food production. It is now clear that such components of global NSS functioning as alternative energy sources, energy-saving systems and technologies, birth rate and mortality, structure of the workforce and population migration, response of nature to anthropogenic impacts, and many other key indicators need to be parameterized and reflected in the model. For instance, in food production the role of aquaculture in the period 1984–1999 distinctly grew. During this period, global aquaculture production grew by almost 400%, from 6,900,000 t in 1984 to 33,300,000 t in 1999. This growth was heterogeneous both in space and in production components. For instance, the proportion of fish increased from 19% in 1990 to 31% in 2001, 68% of which was down to fish farms in China. The global model component responsible for fossil fuels was rather complex, too. There are about 10 well-studied alternative mechanisms for their economical use and substitution. This aspect was present in the CR model in the form of several primitive scenarios.

Clearly, the decrease or stabilization of per capita food production of the early 21st century predicted by the CR model has proved to be abortive. This is understandable because food production is characterized by more complicated cause-and-effect feedbacks than those considered in the models by the Club of Rome and Forrester (1971). It has to be pointed out that although production of grain, meat, and other components of the human diet suffered increases and decreases, this should not be given as a reason for pessimistic predictions for decades to come. As mentioned above, fish production has sharply increased during recent years. Clearly, a global model should consider bioproductive processes in the World Ocean and inland water basins to estimate the limits of their capabilities to produce food. As for grain, at the turn of the century its production decreased and its consumption increased. For instance, per capita grain production in 2001 constituted 299 kg, which was 14% below that reached in 1984. At the same time, if we consider the long-range trend of grain production, between 1950 and 1984 it grew by 38%. This prevailing of grain consumption over its production and the respective decrease in its global-scale supplies are only a short-term fluctuation of the process of food production. In fact, other constituents of the human diet are characterized by positive gradients of their production, though distributed non-uniformly over countries and continents. For instance, meat production (beef, pork, and poultry) in 2001 grew by 2%. From 1950, per capita meat production doubled from 17.2 kg/yr to 38.2 kg/yr in 2000. In 2001, pork and poultry production reached a peak. In 2007, meat production remained steady at an estimated 275,000,000 t. At the expected level of 310,000,000 t, world cereal production would be 3%, or 68,000,000 t, higher than the 2010/2011 outturn. The overall increase comprises a 4.6% (30,000,000 t) rise in wheat production, a 3% (14,000,000 t) growth in the global rice harvest, and a 2.1% (24,000,000 t) rise for coarse grains (FAO, 2010, 2011a, b). At the present time (2011), a total of 925,000,000 people are still estimated to be undernourished, representing almost 16% of the population of developing countries.

Finally, let us consider how the CR treated non-renewable natural resources (one of the key indicators of the NSS state). Although their model predicted continuous reduction and their limiting role in the development of other NSS levels, their predictions again differed from real trends at the turn of the century. The real global-scale extraction of coal, oil, and natural gas grew at a rate of 1–2%/yr (in 2001 it was 1.3%). The volumes of fossil fuel consumption also grew, though non-uniformly over countries and differing in the type of fuel. On the whole, the global increase in the consumption of oil, coal, and gas constituted 0.2%, 3.2%, and 1.2%, respectively. It is expected (Tanaka, 2010) that global oil, gas and coal production will start to decrease in 2012.

What is more, the use of nuclear, solar, and wind energy is growing and energy-saving technologies are rapidly developing. At the end of the 20th century there was a rapid growth in the industrial production of photoelectric cells to transform sunlight into electricity. In this way, the world gets more than 1,140 MW, and there is an increasing trend in the use of solar energy. Therefore, the dependence of global dynamics on energy resources should be parameterized not on the basis of simplified models but by taking into account the whole spectrum of available information on the nature of multiple feedbacks in the NSS and especially the trends of scientific–technical progress.

### 3.1.3 Encouraging view

In the 30 years that have passed since the appearance of the CR predictions and the respective global model, a major advance has been made in the field of global modeling. The new approach proposed in Krapivin and Kondratyev (2002) was based on the idea of the NSS as a self-organizing and self-structuring system, the correlation of whose elements in time and space is ensured by the process of natural evolution. The anthropogenic constituent in this process is aimed at breaking this integrity. Attempts to formally parameterize the process of the co-evolution of nature and humans as elements of the biosphere are connected with a search for a single description of all the processes in the NSS, which would unite the efforts of various branches of knowledge about the environment. Such a synergism serves the basis of many studies on global modeling.

Let us cover the Earth's surface $\Omega$ with a geographical grid $\{\varphi_i, \lambda_j\}$ with digitization steps $\Delta\varphi_i$ and $\Delta\lambda_j$ by latitude and longitude, respectively, so that within a cell of land surface $\Omega_{ij} = \{(\varphi, \lambda) : \varphi_i \leq \varphi \leq \varphi_i + \Delta\varphi_i; \lambda_j \leq \lambda \leq \lambda_j + \Delta\lambda_j\}$ all the processes and elements are considered as homogeneous and are parameterized by point models. When water surfaces are in the territory of the cell $\Omega_{ij}$ the water masses are stratified into layers $\Delta z_k$ thick; that is, 3-D volumes are selected $\Omega_{ijk} = \{(\varphi, \lambda, z) : (\varphi, \lambda) \in \Omega_{ij}, z_k \leq z \leq z_k + \Delta z_k\}$ inside which all elements of the ecosystem are distributed uniformly. Finally, the atmosphere over the site $\Omega_{ij}$ to a height $h$ is digitized either by levels of atmospheric pressure or by layers $\Delta h_s$ thick.

NSS interactions are considered to be those interactions between natural and anthropogenic components within these spatial structures and between them. The complex model of the NSS realizes the spatial hierarchy of hydrodynamic, atmospheric, ecological, and socioeconomic processes with the division of the whole volume of the environment into structures $\Omega_{ij}$ and $\Omega_{ijk}$. The cells of this division make up the supporting grid in numerical schemes for solutions based on dynamic equations or on the synthesis of data series in evolutionary-type learning procedures.

The cells $\Omega_{ij}$ and $\Omega_{ijk}$ have heterogeneous parameters and functional characteristics. Through this heterogeneity the global model can be referenced to databases. Moreover, to avoid a global model becoming unwieldy, it is supposed *a priori* that all elements taken into account in the model and all NSS processes have a characteristic spatial digitization. Any ambiguity in the spatial resolution of various units of the global model is removed at the algorithmic level of agreement of data fluxes from the monitoring system. As a result, the model's structure is independent of the structure of the database and, hence, does not change if the latter changes. A similar independence between the model's units is also provided. This is realized by data exchange between them only through inputs and outputs under control of the basic data highway. If one or several units are turned off their inputs are identified with corresponding inputs to the database. Then, the model operating in the regime of a simulation experiment can be schematically represented by the process, where the user can choose a spatial image of the modeled medium and of the regime controlling the simulation experiment. Of course, in this case the user should have a certain knowledge base and know how it is structured. For instance, the user can employ a list of key problems of global ecology or a list of the NSS elements recommended for studies.

The spatial structure of the global model is determined by the database. The simplest version of the point model can be brought about using initial information by averaging over land surfaces and the whole World Ocean. Spatial heterogeneity is considered by means of various forms of space digitization. The base form of the spatial division of land and oceans takes the form of a heterogeneous grid $\Delta\varphi \times \Delta\lambda$. Bringing about a real version of the use of the model is achieved by integrating cells $\Omega_{ij}$ such that various forms of the spatial structure of the elements and biospheric processes under consideration are present in each unit. Such a flexible setting of the spatial structure of the biosphere makes it possible to easily adapt the model to heterogeneities in databases and to carry out simulation experiments such that individual regions can be actualized.

Depending on the natural process under consideration, regional division can be identified according to climatic zones, continents, latitudinal belts, socio-administrative structure, and natural zones. For climatic processes, many scientists lean towards regions with dimensions $\Delta\varphi = 4^\circ$ and $\Delta\lambda = 5^\circ$, biogeocoenotic processes are studied at $\Delta\varphi = \Delta\lambda = 0.5^\circ$, the socioeconomic structure is represented by nine regions, atmospheric processes in the biogeochemical cycles of long-lived elements are approximated by point models, functioning of ocean ecosystems is described by heterogeneous digitization of the shelf zone into cells $\Omega_{ij}$ by selecting one of the four pelagic zones of the World Ocean.

The division of the Earth's surface into regions $\Omega_{ij}$ is a feature of all enumerated versions. This means that the general scheme used to digitize processes in the NSS foresees a hierarchy of levels including the global, continental, regional, landscape, and local (there are others). The scheme that independently combines units at all these levels by means of parametric interfaces does not prevent the number of model units from increasing as a result of the introduction of new components specifying models of the processes under consideration. A model of the upper level can serve as an information base for a model of the lower level and *vice versa*. The results of modeling at the lower level can be used to form the information base for models of higher levels. This mechanism of information exchange between models of various levels reduces the requirements of the global database and broadens the capabilities of the NSS model.

The structure of the global model includes auxiliary units that enable the user to interact with the model and operate with the database. In particular, these are units that activate algorithms to spatiotemporally interpolate or coordinate the user's actions with a bank of scenarios. Note that some scenarios can be transformed by the user into model units. Such a duality (excessiveness) is characteristic of scenarios of climate, demography, anthropogenic activity, scientific/

technical progress, and agriculture. The user interface makes it possible to select a structure ($\Omega_{ij}$) in default mode or the needed spatial structure can be formed from base elements by averaging and interpolation.

Thus, synthesis of a global model requires preliminary analysis of the present situation of global databases and knowledge bases. Researchers face a number of principal difficulties here, the main one being the absence of adequate knowledge of climatic and biospheric processes or of a coordinated database on global processes on land, in the atmosphere, and in the oceans.

Another principal difficulty is the inability of science at the moment to formulate the requirements necessary for global databases to reliably assess the state of the environment and give reliable forecasts of how the environment will develop over the long term. Moreover, no database-forming technology has been set up with the aim of creating a global model.

Many scientists have made attempts to address these difficulties (Tianhong *et al.*, 2003). An efficient way to do so would be a single planetary adaptive GIMS, which has a hierarchical structure for data collection and forms a multi-level global database. The adaptive character of such a system is provided by correcting the regime of data collection and by changing the parameters and structure of the global model.

A global GIMS can only be created by taking the existing structure of databases into account. Databases continue to be developed within the framework of the IGBP and a number of national ecological and nature conservation programs. A developed system of world data centers would favor rapid use of accumulated information about global processes and simplify GIMS synthesis. However, significant progress in this direction connected with large economic expenses cannot in isolation lead to successful solution of the problem of global environmental control. Though, this phase cannot be avoided.

According to Kondratyev (1998a, b), regular observations of specific key variables are needed to control the global geobiosystem of the Earth. With the increasing probability of dramatic global changes, the spectrum of these variables will vary and, as a consequence, the global predicting system should be constantly modernized. Methodological substantiation of the information content needed by variables of the monitoring system can only be objective when the GIMS is operational. Many variables can be calculated using respective models, hence, there is no need to measure them. However, at the moment measurements are planned in line with model development, and there are no outcomes from global experiment planning that currently offer a glimmer of hope. However, some studies (Kondratyev and Galindo, 2001; Kondratyev *et al.*, 2002) indicate the bases of global knowledge and data necessary to make it possible to synthesize and develop a GIMS series of models (Kondratyev *et al.*, 2003b).

Including a global model in the GIMS structure enables it to be considered an expert system. This means that complex analysis of numerous elements of the NSS when reconstructing hypothetical situations, be they natural or anthropogenic, is possible. Figure 1.23 reflects the basic elements taken into account in the NSS global model (GMNSS). Concrete realization of each unit of the GMNSS is determined by the level of knowledge of processes reflected in the unit. The units responsible for modeling biogeochemical and biogeocenotic processes are described using balance equations.

Let $\psi_S(t)$ be the information content of an element $\psi$ in a medium $S$ at a moment $t$. Then, following the law of preservation of matter and energy, we write the following balance equation

$$\frac{d\psi_S}{dt} = \sum_j H_{jS} - \sum_i H_{Si},$$

where fluxes $H_{jS}$ and $H_{Si}$ are, respectively, incoming and outgoing fluxes with respect to the medium $S$. Summing is carried out by external media $i$ and $j$ interacting with $S$. A variety of functional parameterizations of fluxes $H_{pq}$ are determined by the level of knowledge of the physical, chemical, and biological features of the element $\psi$.

Parameterization of the processes of photosynthesis, dying off, and respiration of plants in land ecosystems needs the researcher to be well versed in phytocenology, a field of study that includes information about external and internal connections of the vegetation community such as the temperature dependence of photosynthesis and the evapotranspiration of plants, gas exchange processes between plants and the atmosphere, impacts of solar radiation on the processes of growth and exchange, relationships between plants and processes in the soil, interaction of vegetation covers with the hydrological cycle.

The GMNSS units responsible for parameterizing climatic and anthropogenic processes are complex; they are partially described by equations of motion and balance and partially by an evolutionary model specially constructed for them based only on observational data.

### 3.1.4 Perspectives and hope

Having considered approaches to assessing the dynamics of the NSS from Forrester (1971) and Meadows *et al.* (1972) to recent publications (Krapivin and Varotsos, 2007, 2008), we conclude that much progress in finding ways to reach global sustainable development can be made by adopting a systematic approach to multi-functional monitoring of the NSS. Thanks to the CR authors, more than 30 years ago a contradiction between the growth of population size and limited natural resources was put in the limelight. For the first time since Vernadsky (1944) an attempt was made to use numerical modeling to study evolution of the NSS. Of course, the CR model oversimplified the real internal connections in the NSS, describing interactions between its elements as averaged indirect relationships, without direct account of economic, ecological, social, and political laws. The possibility of taking account of these laws appeared later in the field of simulation and evolutionary modeling and the theory of interaction optimization for complex systems resulting in the creation of methods and algorithms for prognostic estimation of dynamic processes under conditions of *a priori* uncertainty. However, the problem of creating a global model sufficiently representative of the real world still cannot be solved even today (2010). Complete consideration of all NSS parameters leads to insurmountable multi-variance and information uncertainty with irremovable problems. Moreover, in such fields as the physics of the ocean, geophysics, ecology, medicine, sociology, etc., adequate parameterization of real processes will always be problematic because it is impossible to have a complete database. Nevertheless, a search for new efficient ways to synthesize a global system of NSS control based on adaptive principles in the use of the global model and renewed databases seems to be the perspective of many to get reliable forecasts of the dynamics of the NSS. Preliminary calculations using the GMNSS have shown that the role of biotic regulation in the NSS has been underestimated and the forecasts, for instance, of the greenhouse effect have been overestimated (Streets *et al.*, 2001). Therefore, in this book we attempt to synthesize the GMNSS taking the earlier experience and accumulated databases into account in addition to knowledge about the environment and human society.

Further improvement of the GMNSS will involve balanced development of studies both in parameterizing the NSS and in modernizing the Earth observation systems, covering the whole thematic space of the NSS (Kramer, 1995; Kump *et al.*, 2003):

- Sun–land interactions (physical mechanisms of the transport of mass, momentum, and energy in the geosphere);
- atmospheric dynamics (atmospheric chemistry, atmospheric physics, meteorology, hydrology, etc.);
- the dynamics of the World Ocean and coastal zones (winds, circulation, sea surface roughness, color, photosynthesis, trophic pyramids, pollution, fishery);
- lithosphere (geodynamics, fossil fuel and other natural resources, topography, soil moisture, glaciers);
- biosphere (biomass, soil–plant formations, snow cover, agriculture, interactions at interfaces, river runoff, sediments, erosion, biodiversity, biocomplexity);
- climatic system (climate parameters, climate-forming processes, radiation balance, global energy balance, greenhouse effect, long-range climate forcings, delay of climate effects);
- sociopolitical system (demography, geopolitics, culture, education, population migration, military doctrines, religion, etc.).

Parallel with the search for new methods of global modeling, detailed studies of global environmental systems are important for global model development and testing. There are numerous books and papers full of useful data, ideas, and models that can be used to modernize the GMNSS. For example, Turekian (1996) investigates accelerated environmental change, be it natural or anthropogenic, on a global scale and at the same time gives a chronology of global evolution and a detailed description of environmental subsystems and their parameters.

## 3.2 GLOBAL MODEL AND THE GIMS

To create NSS global modes as a component of the GIMS demands synthezing series-specific models capable of describing separate environmental or socioeconomic processes within the NSS. Many scientists have tried to model the global carbon cycle (Alexandrov and Oikawa, 2002; Kondratyev *et al.*, 2002, 2003c; Krapivin and Varotsos, 2007, 2008; Nitu *et al.*, 2000a, b, 2004; Tarko 2001, 2003, 2005). Most models are based on data for the pre-industrial period and predict the $CO_2$ concentration for this century. The number of factors taken into account in the models is constantly growing and the models are becoming increasingly better. One of the first sufficiently complete models of the global $CO_2$ cycle was proposed by Bjorkstrom (1979). This model took into account the dynamic interaction between the carbon reservoirs in the biosphere and fluxes between them. For the first time, a unit for the World Ocean was realistically represented, in which the ocean was considered as a multi-layer composition of uniformly mixed reservoirs, the $CO_2$ exchange between them being described by linear laws. Unit-type models have been analyzed by many experts. They are the reasons for successful division of the Earth biogeosystem into heterogeneous compartments and the derivation of respective box models (Pervaniuk, 2001). The ability of carbon to reside in the atmosphere for a long time makes it possible to use a point model to describe the atmosphere. Bacastow (1981) derived a global four-reservoir model, that approximated $CO_2$ exchange between the northern and southern hemispheres.

Every available model of the global carbon cycle differs in its set of assumptions and, therefore, leans towards selection of some effect. Models allowing mutual parametric correlation can be used in a more complicated model. For instance, a simple numerical model of gas exchange at the ocean–atmosphere boundary under wind-driven roughness at wind speeds up to 7 m/s enables a unit to calculate the established $CO_2$ flux between the water surface and the atmosphere to be included in the global model. Models of the ocean carbonate system described in the literature exemplify the same type. There are also other models of the $CO_2$ cycle in natural systems (Demirchian and Kondratyev, 2004; Dore *et al.*, 2003; Hales *et al.*, 2005; Harvey and Huang, 2001; Lal, 2008; Riedo *et al.*, 2000; Zonneveld, 1998).

All known models of the $CO_2$ cycle should be modified by detailing the spatial distribution of soil–plant formations and by specifying exchange processes both in the ocean and at the atmosphere–ocean boundary. Moreover, the most accurate parameterizations of all known elements of the biogeochemical cycles of carbon should be combined into a single system. Such an attempt was shown as a block diagram of the model described by Krapivin and Kondratyev (2002). The basic sources of $CO_2$ are the functioning of land and marine animals, photochemical reactions, decomposition of dead organic matter, and anthropogenic activity. The time it takes $CO_2$ to migrate into upper-atmospheric layers is substantially less than the time step of the model. The adequacy of estimates of the role land and water ecosystems play in the assimilation and burial of excess amounts of carbon dioxide in the atmosphere is thus problematic.

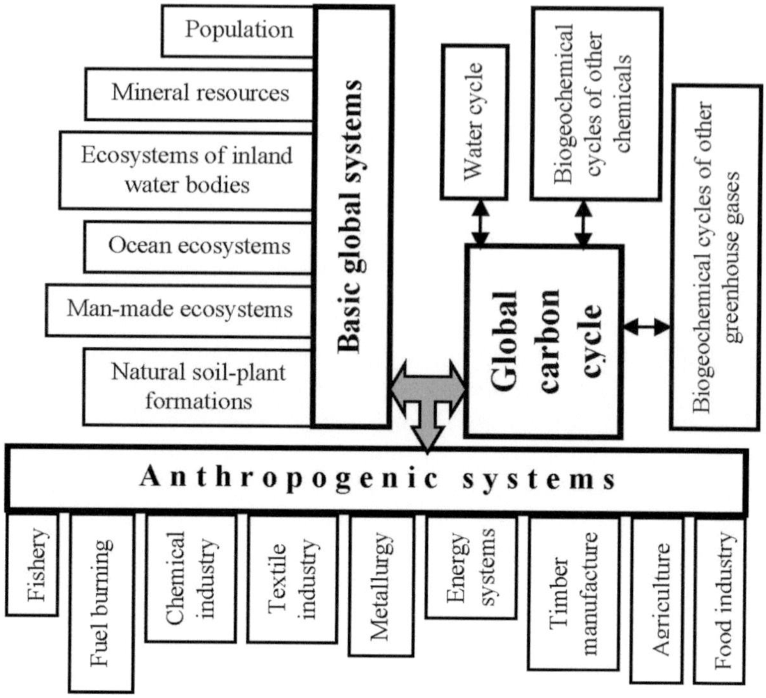


**Figure 3.1.** A conceptual scheme of the carbon cycle in the environment and position of the biogeochemical carbon cycle in the global system of energy exchange.

Thorough analysis of the many publications on the role of increasing $CO_2$ concentrations in the atmosphere leading to an enhanced greenhouse effect and to possible climate warming in the near future shows that predicted respective cataclysms in the environment are merely hypothetical. Researchers have reached these conclusions using rough models of the global biogeochemical $CO_2$ cycle that do not take into account the spatial mosaic of most of the effects and feedbacks considered above. However, resolving this is currently beyond our capabilities because, despite the many studies of the greenhouse effect, there are no global databases on greenhouse gases that reflect the seasonal and spatial structure of their fluxes. Moreover, even the fluxes enumerated in Figure 3.1 have not been modeled and, hence, the global models of $CO_2$ are little more than educational. For example, the results obtained by Alexeev *et al.* (1992), who studied $CO_2$ exchange in the coastal waters of the Caspian Sea, show that the mosaic of $CO_2$ fluxes is characterized by variations of the $\Delta p$ parameter within $-10$ to 234 ppm and of fluxes $H_2$ and $H_3$ within 1.2 to 32.3 mmol/m$^2$/day, even in such a small space. This confirms the need for thorough analysis of local fields $p_m$ and $p_a$ to derive accurate models of gas exchange in the ocean–atmosphere system. Similar conclusions suggest themselves for the soil–plant–atmosphere system. In this case it is necessary to consider the effect of as many types of plants and soils as possible. Available models consider no more than 30 types, whereas 172 species of plants are endemic (Eliasson *et al.*, 1999; Spilsbury and Spilsbury, 2008).

Inaccuracies in global models of the greenhouse effect as a result of $CO_2$ are on the increase, because of the lack of regular reliable data on the biospheric cycles of other greenhouse gases, including ozone and water vapor. This is confirmed by the large scatter between predicted estimates of climate warming by 2075 compared with the pre-industrial period: $1.1°C \leq \Delta T_{\Sigma} \leq 4.2°C$.

Detailed analysis of the causes and consequences of the greenhouse effect was undertaken at the Fourth International Conference on Greenhouse Gas Control held from August 30 to September 2, 1998 in Sweden (Eliasson *et al.*, 1999). The papers presented at the conference questioned the reliability of numerous predictions concerning various aspects of the greenhouse effect. The authors discussed the biological, energetic, chemical, geological, and social aspects of the greenhouse effect. However, a complex consideration of all these aspects that takes account of additional feedbacks reflecting the correlations between them still remains to be done. A solution to this was suggested by Krapivin and Kondratyev (2002). They proposed combining the functioning of several spatially heterogeneous natural–anthropogenic systems and processes in a global model. Such a global model would make it possible, in an adaptive regime, to search for an efficient mechanism to monitor the greenhouse effect and reliably assess the role of land and ocean biocenoses.

Accurate prediction of emissions of anthropogenic $CO_2$ and other greenhouse gases into the atmosphere is only possible by thorough and regular monitoring of the spatial structure of a huge number of NSS parameters. This can be achieved by finding an adaptive mechanism for the global model of this system to use (Kondratyev *et al.*, 2004a). Only an adaptive/evolutionary approach to solution of this problem can reduce uncertainty in the estimates of NSS parameters and guarantee harmonic development of human civilization. Thorough analysis of the publications of numerous experts raises hopes that within the Global Carbon Project (GCP) a constructive step forward will be made in synthesizing a global model that adequately reflects the whole spectrum of direct correlations and feedbacks in the environment and takes into account all developed models of various processes (Huesemann *et al.*, 2002; Lal, 2008; Peng *et al.*, 2002; Porte and Bartelink, 2002; Qi *et al.*, 2002; Sternberg and De Angells, 2002; Yang *et al.*, 2002; Yemshanov and Perera, 2002).

## 3.3 THE NSS GLOBAL MODEL

The present stage of scientific and technical progress, constrained as it is by the realization that the global ecological situation on Earth is characterized by limited supplies of energetic, geological, biogeocenotic, and other resources, highlights the need for information on global ecology (i.e., knowledge of the conditions under which humans and nature have co-evolved). The level of this information resource for thousands of years had been determined by weakly correlated total human activity which up to the beginning of the industrial era had been relatively low. Then, there was a historically rapid approach to the situation when a commercial attitude to the biosphere became the governing strategy of humankind and when ecological deadlock was reached. The significance of the information resource has reached its maximum.

This resource includes not only academic knowledge. Every ecological problem is "open", it is part of today's system of global problems, the main one being the need to preserve the homeostasis of humankind. This means that the threats to the biosphere that dawned on researchers at the end of the 20th century have highlighted the problem of the very survival of *Homo sapiens* and how to guarantee a responsible attitude to nature. Both ecological and moral problems have clashed. The rhetorical question "Why do we live and work?" has acquired a deep social aspect and needs to be answered for a harmonious relationship between society and nature. The fate of the biosphere will depend on how rapidly humankind solves these problems and finds the optimal balance between attitudes to the environment. Model estimates have shown that 90% of humankind needs to become involved for these problems to be resolved. However, at the present time, it is unlikely that such a large proportion of the population can deliberately change from their position as master of nature and develop new harmonious relationships with nature. To reach global harmony, it is necessary to concentrate on negative ecological and socioeconomic changes, so that ecological knowledge can be put into practice in the form of concrete technologies that ensure the decisions made in nature conservation are of sufficiently high quality.

At the present stage of scientific and technical progress, there has been intensive development in this field, an analysis of which reveals the characteristic features of ecological knowledge and the problems facing applied methods to formulate the basic requirements for an efficient information technology. A prerequisite to creating systems to monitor the environmental condition is the availability of data of different quality and, hence, a multitude of numerical models of various types (balance, optimizing, evolutionary, statistical, etc.). Derived on the basis of parameterization and, as a rule, linearization of the laws of natural phenomena, these models include a wide spectrum of determinate and probabilistic descriptions of the geological, ecological, oceanographic, biogeochemical, and biogeocenotic processes of a global, regional, and local nature. Most aim at a theoretical understanding of the specific features of higher living systems on the basis of available knowledge, and only a few are aimed at first steps to objective assessment of the present global ecological situation. Differing in the goals and mathematical apparatus of description, many models have little option but to be rough as a result of limitations, inadequacy, and uncertainty of the information base. What is more, present-day instrumental systems in the field of numerical experiments are lacking. An increase in the number of biospheric parameters and their relationships to improve model adequacy makes them multiparametric (i.e., the "multi-variance curse" rears its ugly head).

Simulation modeling has been considered as the basic instrument to solve these problems. This methodology makes it possible to adjust different data that refer to various mathematical formalisms and remove the excess number of parameters. Such a model is constructed on the basis of empirical information not restricted *a priori* by any mathematical apparatus, which ensures flexible formalization, something that is unavoidable when important laws of phenomena are unknown. In some cases, genetic algorithms are used to model undetermined systems with stochastic elements that have no unique solution. One such algorithm is the genetic algorithm for rule set prediction (Anderson *et al.*, 2003). In addition, there is a multitude of non-traditional algorithms based on the technology of evolutionary modeling (Bukatova *et al.*, 1991) and self-organizing numerical models (Timoshevskii *et al.*, 2003).

The development of simulation modeling by broadening the information base, by combining formal and informal methods in stage-by-stage synthesis of the necessary model, and finally by active dialogue between humans and computers will provide an efficient approach to system ecological modeling. However, it now appears that it is not that easy. If we compare information queries that crop up in ecological problems with existing information provision for their solution (different numerical and simulation models, principles of ecological information processing), we see that not all levels of natural and anthropogenic complexes have a well-developed apparatus to describe them or to construct efficient information approaches to obtain necessary estimates of problematic situations. Difficulties appearing in this connection are not only technical, they also consist in accumulating models of different types. These features manifest themselves especially in global modeling, which reveals the substantial and principally ineradicable inadequacy of our knowledge about natural processes, as manifest both by the fragmentary character of empirical data and by lacking adequate ideas about the laws of natural process evolution. It is now clear that simply having a mechanical set of hierarchies of models and a desire to accumulate empirical databases are an attempt to reanimate primitive reasoning about the integral pattern of the development of biospheric processes without hope of success, without the possibility of explaining the permanent self-organizing ability of living systems, and without substantially improved understanding of the mechanism involved in NSS functioning. The situation is such that it is necessary to use computer technologies that combine the methods of both evolutionary and simulation modeling, as described in Bukatova *et al.* (1991). This makes it possible to take into account the internal dynamics (evolution) of modeled processes and to synthesize models under conditions of inadequate and partially reliable data.

Traditional approaches to global model construction face difficulties in algorithmic description of many socioeconomic and natural processes. This means we are dealing with information uncertainty. Developed approaches to global modeling ignore this uncertainty and, as a result, the models do not adequately consider real processes. Combined use of evolutionary and simulation modeling makes it possible to remove this shortcoming by synthesizing a combined model, the structure of which is adapted on the basis of the prehistory of biospheric and climatic components. The model can be created in various model classes using traditional computer programs and special evolutionary-type processors. The form of such a combination can vary according to the spatiotemporal adequacy of global databases.

Global modeling comes up against many examples of insurmountable difficulties when attempting to find ways to describe scientific and technical progress and human activity in its various manifestations. Moreover, difficulties appear in modeling the climate characterized by superposing processes with different time scales. As for adequate description in a global model, it is impossible here to clearly define the limits of information provision and those for the necessary spatial and structural resolution. Therefore, without dwelling on philosophical analysis of global problems and without trying to recommend a comprehensive approach to global modeling, we shall discuss one possible way of showing how evolutionary modeling by creating a special processor version of the model enables the difficulties mentioned above to be overcome.

Adjusting the evolutionary model by the prehistory of natural rhythms makes it possible to obtain a model that implicitly follows the various regularities of NSS dynamics in the past and to make forecasts in the same temporal rhythm. The special processor version of the model overcomes all algorithmic and digital difficulties caused by multi-dimensionality of the global model and the multitude of parametric uncertainties.

Clearly, it is impossible to completely overcome contradictions between researchers' needs and NSS models because of insufficient information about the laws under which environmental elements and human society function. Therefore, a huge number of model types have appeared, each responding to a certain level of adequacy. Small-sized models mainly based on empirical dependences are most widespread. Among successfully used approaches to overcome information uncertainties are neural networks, which (like evolutionary modeling) enables models to be synthesized when relationships between the characteristics of environmental systems are poorly known (Aitkenhead *et al.*, 2003). In other words, neural networks permit the establishment of cause-and-effect bonds without equations. Of course, any loss of accuracy when modeling the process under study using neural networks takes place not because of any inadequacy of parametric description, but because of other input suppositions about characteristic functions of neural networks (the weights of synoptic connections, node activities, etc.). For instance, Aitkenhead *et al.* (2003) used the activation function $y$ in the form of a curve with saturation $y = 1/[1 + \exp(-cx)]$, where $x$ is the input variable. The neural network approach applied to a small headwater stream called Brocky Burn (27 km in length) in northeast Scotland resulted in more than 50% of predictions having an accuracy of up to 10% and over 90% of predictions an accuracy up to 30%. This local example confirms the efficiency of neural-based models to synthesize some GSM units that have a high degree of functional uncertainty but do have a series of observations of their dynamics.

Synthesis of the GMNSS is based on considering it as a self-organizing and self-structuring system, in which elements are coordinated in time and space by the process of natural evolution. The anthropogenic constituent in this process breaks this integrity. Attempts to parameterize formally the process of co-evolution of nature and humans, as elements of the biosphere, are connected with the search for a single description of all processes in the NSS that would combine all spheres of knowledge in perceiving the laws of the environment. Such a synergetic approach has formed the basis of numerous studies in global modeling (Kondratyev *et al.*, 2003b, 2004a; Krapivin and Kondratyev, 2002; Krapivin *et al.*, 1982).

Synthesis of the global model requires choosing a spatial grid that digitizes climatic, biospheric, and anthropogenic processes. Usually, they take a uniform geographic grid with steps $\Delta\varphi$ by latitude and $\Delta\lambda$ by longitude. The convenience of this choice is connected with a simplified process of calculations but leads to loss of data, excessiveness, and difficulties in correlating accumulated knowledge with available global databases. One of the first attempts to reflect NSS spatial heterogeneity in the model was made by Krapivin (1978), where the NSS was divided into the atmosphere, the World Ocean, and two land regions. NSS input is assumed to be solar radiation and irreplaceable mineral resources. Solar energy is used by photosynthesizing elements of the biosphere (phytoplankton and plants). The flux of this energy is regulated by atmospheric turbidity formed by natural and anthropogenic processes. Atmospheric transparency is determined by the number of dust particles residing in the atmosphere as a result of dust storms, volcanic eruptions, burning of solid and liquid fuels, emitted from the chimney stacks of metallurgical and chemical factories, cement plants, and other industrial factories, and as a result of cloud formation.

Calculations using the first version of GMNSS demonstrated a point of balance of the interacting forces between two regions of land. This result necessitated further digitizing of land territory, which was done in Krapivin *et al.* (1982). It was found that the assumed scheme of human–environment interaction (mutual influence of ecological and production processes) is a classical example of Germeyer's system (Germeyer, 1974). Breaking the balance of biota or irreversible climate change will disrupt humankind's homeostasis, the condition of which can be formulated in terms of indicators of environmental quality. Hence, irrespective of the subjects of the NSS under consideration, among the criteria for subjective decisions there is always a criterion general for all subjects (regions, states, or other groups). One such criterion is the survival of humankind.

Since Vernadsky's (1944) thesis "no living species can exist in a medium consisting of its waste" is currently valid, we shall not discuss the ramifications of creating a completely artificial habitat for the global population. Nevertheless, it is important to preserve the stability of the environment and the general homeostasis of all constituents involved in the combined functioning of nature and humans. In Germeyer's system, the search for persistent collective decisions belonging to the Pareto set is justified. Unfortunately, the latest international forums on the problem of sustainable development have not pursued this path (i.e., have not posed synthesis of the GMNSS in such a way as to find persistent collective decisions in the interaction processes between human activity and the environment). A vivid negative example of this is the Kyoto Protocol, despite the greenhouse effect as a result of carbon dioxide becoming increasingly more urgent every year.

One cause of the observed gap between the suggestions of researchers and the decisions of politicians consists in the absence of an international program that could combine them and lead to scientifically grounded criteria being assumed when estimating developing trends in the NSS and its elements. It is for this reason that the global database continues accumulating but does so in a way that is not methodical (Forest *et al.*, 2002; Lovett, 2002; McCarl and Schneider, 2001; Milford *et al.*, 2001).

The ISGGM proposed by Kondratyev *et al.* (2002) has the potential to be an efficient mechanism for facilitating our search for stable NSS conditions. It is in this way that a single planetary adaptive evolutionary geoecoinformation service can be created, which would have a hierarchical structure of data accumulation and form a multi-level global database (Gorshkov *et al.*, 2000, 2002; Ivanov-Rostovtsev *et al.*, 2001; Kondratyev, 1999, 2000a, b; Marchuk and Kondratyev, 1992).

As mentioned above, existing difficulties in synthezing a global NSS model consist mainly in parameterizing energy fluxes at the atmosphere–land and atmosphere–ocean boundaries. A series of studies by Sellers *et al.* (1995, 1996, 1997) threw up the possibility of combined use of climate and biosphere models. The SiB1 and SiB2 models lean toward describing the energy and heat balance of the Earth's surface. In these models the processes of photosynthesis and evapotranspiration are parameterized having regard to plant physiology and local conditions for the formation of vegetation cover. It is also important that these types of global models are orientated towards making their parameters compatible with satellite monitoring. Since 1987, a global series of NDVI has been created over a $1^\circ \times 1^\circ$ monthly averaged geographical grid by means of satellites of the NOAA series and the AVHRR. Moreover, databases have been formed of the types of soils, types of vegetation, absorbed radiation, albedo, cloudiness, soil moisture, etc. These databases can only be justified from a research perspective.

Finally, there is the ICLIPS (Integrated Assessment of Climate Protection Strategies) model, a potential unit of the global model that can be used to simulate the global socioeconomic structure (Leimbach and Toth, 2003). This model divides the economic section of the NSS into 11 regions:

- sub-Saharan Africa;
- China, Mongolia, Vietnam, Cambodia, Laos;
- Eastern Europe;
- former Soviet Union;
- Latin America and the Caribbean;
- Middle East and North Africa;
- North America;
- Pacific OECD (Japan, Australia, New Zealand);
- South Asia (mainly India);
- Western Europe.

## 3.4 MODEL-BASED METHOD TO ASSESS GLOBAL CHANGE IN THE NSS

### 3.4.1 Backdrop

As mentioned above, many problems arising from the interaction between nature and society have been considered by various authors. The growing number of publications dedicated to global environmental change demonstrates just how urgent protection of the natural environment has become. Ascertaining the principles that underlie co-evolution of humans and nature is being pursued with ever-increasing persistence. Scientists in many countries are making attempts to find ways of formulating the laws that govern human processes acting on the environment. Many national and international programs of the biosphere and climate studies have contributed to the quest for a means of resolving the conflict between human society and nature. However, attempts to find efficient ways to regulate human activity globally have encountered many difficulties. The major difficulty is the absence of an adequate knowledge base pertaining to climatic and biospheric processes along with the largely incomplete state of databases concerning global processes occurring in the atmosphere, in the ocean, and on land. Another difficulty is the inability of modern science to formulate the requirements that must be met by global databases to reliably evaluate the state of the environment and forecast how it will develop over the long term.

Many scientists are trying to find answers to the above questions. The majority suggest creation of a unified planetary-scale adaptive GIMS as an efficient way of resolving the conflicts between nature and man. Based on regenerated knowledge bases and global datasets, the adaptive nature of such a system should be provided by constantly updating the data-acquisition mode and by varying the parameters and structure of the global model. The main idea of this approach to studying the NSS was developed in detail by Kondratyev *et al.* (2002). This book continues to develop this approach.

This section describes schematically a simulation model of NSS dynamics, considering different aspects of global ecodynamics and globalization processes. The main idea behind this model is the interactivity concept developed by Krapivin *et al.* (1982). The model is constructed of blocks parameterizing natural and anthropogenic processes. The various blocks describe: biogeochemical cycles of greenhouse gases; the global hydrologic cycle in liquid, gaseous, and solid phases; productivity of soil–plant formations and definition of the many types; photosynthesis in ocean ecosystems with depth and surface heterogeneity taken into account; demographic processes and anthropogenic changes. The model makes it possible to compute the dynamics of industrial $CO_2$ distribution between the ocean, terrestrial biota, and the atmosphere. The ocean is described by a spatial four-layer model with due regard for water chemistry. The model is specially designed for connection to a global climate model. Input model data are combined from existing global databases and the model-oriented environmental monitoring is proposed to be adaptive to simulation model input. Examples are presented of using the model to estimate the state of the NSS and its subsystems. The respective roles played by vegetation and the global ocean in climate change are evaluated. Different hypotheses of the causes of global change are also considered.

### 3.4.2 A new type of global model

Approaches to synthesis of a global model include the need to describe all aspects of human interactions with the environment and its physical, biological, and chemical systems. One such application has its origin in studies that took place at the Computer Center of the Russian Academy of Sciences in Moscow (Krapivin *et al.*, 1982). This type of global model is formulated on the basis of detailed description of the climate system by considering a small set of biospheric components. Such a strategy for global modeling is followed by the Potsdam Institute for Climate Impact Research studies (Boysen, 2000), where prototypes of the Moscow Global Model were developed. More than 30 climate models are currently being developed in different countries in an attempt to generate new trends in the science of global change. Unfortunately, global and regional studies using this approach to assess the processes and impacts of global change have not produced sufficiently acceptable results. This is the reason another approach to global modeling, known as evolutionary modeling, has been developed by many authors (Kondratyev *et al.*, 2004a; Krapivin, 1993; Sellers *et al.*, 1996).

Traditional approaches to building a global model face some difficulties in algorithmic description of many socioeconomic, ecological, and climatic processes (Kondratyev, 1999), such that we have to deal with information uncertainty. These approaches to global modeling simply ignore such uncertainty and, consequently, the resultant models do not adequately reflect real processes. Evolutionary modeling makes it possible to remove this drawback by synthesizing a combined model whose structure allows the background history of a system to be adapted to the biosphere and climate components. The implementation of such a model can also be combined in various classes of models using conventional software and hardware and special purpose evolutionary-type processors. The form of such a combination is dependent on the spatiotemporal completeness of the databases (Rochon *et al.*, 1996).

Global modeling abounds with examples of insoluble problems that are encountered when looking for ways to describe scientific and technological advances and human activity in their diverse manifestations. No less difficulty arises when modeling climate described by superimposing processes with different temporal variability rates. As to the inclusiveness of description in the global model, it is impossible to clearly delineate the bounds of information availability and the extent of required spatial and structural detail. Therefore, without going into natural–philosophical analysis of global problems and skirting the issue of the ultimate solution to global modeling, we will confine ourselves to discussing just one possible approach. This approach will demonstrate the way in which evolutionary modeling, developed in special processors, can help overcome some of the difficulties of global modeling, such as those inherent in computing and algorithmic variations. All of this implies that a search for effective traditional-type models may well be of value. At present, the building of global biogeocenotic models is not seen as difficult. Many such models have been created, and the gathering of information to support them is under way. The history of the interaction of the biosphere with the climate system and human society is not sufficiently understood, which is one of the obvious hurdles in describing climatic cycles. This is the reason an evolutionary approach is essential to building a global model that accounts for the interaction of the biosphere, climate system, magnetosphere, etc. Such an approach helps overcome uncertainties in describing such a complex interaction. As a result of adjusting such a model to the history of the prescribed cycle, we will obtain a model implicitly tracing various regularities of the dynamics of the biosphere in the past and allowing for forecast assessments to be made in the same temporal cycle. A special processor version of this model completely removes all existing algorithmic and computing hurdles arising from the large dimensionality of the global model and the conditions that bring about irreducible non-parametric uncertainty.

Figure 3.2 shows the key elements of this new type of global model. The data archive is formed here as two structures. Data of the first type for computer models of biosphere processes are stored as climatic maps and as tables of model equation coefficients. It is necessary to fill in all cells on the schematic maps. Data of the second type are represented as fragments recorded disparately (possibly irregularly) in time and space (i.e., $CO_2$ concentration, temperature, precipitation, pressure, population numbers, availability of resources, etc.). Data of this type are used to adjust the evolutionary processor to the given class of models (e.g., finite automata). As a result of this procedure the model is adapted to the history of the prescribed time cycle. As shown by Kondratyev *et al.* (2002), a stable forecast is produced with 75–95% reliability covering several temporal steps. The extent of a forecast is determined by the length of its history. Given the need for a forecast under conditions of change in the trends of human economic activities, an evolutionary processor is adjusted to the assigned scenario, thus automatically providing for simulation of the corresponding response of the biosphere to this change.

The suggested structure of the global model thus ensures traditional and evolutionary-type models are flexibly combined. The proposed approach helps avoid the need to model non-stationary processes (climatic, socioeconomic, demographic, etc.) and allows uncertainty to be overcome. A model of this new type makes it possible to go from learning experiments to assessment of the viability of the biosphere with regard to actual trends of anthropogenic stresses in all regions of the globe.

Departing from established global modeling techniques based on new information technology makes it possible to proceed to creating a global monitoring system that has the global model as a part of system support. A conceptual block diagram of geoinformation monitoring and use of the global model is represented in Figure 3.3. Application of evolutionary computer technology allows

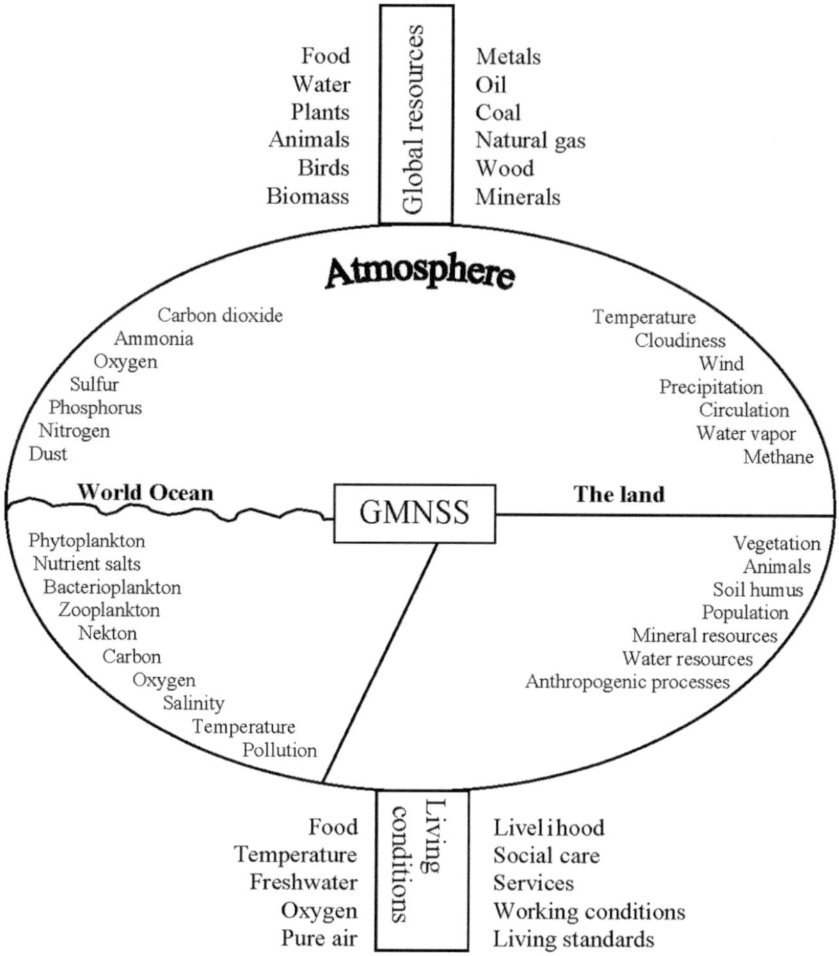


**Figure 3.2.** Key elements of the nature–society system and energy components that need to be taken into account for global ecodynamics forecast in global model use.

categorizing the whole system as a class of subsystems with a variable structure and making it adaptable to changes in the natural process or entity under observation. Furthermore, it becomes possible to detail heterogeneously the natural systems under study in the space of phase variables and to select non-uniform geographical grids in sampling analysis of the planetary surface (i.e., arbitrary insertion of significant regularities at the regional level becomes possible).

The automatic system for processing global information is aimed at acquiring combined models that reflect in real time climatic and anthropogenic changes in the biosphere and is based on its known history (or, rather, its simulation). The system relies on a set of models of biosphere processes and, using the software of other units along with a scenario of anthropogenic behavior formulated at input,

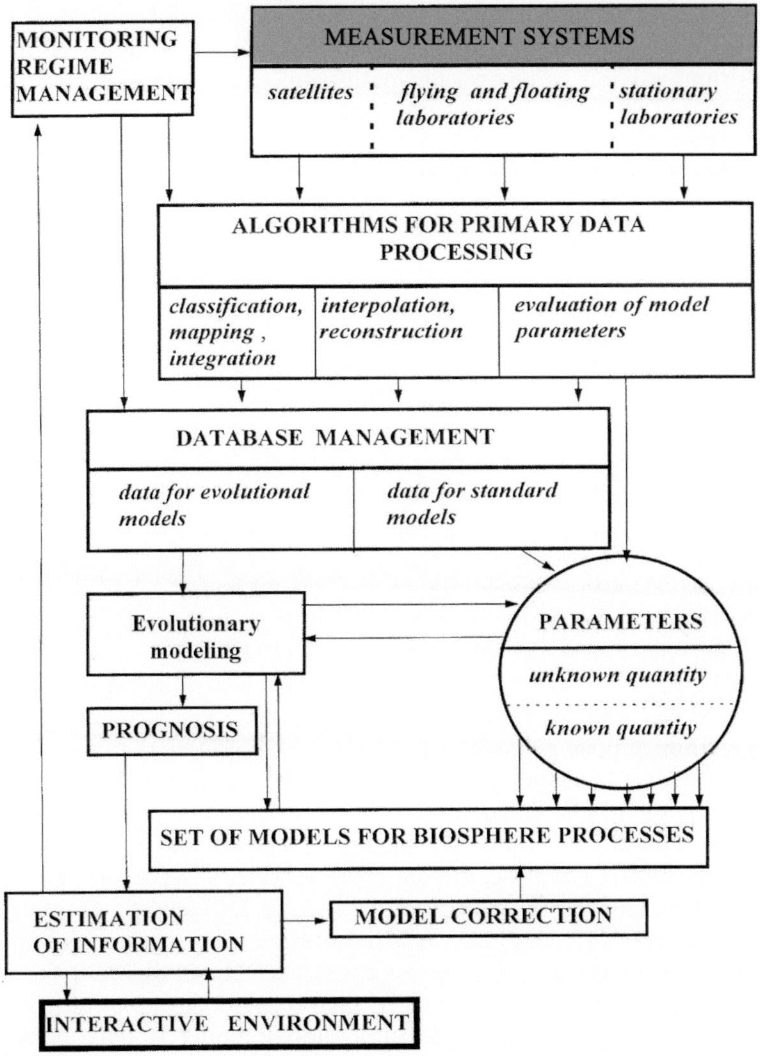


**Figure 3.3.** Block diagram of interactive adjustment of the GMNSS and control of the geoinformation monitoring regime.

provides for prompt assessment of the environmental state and for forecast assessments within the framework of this scenario. The automatic system for processing global information has a further advantage in that it formulates the entry and bound conditions for particular built-in models in the study of regional systems and can be used as a virtual substitute for field measurements of those conditions.

The first version of the global model (Krapivin *et al.*, 1982) was oriented towards rigid spatiotemporal detailing and, therefore, required a large quantity of information. The subsequent development of an automatic system for processing global information made it possible, owing to evolutionary technology, to discard the generally accepted regular geophysical grid in archive development and replace it by algorithms capable of recovering spatiotemporal information.

### 3.4.3 Mathematical model of NSS dynamics

#### *3.4.3.1 General description of the global model*

In connection with the different aspects of environmental change that have taken place during recent decades, experts have put forward numerous concepts for global description of the NSS, and models of various complexity have been developed to parameterize characteristics of the biosphere and climate. The availability of a large database on these characteristics enables the consequences of different scenarios of the development of NSS subsystems to be considered and estimated. Traditional approaches to synthesizing global models are based on considering the totality of balance equations, which include parameters $\{x_i\}$ in the form of functions, arguments, coefficients, and conditions of transition between parametric descriptions of processes taking place in the environment. Moreover, other approaches have been applied that use evolutionary and neural network algorithms. Organization of the global model of NSS functioning is presented as a conceptual scheme in Figure 3.4. This scheme is realized by introducing a geographical grid $\{\varphi_i, \lambda_j\}$ that has step progressions when sampling the land surface and the World Ocean given by $\Delta\varphi_i$ and $\Delta\lambda_j$ in latitude and longitude, respectively, such that within a pixel $\Omega_{ij} = \{(\varphi, \lambda) : \varphi_i \leq \varphi \leq \varphi_i + \Delta\varphi_i, \lambda_j \leq \lambda \leq \lambda_j + \Delta\lambda_j\}$ all processes and elements of the NSS are considered as homogeneous and parameterized by point models. The choice of pixel size is determined by several conditions governed by the spatial resolution of satellite measurements and the availability of an appropriate global database. In the case of water surface, in pixel $\Omega_{ij}$ water mass is divided into layers by depth $z$ (i.e., 3-D volumes $\Omega_{ijk} = \{(\varphi, \lambda, z) : (\varphi, \lambda) \in \Omega_{ij}, z_k \leq z \leq z_k + \Delta z_k\}$ are selected and all elements are distributed uniformly within). Finally, the atmosphere over pixel $\Omega_{ij}$ is digitized by altitude $h$ either by atmospheric pressure levels or by characteristic layers of altitude $\Delta h_s$.

It is clear that the development of a global model is only possible using knowledge and data at the international level. Of the many models available, the most adequate is that described in Kondratyev *et al.* (2004a, b). The block scheme construction of this model is shown in Figure 3.5 and Table 3.2. To synthesize a model on such a scale requires careful consideration of existing models of various partial processes derived from information on climatology, ecology, hydrology, geomorphology, etc.

An adaptive procedure for introducing the global model into a system of geoinformation monitoring was proposed in Kondratyev *et al.* (2003b). This procedure is schematically shown in Figure 3.6.

Note that since the NSS is a part of the Earth system, it can be considered a closed object of the energy exchange with space and part of the Earth system that includes the core and mantle as sources of planetary energy through gravitational

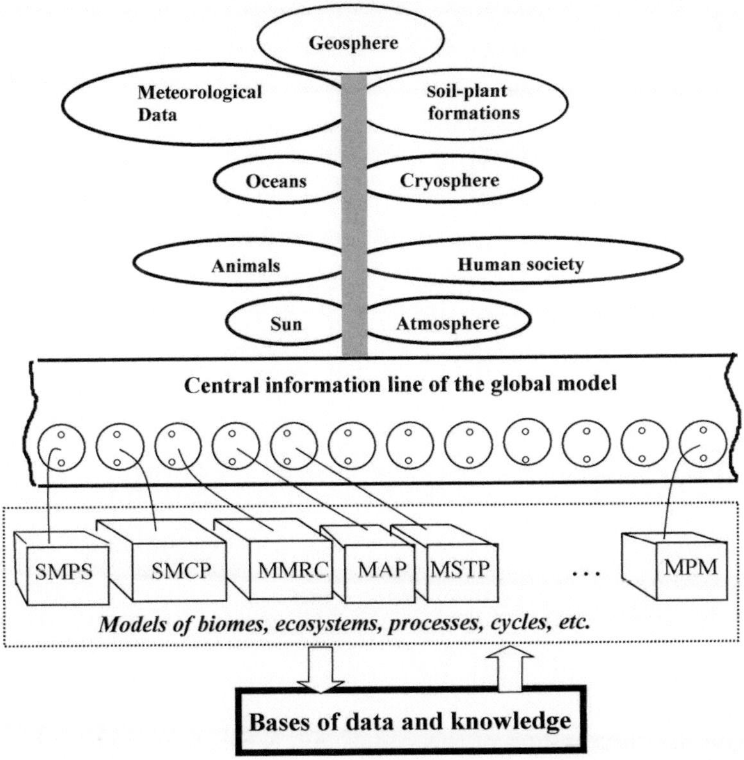


**Figure 3.4.** The information/functional structure of the global NSS model. Notation is given in Table 3.2.

differentiation and radioactive decay. The GMNSS needs to have certain mechanisms that can regulate $(V, W)$ exchange by considering correlations between the biogeochemical cycles of carbon, methane, ozone, water, oxygen, nitrogen, sulfur, and phosphorus. Interdisciplinary studies currently under development should enable elaboration of this series and make it possible to establish functional connections between all spheres of the Earth system. Here an attempt has been made to synthesize the GMNSS elements responsible for simulating the complex of biogeochemical cycles of some important elements.

#### *3.4.3.2 Model of the global biogeochemical cycle of carbon dioxide*

In recent decades, much focus has been on the global carbon cycle (GCC) in terms of the numerous, often speculative, explanations of the role of $CO_2$ in future climate change. Unfortunately, an objective assessment of this role is still absent. Recent studies (Kondratyev and Krapivin, 2004; Kondratyev *et al.*, 2003c, 2004a, b; Krapivin and Kelley, 2009) have summed up the first results from a formalized technology that assesses the greenhouse effect as a result of $CO_2$ with due

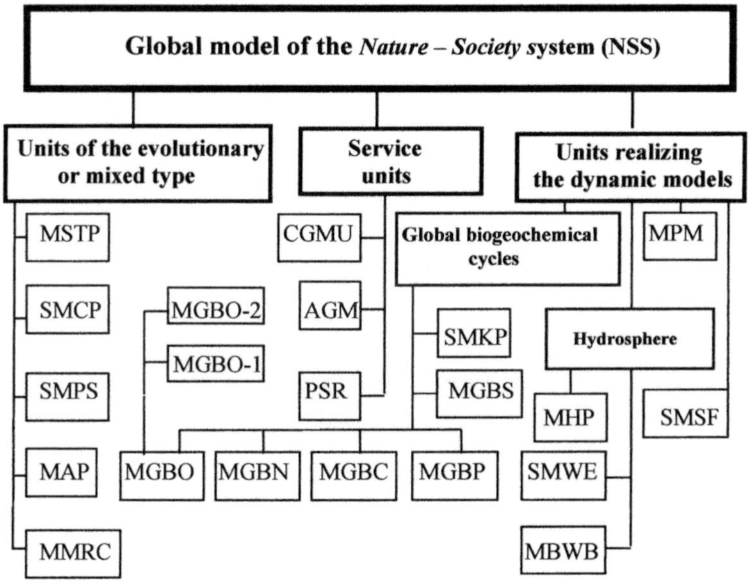


**Figure 3.5.** Block diagram of the GMNSS. The units are explained in Table 3.2.

regard to the role played by land and ocean ecosystems. An interactive connection has been demonstrated between the global carbon cycle in the form of $CO_2$ and climate change. Formalization of this connection is based on synthesizing GMNSS functioning and taking the spatial distribution of the elements of this system into account, which makes it possible to reduce to a single correlated scheme the cause-and-effect connections of carbon fluxes between its different biospheric and geospheric reservoirs.

Objective formalization of the biospheric sources and sinks of $CO_2$ as functions of environmental parameters and consideration of the actual role played by anthropogenic processes is possible as a result of the recent studies of many researchers who have developed models that describe (in varying degrees of detail) spatially distributed carbon fluxes and their interaction with NSS components (Kondratyev *et al.*, 2004a).

In a recently published first report of the GCP investigating the global carbon cycle (Canadel *et al.*, 2003; Yude *et al.*, 2011), a strategy was adopted for interdisciplinary cooperation within a broad spectrum of environmental problems considered in the context of a global system of nature–society interaction, with special emphasis on the need to develop methods and information technologies to analyze the carbon–climate–society system (CCSS). The central goal was consideration of the following five aspects of the global carbon cycle:

- study of the GCC based on integrating natural and anthropogenic components by analyzing the interactions between energy systems based on fossil fuel, the biogeochemical carbon cycle, and the physical climate system;

**Table 3.2.** Characteristics of GMNSS units shown in Figures 3.4 and 3.5.

| *Identifier* | *Characteristics of unit functions* |
|---|---|
| SMPS | A set of models of population size with regard to age structure (Logofet, 2002) |
| SMCP | A set of models of climatic processes with differently detailed consideration of parameters and their correlations (Kiehl and Gent, 2004) |
| MMRC | Model of mineral resources control (Nitu *et al.*, 2004) |
| MAP | Model of agricultural production (Kondratyev and Krapivin, 2001b) |
| MSTP | Model of scientific/technical progress |
| CGMU | Control of the global model units and database interface |
| AGM | Adjustment of the global model to simulation experiment conditions and its control |
| PSR | Preparation of simulation results to visualize other forms of account. |
| MBWB | Model of the biospheric water balance (Krapivin and Kondratyev, 2002) |
| MGBC | Model of the global biogeochemical cycle of carbon dioxide (Kondratyev *et al.*, 2003b; Krapivin and Varotsos, 2007) |
| MGBS | Model of the global biogeochemical cycle of sulfur compounds (Krapivin and Nazaryan, 1997; Krapivin and Varotsos, 2008) |
| MGBO | Model of the global biogeochemical cycle of oxygen and ozone (Krapivin and Kondratyev, 2002; Krapivin and Varotsos, 2007) |
| MGBN | Model of the global biogeochemical cycle of nitrogen (Krapivin and Kondratyev, 2002; Krapivin and Varotsos, 2007) |
| MGBP | Model of the global biogeochemical cycle of phosphorus (Krapivin and Kondratyev, 2002; Krapivin and Varotsos, 2007) |
| SMKP | A set of models of the kinetics of some types of pollutants in different media (Kondratyev *et al.*, 2006; Krapivin and Potapov, 2002) |
| SMWE | A set of models of water ecosystems in different climatic zones (Cracknell *et al.*, 2009; Krapivin and Varotsos, 2007) |
| MHP | Model of hydrodynamic processes (Kondratyev *et al.*, 2002) |
| SMSF | A set of models of soil–plant formations (Kondratyev *et al.*, 2004a) |
| MPM | Model of processes in the magnetosphere (Korgenevsky *et al.*, 1989) |

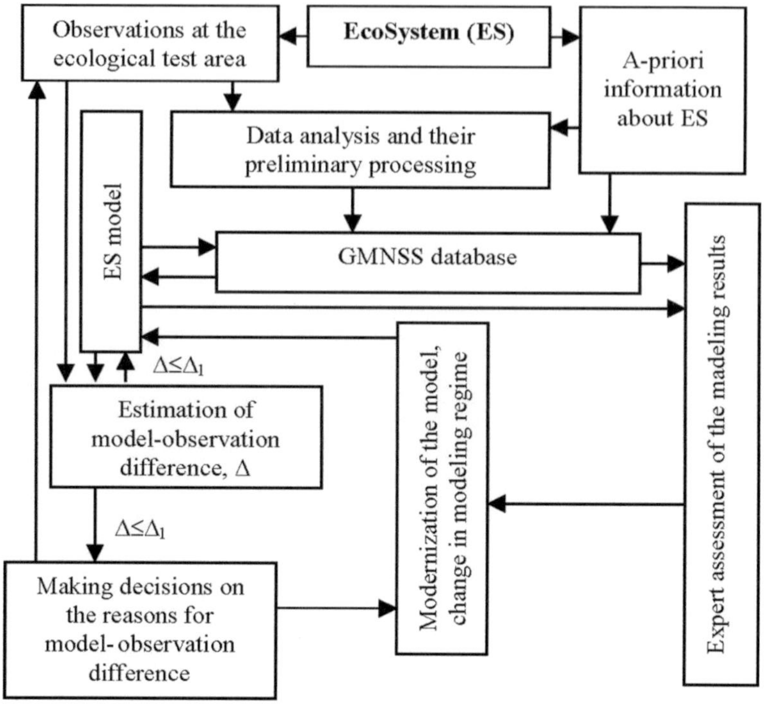


**Figure 3.6.** Principal scheme of the organization of ecological monitoring using an adaptive modeling regime. Notation: $\Delta$ is the integral or subject estimate of difference between modeling and observational data; $\Delta_1$ is the permissible level of difference for $\Delta$ estimation.

- development of new methods of analysis and numerical modeling of an integrated carbon cycle;
- global studies of the carbon cycle carried out with due regard to the results of national and regional research programs that have investigated GCC reservoirs;
- search for a means in regional development of achieving a stabilized concentration of $CO_2$ in the atmosphere;
- classification of all countries into developed and developing status in order that the technologies produced are respectively divided into industrial, economic, and energy sectors of the NSS by their significance as sources of anthropogenic $CO_2$ emissions.

In addition to this, let us enumerate the key directions of developments within the CCSS program (DOE, 2004, 2009):

- What are the specific features of spatial-temporal variability of the sources and sinks of carbon on the continent of North America on time scales from seasonal to centennial, and what processes are attributable to the prevailing effect on carbon cycle dynamics?

- What are the respective features of variability and their determining factors in the case of ocean components (sources and sinks) of the carbon cycle?
- How do local, regional, and global processes on the land surface (including land use) affect the formation of carbon sources and sinks in the past, present, and future?
- How do the sources and sinks of carbon vary on land, in the ocean, and in the atmosphere on time scales from seasonal to centennial, and how can the respective information be used to obtain a better understanding of the laws of GCC formation?
- What future changes can be expected in concentrations of atmospheric $CO_2$, methane, and other carbon-containing GHGs as well as changes in sources and sinks of carbon on land and in the ocean?
- How will the Earth's system and its components respond to various choices of strategy for regulating the carbon content in the environment, and what information is needed to answer this question?

The carbon cycle is closely connected with climate, water, and nutrient cycles and with photosynthesis production on land and in the ocean. Therefore, all studies of GCC that overlook such connections are inevitably doomed to fail and, hence, cannot give even approximately adequate estimates of the consequences of anthropogenic emissions of carbon to the environment. For this reason, many international projects on the greenhouse effect and its impact on climate have failed, such as the Kyoto Protocol's intention to regulate $CO_2$ emissions. The GCP is hopeful of making progress in this sense by planning interdisciplinary studies of the GCC. Such studies can be divided into three directions:

- formation of a strategy for GCC studies and evaluation of its variability;
- analysis of connections between causes and consequences in studies of mechanisms of environmental interaction with natural and anthropogenic $CO_2$ sources and sinks;
- identification and quantitative estimation of evolutionary processes in the CCSS.

The first GCP report (Canadel *et al.*, 2003) formulated the goal of the coming decade of GCC study as ideologically combining earlier isolated programs such as the IGBP, IHDP, and WCRP. The authors of this report substantiated a detailed scheme of the cause-and-effect connections between components of the climate–biosphere system and pointed to the necessity to consider them jointly in order to raise the level of reliability of estimates and forecasts for $CO_2$ climatic impact. All these problems had been discussed in other publications (Cracknell *et al.*, 2009; Krapivin and Varotsos, 2007, 2008). In the first GCP report, the role of other GHGs was underestimated, unfortunately, even though their contribution in the most immediate future looks likely to exceed that of $CO_2$. The list of GHGs, such as methane, nitric oxide, hydrofluorocarbons, perfluorocarbons, and hydrofluoroethers, increases with time. Moreover, these gases contribute more readily to formation of the greenhouse effect than $CO_2$. Their total equivalent emissions in

**Table 3.3.** Potentials of relative global warming due to various greenhouse gases (EPA, 2001). Global warming potential (GWP) is a measure of how much of a given mass of greenhouse gas is estimated to contribute to global warming. GWP is defined as the ratio of time-integrated radiative forcing from the instantaneous release of 1 kg of a trace substance relative to that of 1 kg of a reference gas. Carbon dioxide has a GWP of exactly 1 (since it is the baseline unit with which all other greenhouse gases are compared).

| *Gas* | *Potential* | *Gas* | *Potential* |
|---|---|---|---|
| $CO_2$ | 1 | HFC-227es | 2,900 |
| $CH_4$ | 21 | HFC-236fa | 6,300 |
| $N_2O$ | 310 | HFC-4310mee | 1,300 |
| HFC-23 | 11,700 | $CF_4$ | 6,500 |
| HFC-125 | 2,800 | $C_2F_6$ | 9,200 |
| HFC-134a | 1,300 | $C_4F_{10}$ | 7,000 |
| HFC-143a | 3,800 | $C_6F_{14}$ | 7,400 |
| HFC-152a | 140 | $SF_c$ | 23,900 |

1990 constituted 3.6 Gt $CO_2$, and by the year 2010 the level of 4.0 Gt $CO_2$ was exceeded (as predicted by Bacastow, 1981). In 2010, anthropogenic $CO_2$ emissions reached 6 Gt $C \cdot yr^{-1}$ (as predicted by IPCC, 2007). Table 3.3 compares the potential impact of various GHGs on climate change. According to historical data from the beginning of the industrial revolution (EPA, 2005), the relative levels of anthropogenic emissions of certain GHGs contributing to enhancement of the greenhouse effect consisted of $CO_2$ 55%, $CH_4$ 17%, $O_3$ 14%, $N_2O$ 5%, and others 9%.

All of this testifies to the fact that continuing with inadequate descriptions (primitive in most cases) of the GCC, let alone parameterizing other GHGs, cannot lead to reliable estimates of possible future climate changes due to anthropogenic activity within the NSS. The idea of identifying the locations and impact potential of $CO_2$ sources and sinks on land and in the World Ocean as declared in the GCP report (Canadel *et al.*, 2003) has not been followed up by the serious and substantial motivation needed to develop new information technologies for comprehensive analysis of the Earth's radiation budget.

Nevertheless, it should be recognized that the fundamental postulate of the GCP is to promote a better understanding of the GCC. Studies are based on the concept of combined natural and anthropogenic components, while application is by established analytic methods, algorithms, and models. The main structure of the carbon cycle is determined by fluxes between its basic reservoirs, including carbon in the atmosphere (mainly in the form of $CO_2$), the oceans (surface, intermediate, deep layers, and bottom deposits), terrestrial ecosystems (vegetation, litter, and soil), rivers and estuaries, and fossil fuels. All of these reservoirs should be studied with due regard for their spatial heterogeneity and dynamics as influenced by natural and anthropogenic factors according to such bases of accumulated knowledge as the following observations:

- Anthropogenic carbon emissions have been growing constantly from the beginning of industrial development, reaching levels of 5.2 Gt C in 1980 and 6.3 Gt C in 2002 (IPCC, 2007).
- The content of the main GHGs ($CO_2$, $CH_4$, and $N_2O$) in the atmosphere has increased from the year 1750 by 31%, 150%, and 16%, respectively. About 50% of the $CO_2$ emitted to the atmosphere due to the burning of fuel is assimilated by vegetation and the oceans.
- The observed distribution of atmospheric $CO_2$ and the oxygen/nitrogen relationship show that a land sink of carbon prevails in northern and middle latitudes over the oceanic sink. In tropical latitudes, emissions of $CO_2$ to the atmosphere are substantial as a result of Earth's resources being used.
- Inter-annual oscillations of $CO_2$ concentration in the atmosphere follow changes in the use of fossil fuels. The intra-annual variability of atmospheric $CO_2$ concentration correlates more closely with the dynamics of land ecosystems than with those of ocean ecosystems.
- The regional flow of carbon in 2000, due to production and commercial trade of crops, timber, and paper, constituted 0.72 Gt $C \cdot yr^{-1}$. The global pure carbon flux at the atmosphere–ocean boundary observed in 1995 was estimated at 2.2 Gt C (−19% to +22%), with an intra-annual variation of about 0.5 Gt C. Maximum amplitudes of $CO_2$ flux oscillations in the atmosphere–ocean system are observed in the equatorial zone of the Pacific Ocean.
- An approximate picture of $CO_2$ distribution in ocean sources and atmospheric sinks is known: tropical basins of the oceans are sources of $CO_2$ while high-latitude water basins are $CO_2$ sinks. The role played by rivers is reduced mainly to the transport of carbon to coastal zones of the World Ocean (~1 Gt C $yr^{-1}$).

The most important task of the GCP is global environmental monitoring and the accumulation of detailed information on the productivity of land ecosystems, $CO_2$ fluxes at the atmosphere–ocean boundary, and volumes of anthropogenic emissions. Spaceborne sounding of $CO_2$ using AIRS (Atmospheric Infrared Sounder) on board the spaceborne laboratory EOS-Aqua launched by NASA on March 4, 2002 to an altitude of 705 km and IASI (Infrared Atmospheric Sounder Interferometer) carried on board METOP (Meteorological Operational Polar satellite) plays a special role (Nishida *et al.*, 2003). Other space vehicles, either presently functioning or planned for launch, will be used to evaluate $CO_2$ fluxes from data of indirect measurements of environmental characteristics. In particular, these are the aims of TIROS-N (Television Infrared Observational Satellite-N) and SCIAMACHY (Scanning Imaging Absorption Spectrometer for Atmospheric Cartography). The latter spectrometer launched in 2002 provides high spectral resolution within the absorption bands of GHGs such as $CO_2$, $CH_4$, $H_2O$

(accuracy 1%) and $N_2O$, CO (accuracy 10%) with a surface resolution ranging from 30 to 240 km depending on latitude.

Traditional ground measurements will be continued with the particular goal of substantiating national strategies for compatible use of Earth's resources, including the development of forestry, agriculture, stockbreeding, and cultivation of field crops.

The GCP program foresees extensive study of the physical, biological, biogeochemical, and ecophysiological mechanisms involved in the formation of environmental carbon fluxes. A deeper understanding of these mechanisms and their parameterizations will make it possible to specify GCC models and related climate changes. Broadening the respective base of knowledge will make it possible to specify the following information about these mechanisms:

- Atmosphere–ocean carbon exchange is controlled mainly by physical processes, including mixing between surface and deep layers of the ocean through the thermocline. Biological processes promote the transport of carbon from the ocean surface to deeper layers and farther down to bottom deposits. A biological "pump" functions as a result of phytoplankton photosynthesis.
- A complex of feedbacks controls the interactive exchange of energy, water, and carbon between the atmosphere and land surface, causing a response of these fluxes to such disturbances as transformation of land covers or oil pollution of the World Ocean. Plant communities respond physiologically to changes in temperature and humidity of the atmosphere and soil.
- The carbon sink in the northern hemisphere depends on forest growth, climate change, soil erosion, use of fertilizers, and the accumulation of carbon in freshwater systems. Unfortunately, the processes taking place in the southern hemisphere have been poorly studied, and factual information is practically absent. The significance of the terrestrial carbon sink can increase according to certain dimensions of climate change. When the atmospheric $CO_2$ concentration exceeds a level of 550 ppm, many processes in land ecosystems run short of nutrients and water, and therefore the photosynthetic accumulation of carbon by terrestrial vegetation becomes physiologically saturated.
- The extent of key factors determining the directions and amplitudes of $CO_2$ fluxes between the atmosphere and land ecosystems is limited by several factors:
  - extreme climatic phenomena such as droughts, serious drifts of seasonal temperatures, solar radiation change due to large-scale input of aerosol to the atmosphere (e.g., volcanic eruptions or large-scale fires like those that took place in Iraq in connection with recent military operations);
  - forest wildfires and other fires that introduce large-scale and long-term changes in carbon cycle characteristics (about 5–10% of pure primary production, estimated at $57\,\text{Gt C} \cdot \text{yr}^{-1}$ globally, is emitted to the atmosphere by the burning of wood);
  - land use leading to a change in the boundaries of biomes and in their types (from evergreen forests to coniferous stands, from forests to pastures, from meadows to inhabited territories);

  - reduction of biodiversity and change in the structures of communities, which alters how they impact nutrient, carbon, and water cycles.
- Phenomena of El Niño type or thermohaline circulation in the North Atlantic lead to global instability in the processes involved in energy–matter exchange, which should be reflected in the parameterization of non-linear feedbacks.

The future dynamics of carbon exchange in the NSS will be determined by the strategy adopted to manage the interaction between natural and anthropogenic factors. While this is patently obvious, there remain doubts since the problem of the greenhouse effect within the GCP has no simple solution. The GCP as a program is isolated from other investigative efforts concerning global ecodynamics. Although a broader approach to this problem has been introduced (Kondratyev *et al.*, 2002, 2003b, c, 2004a), these studies were unfortunately ignored by the GCP authors.

In a number of recent studies (Bartsev *et al.*, 2003; Kondratyev *et al.*, 2002, 2003a, 2004b), it has been proposed that the GCP be considered in the context of how it interacts with other processes in the NSS. As shown in Figure 3.1, the carbon cycle correlates with a multitude of natural and anthropogenic factors whose interaction forms key processes in the NSS. For $CO_2$, such processes are exchanges at the boundaries of the atmosphere with land surfaces and sea and ocean basins. It is clear that $CO_2$ dynamics in the biosphere can be analyzed with available data on the spatial distribution of sinks and sources. The present level of knowledge makes it possible to specify and solve the problem of the impact the greenhouse effect has on climate and, thereby, decrease uncertainty in estimates of future climate change. However, the applied GCC model should reflect not only the spatial mosaic of its reservoirs, sinks, and sources, but should also provide a dynamic calculation of the respective influences. Earlier calculations using GCC models have not taken information on the status and classification of land cover into account adequately and have considered even less World Ocean basin variability. Therefore, Figure 3.7 and Table 3.4 are aimed at compensating for the above-indicated shortcomings of other models. The system of balance equations for such a scheme is written as

$$\frac{\partial \alpha_S^i(\varphi,\lambda,z,t)}{\partial t} + V_\varphi \frac{\partial \alpha_S^i(\varphi,\lambda,z,t)}{\partial \varphi} + V_\lambda \frac{\partial \alpha_S^i(\varphi,\lambda,z,t)}{\partial \lambda} + V_z \frac{\partial \alpha_S^i(\varphi,\lambda,z,t)}{\partial z} = \sum_{j \in \Omega_S} H_{jS} - \sum_{m \in \Omega_S} H_{Sm} \quad (i = 1,\ldots,N), \qquad (3.1)$$

where $S$ is the carbon reservoir in the $i$th cell (pixel) of spatial digitization; $\varphi$ is latitude; $\lambda$ is longitude; $z$ is depth; $t$ is time; $H_{jS}$ is the carbon sink from the $j$th reservoir to reservoir $S$; $H_{Sm}$ is the carbon sink from reservoir $S$ to the $m$th reservoir; $\Omega_S$ is the multitude of carbon reservoirs bordering the reservoir $S$; $N$ is the number of carbon reservoirs; and $V(V_\varphi, V_\lambda, V_z)$ is the rate of exchange between reservoirs.

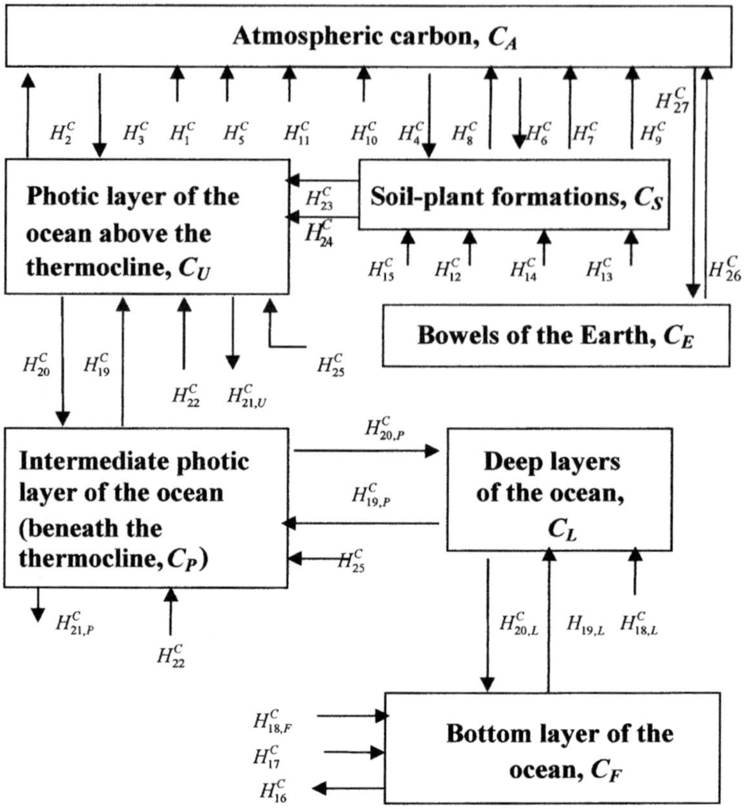


**Figure 3.7.** Block diagram of the global biogeochemical cycle of carbon dioxide (MGBC unit of the GMNSS) in the atmosphere–land–ocean system. $CO_2$ reservoirs and fluxes are described in Table 3.4.

In equation (3.1) rate $V$ and fluxes $H$ are non-linear functions of environmental characteristics. These functions were described in detail by Krapivin and Kondratyev (2002) and so are only specified here. Elements of the biogeocenotic unit of the global model shown in Figures 3.1 and 3.7 should first be designated. This can be visualized by covering the whole land surface $\Sigma$ with a homogeneous grid of geographic pixels $\Sigma_{ij} = \{(\varphi, \lambda) : \varphi_{i-1} \leq \varphi < \varphi_i; \lambda_{j-1} \leq \lambda < \lambda_j\}$ with boundaries in latitude $(\varphi_{i-1}, \varphi_i)$, longitude $(\lambda_{j-1}, \lambda_j)$, and area $\sigma$. The number of pixels is determined by the available database (i.e., by the choice of grid sizes $(\Delta\varphi, \Delta\lambda)$: $i = 1, \ldots, n$; $n = [180/\Delta\varphi]$; $j = 1, \ldots, k$; $k = [180/\Delta\lambda]$). Each pixel can contain $N$ types of surfaces, including the types of soil–plant formations, water basins, and other objects. The dynamics of the vegetation cover of the $s$th type follows the law:

$$\frac{dB_s}{dt} = R_s - M_s - T_s, \tag{3.2}$$

**Table 3.4.** Reservoirs and fluxes of $CO_2$ in the biosphere considered in the simulation model of the global biogeochemical cycle of $CO_2$ shown in Figure 3.7.

| *Reservoirs and fluxes of carbon dioxide* | *Identifier* | *Estimate of reservoir* ($10^9$ t) *and flux* ($10^9$ t/yr) |
|---|---|---|
| Carbon | | |
| Atmosphere | $C_A$ | 650–750 |
| Photic layer of the ocean | $C_U$ | 580–1,020 |
| Deep layers of the ocean | $C_L$ | 34,500–37,890 |
| Soil humus | $C_S$ | 1,500–3,000 |
| Emission in burning | | |
| Vegetation | $H_8^C$ | 6.9 |
| Fossil fuel | $H_1^C$ | 3.6 |
| Desorption | $H_2^C$ | 97.08 |
| Sorption | $H_3^C$ | 100 |
| Rock weathering | $H_4^C$ | 0.04 |
| Volcanic emanations | $H_5^C$ | 2.7 |
| Assimilation by land vegetation | $H_6^C$ | 224.4 |
| Respiration | | |
| Plants | $H_7^C$ | 50–59.3 |
| Humans | $H_{10}^C$ | 0.7 |
| Animals | $H_{11}^C$ | 4.1 |
| Emission | | |
| Soil humus decomposition | $H_9^C$ | 139.5 |
| Plant roots | $H_{15}^C$ | 56.1 |
| Vital activity | | |
| Population | $H_{12}^C$ | 0.3 |
| Animals | $H_{13}^C$ | 3.1 |
| Plants dying off | $H_{14}^C$ | 31.5–50 |
| Bottom deposits | $H_{16}^C$ | 0.1–0.2 |
| Solution of marine deposits | $H_{17}^C$ | 0.1 |
| Detritus decomposition | | |
| Photic layer | $H_{22}^C$ | 35 |
| Deep layers of the ocean | $H_{18}^C$ | 5 |
| Rising with deep waters | $H_{19}^C$ | 4 |
| Lowering with surface waters and due to gravitational sedimentation | $H_{20}^C$ | 40 |
| Photosynthesis | $H_{21}^C$ | 69 |
| Groundwater runoff | $H_{23}^C$ | 0.5 |
| Surface runoff | $H_{24}^C$ | 0.5–0.6 |
| Respiration of living organisms in the ocean | $H_{25}^C$ | 25 |
| Degassing processes | $H_{26}^C$ | 21.16 |
| Sinking to the Earth's bowels | $H_{27}^C$ | 1.3 |

where $R_s$ is photosynthesis; and $M_s$ and $T_s$ are losses of biomass $B_s$ due to dying off and evapotranspiration, respectively.

Components shown on the right-hand side of equation (3.2) are functions of environmental characteristics: illumination, temperature, air and soil humidity, and atmospheric $CO_2$ concentration. There are several methods and forms of parameterizing these functions. An example is the model of Collatz *et al.* (2000), which provided the basis for developing the SiB2 global biospheric model (Sellers *et al.*, 1996). Temperature, humidity, and rate of moisture evaporation in the vegetation cover and soil depend on biospheric parameters and energy fluxes in the atmosphere–plant–soil system. By analogy with electrostatics, the notion of resistance is introduced, and fluxes are calculated from a simple formula: flux = potential × difference/resistance. The SiB2 model takes into account the fluxes of sensible and latent heat through evaporation of water vapor in plants and soil, and $CO_2$ fluxes are divided into classes $C_3$ and $C_4$, which substantially raises the accuracy of parameterizing functions on the right-hand side of equation (3.2). According to Collatz *et al.* (2000), three factors regulate function $R_s$: the efficiency of the photosynthetic enzymatic system, amount of photosynthetically active radiation (PAR) absorbed by cellulose chlorophyll, and the ability of plant species to assimilate and transmit the products of photosynthesis to the outside medium. Application of the Libich principle (Kondratyev *et al.*, 2002; Nitu *et al.*, 2000a) and consideration of the data on distribution of the types of vegetation cover by pixels $\{\Sigma_{ij}\}$, partial pressures of $CO_2$ and $O_2$, temperature and density of the atmosphere, and the level of illumination makes it possible to calculate fluxes $H$ in equation (3.1) for all land pixels.

A model of the carbon cycle in the atmosphere–ocean system was described in detail by Tarko (2005). It is based on the same grid of geographic pixels but is combined with a zonal principle classified by Tarko (2001, 2005). Ocean thickness is considered a single biogeocenosis in which the main binding factor is the flux of organic matter produced in surface layers which then penetrates down to the deepest layers of the ocean. In this medium the carbonate system, a parametric description of which was given in Kondratyev *et al.* (2004a), is a regulator of carbon fluxes.

An important element of $CO_2$ atmosphere–ocean exchange is the role played by hurricanes, which has not been studied in detail. Perrie *et al.* (2004) conducted a study of hurricane influence on local rates of air–sea $CO_2$ exchange. Hurricanes are shown to affect the thermal and physical structure of the upper ocean. Air–sea gas transfer includes processes such as upper-ocean temperature changes and the upwelling of carbon-rich deep water. Observations show that sea surface temperature and $CO_2$ partial pressure can decrease by 4°C and 20 μatm, respectively, as a result of hurricane activity. Perrie *et al.* (2004) proposed a model to parameterize $CO_2$ flux $H_3$ with the following formula:

$$H_3 = k_L \alpha \Delta[CO_2],$$

where $\alpha$ is the solubility of $CO_2$; and $\Delta[CO_2]$ is the difference between its partial pressure in the atmosphere and the upper layer of the sea. Parameter $k_L$ ($cm \cdot h^{-1}$)

is determined by one of the following correlations that depend on wind speed:

$$k_L = \begin{cases} 0.31U_{10}^2(S_c/660)^{-0.5} & \text{for category 1-3 hurricanes;} \\ 0.0283U_{10}^3(S_c/660)^{-0.5} & \text{for category 4-5 hurricanes,} \end{cases}$$

where $S_c$ is the Schmidt number (Hasegawa and Kasagi, 2001, 2005); and $U_{10}$ is the wind speed at an altitude of 10 m ($\text{m} \cdot \text{s}^{-1}$).

Introducing wave spectrum peak frequency $\omega_p$, air–side friction velocity $u_*$, and kinematic viscosity $\nu$, the parameter $k_L$ can be calculated by the formula:

$$k_L = 0.13\left[\frac{u_*^2}{\nu\omega_p}\right]^{0.63}.$$

Parameter $k_L$ is actually formed from two components: $k_L = k_{L_1} + k_{L_2}$, where $k_{L_1}$ and $k_{L_2}$ are the wave-breaking and interfacial terms, respectively. The terms $k_{L_1}$ and $k_{L_2}$ are calculated using the following formulas:

$$k_{L_1} = u_*^{-1}\lfloor\sqrt{\rho_w/\rho_a}(h_w S_{c_w}^{0.5} + \kappa^{-1}\ln\{z_w/\delta_w\}) + \alpha(h_a S_{c_a}^{0.5} + c_d^{-0.5} - 5 + 0.5\kappa^{-1}\ln S_{ca})\rfloor;$$

$$k_{L_2} = fV\alpha^{-1}[1 + (e\alpha S_c^{-0.15})^{-1/n}]^{-n},$$

where $f = 3.8 \cdot 10^{-6}U_{10}^3$; $\alpha$ is gas solubility; subscript $a$ ($w$) denotes air– (water–) side; $\rho$ is density; $z$ is measurement depth; $\delta$ is turbulent surface layer thickness; $\kappa$ is the von Kármán constant; $c_d$ is the drag coefficient; $h \equiv \Lambda\varphi^{-1}R_r^{0.25}$; $\Lambda$ is an adjustable constant; $R_r$ is the Reynolds number of roughness; $\varphi$ is an empirical function that accounts for buoyancy effects on turbulent transfer in the ocean; and $V$, $e$, and $n$ are empirical constants equal to 14, 1.2, and 4,900 $\text{cm} \cdot \text{h}^{-1}$ in the GasEx-1998 field experiment (Perrie *et al.*, 2004) and may need readjustment for other datasets.

What is significant here is the fact that hurricane activity initiates an upwelling zone where air–water gas exchange takes on a different character. Hales *et al.* (2005) studied atmospheric $CO_2$ uptake in a coastal upwelling system located off the Pacific coast of Oregon using high-resolution measurements of the partial pressure of $CO_2$ and nutrient concentrations in May through August 2001. Results showed that the dominance of low-$CO_2$ waters over the shelf area renders the region a net sink during the upwelling season as a result of:

- the presence of upwelled water rich in preformed nutrients;
- complete photosynthetic uptake of these excess nutrients and a stoichiometric proportion of $CO_2$;
- moderate warming of upwelled waters.

It is estimated that:

- the eastern boundary area of the North Pacific can constitute a sink of atmospheric $CO_2$ equivalent to 5% of the annual North Pacific $CO_2$ uptake;

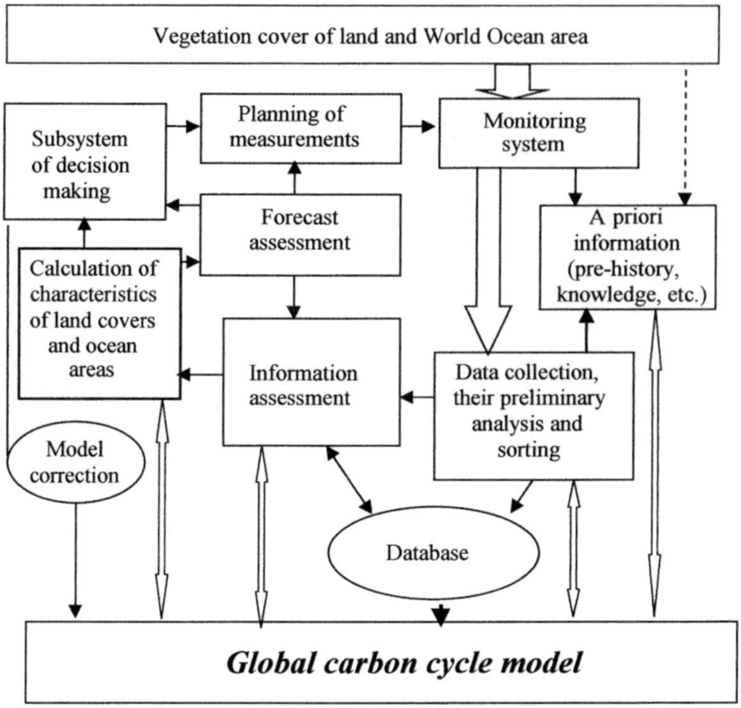


**Figure 3.8.** An adaptive regime of greenhouse effect monitoring to assess the role played by vegetation cover of land and ocean areas.

- by mid-August, the partial pressure of $CO_2$ in subsurface waters increases 20–60%, corresponding to an increase of 1.0–2.3% total dissolved $CO_2$ as a result of settling biogenic debris respiring.

Many parameters of the global carbon cycle model are measured in the satellite-monitoring regime, which makes it possible to apply an adaptive scheme to calculate greenhouse effect characteristics (Figure 3.8). This scheme makes it possible to add information to a model of the continuous regime by correcting its structure and parameters. Satellite measurements in the visible and near-IR provide operational estimates of PAR and vegetation characteristics such as canopy greenness, area of living photosynthetically active elements, soil humidity and water content in elements of the vegetation cover, $CO_2$ concentration on the surface of leaves, etc. The regime of prediction of vegetation cover biomass in each pixel $\Sigma_{ij}$ and comparisons with satellite measurements enables some fragments of the model to be corrected, for instance, by doubling its units or their parametric adjustment to minimize discrepancies between prediction and measurements (Figure 3.8). In particular, to calculate primary production, there are some semi-empirical models that can be used by a sample criterion in different pixels.

**Table 3.5.** Model estimates of excessive $CO_2$ assimilation over Russia. A more detailed classification of soil–plant formations is given in Table 1.2.

| *Soil–plant formation* | *Flux of assimilated carbon as $CO_2$* ($10^6$ t C/yr) |
|---|---|
| Arctic deserts and tundras, sub-Arctic meadows and marshes | 2.2 |
| Tundras | 3.3 |
| Mountain tundras | 3.6 |
| Forest tundra | 2.8 |
| North taiga forests | 10.8 |
| Mid-taiga forests | 31.2 |
| South taiga forests | 22.9 |
| Broad-leaved and coniferous forests | 4.8 |
| Steppes | 3.6 |
| Alpine and subalpine meadows | 1.1 |
| Deserts | 2.2 |

There is a certain freedom of choice when estimating evaporation from vegetation cover (Wange and Archer, 2003).

The key component of the global $CO_2$ cycle is anthropogenic emissions to the environment. What most researchers are interested in here is an assessment of the ability of the biosphere to neutralize an excess amount of $CO_2$. Table 3.5 and Figure 3.9 illustrate the modeling results. It can be seen that 41.3% of the 6.3 Gt C emitted to the atmosphere by industry remains in the atmosphere, while the oceans and land vegetation absorb 20.2 and 38.5%, respectively. Taking the dependence of air temperature changes on $CO_2$ variation as a basis (Mintzer, 1987):

$$\Delta T_{CO_2} = -0.677 + 3.019 \ln[C_a(t)/338.5],$$

for the realistic scenario, we obtain $\Delta T_{CO_2} \leq 2.4°C$. This correlates well with the estimate $\Delta T_{CO_2} \leq 4.2°C$ published by many authors and assumed in the Kyoto Protocol.

As Figure 3.9 shows, the discrepancy between the forms of $CO_2$ distribution in the absorption curve and the vegetation index suggests that the structure of model pixels and their correspondence with observational data in the southern hemisphere should be specified. Nevertheless, introduction to the GCC model of a pixel mosaic has made it possible to evaluate the role played by some types of ecosystems and regions in Russia in regulating the greenhouse effect. Table 3.5

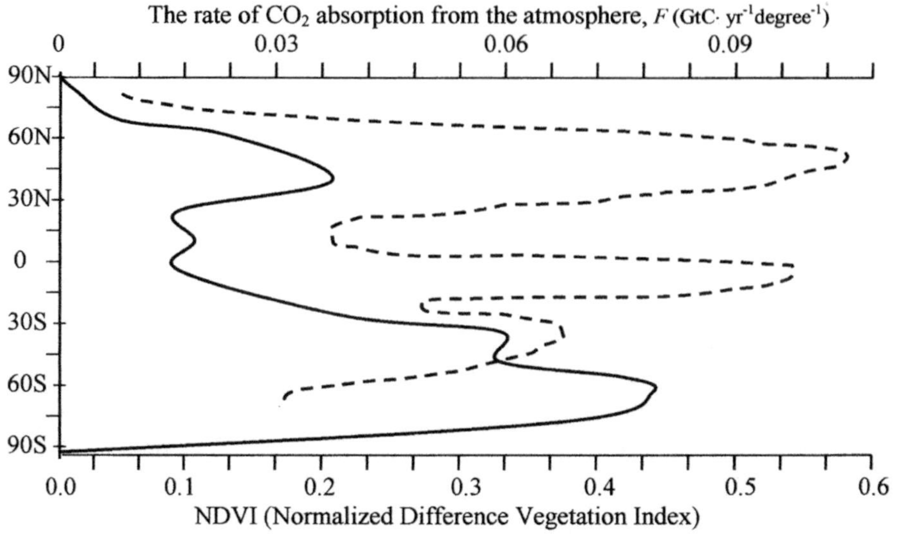


**Figure 3.9.** Latitudinal distribution of the rate $F$ (Gt C · yr$^{-1}$ deg$^{-1}$) of carbon absorption (solid curve) from the atmosphere and vegetation index (dashed curve). Types and spatial distribution of soil–plant formations are determined in Table 1.2 and Figure 1.19. Industrial emissions of $CO_2$ are assumed to be 6.3 Gt C · yr$^{-1}$.

demonstrates the regulatory role played by taiga in Russia. On the whole, the model enables consideration of various scenarios of land cover changes and study of the dependence of $CO_2$ partial pressure in the atmosphere on their structure. For instance, if by 2050 forest areas reduce by 10% with respect to 1970 ($\sim$42,000,000 km$^2$), then by the end of the 21st century the content of atmospheric $CO_2$ could increase by 46.7% with stable anthropogenic emissions of carbon of about 6 Gt C · yr$^{-1}$. In contrast, if forest areas increase in the northern hemisphere by 10% then the anthropogenic impact on the greenhouse effect could reduce by 14.8%.

Global warming as a result of the growth of GHGs is the most important problem facing civilization and sustainable development. The approach proposed here advocates synthesizing accumulated data and knowledge of the GCC and other GHGs into a single monitoring system. Unfortunately, the international program set up to study the GCC (Canadel *et al.*, 2003), like other similar global programs, was not aimed at developing a constructive information technology to substantially raise the reliability of prognostic estimates of future climate change. Nevertheless, the ideas and approaches of Russian researchers (Bartsev *et al.*, 2003; Kondratyev *et al.*, 2002, 2003a, 2004a; Krapivin and Kondratyev, 2002; Nefedova, 1994; Nefedova and Tarko, 1993; Tarko, 2005), as well as models developed by American researchers (Collatz *et al.*, 2000; Sellers *et al.*, 1996) should make it possible, though not within the GCP, to overcome the existing isolation of GCC studies and create a global model operational in the regime of satellite monitoring capable of giving reliable estimates of the roles played by individual regions in the greenhouse effect. Such a model could be used to work out an efficient strategy for land use and to make substantiated (in contrast to the Kyoto Protocol) international decisions.

To get at the key factors of global change by modeling global ecodynamics involves overcoming some difficulties as to choice of form and methods of modeling. The most important of these were discussed in Bartsev *et al.* (2003), Kondratyev *et al.* (2003c), among others. The main difficulty lies in combining the parameters used in models with available data and knowledge bases, as well as in reaching a compromise between the complicated structures of these models and their semi-empirical realizations. The choice of technology used to model and interpret results is arbitrary. For example, a model can be selected on the basis of important parameters of the NSS that affect global ecodynamics (Kondratyev *et al.*, 2004a). However, according to Barenbaum (2002, 2004) and Yasanov (2003), the description of the GCC ignores carbon buried in geological structures and its intake from space. Therefore, when synthesizing the global model of the carbon cycle, it is necessary to consider a more detailed description of the biospheric and lithospheric parts of this cycle. Clearly, the lifetime of carbon in each subcycle needs to be estimated more accurately. The lithospheric part of the GCC includes its transformation in the process of long interactions and conversions, including the transformation to methane, oil, coal-like deposits, etc. Depending on temperature and pressure, hydrocarbons can be oxidized and become the main component of underground fluids and magmas. In this way the carbon cycle correlates well with the cycles of methane and water (Trudinger, 1980).

### 3.4.4 Global model units for other biogeochemical cycles

#### *3.4.4.1 Sulfur unit*

In line with the designations in Figure 2.7 and Table 2.3, the equations of the sulfur unit of the GMNSS are written in the form of balance correlations:

$$\frac{dAH2SL}{dt} = C_1 + C_2 + C_3 + C_{21} - C_4$$

$$\frac{dASO2L}{dt} = C_4 + C_4 + C_6 - C_7 - C_8 - C_9$$

$$\frac{dASO4L}{dt} = C_9 + C_{13} + C_{20} - C_{11} - C_{12}$$

$$\frac{dS}{dt} = C_{17} - C_{16} - C_{19}$$

$$\frac{dSO4L}{dt} = C_{10} + C_{11} + C_{12} + C_{16} - C_3 - C_{13} - C_{14}$$

$$\frac{dFIX}{dt} = C_7 + C_{15} + C_{22} - C_{17}$$

$$\frac{dH2SO4L}{dt} = C_8 - C_{18} - C_{21} - C_{22}$$

$$\frac{dAH2SO}{dt} = H_1 + H_3 + H_4 + H_{26} - H_2$$

$$\frac{dASO2O}{dt} = H_2 + H_5 + H_6 - H_7 - H_8 - H_{24}$$

$$\frac{dASO4O}{dt} = H_8 + H_9 + H_{12} - H_{10} - H_{11}$$

$$\frac{\partial SO4OU}{\partial t} + v_z \frac{\partial SO4OU}{\partial z} + k_z \frac{\partial^2 SO4OU}{\partial z^2} = H_7 + H_{10} + H_{11} + H_{20} + H_{22} + H_{27} + C_{14} - H_{12} - H_{13}$$

$$\frac{\partial H2SOU}{\partial t} + v_z \frac{\partial H2SOU}{\partial z} + k_z \frac{\partial^2 H2SOU}{\partial z^2} = H_{21} + H_{23} - H_4 - H_{22}$$

$$\frac{\partial H2SOD}{\partial t} + v_z \frac{\partial H2SOD}{\partial z} + k_z \frac{\partial^2 H2SOD}{\partial z^2} = H_{17} - H_{18} - H_{21}$$

$$\frac{\partial SO4OD}{\partial t} + v_z \frac{\partial SO4OD}{\partial z} + k_z \frac{\partial^2 SO4OD}{\partial z^2} = H_{18} - H_{19} - H_{20}$$

$$\frac{\partial DU}{\partial t} + v_z \frac{\partial DU}{\partial z} + k_z \frac{\partial^2 DU}{\partial z^2} = H_{14} - H_{15} - H_{23}$$

$$\frac{\partial DD}{\partial t} + v_z \frac{\partial DD}{\partial z} + k_z \frac{\partial^2 DD}{\partial z^2} = H_{15} - H_{16} - H_{17}$$

$$\frac{\partial FI}{\partial t} + v_z \frac{\partial FI}{\partial z} + k_z \frac{\partial^2 FI}{\partial z^2} = H_{13} - H_{14}$$

$$\frac{dBOT}{dt} = H_{16} + H_{19},$$

where $v_z$ is advection velocity (m day$^{-1}$); and $k_z$ is the coefficient of turbulent mixing (m$^2$ day$^{-1}$). Reservoir designations are given in Table 3.6. Functional representations of sulfur flows are given by Krapivin and Kondratyev (2002) and Marchuk and Aloyan (2009).

The discharge speed of $H_2S$ to the atmosphere due to humus decomposition is described by a linear function $C_3 = \mu_1(pH)SO4L \cdot T_L$ where $\mu_1$ is the proportionality coefficient depending on the soil acidity pH (day$^{-1}$ K$^{-1}$) and $T_L$ is soil temperature (K).

It is supposed that flow $H_4$ is a function of the rates of alignment for $H_2S$ oxidation in the photic layer with the vertical velocity of water rising. Therefore, for the description of flow $H_4$ the parameter $t_{H2SU}$ which reflects the lifetime of $H_2S$ in the water is used: $H_4 = H2SU/t_{H2SU}$, where $t_{H2SU}$ is a function of the velocity of vertical advection $u_z$ and of the oxygen concentration $O2$ in the upper

**Table 3.6.** Some estimates of sulfur reservoirs that can be used as initial data.

| *Reservoir* | *Identifier of the GSM* | *Quantitative estimate of the sulfur reservoir* (mg/m$^2$) |
|---|---|---|
| The atmosphere over the ocean | | |
| $H_2S$ | *AH2SO* | 10 |
| $SO_2$ | *ASO2O* | 5.3 |
| $SO_4^{2-}$ | *ASO4O* | 2 |
| The atmosphere over land | | |
| $H_2S$ | *AH2SL* | 36.9 |
| $SO_2$ | *ASO3L* | 17.9 |
| $SO_4^{2-}$ | *ASO4L* | 12.9 |
| Land | | |
| $SO_4^{2-}$ | *SO4L* | 11.2 |
| Biomass | *FIX* | 600 |
| Soil | *S* | 5,000 |
| Photic layer of the World Ocean | | |
| $H_2S$ | *H2SOU* | 1.9 |
| $SO_4^{2-}$ | *SO4OU* | $19 \times 10^7$ |
| Biomass | *FI* | 66.5 |
| MOB | *DU* | 730 |
| Deep layers of the World Ocean | | |
| $H_2S$ | *H2SOD* | $2 \cdot 10^6$ |
| $SO_4^{2-}$ | *SO4OD* | $3.4 \times 10^9$ |
| MOB | *DD* | 13,120 |

layer having the thickness $Z_{H2S}$:

$$t_{H2SU} = H2SOU \cdot O2^{-1} u_z(\theta_2 + O2)(\theta_1 + u_z)^{-1}.$$

The constants $\theta_1$ and $\theta_2$ are defined empirically, while the value of $O2$ is estimated by the oxygen unit of the GSM. Flows $H_2$ and $C_4$ reflect the correlation between the sulfur and oxygen cycles: $C_4 = AH2SL/t_{H2SA}$, $H_2 = AH2SO/t_{H2SA}$, where $t_{H2SA}$ is the lifetime of $H_2S$ in the atmosphere.

The mechanism of $SO_2$ removal from the atmosphere is described by flows $H_7$, $H_8$, $H_{27}$, $C_7$, and $C_9$. These flows are characterized by typical parameters $t_{SO2L}$ and $t_{SO2A1}$, which are the lifetimes of $SO_2$ over land and the water surface, respectively. $SO_2$ is absorbed from the atmosphere by minerals, vegetation, and soil. Dry absorption of $SO_2$ by vegetation from the atmosphere is described by the model $C_7 = q_2 RX$, where $q_2 = q_2' \cdot ASO2L/(r_{tl} + r_s)$, $r_{tl}$ is the atmospheric resistance to $SO_2$ transport over vegetation of $l$th type (day $\cdot$ m$^{-1}$), $r_s$ is surface resistance to $SO_2$ transport over a surface of $s$th type (day $\cdot$ m$^{-1}$), $RX$ is the production of $X$-type vegetation (mg · m$^{-2}$ day$^{-1}$), and $q'_2$ is the proportionality coefficient. Production $RX$ is calculated by the biogeocenotic unit of the GSM.

The process of $SO_2$ washing from the atmosphere is described by the model: $C_8 = q_{1l}$, where $q_{1l}$ is the characteristic parameter for a surface of $l$th type and $W(t, \varphi, \lambda)$ is precipitation intensity. The interaction of acid rain with the land surface was reflected in Figure 2.7 as flows $C_{18}$, $C_{21}$, $C_{22}$, $H_{25}$, $H_{26}$, and $H_{27}$. These flows are parameterized by models: $C_{18} = h_1 \cdot H2SO4L$, $C_{22} = h_2 \cdot RX \cdot H2SO4L$, $C_{21} = h_3 T_a \cdot H2SO4L$, $H_{25} = h_6 \cdot H2SO4O$, $H_{26} = h_4 T_a \cdot H2SO4O$, $H_{27} = h_5 \cdot RFI \cdot H2SO4O$, where $T_a(t, \varphi, \lambda)$ is atmospheric temperature, $h_1 + h_2 \cdot RX + h_3 T_a = 1$, $h_4 T_a + h_5 \cdot RFI + h_6 = 1$, and $RFI$ is the production of phytoplankton.

Similarly, flows $H_8$, $C_9$, $H_7$, and $H_{24}$ are simulated by models: $H_8 = ASO2O/t_{SO2A1}$, $C_9 = ASO2L/t_{SO2L}$, $H_7 = ASO2O/t_{SO2A2}$, and $H_{24} = q_{1l} W \cdot ASO2O$.

The physical mechanisms behind sulfate transportation in the environment are described by the models of Bodenbender *et al.* (1999), Luecken *et al.* (1991), Park *et al.* (1999): $H_{10} = \mu W \cdot ASO4O$, $H_{11} = \rho v_0 \cdot ASO4O$, $C_{11} = b_3 W \cdot ASO4L$, $C_{12} = d_1 v_a \cdot ASO4L$, where $v_0$ and $v_a$ are the rates of dry sedimentation of aerosols over the water surface and land, respectively.

For flows $C_{13}$, $H_{12}$, $C_{14}$, $C_{16}$ we consider the following models: $C_{13} = d_2 RATE \cdot SO4L$, $H_{12} = \theta RATE \cdot SO4L$, $C_{16} = b_2 ST_L$, $C_{14} = d_3 W \cdot SO4L + (C_{11} + C_{12})\sigma$, where $RATE(t, \varphi, \lambda)$ is wind velocity over the surface (m · s$^{-1}$) and coefficient $b_2$ reflects the sulfur content in dead plants.

The terrestrial part of the sulfur cycle correlates with the water part through flows in the atmosphere–hydrosphere–land system. We have: $H_{13} = \gamma RFI$, $H_{14} = bMFI$, $H_{15} = fDU$, $H_{16} = pDD$, $H_{17} = qDD$, $H_{18} = H2SOD/t_{H2SOD}$, $H_{19} = uSO4D$, $H_{20} = a_1 v_D SO4D$, $H_{21} = b_1 v_D H2SOD$, $H_2 = H2SOU/t_{H2SOU}$, and $H_{23} = gDU$, where $MFI$ is the mass of dead phytoplankton, $t_{H2SOU}$ and $t_{H2SOD}$ are the characteristic times for $H_2S$ total oxidation in the photic layer and deep waters, respectively.

### *3.4.4.2 Nitrogen unit*

The MGBN unit (Table 3.2), which simulates the fluxes of nitrogen in the environment, is necessary in the global biospheric model for several indisputable reasons: nitrogen compounds can affect environmental conditions, food quality, the climate, and hydrospheric parameters. The abundant use of nitrates leads to water pollution and deteriorates the quality of food products. It is well known that intensive exploitation of soils without taking into account the consequences of the misuse of nitrogen fertilizers breaks the stability of agro-ecosystems and human health. Moreover, nitrous oxide ($N_2O$), nitrogen dioxide ($NO_2$), and nitric oxide (NO), being minor gas components of the atmosphere, substantially affect the formation of processes that absorb optical radiation in the atmosphere. Small deviations in their concentrations can cause significant climatic variations near the Earth surface.

The nitrogen cycle is closely connected with the fluxes of hydrogen, sulfur, and other chemicals. The global cycle of nitrogen, as one of the nutrient elements, is a mosaic structure of local processes of its compounds formed as a result of water migration and atmospheric processes. The present-day nitrogen cycle is especially vulnerable to anthropogenic impacts manifested through interference with the nitrogen cycle both directly and via the influence on related processes. Therefore, the construction of an adequate model of the nitrogen cycle in nature at present should be based on describing the whole complex of natural processes and those initiated by humans.

Natural sources of nitrogen oxides are associated with the vital functions of bacteria, volcanic eruptions, and several atmospheric phenomena (e.g., lightning discharges). The biogeochemical cycle of nitrogen includes such processes as fixation, mineralization, nitrification, assimilation, and dissimilation. The structural schemes of these processes have been described in detail by many authors (Moir, 2011). Their complexity level is determined by the goal of a given study, the availability of data on the rates at which nitrogen-containing compounds and their supplies are transformed, the level of detail, etc.

Nitrogen transport in the biosphere is driven by a complicated meandering structure of fluxes, including a hierarchy of cycles at various levels of life organization. From the atmosphere, nitrogen enters the cells of microorganisms, from where it enters the soil and then passes to higher plants, animals, and humans. Living organisms return nitrogen to the soil, where it is either taken up by plants and living organisms or emitted to the atmosphere. A very similar scheme of nitrogen oxide cycling prevails in the hydrosphere. The characteristic feature of these cycles is their accessibility to available processes of nitrogen removal from the biospheric balance and subsequent migration into rock formations, from where it returns to the biosphere at a much slower rate than its outbound flux. A global scheme of nitrogen fluxes can be formulated by taking into account the nature of the nitrogen cycle in the biosphere and its reservoir structure.

Comparative analysis of model schemes to diagram the flux of nitrogen compounds in nature as proposed by various experts makes it possible to construct a block scheme like that presented in Figure 3.10. In the diagram the atmosphere, soil, lithosphere, and hydrosphere are considered as nitrogen reservoirs. The first three reservoirs are described by 2-D models, and the hydrosphere is described by a 3-D multi-layer model. The characteristics of nitrogen fluxes between these reservoirs are given in Table 3.7. The equations of the model are written as

$$\frac{\partial N_A}{\partial t} + V_\varphi \frac{\partial N_A}{\partial \varphi} + V_\lambda \frac{\partial N_A}{\partial \lambda} = H_1^N + \begin{cases} H_{20}^N - H_{16}^N, & (\varphi, \lambda) \in \Omega_O \\ H_7^N + H_{19}^N - H_8^N - H_9^N + H_{22}^N - H_2^N - H_{10}^N, & (\varphi, \lambda) \in \Omega/\Omega_O \end{cases}$$

$$\frac{\partial N_{S_1}}{\partial t} = H_8^N + H_6^N - H_3^N$$

$$\frac{\partial N_{S_2}}{\partial t} = H_2^N + H_3^N + H_5^N + H_9^N - H_6^N - H_7^N - H_{11}^N - H_{21}^N$$

$$\frac{\partial N_U}{\partial t} + v_\varphi \frac{\partial N_U}{\partial \varphi} + v_\lambda \frac{\partial N_U}{\partial \lambda} = H_{16}^N + H_{4,U}^N + H_{18,U}^N + H_{11}^N - H_{17,U}^N - H_{20}^N - H_{14,UP}^N - H_{15,UP}^N$$

$$\frac{\partial N_P}{\partial t} + v_\varphi \frac{\partial N_P}{\partial \varphi} + v_\lambda \frac{\partial N_P}{\partial \lambda} = H_{18,P}^N + H_{4,P}^N + H_{14,UP}^N + H_{15,PL}^N - H_{17,P}^N - H_{14,PL}^N - H_{15,UP}^N$$

$$\frac{\partial N_L}{\partial t} = Q_L + H_{12,L}^N + H_{14,PL}^N + H_{15,LF}^N - H_{14,LF}^N - H_{15,PL}^N$$

$$\frac{\partial N_F}{\partial t} = Q_F + H_{12,F}^N + H_{23}^N + H_{14,LF}^N - H_{13}^N - H_{15,LF}^N$$

where $V(V_\varphi, V_\lambda)$ is wind speed; $v(v_\varphi, v_\lambda)$ is current velocity in the ocean; and $Q_L$ and $Q_F$ are functions describing the mixing of the deep waters of the ocean.

To simplify the calculation scheme shown in Figure 3.10, advection processes in the equations can be described by superposing fluxes $H_{14}^N$ and $H_{15}^N$. Computer realization of the equations of the nitrogen unit makes some corrections to them so that variable dimensions conform with the spatial digitization of $\Omega$. Therefore, the estimates of fluxes $H_i^N$ given in Table 3.7 should be corrected according to this criterion.

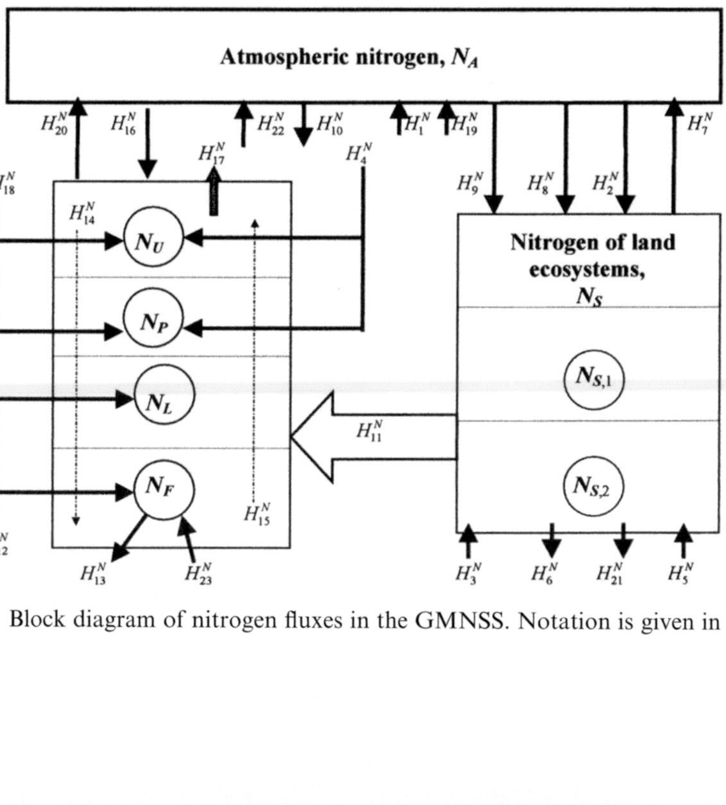


**Figure 3.10.** Block diagram of nitrogen fluxes in the GMNSS. Notation is given in Table 3.7.

**Table 3.7.** Characteristics of reservoirs and fluxes of nitrogen in the biosphere (Figure 3.11).

| *Reservoirs* (Gt) *and fluxes* ($10^6$ t/yr) | *Identifier* | *Estimate* |
|---|---|---|
| Nitrogen supplies | | |
| Atmosphere | $N_A$ | $39 \cdot 10^5$ |
| Soil | $N_S$ | 280 |
| Photic and intermediate layer of the ocean | $N_U + N_P$ | 2,800 |
| Deep and bottom layer of the ocean | $N_L + N_F$ | 36,400 |
| Natural sources of the hydrosphere | $H_1^N$ | 0.392 |
| Anthropogenic accumulation | | |
| Fuel burning | $H_2^N$ | 22.8 |
| Fertilizer production | $H_9^N$ | 41.8 |
| Input from dead organisms | | |
| On land | $H_3^N$ | 42.2 |
| In upper layers of the World Ocean | $H_{18}^N$ | 5 |
| In deep layers of the World Ocean | $H_{12}^N$ | 7.8 |
| Input from organism functioning | | |
| On land | $H_5^N$ | 0.1 |
| In the World Ocean | $H_4^N$ | 0.3 |
| Biological fixation | | |
| On land | $H_6^N$ | 20.3 |
| In the World Ocean | $H_{17}^N$ | 10 |
| In the atmosphere | $H_{10}^N$ | 40 |
| Denitrification | | |
| On land | $H_7^N$ | 52 |
| In the World Ocean | $H_{20}^N$ | 49.8 |
| Atmospheric fixation | | |
| Over land | $H_8^N$ | 4 |
| Over the World Ocean | $H_{16}^N$ | 3.6 |
| Runoff from land into the World Ocean | $H_{11}^N$ | 38.6 |
| Precipitation | $H_{13}^N$ | 0.5 |
| Vertical exchange processes in the oceans | | |
| Descending | $H_{14}^N$ | 0.2 |
| Lifting | $H_{15}^N$ | 7.5 |
| Anthropogenic emissions to the atmosphere | $H_{19}^N$ | 15 |
| Removal of nitrogen from the cycle by sedimentation | $H_{21}^N$ | 0.2 |
| Input of nitrogen to the atmosphere during rock weathering | $H_{22}^N$ | 0.217 |
| Input of nitrogen to the water medium with dissolving sediments | $H_{23}^N$ | 0.091 |

### *3.4.4.3 Phosphorus, oxygen, and ozone units*

Biogeochemical cycles are highly interactive with other environmental processes. The global cycles of phosphorus, oxygen, and ozone were described in detail by Kondratyev *et al.* (2004a). In contrast to nitrogen, the main reservoir of phosphorus in the biosphere is not the atmosphere but rocks and other deposits formed in past geological epochs, which, being subject to erosion, emit phosphates. There are other mechanisms by which phosphorus is returned to the biospheric cycle, but they are not that efficient as a rule. One of these mechanisms is trawler fishing, which returns about $60 \cdot 10^3$ t P/yr to land from the hydrosphere; another is extraction of phosphorus-containing rocks at an estimated rate of $1\text{–}2 \cdot 10^6$ t P/yr. The present cycle of phosphorus terminates when its fluxes reach the bottom deposits of the World Ocean, where it combines with sewage; it may also get involved with coastal and river runoff.

The oxygen cycle in nature is composed of characteristic biogeochemical transitions among the various reservoirs of basic constituents circulating in the biosphere. A block scheme of oxygen exchange would thus be similar to those of sulfur, nitrogen, carbon, and phosphorus. However, oxygen is the most widespread constituent on Earth, which makes it one of the most substantial components of biogeochemical cycles. The proportion of oxygen in the Earth's crust, including the hydrosphere, reaches 49% by mass. The lithosphere (without considering the ocean and atmosphere) contains 47.2% oxygen, and water is comprised of 88.89% oxygen. Oxygen constitutes 85.82% of ocean water, and marine biota account for 65% of oxygen by mass. These estimates testify to the overriding significance of oxygen in the biosphere, whose very appearance and existence are determined by oxygen. Presently, about $39 \cdot 10^{14}$ t $O_2$ circulate in the biosphere.

Oxygen is present in the biosphere as molecular oxygen ($O_2$), ozone ($O_3$), atomic oxygen (O), and as a constituent of various oxides. On the one hand, oxygen maintains life on Earth through the process of respiration and formation of the ozone layer, and yet oxygen is itself a product of organism functioning. This fact confounds simple description of the oxygen cycle, since it requires synthesizing the descriptions of various processes. A model of the global oxygen cycle (MGOC) was described in detail by Kondratyev *et al.* (2003b, c).

In the imminent future many authors believe nothing threatens the stability of the global biogeochemical cycle of oxygen. Such a statement is not valid for ozone, whose concentration and spatial distribution have undergone serious changes over recent decades. According to Kondratyev and Varotsos (2000), available observations of the vertical profile of atmospheric ozone show a very complicated spatiotemporal variability that depends on many characteristics of the nature–society system. The decision to use the MGOC unit to parameterize ozone fluxes follows a numerical model proposed by Aloyan (2004) and Arutiunian *et al.* (2004) (see also Kondratyev and Varotsos, 2000), after making a necessary correction. This correction consists in substituting certain functional dependences for scenarios reflecting the dynamics of change in concentrations of chemicals not described in the global model of the carbon cycle.

### 3.4.5 World Ocean bioproductivity

World Ocean ecosystems can be represented by three trophic structures that characterize:

(1) tropical pelagic zones (long trophic chains);
(2) tropical latitude shelf zones and mid-latitude aquatic zones (medium trophic chains); and
(3) Arctic latitudes (short trophic chains). In each of these structures the water column is considered as a single biogeocenosis.

The major factor ensuring this unity is the flow of organic matter, which is produced in the surface layers and subsequently reaches maximum depths.

The functioning of the trophic pyramid is characterized by consumption intensity for $s$th food variety at the $i$th level:

$$C_{is} = k_{is}\overline{B_s} \Big/ \sum_{j \in S_i} k_{ij}\overline{B_j}$$

where $\overline{B_j}$ is the effective biomass of the $s$th level; $S_i$ is the food spectrum of the $i$th level; and $k_{is}$ is the Ivlev coefficient used in the formula for the $i$th component ration:

$$R_i = k_i \left[1 - \exp\left(-\sum_{j \in S_i} k_{ij}\overline{B_j}\right)\right].$$

The equations used to describe the bioproduction process in the water column have the following forms:

$$\frac{\partial p}{\partial t} = P_p - \tau p - M_p - \beta\frac{\partial p}{\partial z} + \frac{\partial(A\,\partial p/\partial z)}{\partial z}$$
$$- \sum_{i \in \Gamma_p} k_{ip}\bar{p}R_z^i \Big/ \left(k_{ip}\bar{p} + k_{id}\bar{d} + \sum_{s \in S_i} k_{is}\overline{Z_s}\right)$$

$$\frac{\partial Z_i}{\partial t} = (1 - h_Z^i)R_Z^i - \tau_Z^i Z_i - M_Z^i - \beta_Z\frac{\partial Z_i}{\partial z}$$
$$- \sum_{j \in \Gamma_i} k_{ji}\overline{Z_i}R_Z^i \Big/ \left(k_{jp}\bar{p} + k_{jd}\bar{d} + \sum_{s \in S_j} k_{js}\overline{Z_s}\right)$$

$$\frac{\partial d}{\partial t} + \beta\frac{\partial d}{\partial z} + A\frac{\partial^2 d}{\partial z^2} = M_p + \sum_{i=1}^{m} M_Z^i - \mu_d + \sum_{i=1}^{m} h_Z^i R_Z^i$$

$$\frac{\partial n}{\partial t} + \beta\frac{\partial n}{\partial z} + A\frac{\partial^2 n}{\partial z^2} = \lambda_0 d - \delta P_p + \rho\sum_{i=1}^{m} \tau_Z^i Z_i,$$

where $M_\omega^i = \mu_\omega \max\{0, \omega_i - \omega_{i,\min}\}^{r_Z^i}$ represents the speed at which element $\omega$ $(p, Z_1, \ldots, Z_m)$ dies off; $p$ is phytoplankton biomass; $Z_i$ $(i = 1, \ldots, m)$ is the biomass of the $i$th component of zooplankton; $d$ and $n$ are the concentrations of detritus and nutrients, respectively; $\tau_\omega$ is the index of energy inputs of component $\omega$; $A$ is the turbulent diffusion coefficient; $\beta_Z^i$ is the mobility index of the $i$th component of zooplankton in vertical migrations; and $\beta$ is the upwelling velocity of the water.

### 3.4.6 Units of biogeocenotic, hydrologic, and climatic processes

As shown in Figures 1.19 and 1.20, the GMNSS comprises up to 30 models for soil–plant formations. When synthesizing these models, use was made of results obtained by Holmberg *et al.* (2000), Krapivin and Kondratyev (2002), Papakyriakou and McCaughey (1991), Peng (2000), Wirtz (2000), and Yokozawa (1998). They are all based on the biomass balance equation $X(t, \varphi, \lambda) : \partial X/\partial t = \xi - \omega_X - \tau - \Sigma$, where $\xi$ is actual plant productivity, $\omega_X$ and $\tau$ are the number of deaths and the outlays for energy exchange with the environment, and $\Sigma$ are biomass losses for anthropogenic reasons. These functions are described in detail by many authors (e.g., those cited above). In the GSM the value of $\xi$ is approximated as follows:

$$\xi = \delta_c \delta_o (1 + \alpha_T \,\Delta T/100) \exp(-\beta_1/X) \min\{\delta_e, \delta_Z, \delta_W, \delta_{\mathrm{N}}, \delta_{\mathrm{S}}, \delta_{\mathrm{P}}\}, \qquad (3.3)$$

where $\alpha_T$ and $\beta_1$ are indexes corresponding to the dependence of production, respectively, on temperature and biomass; $\delta_e$ is the index of production limitation by the $\theta$ factor ($e$ = illumination, $Z$ = pollution, $W$ = soil moisture, N, S, and P are nitrogen, sulfur, and phosphorus in soil, respectively).

Formula (3.3) was chosen after carrying out a number of computational experiments that took various options for the limiting factor dependence of plant productivity into account. The $\delta_\theta$ functions actually used were calculated based on data published in the literature. Thus, the role played by $C_A$ in photosynthesis was described by the relation $\delta_c = bC_A/(C_A + C_{0.5})$, where $C_{0.5}$ is $CO_2$ concentration for which $\delta_c = b/2$. The influence of solar radiation intensity $e(t, \varphi, \lambda)$ on photosynthesis was parameterized by $\delta_e = \delta^* \exp(1 - \delta^*)$, where $\delta^* = e/e^*, e^*$ is optimal illuminance. In the soil–plant formation unit for which the maximum photosynthesis value $d_1$ and the initial slope of the photosynthesis curve $m_1$ are known, use was made of the relation $\delta_e = d_1 e/(d_1/m_1 + e)$. The limiting of photosynthesis by pollution was defined by the exponential dependence $\delta_Z = \exp(-fZ)$, where $f$ is a constant. The effect of soil moisture on photosynthesis was expressed by the function $\delta_W = 1 - \exp(-gW)$, where $g$ is a constant. The biogenic element dependence of plant production was represented as $\delta_\theta = \theta/(\theta + \theta_A)$, where $\theta_A$ is the $\theta$ element concentration in soil for which $\delta_\theta = 0.5$.

Water is responsible for the interrelationship channels that exist between natural systems in the biosphere. The water cycle in the biosphere includes the exchange of water in its various phase states between the hydrosphere, atmosphere, and living organisms. The reserves of water in their various forms are described in the literature at great length; therefore, there is every possibility of constructing a mathematical model of the global water cycle. Such a model was suggested in the study by Krapivin *et al.* (1982) and made to conform with a diagram of the water balance. In this version of the model, atmospheric water circulation is simulated by a simplified diagram of stable transports. In reality the process of atmospheric circulation is far more complex in space and time. It can only be roughly characterized by alternating zonal and meridional motions.

Using satellite systems to measure environmental parameters allows rapid acquisition of data pertaining to water content in various biospheric reservoirs; this is particularly so for atmospheric moisture content. This information may be obtained simultaneously with synoptic data on temperature, velocity and direction of wind, atmospheric pressure, content of pollutants in the atmosphere, and ground surface heat–flux balance. Such measurements and published data on biospheric water distribution have made it possible to construct a flow chart of global water balance. These flow charts are based on balance equations. The form in which the latter are written is illustrated by the following example:

$$\frac{dW_{iH}}{dt} = W_{SiH} - W_{iHG} - \sum_k W^k_{HiO} \quad (i = 1, \ldots, n),$$

where $W_{iH}$ is the level of underground waters; $W_{SiH}$ is infiltration; $W_{iHG}$ is irrigation waters; $W_{HiO}$ is runoff to the oceans; and $n$ is the number of land regions.

The precipitation formation regime is represented by a threshold algorithm: rainfall for $T > T^*$, snowfall for $T < T^*$, and snowmelt for $T > T_W$, where $T^* = 0°\text{C}$ and $T_W = 5.5°\text{C}$.

The most important climatic factor representative of human activity in the various regions appears to be atmospheric temperature. A change in atmospheric temperature leads to changes in the intensity of biological processes on land and causes disturbances in biogeochemical cycles (Chen *et al.*, 2000; Power, 2000).

Atmospheric temperature is a function of carbon dioxide $C_A$ and water vapor $W_A$ content in the near-surface layer of the Earth: $T = T_D + \Delta T(C_A, W_A)$, where $T_D$ is temperature in the period when anthropogenic processes were negligible. Estimations of $\Delta T$ are calculated by $\Delta T = \Delta T_C + \Delta T_W$, where $\Delta T_C$ and $\Delta T_W$ reflect changes in $T$ caused by fluctuations in $CO_2$ and $W_A$ concentrations in the atmosphere, respectively. The value of $\Delta T_C$ has two components: $\Delta T_C = 0.5(\Delta T_{C_1} + \Delta T_{C_2})$. The values of $\Delta T_{C_1}$ and $\Delta T_W$ are calculated using a climate model. The spatial distributions of $\Delta T_{C_1}$ and $\Delta T_W$ are calculated as functions of $C_A$ and $W_A$, respectively. The value of $\Delta T_{C_2}$ is calculated using:

$$\Delta T_{CO_2} = \begin{cases} L_1 & \text{when } X \geq 1 \\ L_2 & \text{when } X < 1, \end{cases}$$

where

$$X = C_A(t)/C_A(1900),$$

$$L_1 = -0.847 + 4.528 \ln X - 1.25 \exp\{-0.82(X - 1)\},$$

$$L_2 = -2.63X^2 + 6.27X + 1.509 \ln X - 3.988.$$

### 3.4.7 Demographic unit

The effect of numerous environmental and social factors on population dynamics in the $i$th region $G_i$ is reflected in the birth rate $R_{G_i}$ and death rate $M_{G_i}$:

$$\frac{dG_i}{dt} = R_{G_i} - M_{G_i} \quad (i = 1, \dots K).$$

In each $K$ region, the birth and death rates depend on food supply and quality, environmental contamination, gas composition of the atmosphere, the standard of living, power resource sufficiency, and population density as follows:

$$R_{G_i} = (1 - h_{G_i})K_{G_i}G_i \min\{H_{GV_i}, H_{GG_i}, H_{GM_i}, H_{GA_i}\},$$
$$M_{G_i} = \mu_{G_i}G_i \max\{H_{\mu A_i}, H_{\mu G_i}, H_{\mu V_i}\},$$

where $h_{G_i}$ is a quality coefficient representing the lack of nutrition in food consumed by the population (food inassimilability); $H_{G_i} = 0$ under ideal conditions; $K_{G_i}$ and $\mu_{G_i}$ are constants of the birth rate and death rate, respectively; and the functions

$$H_{GV_i}(H_{\mu V_i}), \quad H_{GG_i}(H_{\mu G_i}), \quad H_{GM_i}, \quad H_{GA_i}(H_{\mu A_i})$$

indicate the effect on birth rate (death rate) of various factors, respectively, such as food supply, population density, standard of living, and environmental quality. Functional descriptions of these factors are related to effects on human ecology. Thus, the function $H_{GV_i}$ is represented as $H_{GV_i} = 1 - \exp(-V_{G_i})$, where $V_{G_i}$ is the effective food amount determined as a weighted sum of components in the food spectrum of *Homo sapiens*:

$$V_{G_i} = K_{Gp_i}p + K_{GF_i}\left(F_i + \sum_{j\neq i} a_{Fj_i}F_j\right) + K_{Gr_i}I_i(1 - \theta_{Fr_i} - \theta_{Ur_i})$$
$$+ K_{GX_i}\left[(1 - \theta_{FX_i})X_i + (1 + \nu_{FX_i})\sum_{j\neq i} a_{Xj_i}X_j\right],$$

where coefficients $K_{Gp_i}$, $K_{GF_i}$, $K_{Gr_i}$, and $K_{GX_i}$ are defined according to Krapivin (1996); $a_{Fj_i}$ and $a_{Xi_j}$ are food of animal and plant origin, respectively, for the $i$th region; $\theta_{FX_i}$ and $\nu_{FX_i}$ are food of plant origin used as cow feed produced in and imported to the $i$th region, respectively; and $\theta_{Fr_i}$ and $\theta_{Ur_i}$ are food of fish origin $I_i$ used for cow feed and fertilizer production, respectively.

It is assumed that as food supply increases the death rate of the population diminishes at a rate $\rho_{G_i}$ to a certain level determined by the constant $\rho_{\mu_i}$ so that $H_{\mu H_i} = \rho_{\mu_i} + \rho_{G_i}/F_{RG_i}$, where

$$F_{RG_i} = V_{G_i}/G_i.$$

Similarly, it is assumed that the dependence of birth rate on the standard of living $M_{SG_i}$ is described by a saturating function such that the birth rate is maximum for low $M_{SG_i}$ values and falls as $a_{G_i}$ increases, down to a certain level

$a_{GM_i}$. The rate at which the transition between maximum and minimum birth rate levels occurs is denoted by $\alpha_G$ and defined by the equation

$$H_{GM_i} = a_{G_i} + a_{GM_i} \exp(-\alpha_G M_{SG_i}).$$

The dependence of birth rate on population density is approximated by the relationship

$$H_{GG_i} = g' + g^* \exp(-g'' G_i).$$

In general, the demographic unit has a branching structure permitting the use of different parameters to describe population dynamics within the framework of a concrete computational experiment. The option used in the GMNSS is described in detail by Krapivin (1978) and Krapivin *et al.* (1982). The demographic unit includes a matrix model comprising three population age groups (0–14, 15–64, and 64 and older) and a disabled group. The unit structure also permits the use of different scenarios in describing both population dynamics as a whole and its parts.

### 3.4.8 Global simulation experiments

A special feature of today's global ecodynamics is the intensive development of science and technology; this is the reason that the environment and human society are changing at a higher rate than, say, 50 years ago and that the role played by interdisciplinary studies of the NSS using accumulated knowledge in mathematics, ecology, sociology, medicine, chemistry, geophysics, etc. has grown. A global simulation experiment based on employing the latest computer technologies has become a precursor to decision-making strategies in the field of natural resource management and, likewise, the concept and understanding of the experiment have changed as the process has evolved as a result of real natural observations. Of course, problems come up when assessing how adequate the results of simulation experiments are; ecoinformatics solves these problems (Kondratyev *et al.*, 2002). The late 20th to early 21st centuries witnessed a new mentality—a transition to experiments using models in which real natural components were not included. How, for instance, is it possible to carry out a field experiment of such unique systems as oceans, continental ecosystems, and the biosphere using such a model? Such experiments are obviously very dangerous and should be given short shrift.

The GMNSS allows a wide range of simulation experiments to be carried out within a multi-dimensional phase window of the NSS and the development of the NSS for this century to be predicted. Longer term predictions are not possible at the present level of knowledge and global databases. The uncertainties that have appeared have been analyzed and found to be closely connected with climate modeling. Unfortunately, there is no universal model at the moment capable of reliably simulating the way in which the Earth's climate system functions and reflecting interactions between the atmosphere, oceans, land, and human society. Therefore, to use the GMNSS some *a priori* assumptions need to be made:

- up to 2100, anthropogenic emissions of $CO_2$ follow the model of Demirchian *et al.* (2002), and emissions of other GHGs until 2050 follow the pattern of 2000, after which they decrease by 25%;
- extraction and consumption of non-renewable fossil fuels in the period up to 2050 grow by 0.5%/year in developed countries and by 0.7%/year in developing countries, after which time they stabilize;
- per capita gross domestic product increases in developed and developing countries at constant rates of 2.5%/year and 1.5%/year, respectively;
- medical services in developed and developing countries reduce mortality by 90% and 50%, respectively, as a result of environmental changes, and in 2020 a remedy for AIDS is found;
- the atmospheric temperature regime follows the model of Mintzer (1987);
- agricultural investments depend on providing the world population with enough food (Krapivin *et al.*, 1982);
- the success of countries in the battle against environmental pollution grows by 0.2%/year until 2050, after which there is no further increase;
- processes of deforestation and replanting reach a balance in 2010 so that the total area of forests existing in 2000 is preserved, and the area of tropical forests decreases by no more than 5%;
- oil pollution of the World Ocean in the 21st century follows the pattern of the last decade of the 20th century;
- agricultural productivity grows steadily, reaching 80% in 2100.

Having made these assumptions, the dynamics of the NSS can be characterized by the following indicators:

- the ecodynamic pattern in which the NSS functions, as characterized by derivatives of functions describing the state of NSS components, is never equal to zero for time periods longer than the time step of the model;
- population size by 2100 will reach 12.7 billion, with the ratio between age groups characterized by the following average values: 0–15 years 32%; 16–65 years 53%; and >65 years 15%; and the disabled will constitute 17%;
- provision of the population with healthy food will increase by 6.8% by the year 2030 (with the World Ocean contributing 12.1%) and then decrease by 3.2% with respect to 2000 (the share of the World Ocean carries on increasing to 22.1%);
- $CO_2$ concentration in the atmosphere will reach 486 ppm with a respective increase of 0.87°C in average planetary temperature;
- the spatial distribution of atmospheric $CO_2$ sinks are characterized by the growing role of the World Ocean and land ecosystems, whose proportions in the middle of the 21st century will constitute 31% and 19%, respectively; then, by the end of the century, the proportion of the World Ocean will decrease to 26.7%, with the role of land ecosystems increasing to 24.4%;
- the conversion of 5% of the tropical forest area to urban ecosystems will lead to a 1.2% reduction in the sink of total $CO_2$;
- complete conversion of wet tropical forests to a grass ecosystem by 2050 will later

take its toll on global climate change by interfering with the regional moisture cycle on the atmosphere–land interface, by increasing outgoing longwave radiation flux from deforested territory by 4.3 W/m$^2$, by increasing this territory's albedo by 5.7%, by decreasing absorbed shortwave radiation flux by 4.1%; by increasing soil temperature by 1.3°C, and by reducing rain and evaporation by 0.7 and 0.9 mm/day, respectively.

For educational purposes, it is of interest to consider hypothetical versions of the interaction between humans and nature. Looking at just a few of the situations hypothesized to occur as a result of changes in the most important anthropogenic factors, it should be noted that, in contrast to many other models, the GMNSS calculates the size of the population as a function of environmental parameters.

Calculations show that the mechanisms involved in demographic process formation depend strongly on the strategies of natural resource use. So, a 50% increase in food provision, which is possible with the introduction of efficient technologies in agriculture and the development of aquaculture systems, should result in a 14% decrease in population. In contrast, a 10% decrease in the rate of mineral resource consumption will cause a 3% increase in population. The NSS is very sensitive to changes in the global structure of soil–plant formations. So, if forest areas decrease steadily by 2050 by only 10% with respect to 2000, the atmospheric $CO_2$ concentration in 2100 will reach 611 ppm (i.e., it will almost double). But if forest areas increase by 15% by 2050, then in 2100 the partial pressure of $CO_2$ will constitute 475 ppm.

These modeling results of the NSS response to possible anthropogenic changes in its parameters show that the suggested method allows various components of this system to be assessed according to hypotheses on the possible rates of these changes. An application of this method to reconstruct the dynamics of the NSS as a whole by some of its parameters for the period 1970–2000, on the basis of data published in Houghton *et al.* (2001), Kondratyev and Losev (2002), and Watson *et al.* (2000), has shown that the prediction error for 10 and 20 years does not exceed 15% and 25%, respectively. Hence, GIMS technology can be recommended as a constructive approach to solving the problems of global ecodynamics formulated by Kondratyev (1999). Simulation experiments using the GMNSS, based on the reliable data available on NSS prehistory, brings to light those conditions that constrain the development of civilization and facilitates the finding of mechanisms for its sustainable development. For this to occur, it is necessary to bring together the efforts of international and national scientific programs when synthesizing the GMNSS database. Kofi Annan, U.N. Secretary-General, emphasized this at his press conference in Johannesburg on the last day of the World Forum (September 4, 2002). He pointed out that governments had come to an agreement to join efforts on biodiversity preservation and solution of global problems regarding water supply, energy, medicine, and agriculture. Speaking about possible ways to solve the global problems of the present civilization, he was fully confident of a way being found that will reduce poverty while preserving the environment.

GIMS technology is one such way. However, for it to be brought about requires the knowledge built up during the development of the GMNSS to be concentrated, its units to be improved, but, mainly, the factors and dynamic strategies involved in social evolution to be discovered and formalized.

### 3.4.9 Concluding remarks

Lomborg (2001, 2004) is of course right to reject apocalyptic predictions of global ecodynamics based on exaggerated fear of limited natural resources and the environmental state. Lomborg's assessment is confirmed particularly well by the data compiled by Holdren (2003), which characterize both real and potential global energy resources. Energy units are expressed in Holdren's work (in the case of non-renewable energy sources) in terawatt years, which is equivalent to 31 exa-J ($1\,\mathrm{TW} = 1\,\mathrm{TW\text{-}yr} \cdot \mathrm{yr}^{-1} = 31.5\,\mathrm{exa\text{-}J} \cdot \mathrm{yr}^{-1}$). It should be added that in 2000 global energy consumption constituted about 15 TW or $15\,\mathrm{TW\text{-}yr} \cdot \mathrm{yr}^{-1}$ with an assumed increase up to $60\,\mathrm{TW\text{-}yr} \cdot \mathrm{yr}^{-1}$ by 2100.

Despite optimistic data, the absence of long-range prospects for development of the present consumer society, as illustrated by global ecodynamics estimates, (Cracknell *et al.*, 2009) is not in doubt. Therefore, at the World Summit on Sustainable Development held in Johannesburg in 2002, emphasis was put on accomplishing 10-year programs to achieve stable production and consumption, and the following recommendations were passed (Starke, 2004):

- developed countries should take the leading role in providing stable production and consumption;
- goals should be achieved on the basis of common but differentiated responsibility;
- stable production and consumption should be paramount;
- the young must get involved in bringing about sustainable development;
- the "polluter pays" principle should be practiced;
- control of the complete cycle of product evolution should be considered—from production to consumption to waste—in order to raise production efficiency;
- support should be given to politics that favor the output of ecologically acceptable products and the delivery of ecologically adequate services;
- ecological and effective methods of energy provision need to be developed and energy subsidies need to be removed;
- support should be given to free-will initiatives of industry aimed at raising social and ecological responsibility;
- study and introduction of ecologically pure production, especially in developing countries, as well as in small and medium-size businesses.

Never mind these recommendations being rather declarative, they are still clearly oriented toward the necessity of changing the paradigm of socioeconomic development (this mainly refers to developed countries) from being a consumer society to a conservation society. Concrete analysis of the means to bring about such development requires the participation of specialists in the field of social sciences. Some related opinions were expressed by Corcoran (2005) in the context of Earth Charter problems. Therefore, the question of whether humans are capable of changing the climate requires further study (Borisov, 2005; Kochergin, 1978).

Finally, the following conclusions can be drawn from the above:

- existing climate models should not be used to make decisions about and assessments of the risk of various anthropogenic scenarios being accomplished;
- the level of uncertainty in climate forecasts can be reduced if global models take on board interactive bonds in the NSS, the mechanism of biotic regulation of the environment, and improvement of the global monitoring system;
- the use of hydrocarbon energy sources in the 21st century will not lead to catastrophic climate change as long as the Earth's cover is preserved and the World Ocean is protected from pollution.

# 4

# GIMS-based study of ocean ecosystems

## 4.1 THE WORLD OCEAN AS A COMPLEX HIERARCHICAL SYSTEM

Investigations of the structure and functioning of oceanic ecosystems have become one of the important and rapidly developing areas of sea biology and environmental studies. Different aspects of relevant problems have been studied in the framework of many international biological and environmental programs. One task is substantiation of the possibilities to forecast system behavior under the impacts of its variable parameters (Osterberg, 1985; Smirnova, 1988; Trites *et al.*, 1999). Because of the unique and enormous spatial extension of World Ocean ecosystems it is difficult to assess their parameters at different times and for the various aquatories. This is the reason modeling technology is effective at studying and solving various problems.

The World Ocean, occupying about 71% of the Earth's surface, provides a substantial amount of the food consumed by people. At the same time the full volume of organic matter produced by the oceans is approximately equal to the productivity of land ecosystems. The total biomass of nekton is estimated at 5.3 Gt. The extraction of commercial food elements from the World Ocean is about $70 \cdot 10^6$ t per year. This includes approximately 20% of the proteins needed by humanity. At the present time (2011) traditional fishing is near to its acceptable limit ($\approx$90–100,000,000 $\mathrm{t \cdot yr^{-1}}$). There is no limit to commercial exploitation of ocean ecosystems in general, since biological objects have many needs that are rarely used by people.

The disparity between the role played by land and ocean ecosystems in food production is mainly explained by the fact that on land intensive agriculture is well developed while in the oceans it is almost completely the opposite. This is the reason that finding a means of increasing ocean ecosystem productivity has become a pressing issue. Humanity mainly uses the upper trophic chains of natural ecosystems. It is known that ocean ecosystems have long trophic chains

V.F. Krapivin and A.M. Shutko, *Information Technologies for Remote Monitoring of the Environment*, Springer Praxis Books, DOI 10.1007/978-3-642-20567-5_4, with transfer coefficients between trophic levels equal to about 0.1. From the oceans and seas people extract production from the three to five upper levels (fish, crabs, whales, etc.). These levels and phytoplankton constitute more than 10 trophic levels. Therefore, optimal exploitation of oceanic ecosystems is possible by adopting a complex simulation approach to the study of global change. Artificial improvement of the ocean's biological community is a consideration here.

Sea communities are complex biological systems comprised of populations of different species whose interaction represents the state of dynamic interaction within the community. The spatial structure of the community is largely dependent on the composition of numerous biotic and abiotic factors that depend on the set of oceanological fields. The latter are defined by the general ocean water circulation which includes ebbs and flows, convergence and divergence zones, wind and thermocline, etc.

During the last 10 years the need to forecast oceanic systems under growing anthropogenic influence has grown. Additionally, it has become important to estimate the anthropogenic role in biosphere dynamics. In recent investigations of global change it has been shown, for instance, that the role played by ocean ecosystems in formation of the greenhouse effect has been underestimated, especially relating to the Arctic Basin (Bobylev *et al.*, 2003). In connection with many catastrophic events such as Hurricanes Katrina and Rita, which took place in summer 2005, being able to forecast the oceanic system has become a priority. Understanding the processes of heat exchange between the atmosphere and ocean is basic to the study of climate change.

Microwaves emitted from the sea surface carry information about the physical and chemical properties of seawater and sea surface state. This is the reason passive microwave remote sensing plays an important role in oceanic investigations of seawater temperature, salinity, sea surface roughness, sea surface wind, sea surface oil film and slick, water vapor and cloud water quanty in the oceanic atmosphere, sea ice temperature, thickness, surface state, age, porosity, salinity, and distribution.

The Marine Observation Satellite-1 (MOS-1) was launched by Japan in 1987 and had a payload of two-frequency microwave radiometers (23.8 and 31.4 GHz). Sasaki *et al.* (1987b) studied the dependence of sea surface microwave emissive properties on oceanic environmental parameters, phenomena, and observational conditions. Wind speed $U(z)$ ($\mathrm{m\,s^{-1}}$) is calculated by:

$$U(z) = (u_*/k)\ln(z/z_0),$$

where $z$ is the height above sea level (meters); $z_0$ is the roughness parameter of the sea surface (meters); $u_*$ is the friction velocity ($\mathrm{cm\,s^{-1}}$); and $k$ is the von Kármán constant.

The influence of oceanic ecosystems on the intensity of biogeochemical cycles occurs at the atmosphere–water boundary and is usually parameterized using observational data. However, the vertical structure of oceanic processes needs to be taken into account in this context. Arctic latitudes show the greatest correlation between state of the atmosphere–water boundary and vertical hydrophysical processes. Investigation of this correlation is important in global change studies (Monin and Krasitskyi, 1985).

The World Ocean has an irregular structure consisting of temperature, salinity, nutrients, insolation, and hydrophysical parameters. This irregularity of the water surface is reflected in phytoplankton production. The irregular variability of the topology correlates with the vertical distribution of the biomass of phytoplankton. Observations of these correlations give only partial descriptions of the actual abiotic, biotic, and hydrophysical processes. To get a global description of the dynamics of the World Ocean ecosystem it is necessary to develop elaborate mathematical models. This chapter describes a set of such models.

## 4.2 THE SYNTHESIS OF OCEAN ECOSYSTEM MODELS

Each element of the oceanic ecosystem $A$ can be described by means of a set of parameters: $x(t) = \{x_j(t), j = 1, \dots, N\}$, where $t$ is time. Ecosystem $A$ is characterized by its structure $|A(t)|$ and behavior $A_B(t)$:

$$A(t) = \{|A(t)|, A_B(t)\} = F(x(t)). \tag{4.1}$$

Consequently, according to equation (4.1) system $A$ has a trajectory in $(N+1)$ space. The abstract formation

$$A_M(t) = F_M(x^M(t)) \tag{4.2}$$

is called the model of ecosystem $A$, where $M \le N$, $\{x^M\} \subset \{x\}$. Parameter $M$ defines the divergence level between the trajectories of $A(t)$ and $A_M(t)$ ecosystems. Let us introduce a goal function to the ecosystem trajectory

$$V = Q(\{x_i(t)\}). \tag{4.3}$$

The form of $Q$ is defined by the character of the system's goal. It is considered that the natural evolution process leads $A$ to the optimal system $A_{\text{opt}}$. Therefore, model $A_{M,\text{opt}}$ giving an extremal value of $Q$ is called the optimal model of ecosystem $A$. The value of divergence between the trajectories $A$ and $A_{M,\text{opt}}$ is a function of the correspondence between $V$ and the real goal $\underline{A}$ of ecosystem $A$. The set $\underline{G} = \{\underline{A}_1, \dots, \underline{A}_r, \dots, \underline{A}_m\}$ of possible real goals $\{\underline{A}_r\}$ of ecosystem $A$ can be formed on the basis of oceanological information. Then, defining

$$A_{M,\text{opt},r} = g_r(\underline{A}_r), \quad \underline{A}_r \in \underline{G} \tag{4.4}$$

we proceed to the restricted set of possible optimal systems $A_{M,\text{opt},r}$ $(r = 1, \dots, m)$, the trajectories of which, together with the $A$ trajectory, are in the space of potential trajectories. Defining $A_{M,\text{opt},r_0}$ with minimal divergence from $A(t)$ we can obtain the value for the most probable ecosystem goal: $\underline{A}_{r_0} = g_r^{-1}(A_{M,\text{opt},r_0}(t))$.

According to the above principles, construction of a mathematical model for the oceanic ecosystem $A$ demands either the formation of a detailed description of the space for all its states or the creation of a full set of mathematical models describing the occurrence in $A$ of processes of energy exchange between trophic levels and of interactions having biotic, abiotic, and hydrophysical origin. Of course, it is supposed that there exists an appropriate set of assumptions about the nature of balance correlations in the ecosystem.

The supposition that solar energy $E$ is the unique source of energy for all forms of life is assumed to be the basic criterion. According to many theoretical and experimental investigations the law of insolation distribution across ocean layers has the exponential form:

$$E(t,\varphi,\lambda,z) = uE_0 \exp\left[-\int_0^z U(x)\,dx - \alpha z\right] + (1-u)E_0 \exp(-\zeta z), \qquad (4.5)$$

where $U(x) = \delta p(t,\varphi,\lambda,x) + \beta d(t,\varphi,\lambda,x) + \nu Z(t,\varphi,\lambda,x)$; $E_0 = E(t,\varphi,\lambda,0)$ is surface illumination; $\alpha$ is the absorption coefficient of seawater; $\delta$, $\beta$, and $\nu$ are relaxation coefficients as a result of attenuation by phytoplankton $p$, detritus $d$, and zooplankton $Z$, respectively; $u$ and $\zeta$ are the parameters selected in the concrete situation to find the shortest distance between $E(t,\varphi,\lambda,z)$ and real data. Note that the influence of other trophic levels on water clarity are ignored.

The insolation level influences photosynthesis $R_p$. It is known that $R_p$ as a function of $E$ reaches a maximum under some optimal value $E_{\max}$ decreasing with either increase or decrease of illumination from this value. The maximum $R_p$ for various latitudes $\varphi$ is located at depths that change with the time of year. This variability of maximal photosynthesis with depth is most substantial in tropical zones. The average location of maximal photosynthesis there is situated at 10 to 30 m, where $E_{\max} = 65$–$85\,\mathrm{cal\cdot cm^{-2}\,day^{-1}}$. Beginning with depths where $E = 20$–$25\,\mathrm{cal\cdot cm^{-2}\,day^{-1}}$ photosynthesis decreases proportionally to $E$. Photosynthesis depression as a result of illumination is actually observed when $E > 100\,\mathrm{cal\cdot cm^{-2}\,day^{-1}}$. Such estimates are different for northern latitudes where maximal photosynthesis is usually found near the water surface.

The variation of photosynthesis on depth $z$ depends on water temperature $T_W$, nutrients $n$, and phytoplankton biomass $p$ as well as on some other factors that are less important and are not taken into consideration. To simulate this dependence different equations are used to take into account the limiting role of elements $E$, $n$, and $p$. Considering that $\partial p/p\,\partial z \to 0$ when $n \to 0$ and $\partial p/p\,\partial z \to \mathrm{const}$ with increasing $n$ the following function is used as the basic correlation to simulate photosynthesis:

$$R_p(t,\varphi,\lambda,z) = k_0(T_W) K_T f_2(p) f_3(n), \qquad (4.6)$$

where

$$\left.\begin{aligned} K_T = Af_1(E), \quad A = kA_{\max}/E_{\max}, \quad f_1(E) = E\cdot\exp[m(1-E/E_{\max})], \\ f_2(p) = [1-\exp\{-\gamma_1 p\}], \quad f_3(n) = [1-\exp\{-\gamma_2 n\}]^{\theta}; \end{aligned}\right\} \qquad (4.7)$$

$k$ is the proportionality coefficient; $k_0(T_W)$ is the function characterizing the dependence of photosynthesis on water temperature $T_W$; $A_{\max}$ is the assimilation number in the area of maximal photosynthesis (increase in phytoplankton species per unit weight); $\gamma_1$, $\gamma_2$, $\theta$, and $m$ are constants whose selection determines the specific characteristics of the type of phytoplankton. For $A_{\max}$ the following estimate is valid: $A_{max} = 5.9E_{max}$ for the area of maximal photosynthesis and $A_{max} = 2.7E_{max}$ for other areas. Corresponding to this value, the assimilation number of tropical phytoplankton in the area of maximal photosynthesis equals 11–12 mg $C \cdot h^{-1}$. For instance, for the Peruvian upwelling $A_{max} = 6.25$ mg $C \cdot h^{-1}$. Under this the light saturation of photosynthesis in equatorial latitudes takes place at the rate of $E_{max} = 9\,\text{cal} \cdot \text{cm}^{-2}\,\text{day}^{-1}$.

As to the function of $k_0(T_W)$, it is known that specific variation of photosynthesis intensity with rising temperature at first increases until it achieves a maximal value for some optimal $p$ temperature and then starts to fall with subsequent increase in temperature. Many authors use the following approximation: $\ln k_0(T_W) = (T_W - T_{W,\text{opt}}) \ln \theta_0$, $0 < \theta_0 \leq 2$.

The exponential dependence of the photosynthesis rate on biogenic elements $n$ (phosphorus, silicon, nitrogen, etc.) as expressed in formula (4.6) by $R_p$ (mg $C \cdot m^{-3}\,\text{day}^{-1}$) is in reality, of course, more complex. Nutrients are one of the most important elements of the ecosystem since they regulate energy fluxes. Nutrients are consumed during photosynthesis at the rate $R_n$ usually approximated by the expression $R_n = \delta R_p$ where $\delta$ is the proportionality coefficient. The replenishment of nutrients occurs at the expense of deep waters where the stored nutrients are formed as a result of the chemical processes involved in decomposition of dead organic matter. It is assumed that nutrients stored in deep ocean waters are unlimited.

The process of dead organic matter decomposition is controlled by a range of abiotic conditions that are specific to the different climatic zones of the World Ocean. The vertical flow of nutrients is determined by the conditions that prevail for water mixing. In tropical zones where the vertical water structure has a pronounced three-layer configuration with a layer that experiences sharp temperature change (the thermocline) the vertical movement of nutrients is restricted by this layer. In those aquatories where the thermocline lies at a depth of 40–100 m, the upper layer is usually made poorer and the flow to this layer takes place only in upwelling zones. In this case the average speed of vertical water transport under the thermocline fluctuates in the range of $10^{-3}$ to $10^{-2}\,\text{cm}\,\text{s}^{-1}$. In upwelling zones, the vertical water transport achieved is of the order of $0.1\,\text{cm}\,\text{s}^{-1}$.

## 4.3 EQUATIONS DESCRIBING THE DYNAMICS OF OCEAN ECOSYSTEMS

All depths of oceanic water are considered unique regions of biogeocenosis in which the basic connecting factor is the flow of organic matter produced in upper layers that then spreads to maximal ocean depths. It is assumed that all model parameters change according to geographical coordinates and the season and that their parametrical description is realized by means of deterministic models. Food interactions between trophic levels are proportional to their effective biomass.

A significant role in the trophic chains of the ocean ecosystem is played by bacterioplankton $b$. According to many investigations the biomass of bacterioplankton biomass can form condensed bodies that are consumed by zooplankton $Z$. This fact is important since the bacterioplankton production of many ocean aquatories occurs at the same level as phytoplankton production. Bacterial plankton have a changeable energy balance depending on the available food supply. The production of bacterial plankton $R_b$ is limited by food abundance (detritus $d$ and dissolved organic matter $g$ excreted during photosynthesis):

$$R_b = k_b b[1 - \exp(-k_{1,d} d - k_{1,g} g)], \tag{4.8}$$

where $k_b$, $k_{1,d}$, and $k_{1,g}$ are empirical coefficients.

The equation describing the dynamics of bacterioplankton biomass has the following form:

$$\partial b/\partial t - U_b + \sum_{s \in \Gamma_b} C_{bs} R_s = 0, \tag{4.9}$$

where

$$U_b + V_\varphi\, \partial b/\partial\varphi + V_\lambda\, \partial b/\partial\lambda + V_z\, \partial b/\partial z$$
$$= R_b - T_b - M_b + k_{2,\varphi}\, \partial^2 b/\partial\varphi^2 + k_{2,\lambda}\, \partial^2 b/\partial\lambda^2 + k_{2,z}\, \partial^2 b/\partial z^2;$$

$T_b$ and $M_b$ are the losses of the bacterioplankton biomass at the expense of the rate of exchange and mortality, respectively; $\Gamma_b$ is the set of trophic subordinates of bacterial plankton (in a typical case $\Gamma_b$ consists of $Z$); $C_{bs}$ is the consumption coefficient of the $s$th trophic level as regards bacterioplankton. Functions $T_b$ and $M_b$ are described by the following formulas (Nitu *et al.*, 2000a):

$$T_b = t_b b \tag{4.10}$$

$$M_b = \max\{0, \mu_b(b - \underline{B}_b)^\xi\}, \tag{4.11}$$

where $t_b$ is the specific loss on exchange; $\mu_b$ is the mortality coefficient; $\underline{B}_b$ and $\xi$ are the constants defining the dependence of bacterial plankton mortality at the expense of its biomass. Coefficients $k_2 = (k_{2,\varphi}, k_{2,\lambda}, k_{2,z})$ determine the process of ocean water turbulence.

The dynamic equation to describe the biomass balance of phytoplankton is

$$\partial p/\partial t - U_p + \sum_{s \in \Gamma p} C_{ps} R_s = 0, \tag{4.12}$$

where

$$U_p + V_\varphi\, \partial p/\partial\varphi + V_\lambda\, \partial p/\partial\lambda + V_z\, \partial p/\partial z$$
$$= R_p - T_p - M_p + k_{2,\varphi}\, \partial^2 p/\partial\varphi^2 + k_{2,\lambda}\, \partial^2 p/\partial\lambda^2 + k_{2,z}\, \partial^2 p/\partial z^2;$$

$\Gamma_p$ is the set of trophic subordinates of phytoplankton (in a typical case $\Gamma_p$ consists of $Z$); $C_{ps}$ is that part of the phytoplankton biomass consumed by the $s$th trophic level; $T_p$ is the rate of exchange; and $M_p$ is the mortality of phytoplankton cells. Similarly to equations (4.10) and (4.11) the functions $T_p$ and $M_p$ are represented by the following correlations:

$$M_p = \max\{0, \mu_p(p - \underline{p})^\theta\} \tag{4.13}$$

$$T_p = t_p p, \tag{4.14}$$

where $t_p$ is the specific loss of phytoplankton biomass at the expense of exchange with the environment; $\mu_p$ is the mortality coefficient; $\underline{p}$ and $\theta$ are coefficients characterizing the dependence of phytoplankton cell mortality on their biomass.

Zooplankton are an important element of ocean ecosystems. They are mainly represented by the unique integrated level $Z$, whose interior has many sub-levels: small-sized herbivores, large-sized herbivores, omnivores, carnivores, etc. Zooplankton consume phytoplankton and bacterial plankton. The trophic scheme of the ocean ecosystem reflects the set of trophic dependence of zooplankton (in a typical case $\Gamma_Z$ consists of $Z$, $r$, and $D$). Zooplankton production $R_Z$ is approximated by an exponential law:

$$R_Z = k_Z(1 - \exp[-\nu B_a]), \tag{4.15}$$

where $k_Z$ is maximal zooplankton production; $\nu$ is the coefficient reflecting the hunger level; $B_a = \max\{0, B - B_{\min}\}$, where $B_{\min}$ is the minimal biomass of food consumed by zooplankton. Maximal production is defined by the relation: $k_Z = T_1 u(1 - q_{2,\max})$, where $q_2 = P_1/(P_1 + T_1)$, $1/u$ is the efficiency of assimilation, $P_1$ is the maximum rate of production under the given rate of exchange $T_1$, and $q_{2,\max} = \max q_2$.

Formula (4.15) explains that under conditions when the store of food is limited the zooplankton ration increases proportionally to $B_a$ and then as the ration approaches the maximal value of $k_Z$ it depends on $B_a$ as little as possible. Expression (4.15) also reflects the absence of full extermination of trophic levels.

Thus, the change in zooplankton biomass obeys the differential equation:

$$\partial Z/\partial t - U_Z + \sum_{s \in \Gamma_Z} C_{ps} R_s = 0, \tag{4.16}$$

where

$$\left.\begin{aligned} &U_Z + V_\varphi\, \partial Z/\partial\varphi + V_\lambda\, \partial Z/\partial\lambda + V_z\, \partial Z/\partial z \\ &\quad = R_Z - T_Z - M_Z - H_Z + k_{2,\varphi}\, \partial^2 Z/\partial\varphi^2 + k_{2,\lambda}\, \partial^2 Z/\partial\lambda^2 + k_{2,z}\, \partial^2 Z/\partial z^2, \\ &\qquad H_Z = h_Z R_Z, \quad T_Z = t_Z Z, \quad M_Z = (\mu_Z + \mu_{Z,1} Z)Z, \end{aligned}\right\} \tag{4.17}$$

The coefficients $h_Z$, $t_Z$, $\mu_Z$, and $\mu_{Z,1}$ are determined empirically.

It follows from formula (4.16) that zooplankton are considered a passive ecosystem element subject to the physical processes of water mixing. However, it is known that zooplankton have vertical migrations. The effect of this vertical migration on trophic relationships can be simulated in the model in such a way that the food requirements of the zooplankton inhabiting the 0 to $z_0$ layer are supplemented by some of the total food requirements of components of the same community occurring in deeper layers ($z_0$ to $H$).

The coefficients $C_{as}$ $(a = p, Z)$ in formulas (4.12) and (4.16) can be determined by assuming that the consumption of different food types at the $s$th trophic level is proportional to their effective biomasses:

$$C_{as} = k_{sa}B_a \Big/ \sum_{a \in S_s} k_{sa}B_a, \tag{4.18}$$

where $B_a$ is the effective biomass of the $a$th food; $S_s$ is the food spectrum of the $s$th component; and $k_{sa}$ is the index of satisfaction of the nutritive requirements of the $s$th component at the expense of the $a$th component biomass.

The equations to describe the biomass dynamics of nekton, detritophages, detritus, dissolved organic matter, and nutrient salts have the following form:

$$\frac{dr}{dt} = R_r - H_r - T_r - M_r - \sum_{s \in \Gamma_r} C_{rs}R_s, \tag{4.19}$$

$$\frac{dD}{dt} = R_D - H_D - T_D - M_D - \sum_{s \in \Gamma_D} C_{Ds}R_s, \tag{4.20}$$

$$\begin{aligned}\frac{\partial d}{\partial t} + V_\varphi \frac{\partial d}{\partial \varphi} + V_\lambda \frac{\partial d}{\partial \lambda} + V_z \frac{\partial d}{\partial z} &= M_b + M_D + M_r + M_p + M_Z + H_Z + H_r + H_D - \mu_d d \\ &\quad - C_{dD}R_D + k_{2,\varphi}\frac{\partial^2 d}{\partial \varphi^2} + k_{2,\lambda}\frac{\partial^2 d}{\partial \lambda^2} + k_{2,z}\frac{\partial^2 d}{\partial z^2}\end{aligned} \tag{4.21}$$

$$\frac{\partial n}{\partial t} + V_\varphi \frac{\partial n}{\partial \varphi} + V_\lambda \frac{\partial n}{\partial \lambda} + V_z \frac{\partial n}{\partial z} = \mu_d d - \delta R_p + k_{2,\varphi}\frac{\partial^2 n}{\partial \varphi^2} + k_{2,\lambda}\frac{\partial^2 n}{\partial \lambda^2} + k_{2,z}\frac{\partial^2 n}{\partial z^2} \tag{4.22}$$

$$\begin{aligned}\frac{\partial g}{\partial t} + V_\varphi \frac{\partial g}{\partial \varphi} + V_\lambda \frac{\partial g}{\partial \lambda} + V_z \frac{\partial g}{\partial z} &= T_p + T_b + T_r + T_D + T_Z - C_{gb}R_b + k_{2,\varphi}\frac{\partial^2 g}{\partial \varphi^2} \\ &\quad + k_{2,\lambda}\frac{\partial^2 g}{\partial \lambda^2} + k_{2,z}\frac{\partial^2 g}{\partial z^2},\end{aligned} \tag{4.23}$$

where $H_a = (1 - h_a)R_a$ is non-assimilated food by the $a$th component $(a = r, D)$; $T_a = t_a a$ is the rate of exchange; $M_a = (\mu_a + \mu_{a,1}a)$ where $a$ is mortality; $\rho_g$ is the rate indicator to replenish nutrient salts at the expense of dissolved organic matter; and $\delta$ is the coefficient of nutrient consumption by photosynthesis.

It follows from equations (4.19) and (4.20) that the model does not consider the possible spatial motion of nekton and detritophages with moving water. It is considered that these components migrate in the water space independently of hydrophysical conditions. There are two possible ways to model the migration process. The first consists in adding terms describing the turbulent mixing process with the coefficient $k_2^* > k_2$ to the right-hand sides of equations (4.19) and (4.20). In other words, this scheme of the migration process identifies fish migration with intense turbulent diffusion. In this case it is necessary to include stochastic elements in the model.

The investigations of many authors have shown that the process of fish migration can outwardly be recognized by purposeful behavior. According to the biological principle of adaptation, the fish migration process is subject to composite maximization of the effective nutritive ration under conservation of the environmental parameters within the limits of the living environment of the fish. Consequently, migrating species travel at characteristic velocities toward the maximum gradient of effective food, subject to temperature and the restrictions of other environmental parameters (salinity, dissolved oxygen, illumination, pollution, and other abiotic conditions).

## 4.4 REAL OCEAN ECOSYSTEMS AND THEIR STUDY

### 4.4.1 Arctic Basin and problems facing its study

The Arctic region plays a unique role in global environmental processes, forming numerous adverse correlations with the Earth's climatic system. Low temperatures, a high level of atmospheric circulation, and the availability of large ice-covered aquatories are characteristic of high latitudes, quite unlike other areas of the globe.

The intensive industrial development of northern countries has led to significant environmental change in these regions, especially in Russia. Oil and gas extraction on the Yamal and Taimir peninsulas in northwestern Siberia, coal and gold extraction in Yakutia and Chukot, as well as mining on the Kola Peninsula are evidence of strong anthropogenic intervention in the natural environment. Plant cover has been damaged over huge regions, the area and productivity of reindeer pastures diminished, and the hydrologic regime of rivers disturbed. Most pollutants are brought to the northern coast of Russia by rivers, where they damage the ecosystems of the northern seas.

Further adverse influence on Arctic ecosystems can lead to increasingly serious regional or even global-scale consequences. Thus, how development of the northern territories of Russia, the U.S.A., Canada, and Scandinavia is further carried out will involve careful analysis of all types of ecosystem dynamics by real-time data collection and formation of national databases and by defining effective ways to coordinate the development of both natural and anthropogenic processes.

The purpose of this chapter is to develop and investigate a model to simulate pollution dynamics in the Arctic Basin. There are many experimental and theoretical results that give estimates of the growing dependences between pollution dynamics in the World Ocean and the state of the continental environment. Arctic Basin pollution is the cause of most anxiety to investigators (Bobylev *et al.*, 2003; Doronin and Heisin, 1975; Kochergin, 1978; Kochetov, 1995; Krapivin *et al.*, 1997a; Ries and Hibler, 1989; Rovinsky *et al.*, 1995). It is known that the ecosystems of the Arctic seas are much more vulnerable than those of other seas. Processes that clean the Arctic Ocean are slower and marine organisms of the Arctic ecosystem live in the polar climate where the vegetation period is restricted.

Some feedback mechanisms operate with significant time delays and the capacity to neutralize the effects of human activity is weak. What is more, the Arctic ecosystem has specific boundary conditions connected with the sea ice ergocline which reduces its survivability level (Legendre and Demers, 1985; Legendre and Legendre, 1998).

As a result of this, the Arctic Basin has become the object of investigations within the frameworks of many national and international environmental programs, such as the International Geosphere–Biosphere Program, U.S. Global Change Research, the international Arctic System Science (ARCSS) Program (McCauley and Meier, 1991), the U.S. Arctic Nuclear Waste Assessment Program (ANWAP), and the International Arctic Monitoring and Assessment Program (AMAP). The research strategies of these programs include the theoretical and experimental study of tundra ecosystems, Siberian rivers, and near-shore and open Arctic waters. The main task consists of determining a set of principle matters that first need to be studied. These include the following:

(1) *Transport modeling of pollutants in Arctic ecosystems*. Based on an experimental database, it is necessary to prepare a complete set of models and computer simulations to describe the processes of transfer and transformation of pollution substances in Arctic natural ecosystems. This set would include such models as:
   - a model of how organic pollution affects the ecosystems of freshwater basins and streams;
   - models of self-cleaning processes for oil, radionuclides, heavy metals, and other pollutants;
   - a model of radionuclide and heavy metal accumulation in river ecosystems of the Far North;
   - a model of the transport of radionuclide, heavy metal, and organic pollution by rivers into near-shore Arctic waters;
   - a model of pollution washout during spring in tundra and forest tundra zones;
   - a model of the kinetics and transformation mechanisms for biospheric elements in water systems;
   - a kinetics model of radionuclides and heavy metals in the food chains of land ecosystems in boreal zones;
   - a model of the surface flow of chemical elements and compounds from territories where open-cast mining takes place under the climatic conditions of the Far North;
   - a model of the seasonal influence of pollution on phytoplankton and primary production in northern seas.

(2) *Modeling the exchange processes of carbon dioxide and methane between tundra ecosystems and the atmosphere*. The global interaction of Arctic ecosystems with the biosphere and with the Earth's climatic system is mainly carried out by the influence of $CO_2$ and methane on biogeochemical cycles. Existing models of the global circulation of these greenhouse gases are incomplete in that they do not take this interaction into account. Present estimations of the gas exchange

between Arctic ecosystems and the atmosphere confirm the importance of taking them into account. To create a model set related to gas exchange in Arctic reservoirs it is necessary to compile a catalog of soil–plant formations, ice fields, and land-based and oceanic reservoirs. Moreover, it is necessary to have estimations of evapotranspiration, dead vegetation decomposition rate, and the productivity of vegetation communities readily available in the database. This model set will facilitate evaluation of the role of tundra ecosystems in forming the greenhouse effect.

(3) *Modeling the hydrological regime and estimating pollutant flows in the Arctic Basin.* It is necessary to prepare a set of models to describe the dynamics of the separate aquatories and of the whole hydrosystem of the Arctic Ocean, including:
- a complex model of water circulation in the Arctic Basin;
- regional models of water circulation in Arctic seas;
- a model of the kinetics of radionuclide, heavy metals, and organic pollutants in the trophic structures of Arctic marine ecosystems;
- a model of pollutant concentration spreading from a point-like source in near-coastal zones of the Arctic Basin;
- a model of thc transfer of radionuclide, heavy metals, and organic pollutants as a result of vertical mixing of Arctic waters;
- a model of the conservation and release processes as a result of the freezing and thawing of the ice cover.

(4) *Modeling Arctic ecosystems impacted by anthropogenic activity.* Anthropogenic influence in the Arctic Basin and on adjacent territories is connected with local, regional, and global activities. Therefore, it is necessary to create such models as:
- models showing the influence that radionuclides, heavy metals, and oil hydrocarbons have on the dynamics of marine ecosystems under Arctic climate conditions;
- models showing how vegetation cover is restricted as a result of the different types of pollution brought to land ecosystems with precipitation and surface flows;
- models for the dynamics of vegetation covers subjected to physical influence;
- models/scenarios for town and infrastructure development;
- models/scenarios for changes in the traditional seasonal regions of activity of nomadic peoples;
- models/scenarios for the social development of scattered peoples in the Far North.

(5) *Modeling the biogeochemical carbon cycle in the atmosphere–Arctic Ocean system.* Experience in modeling the $CO_2$ global cycle has shown that estimates of the role played by the World Ocean in redundant carbon absorption are, to put it mildly, rough (Kahl *et al.*, 1988; Kelley *et al.*, 1999a, b; Krapivin, 2000a; Krapivin and Vilkova, 1990; Nitu *et al.*, 2000a). For models to be more precise they need to be reinforced by more reliable parameterizations of the physical processes involved in the interaction between bordering layers of the atmosphere and the Arctic aquatories. According to many laboratory and natural observations, the direc-

tions these processes take depends on many factors. The most significant are wind speed, the presence of ice cover, and the vertical distribution profile of water temperature. Complex composition of these factors determines the variety of possible models and their details. A significant task here is to model the explosive growth of seaweed during the spring season and, hence, construct a parameterization system for the dynamics of photosynthetic processes under conditions of snow and ice covers and at the time of the spring thaw.

(6) *Development of a complex model of how hydrologic and biogeochemical systems function in the Arctic*. In addition to the model set intended for local and fragmentary processes in Arctic ecosystems and for understanding their global role, it is necessary to synthesize a single model for the whole complex of biogeochemical, biogeocenotic, and hydrologic processes that occur in boreal systems. Creation of such a model will facilitate development of database requirements and will make it possible to obtain a means of estimating the consequences of anthropogenic projects. Armed with such a model it will be possible to estimate the consequences of forest felling and fires, of the broadening of zones with disturbed land cover, of land and basin oil pollution, of hydrogeological changes in separate territories as a result of landscapes merging, of territorial pollution with waste materials from the mining industry, etc.
(7) *Estimation of the stability of Arctic systems under variable global climate conditions*. Human activities in the delicate ecosystems of the Far North need to be conducted with great care. How can the natural balance be safeguarded from the rising influence of human civilization and its conglomerates? How can the survival of these ecosystems be evaluated under all manner of situations? These and other questions need to be answered before any such development takes place.

An understanding of the environmental processes in Arctic regions, a prerequisite for finding scientific solutions to the problems arising there, can only be found by combining many disciplines, including ecology, oceanography, mathematical modeling, and system analysis. Kondratyev *et al.* (2003a) synthesizes many data sources and knowledge from various scientific fields in the Spatial Simulation Model of the Arctic Ecosystem (SSMAE). Separate blocks of the SSMAE were created earlier by many authors (Krapivin, 1995; Legendre and Krapivin, 1992; Mcintyre, 1999; Muller and Peter, 1992; Riedlinger and Preller, 1991). The sequence of these blocks in the SSMAE structure and its adaptation in the Spatial Global Model (SGM) provide a technological base for computer experiments (Kondratyev *et al.*, 2004a; Krapivin, 1993). The basic blocks of the SSMAE are oriented such that they describe the dynamics of any given pollutant. For consideration of a specific pollutant it is necessary to include in the SSMAE an additional block describing its physical and chemical characteristics.

Krapivin and Phillips (2001) demonstrate examples of blocks simulating the characteristics of radionuclides, heavy metals, and oil hydrocarbons. Consideration of these pollutants is restricted to elements whose properties can be averaged.

The Arctic Basin aquatory $\Omega$, which is studied, has boundaries that include peripheral Arctic seas as well as the coastline and southern boundaries of the Norwegian and Bering Seas. The discharge of human-made radionuclides, heavy metals, and oil carbohydrates into the Arctic Basin has been of international concern since the 1992 release of information on Soviet dumping of nuclear reactor and solid industrial waste. There are many experimental and theoretical results giving estimates of the growing dependences between the pollution dynamics in the World Ocean and the state of the continental environment. Let us highlight what we stated earlier, Arctic Basin pollution causes the most anxiety to investigators (Krapivin, 1995). The ecosystems of the Arctic seas are much more vulnerable than those of other seas. This is the reason the Arctic Basin is the object of investigations within the frameworks of many national and international biospheric programs (McCauley and Meier, 1991). The research strategy of these programs includes the theoretical and experimental study of tundra ecosystems, Siberian Rivers, and near-shore and open Arctic waters (Wielgolaski, 1997).

Let us again reiterate, the solution to environmental problems in the Arctic regions is possible via a framework of interdisciplinary approaches to the parameterization of different natural, and anthropogenic, processes. The SSMAE as the GIMS unit generates one such possibility.

Clearly, the processes at play in the Arctic Basin are much broader than those mentioned above. The principal objectives of many investigations can be summarized as:

i. Coordination of theoretical, experimental, and applied research activities focused on certain aspects of the global GHG cycle that take the correlation between climate change and Arctic and subarctic environments into account. Specific fields of interest include:
   - Setting up general principles for the synthesis of an Arctic and subarctic environment model that considers the complexity and interrelationships of all processes involved in the ocean–atmosphere–land system, especially the correlations between various factors regulating the intensity of energetic relationships.
   - Modeling the global biogeochemical cycles of $CO_2$, methane, and nitrogen by considering the spatial structures of land and oceanic ecosystems.
   - Integrating existing data about gas and energy exchanges between the atmosphere and land in northern territories with special focus on permafrost regions.
   - Parameterizing carbon–climate feedbacks at high latitudes to draw attention to shifts in surface cover.
   - Developing a model to simulate the dynamics of permafrost areas by employing ideas and algorithms for processing data as a result of monitoring Arctic and subarctic regions and by applying mathematical models and scenarios of natural and anthropogenic processes.

- Parameterizing changes in the snow cover of boreal regions by considering their role in regulating global climate through complex interactions among different elements of the Arctic and subarctic environment system.

ii. Setting up simulation experiments to estimate the consequences of possible climate changes in the dynamics of permafrost and relevant processes under various *a priori* suppositions about the intensities of flows of pollutants and about other anthropogenic impacts on ecosystems in this region. Possible situations and problems include the following:
   - An assessment of the role that different interactions play in the dynamics of Arctic and subarctic environments.
   - Study of the role anthropogenic factors play in the biospheric cycles of $CO_2$, nitrogen, and methane with special focus on long-range transport of chemical contaminants to northern latitudes.
   - An analysis of correlations between the dynamics of permafrost and biogeochemical cycles in Arctic and subarctic environments.
   - An estimation of land cover shifts as a result of climate change under realistic scenarios.
   - Detection of variations in the thickness of the permafrost layer during the last century to facilitate understanding the laws that manage the energy balance in Arctic and subarctic environments.
   - Calculation of the characteristics of spatiotemporal variations in surface heating between the boreal forest and adjacent tundra.
   - Revealing changes in interactions within Arctic and subarctic environments that in turn affect the dynamics of regional permafrost, biogeochemistry of terrestrial ecosystems, and GHG dynamics.
   - An assessment of the likely direction that cryogenic processes will take under climate change.

iii. Making recommendations to international organizations, governments, and other institutions authorized to take decisions about environmental affairs related to the protection of Arctic and subarctic environments. They will include:
   - ranking of hazardous factors;
   - priorities in remediation and protection measures;
   - assessment of technological impacts connected to human-induced or human-initiated negative processes.

Solution of these and others problems related to Arctic and subarctic environments is possible by employing the Multi-Functional Information Modeling System (MFIMS) (Soldatov, 2010, 2011; Soldatov *et al.*, 2010). MFIMS undertakes the following functions:

- examination of the role played by changes in hydrological cycles and land covers to determine climate responses;
- calculation of the correlations between natural climate variability and anthropogenic changes in the northern hemisphere;

- forecasting changes in Arctic and subarctic environments caused by global climate changes;
- making projections of the optimal data structure and assimilation process necessary to ensure modeling results are of the highest reliability;
- identification of controlling factors to determine distinctive characteristics of climate change in Arctic and subarctic regions.

### 4.4.2 Upwelling ecosystem

Deep-water upwelling zones in the World Ocean are highly productive. Upwelling results from surface waters being wind-driven from the shore, dispersing currents, or water retreat from the shore for other reasons. The rates at which water lifts and upwelling stability are determined by several synoptic parameters. The most characteristic value for the vertical velocity of water lifting in an upwelling zone is $0.77 \cdot 10^{-3}$ cm/s. The depths at which water flow motion starts vary within 200 m.

Let us consider the ocean ecosystem in the upwelling zone, beginning with the concept of successive development of a community from the moment it forms in deep waters lifting to its final state in the oligotrophic region of convergence. Let us derive a model of the upwelling ecosystem (MUE). During these two states the system develops in time and, respectively, moves along with the water flow. In this case changes take place both in the total energy supply in the community and in its structure (spatial, trophic, and specific). The basic moments of these changes have generally been studied in numerous field investigations. Therefore, how well model calculations correspond to these ideas can be one of the criteria for model adequacy.

Let us suppose the ecosystem from the zone of upwelling develops uniformly horizontally, so that changes are observed both in depth $z$ with a step $\Delta z = 10$ m and in time with a step $\Delta t = 1$ day. The horizontal velocity of water flow from the upwelling zone is assumed to be constant and equal to $V = V_\varphi = V_\lambda$, so that the distance of a unit volume of water from the upwelling zone is equal to $\Delta r = (\Delta\varphi^2 + \Delta\lambda^2) = V\Delta t$.

The state of the ecosystem on each horizon $z = \text{const}$ is determined by illumination $E(z,t)$, concentration of nutrient elements $n(z,t)$ and detritus $d(z,t)$, and the biomasses of phytoplankton, $p(z,t)$, bacteria $b(z,t)$, protozoa $Z_1(z,t)$, microzooplankton $Z_2(z,t)$, small filter feeders $Z_3(z,t)$, large filter feeders $Z_4(z,t)$, small predatory cyclopoids $Z_5(z,t)$, predatory calanoids $Z_6(z,t)$, large predatory chaetognaths and polychete $Z_7(z,t)$. Protozoa include infusorians and radiolarians and microzooplankton include the naupliuses of all copepods. Based on studies of the food tropical plankton eat, small filter feeders (<1 mm in size) include plankton such as *Oikopleura*, copepods *Clausocalanus*, *Paracalanus*, *Acartia*, *Lucicutia*, and the small ostracod *Conchoecia*. Large filter feeders include copepods larger than 1 mm, such as *Undinula*, *Pleuromamma*, *Centropages*, *Temora*, *Scolecythrix*, and others like baby euphausiids, pteropods, etc. In the trophic pyramid the impact of nekton is considered negligibly small.

It is assumed that 30% of the mass of bacteria are in natural aggregates >3–5 μm in size, which can be eaten and assimilated by filter feeders ($Z_3, Z_4$); naupliuses ($Z_2$), protozoa ($Z_1$), and small filter feeders ($Z_3$) can also feed on non-aggregated bacterioplankton.

The trophic bonds between elements are described according to the energy principle. Biomass, rates of production and exchange, and rations are expressed in energy units (cal/m$^3$ or cal/m$^2$).

As in the general model of the ocean ecosystem, the original source of energy and substance in the community is the primary production of phytoplankton ($R_p$). Solar radiation energy ($E$) and nutrient elements ($n$) are additional sources. The vertical structure of the water medium is described by a three-layer model with one layer above the thermocline, one layer of sharp gradients, and one layer of lower gradients beneath the thermocline. The upper ($z_b$) and lower ($z_l$) boundaries of the layer with sharp gradients are assumed to be located (depth in meters):

$$z_b = \begin{cases} 10 + 2,2t & \text{for } 0 \le t \le 50 \text{ days,} \\ 120 + 0,6(t-50) & \text{for } t > 50 \text{ days,} \end{cases}$$

$$z_l = \begin{cases} 30 + 2,4t & \text{for } 0 \le t \le 50 \text{ days,} \\ 150 + 1,4(t-50) & \text{for } t > 50 \text{ days.} \end{cases}$$

Thus, the thermocline at an initial time moment $t = 0$ is located in the layer 10–30 m, descending gradually down to 120–150 m on the 50th day and to 150–190 m on the 100th day.

The supply of nutrient elements in the 0–200 m layer is replenished at the expense of detritus decomposition and emission of unassimilated food by living elements of the community, as well as at the expense of the input from deep layers ($z > 200$ m) due to turbulent mixing (coefficient $k_2$) and vertical advection (velocity $V_Z$). The content of nutrients in organic matter is constant and constitutes 10%.

The following function is used as the basic correlation to simulate photosynthesis (Kondratyev *et al.*, 2004a):

$$R_p = k_T[1 - 10^{-0.25p\gamma(t)}][1 - 10^{-0.1n}]^{0.6}, \tag{4.24}$$

where the function $\gamma(t)$ characterizes the time dependence of the production/biomass ratio (P/B coefficient) for phytoplankton. A maximum P/B for phytoplankton at the point of upwelling ($t = 0$) is assumed to be 5, then the P/B coefficient decreases and on the 15th day reaches 1, after which it remains constant. Hence the equation describing the dynamics of the phytoplankton biomass is written in the form (Kondratyev *et al.*, 2002):

$$\partial p/\partial t = R_p - t_p p - \mu_p p - \sum_{j \in \omega_0 \setminus p} C_{pj} R_j + k_2\, \partial^2 p/\partial z^2 + (V_z - w_p)\, \partial p/\partial z,$$

where the coefficient $w_p$ (cm/day) describes the process of gravitational sedimentation and is estimated at

$$w_p = \begin{cases} 50 & \text{for } 0 \leq z \leq z_b, \\ 10 & \text{for } z_b < z < z_l, \\ 30 & \text{for } z \geq z_l. \end{cases}$$

Bacteria feed on detritus and dissolved organic matter emitted by phytoplankton. Allochthonous dissolved organic matter is neglected and it is supposed that the following restriction exists:

$$T_p + T_b + T_Z + T_r + T_D = 0.3R_p.$$

It is also assumed that $\mu_b = 0.01$, $t_b = 0.75\lfloor 1 - 10^{\zeta} \rfloor$, $R_p = 3b\lfloor 1 - 10^{\zeta} \rfloor$, $\zeta = -0.2d - 0.3R_p$.

It has been established that bacterioplankton can consume no more than 10% of all the detritus located in the same water layer in a 24 h period.

By making these assumptions, the equation for biogenic salts can be written as:

$$\partial n/\partial t = -0.1R_p + 0.1d + 0.05 \sum_{i \in \omega_0} t_i B_i + k_2\, \partial^2 n/\partial z^2 + V_z\, \partial n/\partial z,$$

where $\omega_0 = \{p, b, Z_j \ (j = 1\text{–}7)\}$; and $B_i$ is the biomass of the $i$th element.

The coefficients $k_2$ (cm$^2$/day) and $V_z$ (cm/day) are assumed to be:

$$V_z = \begin{cases} 5 & \text{for } 0 \leq z \leq z_b, \\ 0.1 & \text{for } z_b < z < z_l, \\ 1 & \text{for } z \geq z_l, \end{cases} \qquad k_2 = \begin{cases} 200 & \text{for } 0 \leq z \leq z_b, \\ 50 & \text{for } z_b < z < z_l, \\ 150 & \text{for } z \geq z_l. \end{cases}$$

The formula describing the penetration of sunlight into deep layers of the ocean follows an exponential law:

$$E(z_n, t) = E_0 \cdot 10^{-\alpha z_n},$$

where $z_n$ is the depth of the $n$th layer (meters);

$$\alpha = 0.01 + 0.001 \sum_{k=0}^{n} [p(z_k, t) + d(z_k, t)]/z_n.$$

A marked suppression of phytoplankton is observed at $E > 100$ cal/cm$^2$ day. Hence, the formula for $k_T$ in equation (4.24) is written:

$$k_T = 0.041E \cdot 10^{0.25[1 - E/E_{\max}]},$$

where $E_{\max} \approx 70$ cal/cm$^2$ day.

For this case the equation for the dynamics of bacterioplankton biomass is written in the form:

$$\partial b/\partial t = R_b - \mu_b b - t_b b - \sum_{j \in \omega_0 \backslash (p,b)} C_{bj} R_j + k_2\, \partial^2 b/\partial z^2 + (V_z - w_p)\, \partial b/\partial z.$$

The rate of change of the biomass of each element of zooplankton $\partial Z_i/\partial t$ ($i = 1$–7) is determined by the level of consumption (ration) $R_{Z,i}$, food assimilation

$1/u_{Z,i}$, expenditure on energy exchange $T_{Z,i} = t_{Z,i}Z_i$, the rate of dying off $\mu_{Z,i} = 0.01Z_i$, consumption of the $i$th element by the $j$th element with the coefficient $C_{ij}$, and the age transition of naupliuses into copepod stages of filter feeders and predators. Thus

$$\partial Z_i/\partial t = \Omega_{m,i} + (1 - h_{Z,i})R_{Z,i} - (t_{Z,i} + \mu_{Z,i})Z_i - \sum_{j \geq i}^{7} C_{ij}R_{Z,j} \quad (i = \overline{1,7}),$$

where $h_{Z,i} = 1 - 1/u_{Z,i}$; and $\Omega_{m,i}$ characterizes the age transition of naupliuses to another category of zooplankton. The transition of naupliuses $Z_2$ to copepods of filter feeders $(Z_3, Z_4)$ takes place with an intensity

$$\Omega_{m,3} = Z_2Z_3(15Z^*)^{-1},$$

$$\Omega_{m,4} = Z_2Z_4(20Z^*)^{-1},$$

and for predatory copepods $(Z_5, Z_6)$ with an intensity

$$\Omega_{m,5} = Z_2Z_5(15Z^*)^{-1},$$

$$\Omega_{m,6} = Z_2Z_6(20Z^*)^{-1},$$

where $Z^* = Z_3 + Z_4 + Z_5 + Z_6$.

To calculate the food ration for zooplankton elements, the coefficient $\nu$ is determined from the condition $\nu = 0.01/B_{\min}$. When calculating the coefficients $C_{ij}$, it is assumed that food requirements refer to various food objects in proportion to their biomasses after allowing for food selectivity. The coefficients $C_{ij}$ for all elements of the ecosystem are given in Figure 2.4.

The daily migrations of zooplankton are simulated by adding to the daily ration of elements $Z_j$ $(j = 4\text{–}7)$, located in the layer 0–50 m, some share $(k_Z)$ of the total food requirements of the same elements of the community but in the layer 50–200 m. The coefficient $k_Z$ is considered to be time dependent:

$$k_z = \begin{cases} 0.02 + 0.0016t & \text{for } t \leq 50 \text{ days,} \\ 0.1 & \text{for } t > 50 \text{ days.} \end{cases}$$

Various scenarios are possible here for both the migration and nutrition of zooplankton elements. It is only known for certain that some share of the zooplankton elements from deep layers satisfy the needs of others of their kind for food in the upper layers of the ocean.

In this case the equation for detritus is written as:

$$\partial d/\partial t = \sum_{i \in \omega_0} (H_{Z,i} + M_{Z,i}) - \sum_{j \in \omega_0 \backslash p} C_{dj}R_j + k_2\, \partial^2 d/\partial z^2 - \mu_d d + (V_z - w_d)\, \partial d/\partial z,$$

where the coefficient of the rate of gravitational sedimentation of detritus is taken as (cm/day):

$$w_d = \begin{cases} 25 & \text{for } z_b \leq z \leq z_l, \\ 50 & \text{for } z < z_b, z > z_l. \end{cases}$$

Following equation (2.64) the biocomplexity indicator in this case is defined as:

$$\xi(\varphi, \lambda, z, t) = \sum_{i=1}^{9} \sum_{j=1}^{14} x_{ij} C_{ij}.$$

To accomplish these calculations it will be necessary to assume that all ecosystem components at the upwelling point ($t = 0$) are uniformly distributed by depth within the water column: $n(z,0) = 250\,\mathrm{mg \cdot m^{-3}}$, $d(z,0) = 0$, $b(z,0) = 1$, $p(z,0) = 0.5$, $Z_i(z,0) = 0.01$–$0.5$ ($i = 2$–$7$), $Z_1(z,0) = 0.0001\,\mathrm{cal \cdot m^{-3}}$. Note that variations of these estimations that decrease or increase by a factor of 50 have little effect on the character of system dynamics in the future. The system really does "forget" variations of the initial concentrations during the first 50 days of its development.

Zero gradients for all components except $n$ and $d$ are given on the upper ($z = 0$) and lower ($z = 200\,\mathrm{m}$) boundaries. The boundary conditions for $n$ and $d$ components have the values:

$$\left.\frac{\partial n}{\partial z}\right|_{z=0} = 0; \qquad \left.\frac{\partial d}{\partial z}\right|_{z=0} = 0; \qquad \left.\frac{\partial d}{\partial z}\right|_{z=200} = -0.5; \qquad n(200, t) = 250\,\mathrm{mg \cdot m^{-3}}.$$

When estimating changes in the system over time, it was assumed that water takes more than 60 days to cover the distance from the upwelling zone to the oligotrophic zone of planetary convergence. Figure 2.5 gives a representation of how biomasses change as a function of time. We see that the biomass of phytoplankton increases most rapidly here, reaching its maximum ($\approx 4{,}500\,\mathrm{cal \cdot m^{-2}}$) on the 5–10th day of the existence of the system. After this, the biomass decreases. The maximum biomass of bacterioplankton is reached on the 10–15th day. The small herbivores lag somewhat behind phytoplankton in development, and their biomass reaches its maximum only on the 30th day. Nevertheless, its joint influence with the decrease in nutrient salts leads to a sharp drop in the biomasses of phytoplankton and bacterioplankton. For example, the community's inverse chains give weak contributions to $R_p$ and $R_b$. After the 40th day phytoplankton mainly function at the expense of biogenic elements rising in the eutrophic zone across the thermocline from deeper layers. Subsequently, the biomass of phytoplankton decreases slowly. This is the time that stability factors begin to influence the community at the expense of exterior energy flows.

The carnivores prove to be still more inertial than the herbivores as their biomass attains its maximum only on the 35–50th day. At the end of this time range (i.e., the 50–60th day), the system reaches a quasi-stationary state characterized by low concentrations of all living components.

Thus, the present model demonstrates the spatial disconnection and alternation of biomass maxima of phytoplankton, herbivores, and carnivores as water moves farther away from the upwelling zone. System degradation observed 60 days after its development is reported by many authors. From Figure 2.5 we see that indicator $\xi$ describes the ecosystem state with high precision. It shows the ecosystem reaching its maximal complexity on the 30–45th day of development.

**Table 4.1.** Comparison of model results with experimental assessment of components of the upwelling ecosystem (cal/m$^2$).

| *Ecosystem component* | *Community in the middle stage (30 to 40 days)* | | | | *Mature community (60 to 80 days)* | | | |
|---|---|---|---|---|---|---|---|---|
| | *Model* | | *Observation* | | *Model* | | *Observation* | |
| | *Day 30* | *Day 40* | *Day 30* | *Day 40* | *Day 70* | *Day 80* | *Day 70* | *Day 80* |
| Phytoplankton, $p$ | 1,319 | 1,092 | 2,000 | 1,109 | 827 | 734 | 900 | 667 |
| Bacteria, $b$ | 1,673 | 864 | 4,100 | 3,905 | 564 | 465 | 2,180 | 701 |
| Microzoa, $Z_2$ | 394 | 303 | 321 | 298 | 300 | 378 | 411 | 362 |
| Small-sized herbivores, $Z_3$ | 1,338 | 612 | 525 | 465 | 290 | 287 | 74 | 65 |
| Large-sized herbivores, $Z_4$ | 1,416 | 726 | 420 | 376 | 252 | 231 | 164 | 114 |
| $Z_3 + Z_4$ | 2,754 | 1,338 | 945 | 858 | 542 | 482 | 238 | 223 |
| Small-sized predators, $Z_5$ | 624 | 491 | 495 | 493 | 203 | 179 | 236 | 187 |
| Predators (calanoids), $Z_6$ | 288 | 600 | 610 | 609 | 191 | 189 | 175 | 178 |
| Large-sized predators, $Z_7$ | 184 | 183 | 15 | 12 | 102 | 54 | 51 | 44 |
| $Z_5 + Z_6 + Z_7$ | 796 | 1,274 | 1,110 | 1,114 | 496 | 422 | 462 | 409 |

It is very interesting to compare the quantitative estimations of any values obtained from the model with expeditional investigations. Table 4.1 gives an example of such a comparison. The visible discrepancy between the model and experimental estimations of the bacterioplankton biomass is the result of inexact description of its food chain.

The vertical distribution of the biomass component is also subject to essential change in time. It is seen that at times when the system has only just formed ($t \approx 5$ days) and the total biomass of phytoplankton is near its maximum, the biomass of phytoplankton has a uniform vertical distribution in the 0–50 m layer. All other living components have more or less pronounced maxima narrow enough in depth to be connected to the thermocline. However, after 10 days, biogenic salt storage in the upper layer is almost exhausted. In this state it is possible for the biomass of phytoplankton maximum to be reached only at a depth of 10–20 m. A second maximum is formed deeper at the expense of biogenic salts moving across the thermocline. A similar structure having two maxima is formed in the vertical distribution of other components of the ecosystem. It arises at the 20–30th day and is particularly clear when both maxima are well expressed and, owing to the thermocline descending by up to 55–75 m, they are clearly separated by depth.

The upper maximum is formed mainly as a result of the horizontal transport of nutrient salts in the production–destruction cycle of the community. As can be seen from simulation results, the upper maximum is at first more distinctly pronounced than the lower one. As the nutrient salts of the surface layer are consumed, vertical transport across the thermocline begins to play a significant role and, owing to this, the lower maximum is now much larger than the upper one. Subsequently, under the thermocline descending by 80–100 m (sometimes more), illumination becomes the limiting factor for photosynthesis. This situation appears when the lower maximum moves away from the thermocline and its position is defined by the illumination limit and by the flow of nutrient salts from deeper layers. This picture of the vertical distribution of phytoplankton is consistent with the patterns obtained from observations in the World Ocean.

Under a more mature state of the system ($t > 50$–$60$ days) the complete disappearance of the upper maximum is typical for oligotrophic and ultra-oligotrophic waters. This picture exists mainly for the vertical distribution of the biomass of phytoplankton. Other living components of the ecosystem can have a small upper maximum as a result of environmental processes that the model does not take into account.

The results of simulation modeling do not contradict the existing consensus of the structure and behavior of upwelling ocean ecosystems. It is impossible to demand exact numerical correspondence between model results and the real state of the ecosystem. Nevertheless, we can consider in this context more general characteristics such as the stability of the ecosystem. The ecosystem is considered stable in the time interval $(0, T)$ if its state variables (in this case $B_i$, $i = 1 - n$) are within the interval:

$$B_{i,\min} \le B_i \le B_{i,\max} \quad (i = 1 - n).$$

Actually, the right-hand side of this condition is unnecessary. Namely, when one ecosystem component has $B_i \to \infty$, there is always another with $B_j \to 0$, as follows from the data in Figure 2.4. Therefore, the stability condition for an upwelling ocean ecosystem can be rewritten as:

$$B_{\min} \le \sum_{i=1}^{n} B_i,$$

where $B_{\min}$ is the total minimal value of the biomass of the ecosystem (specifically, $B_{\min} = B_{1,\min} + \cdots + B_{n,\min}$).

This criterion was used to estimate model stability under alteration of various parameters. Some results are given in Table 4.2. From these data we see that a deviation in food assimilation ($u^{-1}$) by $\pm 20\%$ leads to a variation of $Z_3 + Z_4$ by 48% and of $Z_2$ by 70%. Variations in respiration coefficients $t_Z$ are not very strong. The same variations in the rate of energy assimilation $T$ and non-assimilated food $H$ cause equivalent changes in the vertical distributions of zooplankton components. Small zooplankton and herbivores are more sensitive to variations in $T$ and $H$ than are predators.

**Table 4.2.** Dynamics of the ecosystem structure as a result of change in the initial biomass of predators ($cal \cdot m^{-2}$). Basic initial biomasses are

$$Z_5(z, t_0) = Z_6(z, t_0) = Z_7(z, t_0) = 0.1 \text{ cal} \cdot \text{m}^{-3};$$

$$Z_0 = \int_0^{200} \{Z_5(z, t_0) + Z_6(z, t_0) + Z_7(z, t_0)\}\, dz.$$

Notation is given in Table 4.1.

| *Component* | *5th day* | | | *30th day* | | |
|---|---|---|---|---|---|---|
| | $Z_0$ | $100\ Z_0$ | $10^3\ Z_0$ | $Z_0$ | $100\ Z_0$ | $10^3\ Z_0$ |
| $b$ | 1,789 | 2,263 | 2,263 | 2,269 | 2,159 | 1,995 |
| $p$ | 3,264 | 3,462 | 3,794 | 3,971 | 2,086 | 2,614 |
| $Z_1$ | 31 | 12 | 3 | 18 | 3 | 2 |
| $Z_2$ | 178 | 106 | 101 | 109 | 265 | 465 |
| $Z_3$ | 70 | 85 | 80 | 80 | 201 | 415 |
| $Z_4$ | 67 | 14 | 74 | 145 | 156 | 191 |
| $Z_5$ | 18 | 14 | 9 | 114 | 119 | 53 |
| $Z_6$ | 1 | 25 | 41 | 14 | 25 | 22 |
| $Z_7$ | 36 | 234 | 183 | 213 | 437 | 821 |

It follows from Table 4.2 that large variations in the initial data do not significantly influence ecosystem dynamics. More detailed calculations show that the ecosystem relaxes these variations betweeen the 20–25th day when its interior energetic interactions are balanced. Stable correlations between the trophic levels are set during the beginning period near the upwelling area.

Many experimental measurements show how variations in illumination influence the structure of the vertical ecosystem (Sorokin and Mikheev, 1979; Vinogradov *et al.*, 1975). We see that an increase in illumination at the ocean surface leads to insignificant lowering of the lower maximum and more rapid decrease of the biogenic salt concentration above the thermocline. When $E_0 = 2{,}000 \text{ kcal} \cdot \text{m}^{-2} \text{ day}^{-1}$, maximal photosynthesis is at the surface. Variations in $E_0$ from 2,000 to $7{,}000 \text{ kcal} \cdot \text{m}^{-2} \text{ day}^{-1}$ in the absence of biogenic limitation lead to a change in the vertical distribution of the biomass of the ecosystem.

Figure 4.1 demonstrates how ecosystem biocomplexity changes depending on illumination variations. We see that ecosystems change complexity with depth as they adapt to the illumination level. This once again shows that the introduced biocomplexity indicator really does reflect ecosystem dynamics.

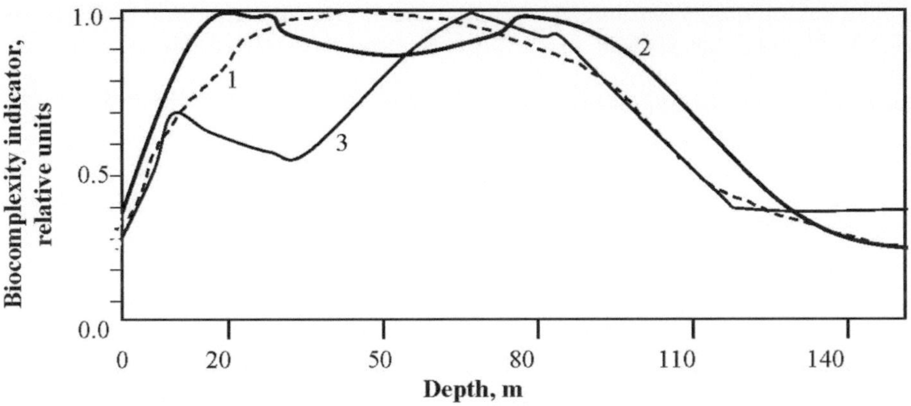


**Figure 4.1.** The dependence of biocomplexity indicator $\xi^* = \xi/\xi_{max}$ on depth $z$ under solar illumination $E_0$ change: 1, $E_0 = 2{,}000\ \mathrm{kcal \cdot m^{-2}\ day^{-1}}$; 2, $E_0 = 3{,}000\ \mathrm{kcal \cdot m^{-2}\ day^{-1}}$; and 3, $E_0 = 7{,}000\ \mathrm{kcal \cdot m^{-2}\ day^{-1}}$.

Model results show how change in the initial biogenic concentration can influence the biomass dynamics of phytoplankton. It follows that variations in nutritive salts above $100\ \mathrm{mg \cdot m^{-3}}$ at moment $t_0$ have practically no influence on the ecosystem state in distant days but only at the initial period of ecosystem development.

The P/B coefficients of phytoplankton and other ecosystem components are important model parameters. For example, if we suppose that the P/B coefficient of phytoplankton decreases over a 15-day period by up to 0.6, the ecosystem practically becomes extinct by the 35th day. Increase in the P/B coefficient of phytoplankton leads to rapid improverishment of the biogenic reserve in the upper layer. This precipitates the disappearance of the upper maximum. As follows from calculations, under $R_p/p \equiv 3$ the biomass of phytoplankton fluctuates widely and an unreal situation arises. This says that in reality the dependence of the P/B coefficient of phytoplankton on time is defined by a diminishing function. Investigation of this function is an important stage for the modeling procedure.

The model of the upwelling ecosystem as considered here raises many questions concerning interactions between trophic levels and another about the aggregation class for trophic levels. Simulation experiments with some variations in the trophic pyramid of the ecosystem show that in the absence of predators the biomass of herbivores increases without limit and the ecosystem loses its stability by the 50th day. In other words, the relations between trophic levels and the disaggregation of the ecosystem are certainly important to achieve the best correlation between model results and field observations. Undoubtedly, further refining of simulation techniques and incorporation of components of higher trophic levels—primarily of fish—will make the model more complete and representative.

The set of model parameters, whose variations are important for ecosystem stability, includes the vertical speed $v_z$ of the motion of water. Model calculations show that the ecosystem is dynamically stable for $v_z \in [10^{-4}, 10^{-1}]\,\mathrm{cm \cdot s^{-1}}$. When $v_z$ is faster or slower, the vertical structure of the ecosystem is changed to a great extent and its stability is broken.

### 4.4.3 Peruvian Current ecosystem

The Peruvian Current moves northward along the coast of South America where it causes an upwelling of cold, nutrient-rich, and oxygen-rich water. This current belongs to a class of World Ocean currents where complex interrelated physical, biological, and chemical processes are taking place within a relatively small area, characterized by a high concentration of nutrients and large quantities of matter for the production of phytoplankton and other living elements (Sorokin 1977). In this region negative environmental conditions occur as a result of El Niño events. During an El Niño warm water appears at the surface of the ocean off South America and kills off marine life. In addition, there are many other global impacts owing to the resultant climate variability (e.g., in World Ocean pollution and fishing). Anthropogenic pressure on the Peruvian Current ecosystem (PCE) is the result of increased fish production.

Off the coasts of Peru and Chile the production of organic matter is so large that there is insufficient oxygen for its oxidation. This results in the emergence of an oxygen-free layer at the 100–800 m depth. The occurrence of such a layer is accompanied by the fauna dying off. Off the coast of Peru there occurs the globe's largest upwelling of abyssal waters enriching the photic layer of the adjoining ocean area with nutrients. For a description of this process we will use the model described in Section 4.2. We will also consider the nutrients as a common element of the PCE and follow the hypothesis that in the PCE area the correlation between nitrogen, phosphate, and silicate is stable with nitrogen as the limiting nutrient.

The investigation of bio-producing processes in the Peruvian Current and its immediate environs was carried out on board research vessels from various countries (Vinogradov *et al.* 1977). The results of these investigations were taken into account when elaborating the simulation unit of the GMNSS that describes bio-oceanographic fields in the Peruvian Current (Krapivin, 1996) and when guiding its further improvement and investigations.

This section considers the PCE area $\Omega$ as part of the East Pacific bounded by latitudes $0 \le \varphi \le 45°\mathrm{S}$ and longitudes $\lambda \le 90°\mathrm{W}$. In contradistinction to earlier-created models of the PCE this unit simulates the trophic pyramid in detail and gives the spatial distributions of PCE elements over the wide area represented by $\Omega$. In the hierarchical composition of the PCE trophic pyramid it is possible to identify the components shown in Figure 4.2. What marks this block diagram is that it takes into account the element of interaction between land and oceanic biocenoses, as expected by the presence of trophic relationships between $B_5$, $B_7$, $B_9$, and $B_8$. Introduction of this relationship appears to be necessary because sea-

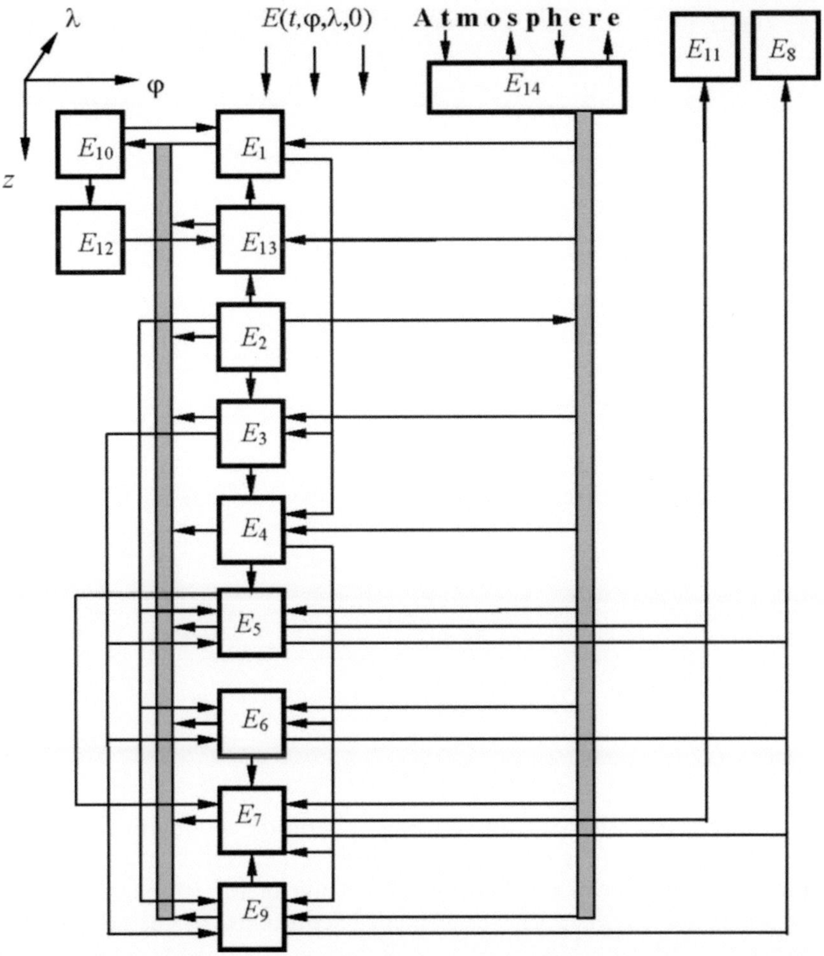


**Figure 4.2.** Block diagram of the PCE. Notation: $E_1$, bacterioplankton; $E_2$, phytoplankton; $E_3$, phytophages; $E_4$, predatory zooplankton; $E_5$, commercial anchovy; $E_6$, anchovy larvae; $E_7$, predatory fish; $E_8$, birds; $E_9$, young anchovy; $E_{10}$, detritus; $E_{11}$, biomass of fished anchovy; $E_{12}$, nutrients; $E_{13}$, dissolved organic matter; and $E_{14}$, dissolved oxygen.

birds are responsible for the perceived trophic pressure on the above components. Anchovy accounts for 80 to 90% of the diet of seabirds. In Figure 4.2 a single relationship describes the consumption of anchovy by other fishes. It is assumed that $B_7$ mainly refers to the Pacific bonito (*Sarda chiliensis*), though anchovy is part of the diet of other animals as well.

The block diagram reflects elements of the interaction between the atmosphere and the ocean reduced to reciprocal oxygen exchange at incoming and outgoing rates of the current. Input to the entire system is solar radiation energy

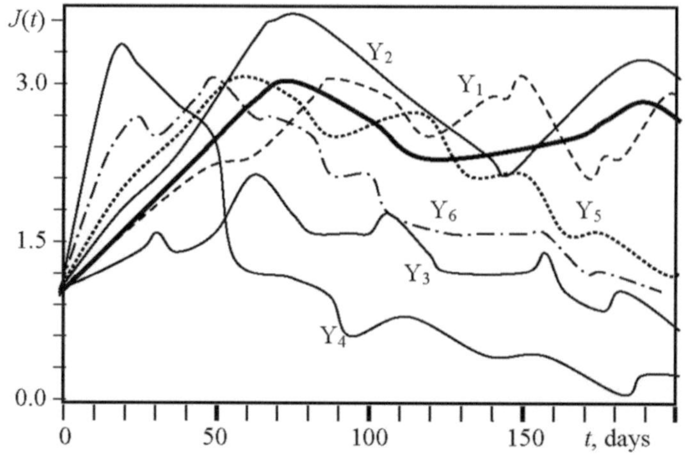


**Figure 4.3.** Dynamics of the survivability function $J(t)$ versus time for the experiments in variation in the food spectra $Y_s$ ($s = 1$–$6$) as given in the text. The thick curve corresponds to the model result. Function $J(t)$ is defined as:

$$J(t) = \frac{\sum_{i=1}^{9} \iint_{(\varphi,\lambda)\in\Omega} \int_z B_i(t, \varphi, \lambda, u)\, d\varphi\, d\lambda\, du}{\sum_{i=1}^{9} \iint_{(\varphi,\lambda)\in\Omega} \int_z B_i(t_O, \varphi, \lambda, u)\, d\varphi\, d\lambda\, du}.$$

$E(t, \varphi, \lambda, z)$ whose intensity depends on time $t$, latitude $\varphi$, longitude $\lambda$, and depth $z$. The block diagram (Figure 4.2) fixes food chains in the PCE. There are uncertainties in this knowledge. The food spectrum of the $j$th element is a vector $S_j = (\eta_{ji})$, where $\eta_{ji} = 1$ when the $i$th element is consumed by the $j$th element, and $\eta_{ji} = 0$ otherwise. We consider a set of simulation experiments whose results are given in Figure 4.3. These experiments correspond to the following variations in the food spectra:

$$Y_1 = \begin{Vmatrix} 0 & 0 & 0 & 0 & 0 & 0 & 0 & 0 & 0 & 1 & 0 & 1 & 1 \\ 0 & 0 & 0 & 0 & 0 & 0 & 0 & 0 & 0 & 0 & 1 & 0 & 1 \\ 1 & 1 & 0 & 0 & 0 & 0 & 0 & 0 & 0 & 1 & 0 & 1 & 1 \\ 1 & 1 & 1 & 1 & 0 & 0 & 0 & 0 & 0 & 1 & 0 & 1 & 1 \\ 0 & 1 & 1 & 1 & 0 & 0 & 0 & 0 & 0 & 0 & 0 & 0 & 1 \\ 0 & 1 & 1 & 1 & 0 & 0 & 0 & 0 & 0 & 1 & 0 & 0 & 1 \\ 0 & 0 & 0 & 0 & 1 & 1 & 1 & 0 & 1 & 1 & 0 & 0 & 1 \\ 0 & 0 & 0 & 0 & 1 & 0 & 1 & 0 & 1 & 0 & 0 & 0 & 0 \\ 0 & 1 & 1 & 1 & 0 & 0 & 0 & 0 & 0 & 1 & 0 & 0 & 1 \end{Vmatrix},$$

$$Y_2 = \begin{Vmatrix} 0 & 0 & 0 & 0 & 0 & 0 & 0 & 0 & 0 & 1 & 0 & 1 & 1 \\ 0 & 0 & 0 & 0 & 0 & 0 & 0 & 0 & 0 & 0 & 1 & 0 & 1 \\ 1 & 1 & 0 & 0 & 0 & 0 & 0 & 0 & 0 & 0 & 0 & 1 & 1 \\ 0 & 1 & 1 & 0 & 0 & 0 & 0 & 0 & 0 & 0 & 0 & 1 & 1 \\ 0 & 1 & 0 & 0 & 0 & 0 & 0 & 0 & 0 & 0 & 0 & 0 & 1 \\ 0 & 1 & 0 & 0 & 0 & 0 & 0 & 0 & 0 & 0 & 0 & 0 & 1 \\ 0 & 0 & 0 & 0 & 1 & 0 & 1 & 0 & 1 & 1 & 0 & 0 & 1 \\ 0 & 0 & 0 & 0 & 1 & 0 & 1 & 0 & 1 & 0 & 0 & 0 & 0 \\ 0 & 1 & 0 & 0 & 0 & 0 & 0 & 0 & 0 & 0 & 0 & 0 & 1 \end{Vmatrix},$$

$$Y_3 = \begin{Vmatrix} 0 & 0 & 0 & 0 & 0 & 0 & 0 & 0 & 0 & 0 & 0 & 1 & 1 \\ 0 & 0 & 0 & 0 & 0 & 0 & 0 & 0 & 0 & 0 & 1 & 0 & 1 \\ 1 & 1 & 0 & 0 & 0 & 0 & 0 & 0 & 0 & 0 & 0 & 0 & 1 \\ 0 & 0 & 1 & 1 & 0 & 0 & 0 & 0 & 0 & 0 & 0 & 1 & 1 \\ 0 & 1 & 0 & 0 & 0 & 0 & 0 & 0 & 0 & 0 & 0 & 0 & 1 \\ 0 & 1 & 0 & 0 & 0 & 0 & 0 & 0 & 1 & 0 & 0 & 0 & 1 \\ 0 & 0 & 0 & 0 & 1 & 0 & 0 & 0 & 0 & 0 & 0 & 0 & 1 \\ 0 & 0 & 0 & 0 & 1 & 0 & 1 & 0 & 1 & 0 & 0 & 0 & 0 \\ 0 & 1 & 0 & 0 & 0 & 0 & 0 & 0 & 0 & 0 & 0 & 0 & 1 \end{Vmatrix},$$

$$Y_4 = \begin{Vmatrix} 0 & 0 & 0 & 0 & 0 & 0 & 0 & 0 & 0 & 1 & 0 & 1 & 1 \\ 0 & 0 & 0 & 0 & 0 & 0 & 0 & 0 & 0 & 0 & 1 & 0 & 1 \\ 1 & 0 & 0 & 0 & 0 & 0 & 0 & 0 & 0 & 0 & 0 & 1 & 1 \\ 0 & 0 & 1 & 0 & 0 & 0 & 0 & 0 & 0 & 0 & 0 & 1 & 1 \\ 0 & 0 & 1 & 0 & 0 & 0 & 0 & 0 & 0 & 0 & 0 & 0 & 1 \\ 0 & 1 & 1 & 0 & 0 & 0 & 0 & 0 & 0 & 0 & 0 & 0 & 1 \\ 0 & 0 & 0 & 0 & 1 & 1 & 1 & 0 & 0 & 0 & 0 & 0 & 1 \\ 0 & 0 & 0 & 0 & 1 & 0 & 1 & 0 & 1 & 0 & 0 & 0 & 0 \\ 0 & 1 & 0 & 0 & 0 & 0 & 0 & 0 & 0 & 0 & 0 & 0 & 1 \end{Vmatrix},$$

$$Y_5 = \begin{Vmatrix} 0 & 0 & 0 & 0 & 0 & 0 & 0 & 0 & 0 & 1 & 0 & 1 & 1 \\ 0 & 0 & 0 & 0 & 0 & 0 & 0 & 0 & 0 & 0 & 1 & 0 & 1 \\ 1 & 1 & 0 & 0 & 0 & 0 & 0 & 0 & 0 & 0 & 0 & 0 & 1 \\ 0 & 0 & 1 & 1 & 0 & 0 & 0 & 0 & 0 & 0 & 0 & 1 & 1 \\ 0 & 1 & 0 & 0 & 0 & 0 & 0 & 0 & 0 & 0 & 0 & 0 & 1 \\ 0 & 1 & 1 & 1 & 0 & 0 & 0 & 0 & 0 & 0 & 0 & 0 & 1 \\ 0 & 0 & 0 & 0 & 1 & 0 & 1 & 0 & 1 & 1 & 0 & 0 & 1 \\ 0 & 0 & 0 & 0 & 0 & 0 & 0 & 0 & 0 & 0 & 0 & 0 & 0 \\ 0 & 1 & 0 & 0 & 0 & 0 & 0 & 0 & 0 & 0 & 0 & 0 & 1 \end{Vmatrix},$$

$$Y_6 = \begin{Vmatrix} 0 & 0 & 0 & 0 & 0 & 0 & 0 & 0 & 0 & 1 & 0 & 0 & 1 \\ 0 & 0 & 0 & 0 & 0 & 0 & 0 & 0 & 0 & 0 & 1 & 0 & 1 \\ 1 & 1 & 0 & 0 & 0 & 0 & 0 & 0 & 0 & 0 & 0 & 0 & 1 \\ 0 & 0 & 1 & 1 & 0 & 0 & 0 & 0 & 0 & 0 & 0 & 0 & 1 \\ 0 & 1 & 0 & 0 & 0 & 0 & 0 & 0 & 0 & 0 & 0 & 0 & 1 \\ 0 & 1 & 1 & 1 & 0 & 0 & 0 & 0 & 0 & 0 & 0 & 0 & 1 \\ 0 & 0 & 0 & 0 & 0 & 0 & 0 & 0 & 0 & 0 & 0 & 0 & 0 \\ 0 & 0 & 0 & 0 & 0 & 0 & 0 & 0 & 0 & 0 & 0 & 0 & 0 \\ 0 & 1 & 0 & 0 & 0 & 0 & 0 & 0 & 0 & 0 & 0 & 0 & 1 \end{Vmatrix},$$

where $Y_k = \|\eta_{ji}\|$ is a matrix of the food spectrum of the ecosystem. Figure 4.3 demonstrates some calculations that characterize the dependence of $J(t)$ on variations in the trophic graph. We can see that extension of the food spectra in scenarios $Y_1$ and $Y_2$ has little effect on ecosystem dynamics. Another situation arises when correlations in the trophic graph are disturbed. For instance, cutting out high trophic levels (birds and/or predatory fishes) leads the PCE to the unstable state seen in scenarios $Y_5$ and $Y_6$. Therefore, a more detailed investigation of correlations in the trophic graph is an important task. In particular, it is important to study the trophic connections between living elements that show predatory relationships. As can be seen from Figure 4.3 the consumption of detritus by bacterioplankton is one such important relationship. In scenarios $Y_3$ and $Y_4$, no such consumption takes place leading to a decline in the survivability of the system.

## 4.5 ARCTIC BASIN STUDY USING GIMS TECHNOLOGY

### 4.5.1 Arctic Basin pollution problems

Compared with most other global regions, the Arctic Basin remains a pristine wilderness. Nevertheless, it has been polluted as a result of atmospheric transport, river and surface coastal outflow, navigation, and other forms of human activity. Unfortunately, there are currently insufficient data on the sources, sinks, and pathways of contaminants in Arctic ecosystems. Therefore, the spatial simulation model of the Arctic ecosystem (SSMAE) can help to evaluate potentially dangerous levels in the framework of different pollution scenarios based on incomplete datasets.

Let us now turn to the ARCSS Program (Arctic Science System: Land/Atmosphere/Ice Interactions), which was initiated at the U.S. National Science Foundation as part of its contribution to the U.S. Global Change Research Program (McCauley and Meier, 1991). The present chapter describes a simulation system based on sets of computer algorithms designed to process data gathered as a result of monitoring Arctic regions and to apply mathematical models of natural and anthropogenic processes.

The basic blocks of the SSMAE are oriented in such a way that the dynamics of any given pollutant can be described. For a specific pollutant to be considered, it is necessary to include an additional block describing its physical and chemical characteristics in the SSMAE. This procedure can be demonstrated by blocks that simulate the characteristics of radionuclides, heavy metals, and oil hydrocarbons. The consideration of these pollutants is restricted to elements whose properties can be averaged.

The Arctic Basin aquatory $\Omega$, which is studied in this chapter, includes the peripheral Arctic seas as well as the coastline and southern boundaries of the Norwegian and Bering Seas.

### 4.5.2 Application of modeling technology to the study of pollutant dynamics in Arctic seas

#### *4.5.2.1 SSMAE structure*

A conceptual diagram and the block contents of the SSMAE are shown in Figure 4.4 and Table 4.3. The functioning of the SSMAE is supported by the SGM and by the climate model (CM) (Sellers *et al.*, 1996). Inputs to the SSMAE are data about pollutant sources of the near-shore Arctic Basin and ice areas, as

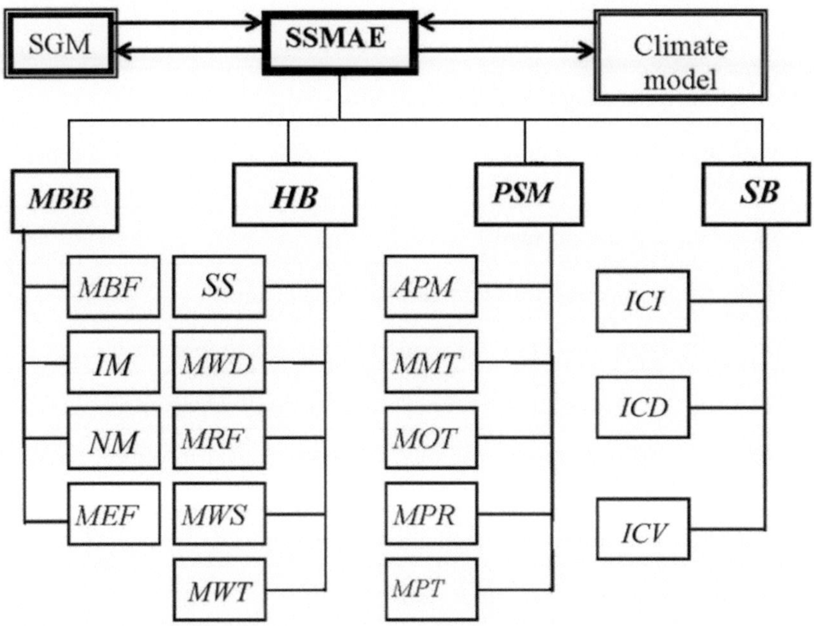


**Figure 4.4.** Block diagram of the SSMAE. Descriptions of the blocks are given in Table 4.3. SGM is the Spatial Global Model (Krapivin, 1993). CM is the Climate Model or the climate scenario.

**Table 4.3.** Description of the SSMAE units in Figure 4.4.

| *Unit* | *Unit description* |
|---|---|
| MBB | Marine biota block containing the set of models for energy flows in trophic chains of the Arctic Basin ecosystem |
| MBF | Model of biota functioning under conditions of energy exchange in the trophic chain of the Arctic Basin ecosystem |
| IM | Illumination model |
| NM | Nutrient model |
| MEF | Model for energy flow transport in the Arctic Basin ecosystem |
| HB | Hydrological block describing water circulation in Arctic seas and the movement of ecological elements |
| SS | Simulator of scenarios describing ice fields, synoptic situations, and change in hydrological regimes |
| MWD | Model for the water dynamics of the Arctic Basin |
| MRF | Model of river flow to the Arctic Basin |
| MWS | Model of water salinity dynamics |
| MWT | Model for calculating the water temperature |
| PSM | Arctic Basin pollution simulation model including a set of anthropogenic scenarios |
| APM | Air pollution transport model |
| MMT | Model for heavy metal transport through food chains |
| MOT | Model for the process of oil hydrocarbon transport to food chains |
| MPR | Model for the process of radionuclide transport to food chains |
| MPT | Model for pollution transport through water exchange between the Arctic Basin and the Atlantic and Pacific Oceans |
| SB | Service block to control the simulation experiment |
| ICI | Interface for control of the identifiers |
| ICD | Interface for control of the database |
| ICV | Interface for control of visualization |

well as current maps. The SSMAE contains three types of blocks: mathematical models of natural, ecological, and hydrophysical processes; service software; and the scenario generator. The marine biota block MBB describes the dynamics of energy flows in the trophic chains of the Arctic Basin ecosystem. The hydrological block HB describes the spatial discretization of $\Omega$ on water circulation in the Arctic. The Pollution Simulation Model (PSM) contains the anthropogenic scenarios and the service control block SB provides for control of the simulation experiment.

Let us designate the Arctic Basin aquatory as $\Omega = \{(\varphi, \lambda)\}$, where $\varphi$ and $\lambda$ are latitude and longitude, respectively. Spatial inhomogeneity of the Arctic Basin model is provided by a set of cells $\Delta T_{CO_2}$ with latitude and longitude steps of $\Delta\varphi$ and $\Delta\lambda$, respectively. These cells are the basic spatial structure of $\Omega$ for the realization of computer algorithms. The cells $\Omega_{ij}$ are heterogeneous as to their parameters and functioning. There is a set of cells adjacent to the river mouths ($\Omega_R$) and to the ports ($\Omega_P$), bordering on the land ($\Omega_\Gamma$), in the Bering Strait ($\Omega_B$) and on the southern boundary of the Norwegian Sea ($\Omega_N$). The aquatory $\Omega$ is divided in depth $z$ by step $\Delta z$. The distribution of depths is given as the matrix $H = \|h_{ij}\|$ where $h_{ij} = H(\varphi_i, \lambda_j)$, $(\varphi_i, \lambda_j) \in \Omega_{ij}$. As a result, the full water volume of $\Omega$ is divided into volumetric compartments $\Xi_{ijk} = \{(\varphi, \lambda, z) \mid \varphi_i \leq \varphi \leq \varphi_{i+1}; \lambda_j \leq \lambda \leq \lambda_{j+1}; \; z_k \leq z \leq z_{k+1}\}$ with volumes $\sigma_{ijk} = \Delta\varphi_i \Delta\lambda_j \Delta z_k$. Within $\Xi_{ijk}$ the water body is considered as a homogeneous structure. The water temperature, salinity, density, and biomass of the compartments $\Xi_{ijk}$ are described by box models. Anthropogenic processes that impact the aquatory $\Omega$ are described for the four seasons: $\tau_w$ winter, $\tau_s$ spring, $\tau_u$ summer, $\tau_a$ autumn.

The procedure of spatial discretization is provided by the ICI block of the SSMAE database, which includes a set of identifiers $A_k = \|a_{ij}^k\|$, where $a_{ij}^k$ is a specific symbol to identify a real element of $\Omega_{ij}$ in computer memory. Identifier $A_1$ reflects the spatial structure of the Arctic Basin and adjoining territories: $a_{ij}^1 = 0$ for $(\varphi_i, \lambda_j) \notin \Omega$; $a_{ij}^1 = 1$ for $(\varphi_i, \lambda_j) \in \Omega$ when $(\varphi_i, \lambda_j)$ belongs to the land (islands), and $a_{ij}^1$ equals the aquatory identifier symbol from the second column of Table 4.4 when $(\varphi_i, \lambda_j)$ belongs to a given sea.

Identifier $A_2$ shows the position of cells $\Omega_R$, $\Omega_P$, $\Omega_N$, $\Omega_S$, $\Omega_\Gamma$ and describes the spatial distribution of pollutant sources. Other identifiers $A_k$ are used to describe ice fields ($k = 3$), the spatial distribution of solar radiation ($k = 4$), and the dislocation of upwelling zones ($k = 5$).

A user of the SSMAE in free-running mode may choose different ways to describe the many input parameters. Blocks ICI and ICD bring about online entry to $A_k$ and to the database. For example, if the user has data about the spatial distribution of ice fields in $\Omega$, he can form identifier $A_3$ with $a_{ij}^3 = 0$ for an ice-free water surface, $a_{ij}^3 = 1$ for new ice, and $a_{ij}^3 = 2$ for old ice. In this case block SS enables the input of data from the climate model concerning the ice fields.

The block structure of the SSMAE is realized by means of visual C++ Builder. Each of the blocks from Table 4.3 is a C++ function. The main function provides the interactions between the SSMAE, SGM, and CM. This functional

**Table 4.4.** Initial data on the distribution of pollutants in Arctic aquatories at moment $t_0$.

| *Aquatory* | *Identifier* | *Concentration* | | | |
|---|---|---|---|---|---|
| | | *Radionuclides* (Bq/L) | *Radionuclides* (Bq/L) | *Heavy metals* (mg/L) | *Oil hydrocarbons* (mg/L) |
| | | $^{137}Cs$ | $^{60}Co$ | | |
| Greenland Sea | Γ | 0.05 | 0.05 | 0.5 | 0.2 |
| Norwegian Sea | *N* | 0.05 | 0.05 | 0.7 | 0.4 |
| Barents Sea | *B* | 0.07 | 0.07 | 0.8 | 0.6 |
| Kara Sea | *K* | 0.1 | 0.1 | 1.0 | 0.4 |
| White Sea | ∇ | 0.1 | 0.1 | 1.1 | 0.4 |
| Laptev Sea | Λ | 0.05 | 0.05 | 0.9 | 0.5 |
| E. Siberian Sea | *E* | 0.01 | 0.01 | 0.9 | 0.5 |
| Bering Sea | *S* | 0.02 | 0.02 | 0.8 | 0.7 |
| Chukchi Sea | *X* | 0.01 | 0.01 | 0.8 | 0.6 |
| Beaufort Sea | Φ | 0.05 | 0.05 | 0.7 | 0.2 |
| Central Basin | *U* | 0.00 | 0.00 | 0.1 | 0.1 |

specification supports overlapping output and input streams of SSMAE blocks. In conversational mode the user can toggle data streams between the slave blocks.

The calculating procedure is based on the subdivision of the Arctic Basin into boxes $\{\Xi_{ijk}\}$. This is realized by means of a quasi-linearization method (Nitu *et al.*, 2000a). All differential equations of the SSMAE are substituted in each box $\Xi_{ijk}$ by easily integrable ordinary differential equations with constant coefficients. Water motion and turbulent mixing are realized in conformity with the velocity fields of currents, which are defined on the same coordinate grid as the $\{\Xi_{ijk}\}$ (Krapivin *et al.*, 1998a).

### 4.5.2.2 *Marine biota block (MBB)*

The ice–water ergocline plays an important role in the biological productivity of northern seas. According to the hypothesis of Legendre and Legendre (1998), energetic ergoclines are the preferential sites for biological production in the Arctic Ocean. Primary production in the food chains of Arctic Basin ecosystems is determined by phytoplankton productivity. This is connected with complex varia-

**Table 4.5.** Vertical structure of the Arctic Basin aquatic system.

| *Layer (A)* | *Parameters* | | | | | |
|---|---|---|---|---|---|---|
| | *Layer thickness* | *Temperature* | *Irradiance* | *Turbulence coefficient* | *Irradiance relaxation coefficient* | *Reflection coefficient* |
| Surface | | $T_0$ | $E_0$ | | | |
| Snow | $g$ | $T_g$ | $E_g$ | | $\alpha_g$ | $\beta_g$ |
| Floating ice | $r$ | $T_r$ | $E_r$ | | $\alpha_r$ | $\beta_r$ |
| Submerged ice | $f$ | $T_f$ | $E_f$ | | $\alpha_f$ | $\beta_f$ |
| Water | $z-f$ | $T_w$ | $E_w$ | $k_w$ | $\alpha_w$ | $\beta_w$ |

tions in the meteorological, hydrodynamic, geochemical, and energetic parameters of the sea environment. Parameterizing phytoplankton production in northern seas was studied by Legendre and Legendre (1998). Table 4.5 gives the vertical structure of the seasonal composition of conditions affecting primary production in $\Omega$. This scheme is applied to each $\Xi_{ijk}$.

The MWT block calculates water temperature $T_w$ by averaging the temperatures of mixed water volumes. In addition, the following correlations are applied:

$$T_g = T_r = T_f = \begin{cases} -0.024b + 0.76T_0 + 8.38 & \text{when } b \leq 50 \text{ cm} \\ -0.042b + 0.391T_0 - 0.549 & \text{when } b < 50 \text{ cm,} \end{cases} \tag{4.25}$$

where $b = g + r + f$; $T_0$ is surface temperature; $g$ is snow depth; $r$ is thickness of the floating ice; and $f$ is depth of the submerged ice below the water surface. If we designate by $\rho_g$, $\rho_r$, and $\zeta$ the density of snow, ice, and seawater, respectively, we obtain for the depth of ice beneath the surface:

$$f = (g\rho_g + r\rho_r)/(\zeta - \rho_r).$$

Figures 4.5 and 4.6 are conceptual flow charts of energy in an ecological system. The energy input during time interval $t$ is provided by solar radiation $E_A(t, \varphi, \lambda, z)$ and upward transport of nutrients from deep-sea layers. The concentration of nutrients $B_{6,A}(t, \varphi, \lambda, z)$ at the depth $z$ is determined by photosynthesis $R_{pA}$, advection, and destruction of suspended dead organic matter $B_7$. The role played by hydrodynamic conditions lies in maintaining the concentration of nutrients required for photosynthesis, which occurs via transport from other layers or aquatories of the sea where the concentration is sufficiently high. Taking into

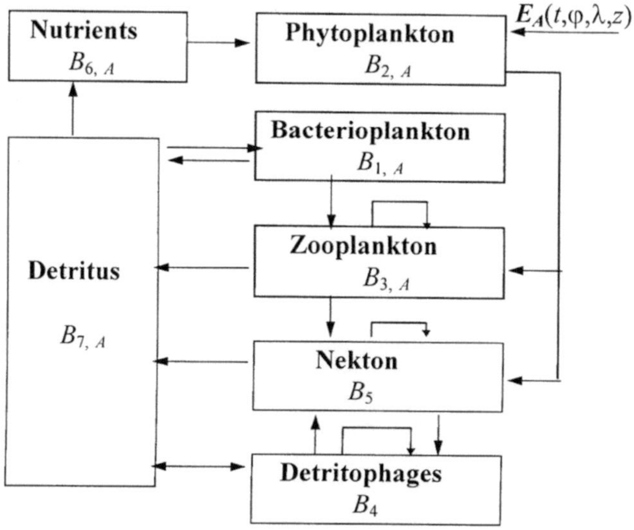


**Figure 4.5.** Block diagram of the energy flows (cal m$^{-3}$ day$^{-1}$) in the trophic pyramid of the Arctic Basin ecosystem. This is realized as block MEF. The boxes with elements denote the generalized trophic levels of the Arctic ecosystem. All of the elements are described by means of averaged parameters for the biological community of the northern seas. It is supposed that this trophic pyramid takes place in each of the Arctic Basin seas. The trophic relations between the elements of the model are described on the basis of the energetic principle (Nitu *et al.*, 2000b). Biomasses, the rates of production and exchange (respiration), and food rations are expressed in energy units. Total nitrogen serves as the nutrient salt prototype in the model.

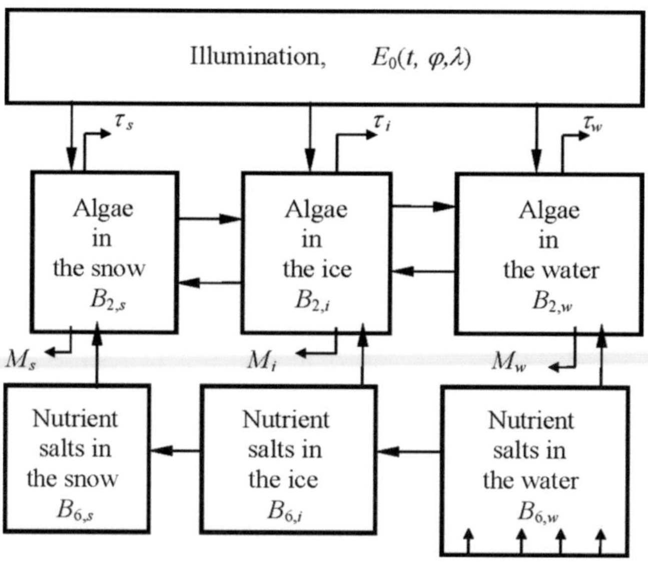


**Figure 4.6.** Block diagram of energy flows (cal m$^{-3}$ day$^{-1}$) at the snow–ice–water interface.

account the designations in Table 4.4 we have:

$$E(\varphi,\lambda,z,t)=\begin{cases} E_0 & \text{for } z=-r-g,\\ E_0\exp(-\alpha_s z) & \text{for } -r-g<z\le -r,\\ E_s(\varphi,\lambda,-R,t)\exp(-\alpha_{i_1}z) & \text{for } -r<z\le 0,\\ E_{i_1}(\varphi,\lambda,0,t)\exp(-\alpha_{i_2}z) & \text{for } 0<z\le r,\\ E_{i_2}(\varphi,\lambda,r,t)\exp(-\alpha_w z) & \text{for } z>r. \end{cases}$$

where the values of $\alpha_A$ ($A=g,r,f,w$) depend on the optical properties of the $A$th medium. Irradiance $E_0$ is that which arrives at the surface $\Omega$. An estimate of $E_0$ can be obtained from monitoring or calculated from the climatic model. The flow of $E_0$ is attenuated by snow, ice, and water according to the scheme of Table 4.5. In each cell $\Omega_{ij}$ the structure of these layers is changed according to the time of year. Within each layer, the attenuation of irradiance with depth is described by an exponential model. The parameters $\alpha_A$ and $\beta_A$ are functions of salinity, turbidity, temperature, and biomass. The form of this dependence can be given as a scenario, otherwise standard functions are used (block IM).

As a basic scheme for the flow of nutrients in water, that proposed by Krapivin (1996) is accepted as correct for conditions in the Arctic Basin by Legendre and Legendre (1998). The spatial distribution of upwelling zones takes into account seasonal variations. Block NUM can be used to realize this scheme in line with the current structure of upwelling regions.

The dynamic equation for nutrients $B_{6,A}$ in the environment, where $A=\{S \text{ snow}, I \text{ ice}, W \text{ water}\}$, is given by

$$\frac{\partial B_{6,A}}{\partial t}+v_\varphi^A\frac{\partial B_{6,A}}{\partial \varphi}+v_\lambda^A\frac{\partial B_{6,A}}{\partial \lambda}+v_z^A\frac{\partial B_{6,A}}{\partial z}$$
$$=Q_A+k_2^W\frac{\partial^2 B_{6,A}}{\partial^2 z}+\beta_V\frac{\partial B_{6,A}}{\partial z}+\rho_1\sum_{i=1}^{5}T_i-\delta_1 R_{pA}+\varepsilon_1^A H_1^\varepsilon, \quad (4.26)$$

where $v_\varphi^A, v_\lambda^A, v_z^A$ are velocity projections of motion in the environment; $Q_A$ is the input of biogenic elements to $A$ resulting from the decomposition of detritus ($Q_A=\delta_n R_D^A$) with $R_D^A=\mu_A B_7$; $\delta_n$ is the content of nutrients in dead organic matter; $\mu_A$ is the rate of decomposition of detritus in environment $A$; $k_2^W$ is the kinematic coefficient of vertical diffusion; $\delta_1$ is the velocity of nutrient assimilation by the photosynthetic process per unit of phytoplankton production; $\varepsilon_1^A$ is the proportional part of the $\varepsilon$th radionuclide that is chemically analogous to $B_{6,A}$ on substrate $A$; $H_1^\varepsilon$ is the rate of input flow of the $\varepsilon$th radionuclide; $T_i$ is the rate of exchange with the environment; $\rho_1$ is the proportion of biomass losses as a result of exchange that transform into nutrients (Legendre and Legendre, 1998); and $\beta_V$ is upwelling velocity. Equation (4.25) is the basic element of block NM.

The net ecosystem production of Arctic ecosystems mainly depends on the seasonal shift of many factors. Figures 4.5 and 4.6 are block diagrams giving the possibility of parameterizing the processes of temporal fluctuations in these factors while taking into account the physical gradients that determine the structure and function of the ecosystems. Phytoplankton is the basic indicator of Arctic ecosystem response to changes in these factors.

Phytoplankton production $R_{pA}$ in environment $A$ is a function of solar radiation $E_A$, the concentration of nutrients $n_A$, temperature $T_A$, phytoplankton biomass $p_A$, and the concentration of pollutants $\xi_A$. There are many models describing the photosynthesis process (Legendre and Krapivin, 1992; Legendre and Legendre, 1998). To describe this function in the present study, an equation of Michaelis–Menten type is used (block MBF):

$$R_{pA} = a_A k_I^A p_{A,\max}/(E_A + k_I^A), \tag{4.27}$$

where $k_I^A$ is the irradiance level at which $R_{pA} = 0.5R_{pA,\max}$; and $p_{A,\max}$ is the maximum quantum yield (Legendre and Legendre, 1998). The coefficient $a_A$ reflects the dependence of phytoplankton production on environment temperature $T$ and the concentration of nutrients $B_{6,A}$. The MBF block realizes the following equation for the calculation of $a_A$:

$$a_A = a_1 K_0(T, t)/[1 + B_{2,A}/(a_2 B_{6,A})], \tag{4.28}$$

where $a_1$ is the maximal rate of absorption of nutrients by phytoplankton (day$^{-1}$); and $a_2$ is the index of the rate of saturation of photosynthesis, and

$$K_0(T, t) = a_3 \max\left\{0, \frac{T_c - T}{T_c - T_{\text{opt}}} \exp\left[1 - \frac{T_c - T}{T_c - T_{\text{opt}}}\right]\right\}, \tag{4.29}$$

where $a_3$ is the weight coefficient; and $T_c$ and $T_{\text{opt}}$ are the critical and optimal temperatures for photosynthesis, respectively (°C).

Equation (4.27) fits laboratory data sufficiently well. The relationships (4.28) and (4.29) make the description of phytoplankton production more accurate for critical environmental conditions when the concentration of nutrients and the temperature fluctuate highly. The coefficients of these relationships are defined on the basis of estimates given by Legendre and Legendre (1998).

The dynamic equation for the biomass of phytoplankton in environment $A$ has the following form:

$$\frac{\partial B_{2,A}}{\partial t} + v_\varphi^A \frac{\partial B_{2,A}}{\partial \varphi} + v_\lambda^A \frac{\partial B_{2,A}}{\partial \lambda} + v_z^A \frac{\partial B_{2,A}}{\partial z}$$
$$= R_{pA} - \theta_p^A - M_p^A + k_2^A \frac{\partial^2 B_{2,A}}{\partial z^2} - [k_{Zp}^A R_{ZA}/\xi_Z^A + k_{Fp}^A R_{FA}/\xi_F^A] B_{2,A}, \tag{4.30}$$

where $R_{ZA}(\xi_Z^A)$ and $R_{FA}(\xi_F^A)$ are the production (the food spectrum) of zooplankton $B_3$ and nekton $B_5$, respectively; $M_p^A$ is mortality; and $\theta_p^A$ is the rate of exchange. Balance equations for the other ecological elements of Figure 4.6 are given by Krapivin (1995, 1996).

The energy source for the entire system is solar radiation $E_A(t, \varphi, \lambda, z)$, whose intensity depends on time $t$, latitude $\varphi$, longitude $\lambda$, and depth $z$. The equations that describe the biomass dynamics of living elements are:

$$\frac{\partial B_i}{\partial t} + \xi_i \left[ V_\varphi \frac{\partial B_i}{\partial \varphi} + V_\lambda \frac{\partial B_i}{\partial \lambda} + V_z \frac{\partial B_i}{\partial z} \right]$$
$$= R_i - T_i - M_i - H_i - \sum_{j \in \Gamma_i} C_{ij} R_j$$
$$+ \xi_i \left[ \frac{\partial}{\partial \varphi}\left(\Delta_\varphi \frac{\partial B_i}{\partial \varphi}\right) + \frac{\partial}{\partial \lambda}\left(\Delta_\lambda \frac{\partial B_i}{\partial \lambda}\right) + \frac{\partial}{\partial z}\left(\Delta_z \frac{\partial B_i}{\partial z}\right) + \beta_V \frac{\partial B_i}{\partial z} \right]$$
$$(i = 1, 3, 4, 5); \quad (4.31)$$

where $V(V_\varphi, V_\lambda, V_z)$ are the components of water current velocity ($V_\varphi = v_\varphi^W$, $V_\lambda = v_\lambda^W$, $V_z = v_z^W$); $R_i$ is production; $M_i$ is mortality; $H_i$ is non-assimilated food; and $\Gamma_i$ is the set of trophic dependences of the $i$th component: $C_{ji} = k_{ji} F_i / \sum_{m \in S_i} k_{jm} F_m$; $S_i$ is the food spectrum of the $j$th component; $k_{jm}$ is the index of satisfaction of nutritive requirements of the $j$th component at the expense of the biomass of the $m$th component; $F_i = \max\{0, B_i - B_{i,\min}\}$; $B_{i,\min}$ is the minimal biomass of the $i$th component consumed by other trophic levels; $\Delta(\Delta_\varphi, \Delta_\lambda, \Delta_z)$ are components of the turbulent mixing coefficient (assuming isotropy of vertical mixing in the horizontal plane $\Delta_\varphi = \Delta_\lambda = \nu_H$); and $\beta_V$ is upwelling velocity. Functions $R_i$, $M_i$, $H_i$, and $T_i$ are parameterized according to the models by Krapivin (1996) and Legendre and Krapivin (1992). Equations describing the dynamics of abiotic elements are represented in conformity with Kondratyev and Krapivin (2001c, d). Functions $M_4$ and $M_5$ include biomass losses as a result of fishing. Parameter $\xi_i$ characterizes the degree to which the $i$th component is subject to the current. It is supposed that $\xi_i = 1$ for $i = 1, 2, 3$ and $\xi_i = 0$ for $i = 4, 5$.

Inert components are described by the following equations (Krapivin, 1996):

$$\frac{\partial B_7}{\partial t} + V_\varphi \frac{\partial B_7}{\partial \varphi} + V_\lambda \frac{\partial B_7}{\partial \lambda} + V_z \frac{\partial B_7}{\partial z}$$
$$= \sum_{i=1}^{5} (M_i + H_i) - \mu_W B_7 - (\nu_* - \beta_V) \frac{\partial B_7}{\partial z} - (k_{1,7} R_1 / P_1 + k_{3,7} R_3 / P_3 + k_{4,7} R_4 / P_4$$
$$+ k_{5,7} R_5 / P_5) B_{7,\min} + \frac{\partial}{\partial \varphi}\left(\Delta_\varphi \frac{\partial B_7}{\partial \varphi}\right) + \frac{\partial}{\partial \lambda}\left(\Delta_\lambda \frac{\partial B_7}{\partial \lambda}\right) + \frac{\partial}{\partial z}\left(\Delta_z \frac{\partial B_7}{\partial z}\right), \quad (4.32)$$

where $P_i = \sum_{j \in S_i} k_{i,j} B_{j,\min}$; $\mu_W$ is the velocity of decomposition of detritus per unit of biomass; $\nu_*$ is the velocity of settling as a result of gravity; and $k_{ij}$ is a coefficient of the relation of the $i$th element with respect to the $j$th element of the ecosystem.

Equations (4.26)–(4.32) should only be used for the complete volume when $(\varphi, \lambda, z) \in W$. In all other cases (i.e., the layers of ice or snow) these equations are automatically reduced in accordance with the scheme represented in Figure 4.6.

#### *4.5.2.3 Hydrological block (HBB)*

The circulation of waters in the Arctic Basin is a complex system of cycles and currents that function on different scales. Block HBB simulates the dynamics of Arctic Basin waters by making use of the system of sub-blocks presented in Figure 4.4. Water dynamics in $\Omega$ is presented by flows between compartments $\Xi_{ijk}$. The directions taken by water exchanges are represented at every level $z_k = z_0 + (k-1)\Delta z_k$ according to Aota *et al.* (1992) in conformity with current maps assigned as SSMAE input. The external boundary of $\Omega$ is determined by coastlines, the bottom, the Bering Strait, the southern boundary of the Norwegian Sea, and the water–atmosphere boundary.

Hydrological data are synthesized via a four-level structure according to the seasons (block MWD). Current velocity in the Bering Strait is estimated by the following binary function:

$$V(t) = \begin{cases} V_1 & \text{for } t \in \tau_u \cup \tau_a \\ V_2 & \text{for } t \in \tau_w \cup \tau_s. \end{cases}$$

Water exchange through the southern boundary of the Norwegian Sea is $V_3$. The water temperature in $T^W_{ijk}$ (block MWT) is a function of evaporation, precipitation, river flows, and inflows of water from the Atlantic and Pacific Oceans. Its change with time in $\Xi_{ijk}$ is described by the heat balance equation:

$$\zeta C \sigma_{ijk} \frac{\partial T^W_{ijk}}{\partial t} = \sum_{s,l,m} (W^{ijk}_{slm} + f^{ijk}_{slm}) - W_{ijk}, \tag{4.33}$$

where $\zeta$ is seawater density (g cm$^{-3}$); $C$ is water thermal capacity, (cal g$^{-1}$ grad$^{-1}$); $\sigma_{ijk}$ is the volume of $\Xi_{ijk}$; $W^{ijk}_{slm}$ is heat inflow to $\Xi_{ijk}$ from $\Xi_{slm}$; $f^{ijk}_{slm}$ is heat exchange between $\Xi_{slm}$ and $\Xi_{ijk}$ caused by turbulent mixing; and $W_{ijk}$ is total heat outflow from $\Xi_{ijk}$ to the bordering boxes. Heat exchange with the atmosphere is calculated in accordance with the empirical equation (4.25).

The dissipation of moving kinetic energy, geothermic flow on the ocean bed, the heat effects of chemical processes in the ocean ecosystem, and freezing and melting of the ice are not considered global determinants of water temperature fields. Thus, the SSMAE does not consider these effects.

The dynamics of water salinity $S(t, \varphi, \lambda, z)$ during the time interval $t$ is described by the balance equation as block MWS. Ice salinity is defined by a two-step scale: $s_1$ old, $s_2$ new. It is supposed that $S(t, \varphi, \lambda, z) = s_0$ for $z > 100$ m, $s_2 = k_s S(t, \varphi, \lambda, f)$ for $r + f > H_{\max}$, and $s_1 = k_r s_2 H_{\max}/(r + f)$ for $r + f < H_{\max}$ where coefficients $k_s$ and $k_r$ are determined empirically and $H_{\max}$ is the maximal thickness of new ice. In accordance with estimations by Krapivin (1995), simulations give the values of $H_{\max} = 50$ cm, $k_s = k_r = 1$. River flows, ice fields, and synoptic situations are described by scenarios given to the MRF and SS blocks by the user of the SSMAE.

Snow layer thickness $g(t, \varphi, \lambda)$ may be described using statistical data with given dispersion characteristics: $g = \bar{g} + g_0$ where the value $\bar{g}$ is defined as the mean characteristic for the chosen time interval and the function $g_0(t, \varphi, \lambda)$ represents how $g$ varies for the given time interval.

An alternative description can be made by parameterizing the snow layer dynamics process in the framework of the atmospheric process simulation algorithm (block APM) relating the thickness of the growth and melting of the snow layer to temperature and precipitation:

$$g(t + \Delta t, \varphi, \lambda) = g(t, \varphi, \lambda) + S_F - S_M,$$

where $S_F$ is the proportion of snow precipitated at temperatures close to freezing ($265\,\mathrm{K} \leq T_0 \leq 275\,\mathrm{K}$); and $S_M$ is snow ablation (i.e., evaporation + melting).

Block SS gives the user the possibility to select between these algorithms. When statistical data on snow layer thickness exist the function $g(t, \varphi, \lambda)$ is reconstructed for $(\varphi, \lambda) \in \Omega$ by means of selected algorithms (Krapivin, 2000a; Nitu *et al.*, 2004).

#### *4.5.2.4 Pollution simulation model (PSM)*

The PSM block simulates pollution processes in the territory $\Omega$ as a result of atmospheric transport, river and surface coastal outflow, navigation, and other human activity (Muller and Peter, 1992). The variety of pollutants is described by three components: oil hydrocarbons $O$, heavy metals ($e$ solids, $\psi$ dissolved), and radionuclides $\varepsilon$. It is supposed that pollutants get into living organisms only through food chains.

Rivers contribute considerably to the level of pollution of Arctic waters. The concentration of pollutant $\kappa$ in river $\gamma$ is $\gamma_\kappa$. Pollutant $\kappa$ enters compartment $\Omega_{ijk} \in \Omega_R$ at velocity $c_\gamma$ reflecting mass flow per unit time. Subsequently, the spreading of the pollutant in $\Omega$ is described by other sub-blocks.

The influence of water exchanges between the Arctic Basin and the Pacific and Atlantic Oceans on the pollution level in $\Omega$ is described by block MPT. The watersheds of the Norwegian Sea $\Omega_N$ and the Bering Strait $\Omega_B$ are characterized in the scenario by currents with varying directions.

The atmospheric transport of heavy metals, oil hydrocarbons, and radionuclides is described by many models (Payne *et al.*, 1991; Phillips *et al.*, 1997; Sportisse, 2000). Using these models to reconstruct the pollution distribution over $\Omega$ makes it possible to estimate optimal values of $\Delta\varphi, \Delta\lambda$ and the step in time $\Delta t$. The present level of the database for the Arctic Basin provides for use of a single-level Euler model with $\Delta t = 10$ days, $\Delta\varphi = \Delta\lambda = 1°$ (block APM). It is supposed that pollution sources can be located at the Arctic Basin boundary. Their detailed distributions are given as SSMAE input. The transport of pollutants to the Arctic Basin and the formation of their spatial distribution are realized in conformity with the wind velocity field, which is considered as that given in Krapivin and Phillips (2001).

It is postulated that oil hydrocarbons $O(t, \varphi, \lambda, z)$ are transformed by the following processes (Payne *et al.*, 1991): dissolution $H_O^1$, evaporation $H_O^2$, sedimentation $H_O^3$, oxidation $H_O^4$, biological adsorption $H_O^5$, bio-sedimentation $H_O^6$, and bacterial decomposition $H_O^7$. The kinetics equation to describe the dynamics of oil hydrocarbons in the Arctic Basin is given by

$$\frac{\partial O}{\partial t} + v_\varphi^w \frac{\partial O}{\partial \varphi} + v_\lambda^w \frac{\partial O}{\partial \lambda} + v_z^w \frac{\partial O}{\partial z} = Q_O + k_2^w \frac{\partial^2 O}{\partial z^2} - \sum_{i=1}^{7} H_O^i, \tag{4.34}$$

where $Q_O$ is the anthropogenic source of oil hydrocarbons.

The process of diffusion of heavy metals in seawater depends on their state. The dissolved fraction of heavy metals ($\psi$) takes part in biogeochemical processes more intensively than suspended particles ($e$). However, heavy metals fall out more rapidly to the sediment as suspended particles. A description of the entire spectrum of these processes in the framework of this study is impossible. Therefore, block MMT can be used to describe processes that have estimates. The transport of heavy metals in seawater includes absorption of the dissolved fraction $\psi$ by plankton ($H_Z^\psi$) and by nekton ($H_F^\psi$), sedimentation of the solid fraction ($H_1^e$), deposition with detritus ($H_D^\psi$), adsorption by detritophages from bottom sediments ($H_L^{e\psi}$), and release from bottom sediments owing to diffusion ($H_a^{e\psi}$). As a result, the dynamic equations for heavy metals become:

$$\frac{\partial e_W}{\partial t} + v_\varphi^W \frac{\partial e_W}{\partial \varphi} + v_\lambda^W \frac{\partial e_W}{\partial \lambda} + v_z^W \frac{\partial e_W}{\partial z} = \sum_{i=1}^{3} \alpha_2^i Q_{e\psi}^i - H_1^e + \alpha_1 H_a^{e\psi} \tag{4.35}$$

$$\frac{\partial \psi_W}{\partial t} + v_\varphi^W \frac{\partial \psi_W}{\partial \varphi} + v_\lambda^W \frac{\partial \psi_W}{\partial \lambda} + v_z^W \frac{\partial \psi_W}{\partial z} = (1-\alpha_1) H_a^{e\psi} + k_2^W \frac{\partial^2 \psi_W}{\partial z^2} - H_Z^\psi - H_F^\psi - H_D^\psi - H_a^\psi \tag{4.36}$$

$$\frac{\partial e^*}{\partial t} = H_1^e - \alpha_1 (H_L^{e\psi} + H_a^{e\psi}) \tag{4.37}$$

$$\frac{\partial \psi^*}{\partial t} = H_D^\psi - (1-\alpha_1)(H_L^{e\psi} + H_a^{e\psi}), \tag{4.38}$$

where $e_W, \psi_W$, and $e^*, \psi^*$ are the concentrations of heavy metals in the water and in the bottom sediments as solid and dissolved phases, respectively; $H_a^\psi$ is the output of heavy metals from the sea to the atmosphere by evaporation and spray; $Q_{e\psi}^i$ is the input of heavy metals to the sea from river waters ($i = 1$), atmospheric deposition ($i = 2$), and ship waste ($i = 3$); $\alpha_2^i$ is the proportion of suspended particles in the $i$th flow of heavy metals; and $\alpha_1$ is the amount of heavy metals in solid form in bottom sediments.

Each radionuclide of $\varepsilon$th type is characterized by a half-life $\tau^\varepsilon$, a rate $H_1^\varepsilon$ of input flow to the aquatory $\Omega$, an accumulation $H_\alpha^\varepsilon$ rate in living organisms $\alpha(p_A, B_A, Z, F, L)$, and a removal rate $H_D^\varepsilon$ with dead elements of the ecosystem. As a result, the concentration $Q^\varepsilon$ of the radionuclide $\varepsilon$ in $\Omega_{ijk}$ is described by the following system of equations:

$$\frac{\partial Q^\varepsilon}{\partial t} + v_\varphi^W \frac{\partial Q^\varepsilon}{\partial \varphi} + v_\lambda^W \frac{\partial Q^\varepsilon}{\partial \lambda} + v_z^W \frac{\partial Q^\varepsilon}{\partial z} = \frac{\sigma_{ijk}}{\sigma} H_1^\varepsilon + k_2^W \frac{\partial^2 Q^\varepsilon}{\partial z^2} - H_\alpha^\varepsilon - H_D^\varepsilon - \frac{\ln 2}{T^\varepsilon} Q^\varepsilon + H_*^\varepsilon, \tag{4.39}$$

$$\frac{\partial Q_*^\varepsilon}{\partial t} = H_D^\varepsilon - H_*^\varepsilon - \frac{\ln 2}{T^\varepsilon} Q_*^\varepsilon, \tag{4.40}$$

where $Q_*^\varepsilon$ is the concentration of the $\varepsilon$th radionuclide in bottom sediments; and $H_*^\varepsilon$ is the rate of output flow of the $\varepsilon$th radionuclide from bottom sediments via desorption. The exchange of radionuclides between water layers by migration of living elements is ignored as it has a small value in comparison with the flow $H_D^\varepsilon$.

#### 4.5.2.5 *Simulation results*

*The assumptions.* The SSMAE allows for estimation of the pollution dynamics of the Arctic Basin under various *a priori* suppositions about the intensities of the flows of pollutants and under other anthropogenic impacts on the ecosystems of this region. Here, some of the possible situations are considered. The thermal regime of the Arctic Basin is given by a normal distribution with average temperatures and with dispersions throughout the aquatories as given by SEDAAR (Strategic Environmental Distributed Active Archive Resource). The scheme of transport of pollutants in the atmosphere is adopted from Christensen (1997). The estimates of parameters for the blocks of Table 4.3 are given by literature sources or personal recommendations (see Table 4.6). The vertical distribution of pollutants at the initial moment is taken as homogeneous. The average diameters of solid particles are estimated to be in the range 0.12 μm to 1,000 μm and the vertical velocity of sedimentation is 0.003 m/s. The concentration of nutrients in ice and snow equals 0. Moreover, it is supposed that deep-water temperature $Y(t, \varphi, \lambda) = 0°\text{C}$ and surface ice temperature $f_1(t, \varphi, \lambda) = -3°\text{C}$ for $(\varphi, \lambda) \in \Omega$. It is further supposed that $\varepsilon_1^A = 0$ and phytoplankton productivity in the ice layer is 2.5% of the primary production in the water column $[(R_{p,r} + R_{p,f})/R_{p,w} = 0.025]$.

Let the ratio between the solid and dissolved phases of heavy metals at moment $t = t_0$ equal 1:2 (i.e., $e(t_0, \varphi, \lambda, z)/\psi(t_0, \varphi, \lambda, z) = 0.5$). The flows of heavy metals, $H_Z^\psi$, $H_F^\psi$, $H_D^\psi$, and $H_L^\psi$, are described by linear models, $H_1^e = 0.01e_w, H_a^{e\psi} = 0$. The boundaries of the Norwegian and Bering Seas are approximated by lines with $\varphi_N = 62°\text{N}$ and $\varphi_B = 51°\text{N}$, respectively. Values for the other parameters are defined by Bard (1999), Preller and Cheng (1999), Valette-Silver *et al.* (1999), Wania *et al.* (1998), and Wielgolaski (1997). The initial data are defined in Tables 4.4 and 4.7.

*The dynamics of Arctic Basin radionuclear pollution.* The intensity of external flows across the boundaries of the Arctic Basin and internal flows as a result of dead organisms $H_D^\varepsilon$, sediment $H_*^\varepsilon$, and living organisms $H_\alpha^\varepsilon$ are described by linear models in accordance with Krapivin and Phillips (2001). Some results of the

**Table 4.6.** The values of some parameters used in SSMAE simulation experiments.

| *Parameter* | *Estimate of the parameter* |
|---|---|
| Step of space digitized by | |
| Latitude, $\Delta\varphi$ | $1^\circ$ |
| Longitude, $\Delta\lambda$ | $1^\circ$ |
| Depth, $\Delta z$ | |
| $z \leq 100$ m | 1 m |
| $z > 100$ m | $h - 100$ m |
| Coefficient of ice heat conductivity, $\lambda_1$ | 2.21 W/m/deg |
| Coefficient of water heat conductivity, $\lambda_2$ | 0.551 W/m/deg |
| Coefficient of turbulent mixing, $k_2^W$ | |
| For open waters | $0.5 \times 10^{-4}$ m$^2$/s |
| For ice-covered water bodies | $5 \times 10^{-6}$ m$^2$/s |
| Characteristic heat of ice melting, $q$ | 334 kJ/kg |
| Content of biogenic elements in dead organic matter, $\delta_n$ | 0.1 |
| Intensity of detritus decomposition, $\mu_A$ | |
| $A = g, r, f$ | 0 |
| $A = W$ | 0.01 |
| Velocity of currents in the Bering Strait, $V_i$ | |
| $i = 1$ | 0.2 m/s |
| $i = 2$ | 0.05 m/s |
| Water heat capacity, $C$ | 4.18 kJ/kg/K |
| Ice salinity, $s_i$ | |
| $i = 1$ | 5‰ |
| $i = 2$ | 1‰ |
| Water salinity at $z > 100$ m, $s_0$ | 34.95‰ |
| Area of the Arctic Basin, $\sigma$ | 16,795,000 km$^2$ |
| Half-life of radionuclides, $\tau_g$ | |
| $\varepsilon = {}^{60}C$ | 5.271 years |
| $\varepsilon = {}^{137}Cs$ | 30.17 years |
| Critical temperature for photosynthesis, $T_c$ | $-0.5^\circ$C |
| The content of nutrients in detritus, $\delta_n$ | 1% day$^{-1}$ |

| *Parameter* | *Estimate of the parameter* |
|---|---|
| Density of | |
| Ice, $\rho_r$ | 10% |
| Snow, $\rho_g$ | $0.9\,g\cdot cm^3$ |
| Seawater, $\zeta$ | $0.1\,g\cdot cm^3$ |
| $z \leq 200\,m$ | $1.024\,g\cdot cm^3$ |
| $z > 200\,m$ | $1.028\,g\cdot cm^3$ |
| Reflection coefficient for | |
| Snow layer, $\beta_g$ | 0.8 |
| Floating ice layer, $\beta_r$ | 0.65 |
| Submerged ice layer, $\beta_f$ | 0.6 |
| Water layer, $\beta_w$ | 0.47 |
| Exchange biomass loss that transforms into nutrients, $\rho_1$ | 0.1 |
| Nutrient assimilation velocity coefficient, $\delta_1$ | 0.1 |
| Vertical velocity in the upwelling zone, $\beta_V$ | $0.1\,m\,s^{-1}$ |

**Table 4.7.** The input flows of radionuclides, heavy metals (suspended *e* and dissolved $\psi$), and oil hydrocarbons *O* to $\Omega$ by water flows taken into account in the SSMAE.

| *Source* | *Flow into the basin* ($km^3$/yr) | *Concentration of pollutant* | | | | |
|---|---|---|---|---|---|---|
| | | $^{137}Cs$ (Bq/L) | $^{60}Co$ (Bq/L) | $e$ (mg/L) | $\psi$ (mg/L) | $O$ (mg/L) |
| Rivers | | | | | | |
| Yenisey | 600 | 0.5 | 0.5 | 0.3 | 5.1 | 2.3 |
| Ob′ | 400 | 0.1 | 0.1 | 0.4 | 6.9 | 4.7 |
| Lena | 500 | 0.0 | 0.0 | 1.1 | 8.8 | 6.9 |
| Pechora | 130 | 0.1 | 0.1 | 0.3 | 1.5 | 3.0 |
| Northern Dvina | 100 | 0.0 | 0.0 | 0.2 | 1.1 | 4.0 |
| Other Siberian | 200 | 0.1 | 0.1 | 0.1 | 0.5 | 2.3 |
| North American | 600 | 0.2 | 0.2 | 0.1 | 1.0 | 1.0 |
| Evaporation | 3,500 | 0.0 | 0.0 | 0.0 | 0.0 | 0.0 |
| Precipitation | 5,300 | 0.0 | 0.0 | 0.1 | 0.1 | 0.0 |
| Southern boundary of the Norwegian Sea | 12,000 | 0.1 | 0.1 | 0.6 | 2.2 | 2.4 |
| Bering Strait | 10,560 | 0.0 | 0.0 | 0.5 | 1.9 | 1.9 |

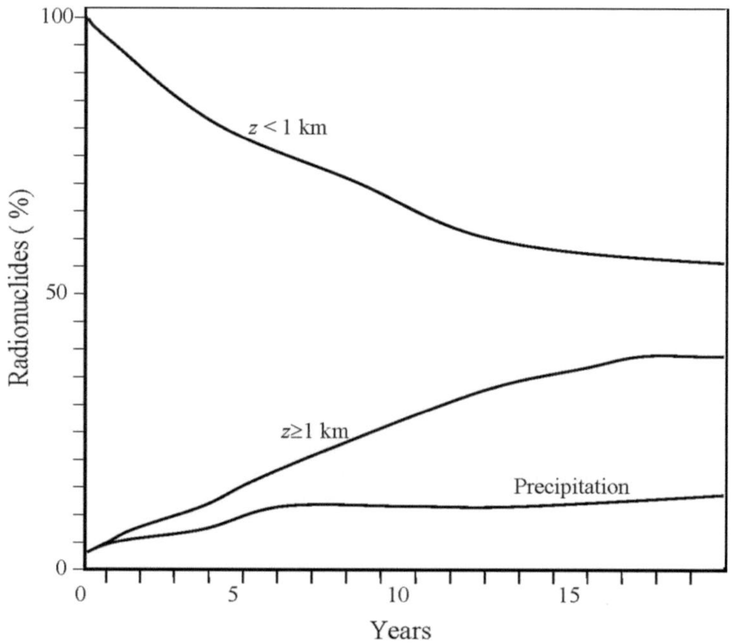


**Figure 4.7.** Dynamics of the distribution of radionuclides in the Arctic Basin. It is assumed that at moment $t_0 = 0$ radionuclide pollution ($^{137}$Cs, $^{60}$Co, see Table 4.7) is in layer $z < 1$ km. The curves show the dynamics of radionuclide pollution in two layers of water thickness and in bottom sediments. The results give the level of pollution by averaging over all Arctic water basins.

simulation experiment are given in Figures 4.7 and 4.8. Figure 4.7 shows the tendency versus time of the average content of radionuclear pollution on the whole Arctic aquatory. Distribution with depth is represented by a three-layer model, upper waters ($z < 1$ km), deep waters ($z > 1$ km), and sediments. The bottom depth is taken as 1.5 km. A more realistic depth representation of the shallow seas and the deeper Arctic Basin will be considered in a future refinement of the model. The curves describe the vertical distribution with time of the radionuclide content in the two water layers and in sediments. The transfer of radionuclides from upper waters to deep waters occurs at a speed that results in a 43.3% reduction in radionuclear pollution in upper waters over 20 years. Distributions for each of the Arctic seas are given in Table 4.8.

Local variations in the vertical distribution of radionuclides are determined by both hydrological and ecological conditions. The correlation between these conditions is a function of the season. Table 4.9 gives estimates of the role played by ecological processes in formation of the vertical distribution of radionuclear pollution in the Arctic seas. These estimates show that the biological community plays a minor role in radionuclide transport from the upper layers to the deep ocean.

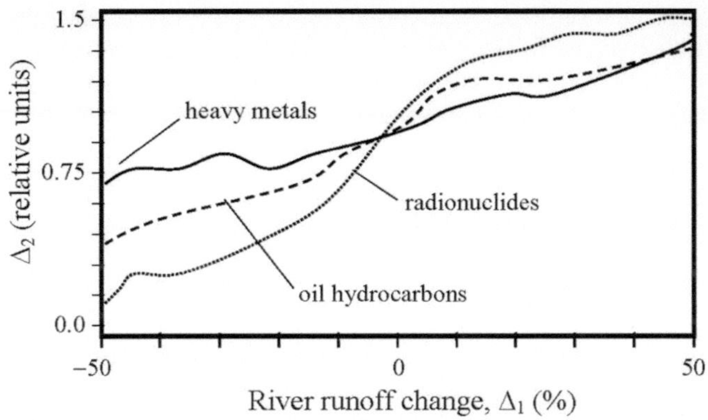


**Figure 4.8.** Influence of variations in river flows on Arctic Basin pollution level. Here $\Delta_1$ is the percentage variation in river flow to $\Omega$ with respect to the value averaged on $\Omega_R$ in the last three years and $\Delta_2$ is the content of pollutant averaged on all rivers in $\Omega_R$ and normalized to the initial data (such that $\Delta_2 = 1$ for $\Delta_1 = 0$).

**Table 4.8.** Distribution of radionuclear pollution in the Arctic aquatories 30 years and 50 years after $t_0$ (%).

| *Parameter* | *Aquatory* (see Table 4.4) | | | | | | | | | | |
|---|---|---|---|---|---|---|---|---|---|---|---|
| | Γ | *N* | *B* | *K* | ∇ | Λ | *E* | *S* | *X* | Φ | *U* |
| 30 years after $t_0$ | | | | | | | | | | | |
| $z \leq 1$ km | 50 | 60 | 69 | 46 | 73 | 44 | 43 | 57 | 58 | 61 | 68 |
| $z > 1$ km | 49 | 39 | 29 | 52 | 26 | 54 | 54 | 39 | 39 | 34 | 29 |
| 50 years after $t_0$ | 1 | 1 | 2 | 66 | 1 | 2 | 3 | 4 | 3 | 3 | 3 |
| $z \leq 1$ km | 65 | 57 | 70 | 27 | 70 | 50 | 49 | 62 | 59 | 58 | 70 |
| $z > 1$ km | 30 | 38 | 24 | 7 | 21 | 47 | 46 | 32 | 37 | 34 | 26 |
| Bottom | 5 | 5 | 6 | 2 | 9 | 3 | 5 | 6 | 4 | 8 | 4 |

The aquatories of the White, Laptev, East Siberian, and Chukchi Seas are subject to visible variations in radionuclear pollution. An accumulation of radionuclides is observed in the central aquatory of the Arctic Basin. The aquageosystems of the Greenland and Kara Seas for some reason lead to a buildup of radionuclear pollution, but in the Norwegian Sea there is even a decrease in the pollution level.

A somewhat stable situation is observed in the vertical distribution of radionuclides. Stability generally takes 5–7 years after initial moment $t_0$ with the exception of the East Siberian, Laptev, and Kara Seas where the stabilization processes of vertical distribution are delayed by 10–12 years compared with the other aquatories of the Arctic Basin.

**Table 4.9.** Some simulation experiment results using the SSMAE to estimate the vertical distribution of radionuclides in the Arctic Basin. The contribution of ecological processes to the vertical distribution of the radionuclide content of water is represented by the parameter $\xi$ (%). The average content of the biomass of phytoplankton is represented by the parameter $p_w$ (g/m$^2$).

| *Aquatory* | *Seasons* | | | | | | | |
|---|---|---|---|---|---|---|---|---|
| | *Winter* ($\tau_w$) | | *Spring* ($\tau_s$) | | *Summer* ($\tau_u$) | | *Fall* ($\tau_a$) | |
| | $p_w$ | $\xi$ | $p_w$ | $\xi$ | $p_w$ | $\xi$ | $p_w$ | $\xi$ |
| Greenland Sea | 3.2 | 2 | 8.4 | 10 | 5.7 | 5 | 6.3 | 5 |
| Norwegian Sea | 2.9 | 2 | 7.8 | 9 | 5.9 | 5 | 6.7 | 6 |
| Barents Sea | 2.1 | 1 | 8.9 | 11 | 6.8 | 6 | 7.1 | 6 |
| Kara Sea | 2.4 | 1 | 9.2 | 12 | 5.3 | 5 | 6.0 | 5 |
| White Sea | 2.2 | 1 | 7.6 | 9 | 6.3 | 6 | 6.4 | 5 |
| Laptev Sea | 0.9 | 1 | 2.4 | 4 | 1.3 | 2 | 1.4 | 2 |
| E. Siberian Sea | 1.3 | 1 | 2.7 | 4 | 1.9 | 3 | 2.1 | 3 |
| Bering Sea | 2.5 | 2 | 7.1 | 9 | 3.9 | 4 | 5.3 | 4 |
| Chukchi Sea | 2.3 | 2 | 6.9 | 8 | 4.1 | 4 | 5.1 | 4 |
| Beaufort Sea | 1.9 | 2 | 5.7 | 7 | 4.8 | 4 | 4.9 | 4 |
| Central Basin | 1.0 | 1 | 1.7 | 2 | 1.5 | 2 | 1.6 | 2 |
| *Average value* | *2.1* | *1.5* | *6.2* | *7.7* | *4.3* | *4.2* | *4.8* | *4.0* |

The results of simulation experiments show that variations of ±100% in the initial data change the stabilization time by no more than 30%, so that distributions take shape in 4–8 years. One unstable parameter is river flow into the Arctic Basin. Figure 4.9 shows how simulation results vary as a result of change in river flow to the Arctic Basin. Radionuclear pollution is reduced by 80% when river flow decreases by 50%. In contrast, when river flow increases by 50%, radionuclear pollution of the Arctic Basin increases by only 58%. Hence, a ±50% error in river flow estimate can cause a <100% deviation of the simulation results for radionuclear pollutants. The other curves in Figure 4.8 show that such deviations are smaller for heavy metals and oil hydrocarbons.

The SSMAE allows a wide spectrum of radionuclear pollution parameters to be estimated; hence, the importance of showing how biological transformation mechanisms depend on initial prerequisites. The biological transport of radionuclides downward out of the mixed layer varies in a wide interval from months to

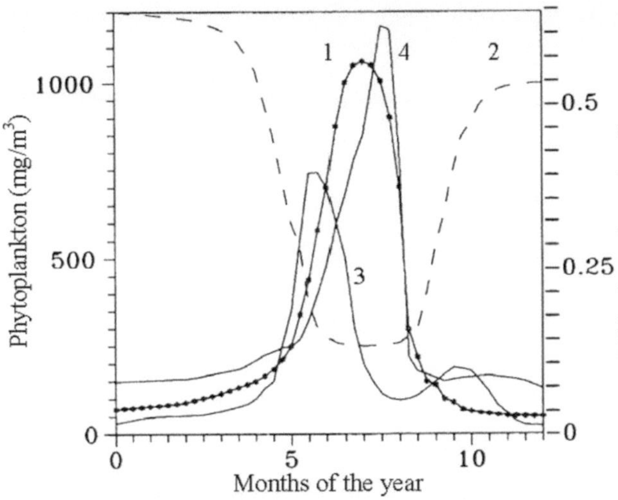


**Figure 4.9.** Influence of the Barents Sea ecosystem on the dynamics of oil hydrocarbons in seawater. Curves 1 and 2 show the simulation results for phytoplankton (solid curve) and oil hydrocarbons (dashed curve), respectively. Curves 3 and 4 show the yearly distribution of phytoplankton in the southwestern and in the northern and northeastern aquatories of the Barents Sea, respectively (Terziev, 1992).

scores of years. Vertical transport by biological elements is split between 11% for migration by living organisms and 89% for transport by dead organisms. Stability is achieved in that the lower trophic levels of the Arctic ecosystem have a greater concentration of radionuclides than the higher trophic levels. However, for block MPR to be more precise the model of biological processes must consider separately each radioactive element and its interaction with plankton. The variant of MPR realized here considers physical processes as major forces.

*The dynamics of Arctic Basin pollution by heavy metals.* The results of the simulation experiment are given in Table 4.10. We can see that the average content of heavy metals on the full aquatory of the Arctic Basin takes 3–5 years to stabilize. Under this stable regime, the concentration of heavy metals in compartments $\Omega_R \cup \Omega_P$ (river mouths and ports) is six times higher than in the central aquatory and two times higher than in $\Omega_\Gamma \cup \Omega_B \cup \Omega_N$ (near-shore waters, the Bering Strait, and the southern boundary of the Norwegian Sea). The concentration of heavy metals in phytoplankton is 18% lower than in zooplankton and 29% lower than in nekton. A process in which heavy metals accumulate in upper trophic levels can be observed; moreover, the relation between the concentration of heavy metals in phytoplankton and their concentration in other trophic levels varies from 0.3 in $\Omega_R \cup \Omega_P$ to 0.5 in the open part of the Arctic Basin.

The spatial features of the distribution of heavy metals in seawater is characterized by more rapid accumulation into aquatories adjoining the west

**Table 4.10.** Simulation experiment results estimating the parameters characterizing the pollution dynamics of Arctic waters by heavy metals.

| *Parameter* | *Estimate of the parameter after* $\Delta t$ (years) | | | | | |
|---|---|---|---|---|---|---|
| | $\Delta t = 1$ | $\Delta t = 3$ | $\Delta t = 5$ | $\Delta t = 10$ | $\Delta t = 15$ | $\Delta t = 20$ |
| Average concentration of heavy metals in the biomass (ppm) | | | | | | |
| Phytoplankton | 0.011 | 0.012 | 0.016 | 0.024 | 0.036 | 0.037 |
| Zooplankton | 0.013 | 0.014 | 0.019 | 0.028 | 0.041 | 0.043 |
| Nekton | 0.015 | 0.017 | 0.022 | 0.040 | 0.070 | 0.070 |
| Detritophages | 0.033 | 0.037 | 0.048 | 0.088 | 0.150 | 0.160 |
| Average content of heavy metals in Arctic waters (ppm) | | | | | | |
| Arctic Basin | 0.022 | 0.027 | 0.036 | 0.037 | 0.038 | 0.038 |
| Flow of heavy metals from the upper layer to deep waters ($\mathrm{mg\,m^{-2}\,day^{-1}}$) | | | | | | |
| Norwegian Sea | 0.71 | 1.07 | 1.14 | 1.17 | 1.19 | 1.22 |
| Barents Sea | 0.72 | 1.08 | 1.25 | 1.19 | 1.24 | 1.16 |
| Greenland Sea | 0.26 | 0.62 | 0.71 | 0.82 | 0.76 | 0.89 |
| White Sea | 0.11 | 0.23 | 0.24 | 0.21 | 0.19 | 0.2 |
| Kara Sea | 0.34 | 0.47 | 0.57 | 0.61 | 0.63 | 0.64 |
| Laptev Sea | 0.55 | 0.78 | 0.81 | 0.89 | 0.74 | 0.77 |
| East Siberian Sea | 0.59 | 0.79 | 0.95 | 0.97 | 1.02 | 1.07 |
| Chukchi Sea | 0.88 | 0.83 | 1.54 | 1.49 | 1.31 | 1.44 |
| Beaufort Sea | 0.34 | 0.67 | 0.66 | 0.81 | 0.74 | 0.69 |

coastline of Novaya Zemlya and situated on the boundary between the Jan Mayen and East Iceland Currents. The central aquatory of the Arctic Basin has a quasi-uniform distribution of heavy metal concentration. The vertical gradients of heavy metals vary from 0.11 to 1.54 $\mathrm{mg\,m^{-2}\,day^{-1}}$. The forms of vertical distributions of dissolved ($\psi_W$) and suspended particles ($e_W$) of heavy metals are not obviously expressed. The average relation of the concentration of heavy metals in sediments and water ($[\psi^* + e^*]/[\psi_W + e_W]$) varies from 1.9 to 5.7 depending on the aquatory of the Arctic Basin. For example, this relation for the Bering Sea is 3.3. The contribution of the biosedimentation process to the vertical distribution of heavy metals is defined by values that vary from 0.23 to 1.24 $\mathrm{mg\,m^{-2}\,day^{-1}}$. The SSMAE allows the characteristics of separate types of heavy metals to be estimated. An example of such calculations is given in Table 4.11.

*The dynamics of oil hydrocarbon pollution in the Arctic Basin.* Spatial distributions were forecast for oil hydrocarbons in the Arctic aquatories for 5 years after $t_0$ under the following restrictions: $O(t_0, \varphi, \lambda, z) = 0$; $H_O^1 = 0.1\,\mathrm{mg\,m^{-3}\,day^{-1}}$; $H_O^2 = 0$ for $g(t) > 0$ and $H_O^2 = 0.01\,\mathrm{mg\,m^{-3}\,day^{-1}}$ for $g(t) = 0$ and $z \geq \Delta z_1$; $H_O^3 = 0.01\,\mathrm{mg\,m^{-3}\,day^{-1}}$; $H_O^4 = 0.01\,\mathrm{mg\,m^{-3}\,day^{-1}}$; $H_O^5 = k_D D_A^{1/3}$ (the adsorption coefficient $k_D = 0$ for $A = g, r, f$ and $k_Z = 0.005\,\mathrm{day^{-1}}$ when $A = W$); $H_O^6 = k_Z Z^{1/4}$ (the biosedimentation coefficient $k_Z = 0$ when $A = g, r, f$ and

**Table 4.11.** Estimates of heavy metal flows to and from the atmosphere.

| *Parameter* | *Heavy metal* | | | | | | | | | | |
|---|---|---|---|---|---|---|---|---|---|---|---|
| | Ag | Cd | Co | Cr | Cu | Fe | Hg | Mn | Ni | Pb | Sb |
| Atmospheric deposition, $Q^2_{e\psi}$ ($\mathrm{mg\,m^{-2}\,h^{-1}}$) | 0.7 | 1.1 | 0.3 | 1.8 | 15 | 599 | 0.6 | 4.2 | 5.5 | 48 | 0.05 |
| Evaporation and spray, $H^{\psi}_{\alpha}$ (t/yr) | 7 | 58 | 5 | 188 | 169 | 894 | 3 | 283 | 60 | 5 | 123 |

$k_Z = 0.004\,\mathrm{day}^{-1}$ for $A = W$); $H^7_O = k_B B_A$ (the bacterial destruction coefficient $k_B = 0.01\,\mathrm{day}^{-1}$ for $A = g, r, f$ and $k_B = 0.05$ when $A = W$). Simulation experiments show that 56% of anthropogenic oil hydrocarbons ($Q_O$) estimated by Mcintyre (1999) are transformed to other forms in the environments of surface snow, floating ice, and submerged ice and 72% in water. It takes 3 years after $t_0$ for the distribution of oil hydrocarbons to stabilize. The average level of oil pollution in the Arctic Basin is 0.005 mg/L. This is lower than the natural level of World Ocean pollution. In the zone of influence of the Gulf Stream and in Pacific waters this level is marginally higher. The Barents and Kara Seas are the most polluted with a concentration of oil hydrocarbons at 0.03 mg/L. The average content of oil hydrocarbons in the Arctic Basin is 65,331 t (metric tonnes) with a dispersion of 32%. The hierarchy of flows $H^i_O$ $(i = 1, \dots, 7)$ (see equation 4.34) is estimated by the set of $H^2_O > H^4_O > H^1_O > H^5_O > H^6_O > H^7_O$. This set is changed for each Arctic sea. Which of the destruction processes $H^i_O$ of oil hydrocarbons prevails is dependent on seasonal conditions. The oxidation process as a result of the evaporation of oil hydrocarbons ($H^2_O$) prevails over other processes in summer.

Oil hydrocarbons that evaporate from the surface of Arctic seas return to the Arctic Basin with atmospheric precipitation. These processes are simulated in blocks APM and PSM. The highest rate at which oil hydrocarbons are destroyed is $0.0028\,\mathrm{g\,m^{-2}\,day^{-1}}$. The flow $H^7_O$ as a result of bacterial decomposition averages $27\,\mathrm{t\,yr^{-1}}$. The values between different seas vary ($\mathrm{t\,yr^{-1}}$): Bering 3.7, Greenland 11, Norwegian 2.2, Barents 3.4, Kara 2.3, White 2.3, Laptev 2.8, East Siberian 2.8, Chukchi 3.4, Beaufort 2.5, Central Basin 0.5.

The role played by the Arctic basin ecosystem in the dynamics of oil hydrocarbons can be estimated for each Arctic sea. As an example, Figure 4.9 gives the results for the Barents Sea. The discrepancy between simulation results (solid curve 1) and the data of curves 2 and 3 (Terziev, 1992) can be explained by the assumption that the trophic structures of different Arctic seas are described by means of the general scheme represented in Figure 4.6 and discrimination between the ecological elements in each of the Arctic seas is not taken into consideration in the SSMAE.

Figure 4.9 indicates that the vegetative period for phytoplankton in the

Barents Sea lasts 4.9 months as shown by the effects on the ecosystem's contribution to self-clearing of oil hydrocarbons (dashed curve.) In this case, the Barents Sea ecosystem neutralizes about 25% of oil hydrocarbons during the vegetative period. The rest of the time this value oscillates near 3%. The dispersion of these estimates with latitude reaches 53%. For example, in the northern part of the Barents Sea the vegetative period lasts from 2.6 to 3.1 months, while in the southern latitudes the variation is 5.3–5.8 months. Consequently, the role played by the ecosystem in sea-cleaning of oil hydrocarbons is 8% and 36% for the northern and southern aquatories, respectively. Such estimates can be calculated for each cell $\Delta\varphi \times \Delta\lambda \times \Delta z$ of the Arctic Basin.

*The dynamics of pollutants in the Arctic Basin.* An important task is determining and understanding the role played by various pollutant sources in forming the pollution levels for different Arctic Basin areas (McCauley and Meier, 1991). A major function of the SSMAE is to estimate the pollution dynamics in each cell $\Omega_{ij} \subseteq \Omega$ as a function of time. The influence pollutant sources have on the Arctic ecosystem occurs across the boundary area $\Omega_R \cup \Omega_P \cup \Omega_B \cup \Omega_N \cup \Omega_\Gamma$ and through the atmosphere. The total picture of the spatial distribution of pollutants is formed from local dynamic processes. The incompleteness of the Arctic database forces the consideration of scenarios within which assumptions necessary to make indefinite Arctic system parameters concrete are admittedly made. From Krapivin *et al.* (1998a) there is the estimation of the flow of pollutants to the Kara Sea based on experimental measurements of radionuclear pollution and heavy metals in the Angara–Yenisey river system. Therefore, the SSMAE can be used to estimate the flow of pollutants from the Kara Sea to the other Arctic Basin aquatories.

The Ob′ and Yenisey rivers are considered to be the main sources of radionuclear pollution, heavy metals, and oil hydrocarbons in the Kara Sea (Table 4.7). Figure 4.10 shows the influence of river flow on the volume of pollutants transported by the Kara Sea aquageosystem to the central basin. As shown by curves 1 and 2, the transfer of heavy metals and radionuclides from the Ob′ and Yenisey rivers through the Kara Sea aquageosystem to the central aquatories of the Arctic Basin amounts to 2.1% when river flow varies from 500 to 1,000 $km^3$/yr and then begins to grow linearly up to 7.6% when river flow reaches 2,000 $km^3$/yr. Hence, there is a critical level of pollution for the Kara Sea ecosystem beyond which it does not have time to dilute the flow of heavy metals and radionuclear pollutants. Similar estimations can be obtained for the other elements of set $\Omega_R$ using the rivers referred to in Table 4.7.

The SSMAE allows the flow of pollutants between the different aquatories of the Arctic Basin to be estimated. For example, the transport of heavy metals and oil hydrocarbons from the Barents Sea to the Kara Sea is 631 kg/yr and 473 kg/yr, respectively. The total flow of pollutants from the Russian coastline to Alaska varies in Table 4.7 between 0.3% and 0.9% of the initial flow. As is evident from curves 3 and 4 of Figure 4.10, the flows of the Ob′ and Yenisey rivers have practically no influence on the pollution level of Arctic waters near Alaska. This effect does not change over time.

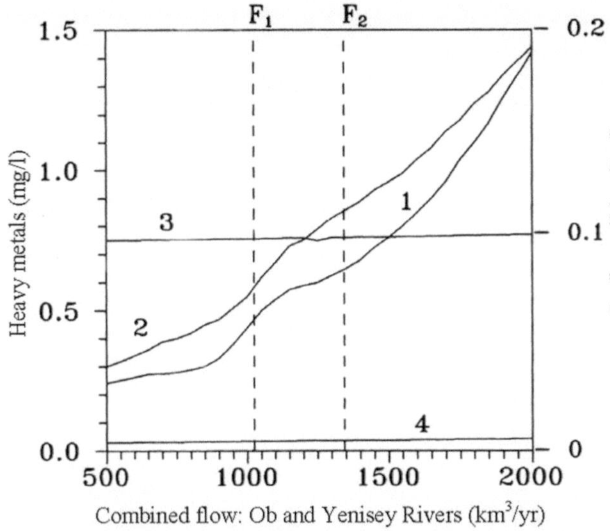


**Figure 4.10.** Concentrations of heavy metals ($\psi + e$) and radionuclides ($\varepsilon = ^{137}$ Cs $+ ^{60}$ Co) at different geographical points as a function of the flow ($F$) of the Ob′ and Yenisey Rivers to the Kara Sea. The interval $[F_1, F_2]$ between the dashed lines corresponds to the range of variations of $F$ in the real world. Curves 1 and 2 show the concentrations of heavy metals and radionuclides, respectively, at the point $\varphi, \lambda = 75°$N, 65°E in the northwestern part of the Kara Sea. Curves 3 and 4 show the concentrations of heavy metals and radionuclides, respectively, at the point $\varphi, \lambda = 72°$N, 150°W above the Alaskan northern coast.

#### *4.5.2.6 Summary and conclusions*

Let us now discuss three aspects of the SSMAE that are very important. The first concerns the incorporation of ecological, hydrophysical, climatic, and biogeochemical relationships in a model simulating the dynamics of Arctic Basin pollution. The main task is parameterizing these relationships to achieve the necessary precision. The second concerns the key problem of database conformity to the model. The task here is to adapt the spatiotemporal scale to the database. The third concerns the user's ability to run the SSMAE in the scenario space.

This investigation shows that simulation experiments can facilitate studying both the common and spatial distributions of pollutant dynamics in the Arctic Basin. The precision of the results is a function both of the scenarios and the forms in which the elements in equations (4.25)–(4.40) are parameterized. The SSMAE is clearly not effective when climatic conditions are varied to a critical state or when anthropogenic impacts are increased to a critical value. But in the SSMAE the Arctic Basin acts as a stabilizing subsystem of the biosphere. When atmospheric temperature is reduced by $-1°$C, there occurs an inverse connection in the water balance of the atmosphere–land–sea system which acts to stabilize estimates. The parameterization of such variations within the framework of the SSMAE is not convenient. Nevertheless, connecting the SSMAE to a global database having estimates of such parameters as ice area, temperature, and albedo distributions, will allow use of the SSMAE in its present form. The effectiveness of the SSMAE will increase with the use of such models as the coupled ice–ocean model described in the paper of Riedlinger and Preller (1991). In general, many different modifications of the SSMAE are possible. Nevertheless, to optimize the structure of the SSMAE depends on greater accuracy of pollutant types, ecosystem structure, water cycle, ice movement, and climate models. The main difficulty lies in synchronizing the optimal modification for every block of the SSMAE.

The pollution level of the Arctic Basin is mainly attributable to river flows. This is the reason the MRF block plays a very important role in the SSMAE. Regular monitoring of water flows and pollutant inputs by rivers to the SSMAE is impossible. Consequently, the study and measurement of these flows during scientific expeditions and the modeling of the results are significant steps in investigating Arctic Basin pollution.

An example of such a step is the U.S./Russian expedition of 1995 to obtain on-site measurements of pollution levels in the Yenisey and its tributary the Angara, to investigate the likely origins of land-based sources contributing to pollution levels in the Yenisey estuary.

Another important task is verification of the SSMAE. However, this should be possible after essential modifications to the SSMAE by incorporating more precise models to account for hydrological, biogeochemical, ecological, and climatic processes. The present structure of the SSMAE can be viewed as a first step toward a new technology of Arctic basin pollution monitoring. Greater accuracy in the SSMAE may be realized by means of simulation experiments in which model parameters are varied over wide intervals. However, a discussion of this is beyond the scope of this chapter.

Let us now look at several results of this investigation. In a nuclear war scenario, for example, the SSMAE shows that Arctic environmental stability is disturbed 3 months after the impact. Other scenario show that variations in the velocity of vertical advection from 0.004 to 0.05 cm/s does not affect the state of the Arctic environment. A 32% error in the ice area estimate leads to simulation results varying by 36%. When this error is more than 32%, simulation results become less stable and can vary manifold. There is also the problem of finding the proper criterion to estimate SSMAE sensitivity to variations in model parameters. Krapivin (1996) showed that a survivability function $J(t)$ reflecting the dynamics of the total biomass of living elements enables this sensitivity to be estimated. In this instance

$$J(t) = \frac{\sum_{i=1}^{5} \iint_{(\varphi,\lambda)\in\Omega} \int_{0}^{H(\varphi,\lambda)} B_i(t,\varphi,\lambda,z)\, d\varphi\, d\lambda\, dz}{\sum_{i=1}^{5} \iint_{(\varphi,\lambda)\in\Omega} \int_{0}^{H(\varphi,\lambda)} B_i(t_0,\varphi,\lambda,z)\, d\varphi\, d\lambda\, dz}.$$

The index $J(t)$ provides an estimate of the uncertainty associated with SSMAE parameters. Although thorough investigation of the influence of SSMAE parameter variations on model results should be done independently by other researchers, we give various estimates here. Preliminary simulation results show that the SSMAE permits initial data to vary between 70% to 150%. The model "forgets" these variations during the first 50 days of its development. Moreover, a large uncertainty (±50%) is permitted in the value of such parameters as $\mu_A, \delta_n, V_i, T_c, T_{opt}, \rho_1, k_{ij}$. The correlation between variations in these parameters and model results is linear. However, high model sensitivity is observed under variations of $\beta_A, \alpha_A, E_0, T_0$. In general terms, the acceptable variation of these parameters is ±20%. What is more, deviation from model results as a result of variations in these parameters is nonlinear. For example, fluctuations of the surface temperature $T_0$ within ±5 K are not hazardous to the system, causing small variations of $J(t)$ by ±10%, but fluctuations of $T_0$ by ±7 K cause much larger (±30%) variations in the value of $J(t)$.

The SSMAE structure falls short of completely describing the processes taking place in Arctic Basin. Optimal extension of SSMAE functions is possible by using environmental monitoring to control parametric and functional model inputs. In such a framework a prognosis of the Arctic aquageosystem state could be realized on the basis of the SSMAE and by processing observed data (Krapivin, 2000b).

### 4.5.3 Interactions in the Arctic system

The Arctic region is a mosaic of freshwater, terrestrial, and marine ecosystems intimately interactive with factors of the NSS. The interactions include many components such as ocean–atmosphere–ice, land–atmosphere–ice, and land–ocean–freshwater. An interpretation and prediction of correlations between the processes occurring in the Arctic environment is only possible only within the framework of a complex approach to the study of these processes. This approach is based on GIMS technology (Kelley *et al.*, 1992b). The interaction between the atmosphere, land, and sea ecosystems under the Arctic climate is characterized by a series of spatiotemporal scales. An understanding of interior correlations at every level is key to environmental monitoring. Each scale has a specific landscape, vegetation cover, topography, character of both hydrophysical and weather structures, and living elements. The discovery of internal and external correlations between these elements and other elements of the NSS makes it possible to form a knowledge base for an Arctic Basin monitoring system.

A methodology for the combined use of measurements and mathematical models is given in Figure 4.11. Such a scheme allows for more precise estimates of the role played by Arctic latitudes in global change: such models make it possible to synthesize a combined picture of the migration of chemical elements in the Arctic Basin and to estimate the consequences of anthropogenic processes in northern territories.

Interactions between Arctic components of the global NSS have been studied by many authors (Kelley and Gosink, 1992; Kondratyev and Johannessen, 1993; McCauley and Meier, 1991; Mertens and Gould, 1979; Preller and Cheng, 1999;

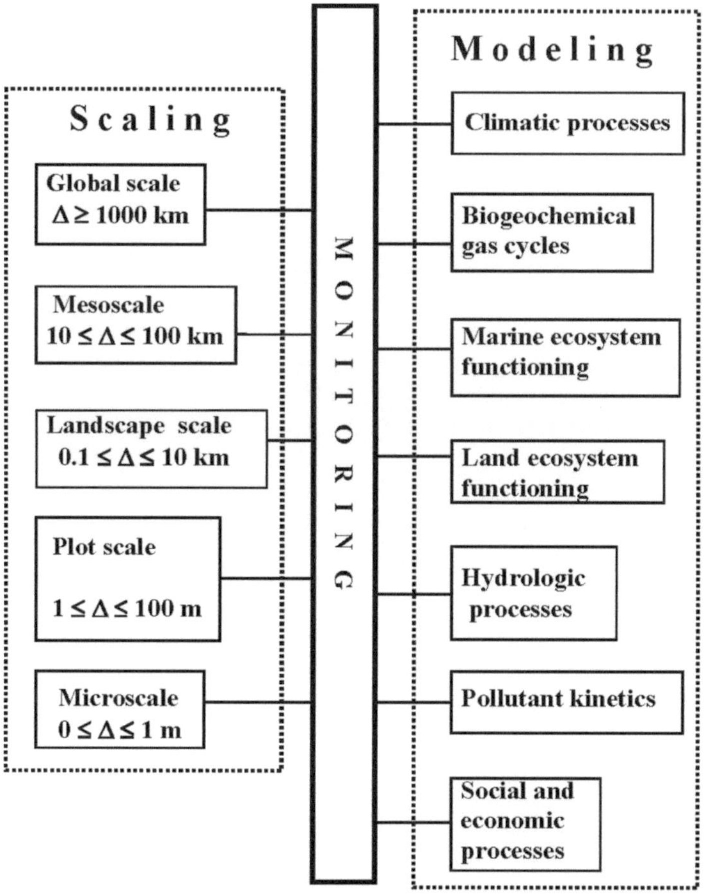


**Figure 4.11.** Conceptual scheme for environmental monitoring of northern latitudes. It reflects correlations between the spatial scales and the problems of investigation that are necessary for understanding the Arctic.

Riedlinger and Preller, 1991). A good example of these interactions is the Angara/Yenisey river system (AYRS) of Siberia. The intensive industrial development of northern Russia has led to significant environmental changes in these regions (Morgan and Codispoti, 1995). As mentioned at the beginning of Section 4.4, anthropogenic intervention in the natural ecosystems of numerous Siberian regions has increased. Most pollutants are brought to the northern coast of Russia by rivers, where they damage the ecosystems of the northern seas. The Angara–Yenisey is one such river system.

The Yenisey river flows northward to the Kara Sea along the boundary between the western Siberian flood plain and the central uplands, draining an area of about 2,600,000 $km^2$ over its 4,100 km length. The flow rate of the Yenisey into the Kara Sea has large seasonal variations, averaging 19,800 $m^3/s$, and reaching as much as 130,000 $m^3$/s during the spring runoff. The Angara, a major tributary, accounts for about one fourth of the total flow. It flows swiftly northward from Lake Baikal for about one third of its 1,850 km length, before turning westward toward its confluence with the Yenisey. Recognizing that major sources of the radionuclides found in the Kara Sea, as well as other environmental pollutants, might lie in the Siberian watersheds of the Yenisey and the Angara, a joint Russian–American expedition was undertaken in July and August of 1995. The region in which the expedition was conducted includes the five hydroelectric dams at Krasnoyarsk and Sayano-Shushenskoye on the Yenisey and at Irkutsk, Bratsk, and Ust-Ilimsk on the Angara. The power output from these facilities has fostered rapid industrial growth in this region. Krasnoyarsk is a major industrial city located on the upper reaches of the Yenisey. Nearby is the nuclear production and processing facility, Krasnoyarsk-26, which is approximately 270 km upstream of the Angara–Yenisey junction. Along the Angara river, there are five other cities involved in major industrial activity: Irkutsk, Angarsk, Usolye-Sibirskoye, Svirsk, and Bratsk. These cities have facilities whose by-products include radionuclides and chemical pollutants both of which likely contribute to pollution in the two rivers.

Samples of Angara river water and sediment were taken beginning near its source at Lake Baikal and continuing at selected sites of interest downstream to its junction with the Yenisey near the lumber-processing village of Strelka. Along the Yenisey sampling was conducted on both sides of the junction beginning upstream from the village of Kazachinskoye and continuing downstream to the town of Lesosibirsk.

Krapivin (1995) showed that comprehensive evaluation of the pollution level in the Arctic Basin as a whole is possible by synthesizing a mathematical model of pollutant transport by rivers from adjacent territories.

#### *4.5.3.1 An AYRS simulation model*

Krapivin (1995) was the first to illustrate the AYRS simulation model (AYRSSM) as a block diagram (see Figure 4.12). The AYRSSM has three levels of blocks. A description of them is given in Table 4.12. The two blocks of the first level are used to control models of different processes and ensure various regimes of computer experiments. The HYD block simulates river flow and considers the correlation between the water regime and ecological, topographical, and synoptic parameters of the territory studied. The CON block is used to set up an informational interface between the user and other AYRSSM blocks. The main function of CON is setting up a database and identifying AYRS elements. The four blocks of the second level analyze input information and synthesize the AYRSSM structure. The 26 blocks of the third level create concrete models and processes during the computer experiment.

Let us suppose that the AYRS watershed has an area $\Omega$. The spatial structure of $\Omega$ is determined by spatial discretization of the AYRS surface using a uniform

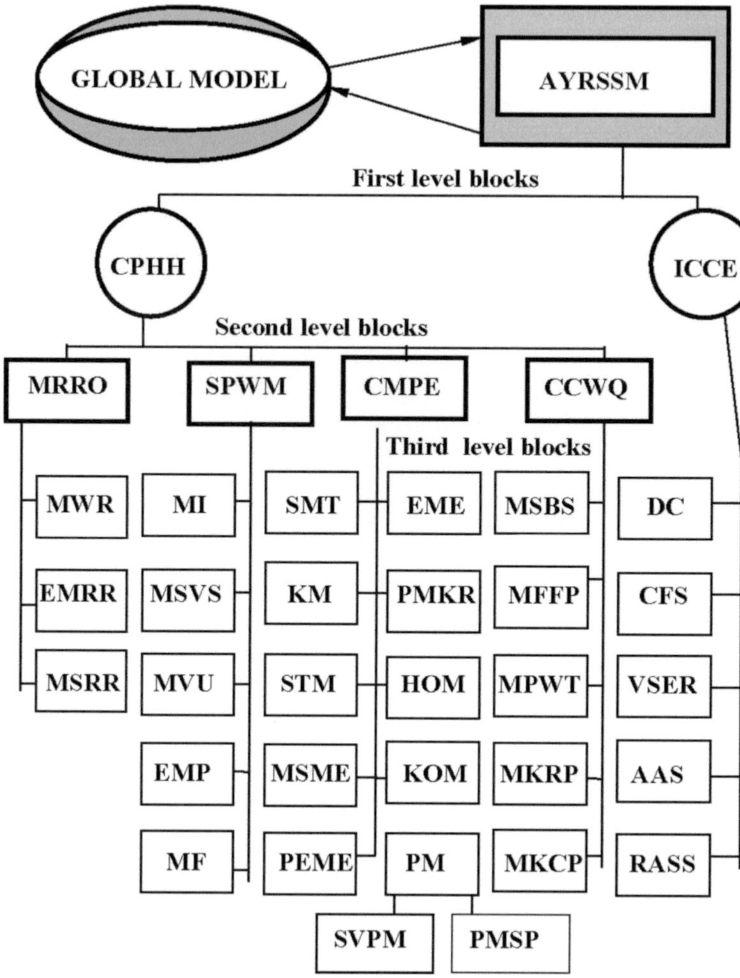


**Figure 4.12.** Structure of the AYRSSM. Identifiers are shown in Table 4.12.

geographic grid with latitude $\varphi$ and longitude $\lambda$, divided into steps of $\Delta\varphi$ and $\Delta\lambda$, respectively. In this study, we suppose $\Delta\varphi = \Delta\lambda = 0.1°$. As a result, the area $\Omega = \bigcup_{i=1}^{N} \Omega_k$, where $\Omega_k$ is that part of $\Omega$ with area $\sigma_k = \Delta\varphi, \Delta\lambda$, $\left(\sum_{k=1}^{N} \sigma_k = \sigma\right)$. The cells $\Omega_k$ are situated along the AYRS beginning with $\Omega_1$ at the source of the Angara river up to $\Omega_N$ in the Yenisey river mouth. The procedure of spatial discretization is provided for by the RASS block by including in the AYRSSM database a set of identifiers $A_k = \|a_{ij}^k\|$, $k = 1, \ldots, 5$ (Borodin *et al.*, 1996; Krapivin *et al.*, 1997a). The hydrology regime of the AYRS is described by the schematic diagram of Figure 4.13. The equations for this scheme can now be

**Table 4.12.** List of blocks of the AYRSSM, whose scheme is presented in Figure 4.12.

| *Identifier* | *Description of the unit* |
|---|---|
| CPHH | Takes control of parameterizing hydrophysical and hydrological processes in the AYRS |
| ICCE | Interface control of the computer experiment |
| MRRO | Model of river runoff |
| SPWM | Simulation procedure that reflects water mass motion in the AYRS |
| CMPE | Choice of model to parameterize the process of evaporation |
| CCWQ | Choice of criterion to assess water quality |
| MWR | Model of the water regime in a water body |
| EMRR | Empirical model of river runoff |
| MSRR | Model to simulate how river runoff spreads beyond the riverbed |
| MI | Model of infiltration |
| MSVS | Model of the sink taking into account the effect of vegetation and soil covers |
| MVU | Model of vertical uplift of ground water in evaporation, feeding, and exfiltration |
| MF | Model of filtration |
| EMP | Empirical model of precipitation |
| SMT | Specified model of transpiration |
| KM | Kuzmin model (Kuzmin, 1957) |
| STM | Simple transpiration model |
| MSME | Model of snow melting and evaporation from snow surface |
| PEME | Parameterization of evaporation using the method of energy balance |
| PM | Penman model (Penman, 1948) |
| PMSP | Penman model as specified by Priestley (Bras, 1990) |
| PMKR | Penman model as specified by Kohler and Richards (1962) |
| SVPM | Simplified version of the Penman model. |
| EME | Empirical models of evaporation |

*(continued)*

**Table 4.12** (*cont.*)

| *Identifier* | *Description of the unit* |
|---|---|
| KOM | Kohler model (Bras, 1990) |
| HOM | Horton model (Bras, 1990) |
| MSBS | Simulation model of sedimentation and biological assimilation of pollutants |
| MFFP | Simulation model of the formation of anthropogenic fluxes of pollutants |
| MPWT | Model of the process of water temperature formation |
| MKRP | Model of kinetics of the radionuclide pollution of water |
| MKCP | Model of kinetics of the chemical pollution of water |
| DC | Database correction |
| CFS | Choice and formation of scenario for the simulation experiment |
| VSER | Visualization of simulation experiment results |
| AAS | Adaptation of the AYRSSM to the scenario of the simulation experiment |
| RASS | Referencing the AYRSSM to spatiotemporal scales |

written in the form of balance correlations on each of the $\Omega_k$ $(k = 1, \ldots, N)$:

$$\sigma_k\left(\frac{\partial W}{\partial t} + \xi_\varphi \frac{\partial W}{\partial \varphi} + \xi_\lambda \frac{\partial W}{\partial \lambda}\right) = V - B\sigma_k + D + T + L \tag{4.41}$$

$$\rho_k \sigma_k \left(\frac{\partial C}{\partial t} + \mu \frac{\partial C}{\partial x}\right) = \rho_k \sigma_k B + J + K - V - U - F - M - R \tag{4.42}$$

$$(1 - \rho_k)\sigma_k \frac{d\Phi}{dt} = U + F + M + N + (1 - \rho_k)\sigma_k B - T - L - K - P \tag{4.43}$$

$$\sigma_k\left(\frac{\partial G}{\partial t} + v_\varphi \frac{\partial G}{\partial \varphi} + v_\lambda \frac{\partial G}{\partial \lambda}\right) = R + P - J - N - D, \tag{4.44}$$

where $\xi_\varphi$ and $\xi_\lambda$ are projections of wind speed; $\rho_k$ is that part of area $\Omega_k$ occupied by the river; $\mu$ is river speed; $v_\varphi$ and $v_\lambda$ are speed projections of ground-water motion; $x$ is the direction of river flow; and $t$ is time.

The functions on the right-hand side of equations (4.41)–(4.44) are described by mathematical expressions in accordance with the papers by Bras (1990) and Krapivin *et al.* (1996). Appropriate models are given in Table 4.12. There are many ways of using these functions. This provides the user of the AYRSSM with

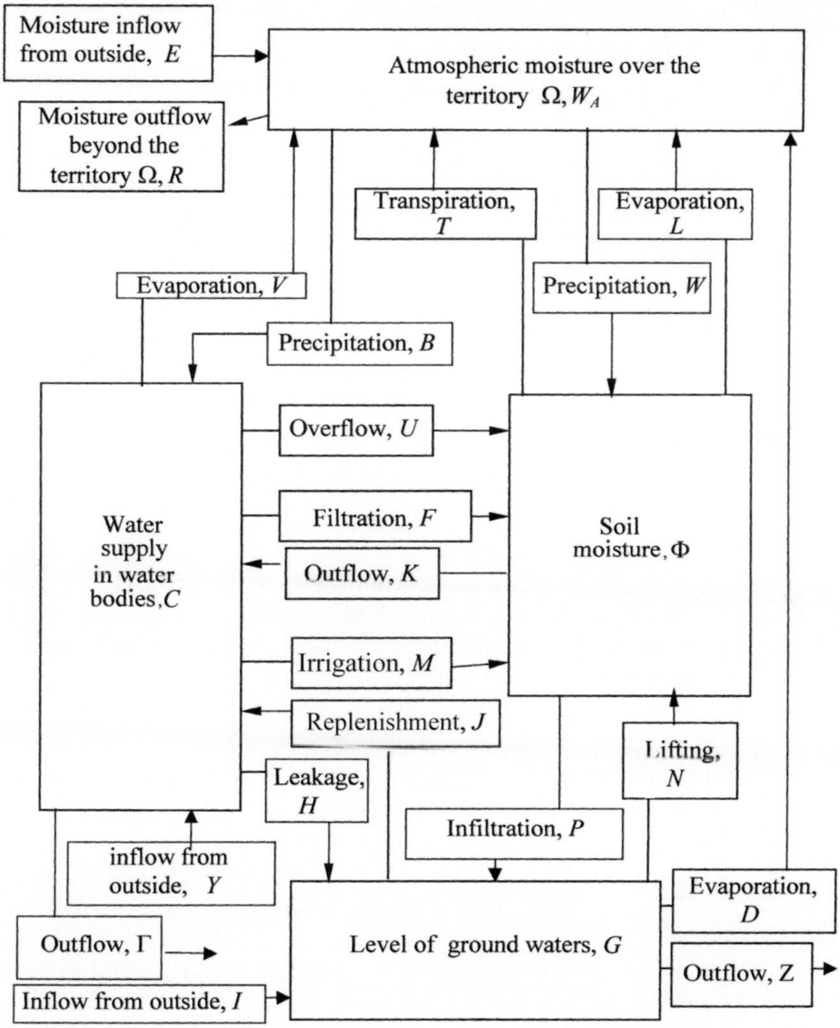


**Figure 4.13.** Block diagram of the AYRS water regime on the area $\Omega_k$ $(k = 1, \ldots, N)$. The functions $W$, $B$, $C$, $G$, and $\Phi$ have a linear dimension (m). All the other functions have a non-linear dimension ($m^3$/day).

the possibility of forming scenarios for computer experiments. Values $\xi$, $\mu$, and $v$ were estimated on the basis of the Irkutsk Scientific Center database. It is possible for the user to vary these parameters during the calculation process. In this study, average values of these parameters are estimated by $\xi = 3.3$ m/s, $\mu = 1.7$ m/s, and $v = 0$. Variations of the parameter $\mu$ are brought about by adapting the left-hand side of equation (4.42) to the empirical data illustrated in Figure 4.14. Boundary conditions for equations (4.41)–(4.44) are formed by the global model (Krapivin,

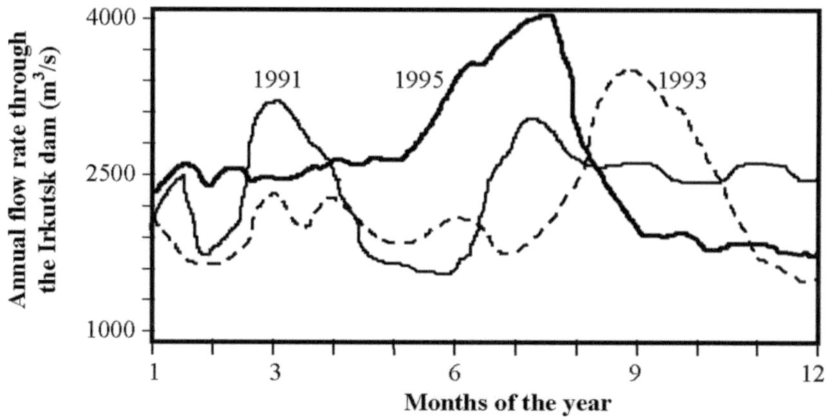


**Figure 4.14.** Annual flow rate through the Irkutsk dam for the years 1991, 1993 and 1995.

1993). Soil moisture transport between the cells $\Omega_k$ is neglected. Synoptic situations are described by a discrete scheme that uses temporal parameters $t_i$ ($i = 1, \ldots, 4$), where $t_1$ is the beginning of the summer period, $t_2$ is the start of winter, $t_3$ is the end of winter, and $t_4$ is the time of the spring thaw when the snow and ice are melting. Between these times the synoptic situation does not change.

The vertical structure of the river aquatory in $\Omega_k$ ($k = 1, \ldots, N$) is commonly described by the MSME block (Krapivin, 1995). A snow layer of thickness $g_k$ forms at the expense of flow $B_k$ according to:

$$\frac{dg_k}{dt} = \begin{cases} 0 & \text{when } t \notin [t_2, t_3] \\ B_k & \text{when } t \in [t_2, t_3]. \end{cases} \tag{4.45}$$

For $t \in [t_3, t_4]$ the value of $g_k(t, \varphi, \lambda)$ is decreased linearly from $g_k(t_3, \varphi, \lambda)$ to $g_k(t_4, \varphi, \lambda) = 0$. A functional representation of the other blocks in Table 4.12 can be realized at times $t_i$ ($i = 1, \ldots, 4$) by values input by the user as a result of scenario realization.

The dynamics of pollutants in the AYRS is determined by the structure of its hydrological regime, which takes into account the transport of pollutants by water motion and their accumulation in sediments, ice, snow, and the living biomass. The pollutant types are divided into radionuclear elements and heavy metals. The set of radionuclear pollutants is described by the index $\nu$. The set of heavy metals is divided into solid particles (index $e$) and dissolved particles (index $\psi$).

The $\nu$th radionuclide is characterized by its half-life $t_\nu$, by the rates $K_{k\nu}$ of input to and output from the area $\Omega_k$ ($k = 1, \ldots, N$), and by the concentrations $Q_\nu$, $E_\nu$, $\Xi_\nu$, and $S_\nu$ in the water, soil, sediments, and ground water, respectively. As a result the main balance equations of the MKRP block are written in the form:

$$\Delta_k\left(\frac{\partial Q_{k,\nu}}{\partial t}+\mu_k\frac{\partial Q_{k,\nu}}{\partial x}\right)=H_{1,\nu}+H_{2,\nu}+H_{3,\nu}-H_{4,\nu}-H_{5,\nu}+H_{6,\nu}-H_{7,\nu}+\rho_kH_{12,\nu}-\frac{\ln 2}{t_\nu}\Delta_kQ_{k,\nu} \tag{4.46}$$

$$\delta_k\frac{\partial E_{k,\nu}}{\partial t}=H_{8,\nu}+H_{4,\nu}+H_{9,\nu}-H_{10,\nu}+(1-\rho_k)H_{12,\nu}-\delta_kE_{k,\nu}\frac{\ln 2}{t_\nu} \tag{4.47}$$

$$\psi_k\frac{\partial \Xi_{k,\nu}}{\partial t}=H_{5,\nu}-H_{6,\nu}-\psi_k\Xi_{k,\nu}\frac{\ln 2}{t_\nu} \tag{4.48}$$

$$y_k\left(\frac{\partial S_{k,\nu}}{\partial t}+v_\varphi\frac{\partial S_{k,\nu}}{\partial \varphi}+v_\lambda\frac{\partial S_{k,\nu}}{\partial \lambda}\right)=H_{7,\nu}+H_{10,\nu}-H_{2,\nu}-H_{9,\nu}-H_{11,\nu}-y_kS_{k,\nu}\frac{\ln 2}{t_\nu}, \tag{4.49}$$

where $H_{1,\nu}$ is radionuclide washout from the soil with flow $K$; $H_{2,\nu}$ is radionuclide input to the river from ground water with flow $J$; $H_{3,\nu}$ is the transport of radionuclides to the area $\Omega_k$ through the boundary of $\Omega$ by the AYRS tributaries; $H_{4,\nu}$ is radionuclide amount carried by flows $U$, $F$, and $M$; $H_{5,\nu}$ is radionuclide sedimentation at the river bottom by gravity mixing with the dead biomass of the river ecosystem; $H_{6,\nu}$ is radionuclide washout from sediments; $H_{7,\nu}$ is radionuclide leakage to ground water with flow $R$; $H_{8,\nu}$ is the anthropogenic source of radionuclides; $H_{9,\nu}$ is radionuclide input to the soil from ground water by flow $N$; $H_{10,\nu}$ is radionuclide washout from the soil to ground water with flow $P$; $H_{11,\nu}$ is radionuclide losses by sedimentation in ground water; and $H_{12,\nu}$ is radionuclide input by rain. The quantities $\Delta_k$, $y_k$, $\delta_k$, and $\psi_k$ are calculated using the following expressions: $\Delta_k=\rho_k\sigma_kC_k$, $y_k=\sigma_kG_k$, $\delta_k=l_k(1-\rho_k)\sigma_k$, and $\psi_k=r_k\rho_k\sigma_k$, where $l_k$ is the thickness of the efficient soil layer in the area $\Omega_k$ and $r_k$ is the thickness of the sediment layer.

The flows $H_{i,\nu}$ ($\underline{i}=1,\ldots,12$) can be parameterized by linear models (Krapivin, 1995; Nitu *et al.*, 2000a). The MSBS and MKCP blocks can be described by similar balance models in analogy with the models described by Somes (1999) and Kram (1999). The AYRS biology can be given in the form of a scenario or described by the model of Legendre and Krapivin (1992). The flows of heavy metals can include an assimilation of dissolved fractions by plankton ($H_\psi^Z$) and by nekton ($H_\psi^F$), the sedimentation of solid fractions ($H_e^1$), absorption from sediments by living elements ($H_{e,\psi}^L$), sedimentation with dead organic matter ($H_\psi^D$), and discharge from sediments ($H_{e,\psi}^a$) by erosion, diagenesis, turbulence, and anthropogenic impacts. The balance equations taking these flows into account have the same form as equations (4.46)–(4.49):

$$\Delta_k\left(\frac{\partial e_w}{\partial t}+\mu_k\frac{\partial e_w}{\partial x}\right)=\sum_{i=1}^{3}\alpha_2^iQ_{e,\psi}^i-H_e^1+\alpha_1H_{e,\psi}^a \tag{4.50}$$

$$\Delta_k\left(\frac{\partial \psi_w}{\partial t}+\mu_k\frac{\partial \psi_k}{\partial x}\right)=(1-\alpha_1)H_{e,\psi}^a-H_\psi^Z-H_\psi^D \tag{4.51}$$

$$\psi_k\frac{\partial e^*}{\partial t}=H_e^1-\alpha_1(H_{e,\psi}^L+H_{e,\psi}^a) \tag{4.52}$$

$$\psi_k\frac{\partial \psi^*}{\partial t}=H_\psi^D-(1-\alpha_1)(H_{e,\psi}^L+H_{e,\psi}^a), \tag{4.53}$$

where $(e_w, e^*)$ and $(\psi_w, \psi^*)$ are heavy metal concentrations in (water, sediments) in their solid and dissolved phases, respectively; $Q_{e,\psi}^i$ is heavy metal input with AYRS tributaries ($i = 1$), by atmosphere precipitation ($i = 2$), and with industrial waste ($i = 3$); $\alpha_1$ is the fraction of solid particles in bottom sediments; and $\alpha_2^i$ is the the fraction of solid particles in the $i$th flow of heavy metals.

Approximate solutions of the initial value problem for equations (4.41)–(4.53) can be found by means of the quasi-linearization method (Nitu *et al.*, 2000b).

#### 4.5.3.2 *In situ measurements*

To obtain data for the AYRSSM database, the joint U.S./Russian expedition to Siberia's Angara and Yenisey rivers was conducted in the summer of 1995 (Krapivin *et al.*, 1998a; Phillips *et al.*, 1997). Figures 4.15–4.18 show the location of the sample sites. Sampling began at the town of Bolshaya Rechka on the Angara near Lake Baikal and continued past the cities of Irkutsk, Angarsk, and Bratsk. Near Angarsk there was a trip along the Kitoy river. The expedition then proceeded to the Yenisey–Angara junction. Sampling began near the village of Kulakovo upstream on the Angara and continued downstream to the junction with the Yenisey at the logging town of Strelka. On the Yenisey, sampling began near the town of Kazachinskoye above the junction and continued downstream past the industrial complex at Lesosibirsk below the junction. Three types of samples were collected:

(1) river sediments;
(2) riverbank and flood plain sediments; and
(3) river water.

Sample locations were geographically recorded using a portable global positioning system (Garman GPS-45) which was generally accurate to within 30 m. River sediment samples were collected from boats in the areas of interest using a 40 kg Phleger gravity corer. This device has a streamlined lead weight with a stainless steel coring tube attached to its nose. During sampling, the corer was dropped and recovered by hand providing river sediment cores up to 30 cm in

**Figure 4.15.** Map of the AYRS section showing sampling locations (indicated by stars) in the area of the Angara–Yenisey junction.

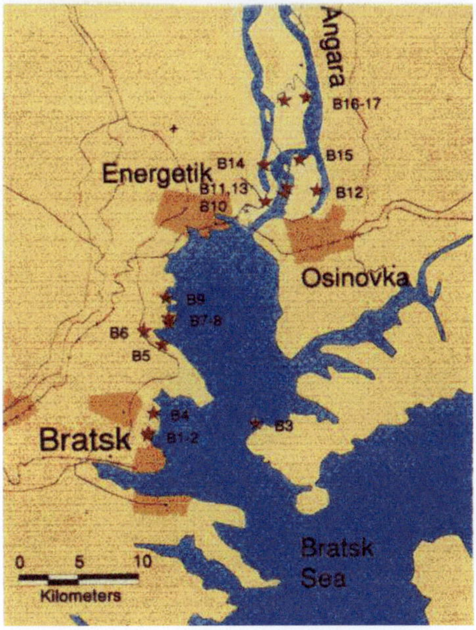


**Figure 4.16.** Map of the AYRS section showing sampling locations (indicated by stars) in the Bratsk area.

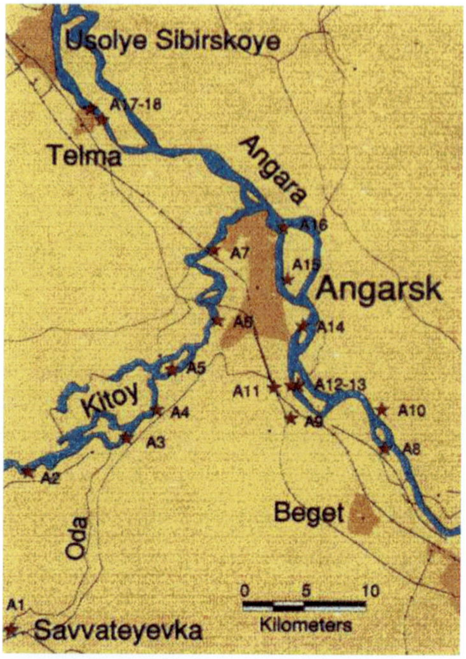


**Figure 4.17.** Map of the AYRS section showing sampling locations (indicated by stars) in the Angarsk area.

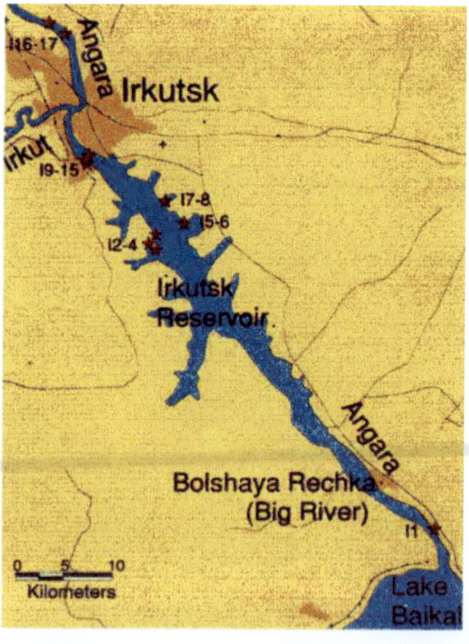


**Figure 4.18.** Map of the AYRS section showing sampling locations (indicated by stars) in the Irkutsk area.

diameter. Riverbank and flood plain sediment samples were taken using standard hand-packed coring procedures. Water samples were collected in small glass vials for chemical analysis only. In addition, detailed chemical analysis was performed on the supernatant water collected above the sediment as part of the gravity corer sampling. The Angara is a swift and extremely rocky river and an ideal location for hard-to-find sediment samples. The leeward side of small islands and backwater inlets provided the best opportunities for sampling sediment. Upstream of the Angara–Yenisey junction, the Yenisey was deeper with a slower current and wider flood plain, which is more conducive to sediment sampling. But downstream of the junction, the Yenisey was rocky and swifter with steeply sloping banks. This is similar in nature to the Angara. The large seasonal fluctuation of flow in the rivers continually alters the shoreline profile and washes away previously deposited sediment. In addition, the water level and flow rate on the Angara River during this period was unusually high compared with previous years, making it a challenge to collect representative samples. Figure 4.14 compares the annual Angara river flow rate through the Irkutsk dam for 1995 with previous years. During the time of this expedition the river flow was at a 5-year high, with a peak flow rate through the Irkutsk dam of 3,800 $m^3/s$.

#### *4.5.3.3 Analysis instrumentation and methods*

An on-site laboratory was set up in the city of Irkutsk. With this laboratory, near real-time results were obtained providing the capability to collect additional samples at sites of particular interest. Three instruments were used on-site to characterize the samples:

(1) a low-background, high-resolution gamma ray spectrometer for radionuclides;
(2) an X-ray fluorescence spectrometer (XRF) for heavy metals; and
(3) a gas chromatograph (GC) for volatile organic compounds.

A mechanically cooled, high-purity, low-background germanium (Ge) spectrometer was used for radionuclide analysis of sediment samples. The detector was oriented vertically, with its face covered by a cylindrical-graded lead–cadmium–cooper shield. This shield reduces the background counting rate by a factor of 100 to less than 2 counts per second. Efficiency calibrations were performed at the Naval Research Laboratory prior to the expedition using a suite of National Institute of Standards and Technology (NIST) traceable standards embedded in a matrix of material of varying thickness designed to simulate sediment densities. The calibration values were checked against a known sample in the field.

As a result of time constraints, only river sediment samples from the top 15 cm of the gravity cores were analyzed in the field. The samples were first dried and then sieved to remove rocks and other debris before they were weighed and packaged in small plastic bags for analysis. The sample masses ranged from about 10 to 60 g.

A personal computer–based multi-channel analyzer system was used for radionuclide analysis. This system included peak search, isotope identification, and quantification capabilities. The data were analyzed in a 4,096-channel spectrum covering the energy range between 40 and 3,000 keV for an average measurement time of 24 hours. Gamma ray detection efficiency was 2.3% for the 662 keV gamma ray line from $^{137}Cs$ with a resolution of 2 keV full width at half maximum (FWHM). The background spectrum of the empty lead shield in the on-site laboratory was measured for more than 31 hours. The activity for each gamma ray, collected for background, is reported in units of Bq/kg of dry sample.

The X-MET 880 XRF system manufactured by Outokumpu Electronics, Inc. (Japan) was used for heavy metal analysis. This is a self-contained unit consisting of a microprocessor-based electronic module. It has a dual-surface analysis probe with a high-resolution, argon-filled proportional counter with a thin beryllium window. The instrument was calibrated prior to the expedition using NIST traceable multi-element standards. The calibration was confirmed in the field by measurement of known pure element standards before and after each day's analysis. Sediment and soil samples were prepared by air drying for 24 hours, sieving to remove foreign debris, and remixing to ensure uniformity. A 2 g aliquot sample was weighed, labeled, and sealed with polypropylene film. The prepared sample was placed at a fixed distance in front of the detector window for analysis. After careful review of the analysis results, it was concluded that this instrument should be used as a screening mechanism in this expedition for determining the presence of heavy metals. Analysis was performed by the Analytical Service Center of Ecology and Environment, Inc. (New York). Analysis was performed on all samples with a Perkin-Elmer Optima-XL Inductively Coupled Argon Plasma Spectrometer (ICP) according to the procedures set forth in USEPA (1983).

A Scentograph Model Plus II portable GC was used to analyze volatile organic compounds in river water samples. This is a self-contained, computer-controlled unit with two GC columns in an oven and two detectors. Sensitivity was determined to sub-parts per billion for volatile hydrocarbons and parts per trillion for chlorinated hydrocarbons using a single-point calibration method at the Idaho National Engineering Laboratory (INEL) prior to the expedition. Calibration was verified on-site with certified standard organic compounds before each analysis. Water samples were collected in a glass container with a sealed lid to prevent the loss of dissolved gaseous components. An aliquot of the water samples was purged with argon gas to transfer gaseous portions of the sample into the GC column for separation and analysis.

#### *4.5.3.4 Summary of expedition results*

Some results from this expedition are given in Tables 4.13–4.16. A transportable radionuclide analysis system consisting of a mechanically cooled germanium gamma ray spectrometer in a lead shield was set up on-site in Irkutsk. The results of this analysis are given in Table 4.13. An X-ray fluorescent spectrometer (XRF) was also set up on-site and used for screening of the samples for heavy metals.

**Table 4.13.** Measurement results of the content of radionuclides in river bottom sediments made in July 1995 (averaged data of isotopic activity, Bq/kg of dry weight). Places where sampling took place are marked on the maps in Figures 4.15–4.18. Errors given are ±2 standard deviations. The sign "<" indicates less than minimum detectable concentration, which is given at the 90% confidence level. Sample J7 was taken in the flood plain of the Yenisey River above the junction.

| *Place of sampling, sample ID* | $^{60}$Co | $^{137}$Cs | $^{152}$Eu | $^{235}$U | $^{238}$U |
|---|---|---|---|---|---|
| Angara river | | | | | |
| Upstream of the Irkutsk dam, I1 | <2.0 | 2.2 ± 1.0 | <3.1 | <1.9 | <280 |
| Downstream of the Irkutsk dam, I2 | <3.5 | < 3.6 | <5.32 | 7.2 ± 4.6 | 720 ± 300 |
| Angarsk (technical canal), A1 | <2.0 | < 1.8 | <3.1 | <1.7 | <320 |
| Upstream of Angarsk, A2 | <4.12 | 5.2 ± 3.1 | <6.0 | <4.2 | <630 |
| Upstream of the Bratsk dam, B1 | <2.3 | < 2.3 | <3.9 | <2.2 | <420 |
| Downstream of the Bratsk dam, B2 | <2.6 | 3.4 ± 1.2 | <4.0 | <2.5 | <440 |
| Before the Angara–Yenisey confluence, J1 | <1.9 | 2.2 ± 1.0 | <2.8 | <1.7 | <290 |
| Yenisey river | | | | | |
| Upstream of Kazachinskoye, J2 | 8.6 ± 1.9 | 22.9 ± 2.3 | 6.7 ± 3.0 | <2.7 | <500 |
| Kazachinskoye, J3 | 241 ± 11 | 392 ± 12 | 151 ± 27 | <8.2 | <1,340 |
| Downstream of Kazachinskoye, J7 | 30.1 ± 3.0 | 203 ± 5 | 42.3 ± 9.5 | <3.8 | <680 |
| Angara–Yenisey confluence (at Strelka), J4 | 96.9 ± 5.1 | 211 ± 6 | 55.7 ± 13.5 | <4.5 | <840 |
| Downstream of Lesosibirsk, J6 | <5.7 | 27.1 ± 3.8 | <8.2 | <6.4 | <860 |

After the expedition, samples were sent to a commercial laboratory for elemental analysis. The results of this analysis are given in Tables 4.14 and 4.15. The results for both categories of pollutants are discussed below.

*Radionuclides in river sediment*. The manmade radioisotope $^{137}$Cs (30-year half-life) was detected in all samples analyzed above the Irkutsk dam, with concentrations ranging from 2 to 12 Bq/kg (dry weight). These values are consistent with background levels that can be expected due to global fallout as a residual from atmospheric nuclear testing. Below the Irkutsk dam in the vicinity of the cities of Irkutsk and Angarsk, the measured $^{137}$Cs concentrations ranged from <4 Bq/kg to 30 Bq/kg, indicating that some of the samples contained $^{137}$Cs concentrations significantly higher than the average global background. For samples taken near Bratsk, only background levels of $^{137}$Cs activity were detected.

The $^{137}$Cs concentrations in samples taken from undeveloped reaches of the Angara river upstream of the AYRS junction at Strelka were determined to be at low background levels, about 2 Bq/kg of dry river sediment. Downstream of the junction with the Yenisey, the $^{137}$Cs concentration in the samples ranged from 3

**Table 4.14.** Laboratory analysis of concentrations of heavy metals in sediments and in water measured in July 1996 during the U.S.–Russian hydrophysical expedition (Krapivin *et al.*, 1998a). Places where samples were taken are marked on the maps in Figures 4.15–4.18. Measurement error did not exceed 1 $\sigma$.

| *Place of sampling, map ID* | As | Cd | Cr | Cu | Ni | Pb | Zn |
|---|---|---|---|---|---|---|---|
| ***River sediment samples (ppm)*** | | | | | | | |
| Bol′shaya Rechka, I1 | 7.2 | <0.52 | 26 | 19 | 25 | 15 | 64 |
| Irkursk, above Dam, I7 | 5.0 | <0.51 | 41 | 22 | 44 | 11 | 55 |
| Irkutsk, below Dam, I11 | 4.5 | <0.51 | 20 | 11 | 21 | 9 | 17 |
| Irkutsk, below Dam, I10 | 8.1 | <0.52 | 54 | 42 | 60 | 20 | 88 |
| Angarsk (technical ditch 1), A11 | 0.9 | <0.50 | 7.0 | 11 | 9.3 | 7.9 | 31 |
| Angarsk (technical ditch 2), A9 | 1.8 | <0.50 | 16 | 9.3 | 23 | 3.2 | 17 |
| Above Angarsk, A13 | 4.3 | <0.51 | 48 | 38 | 40 | 14 | 86 |
| Kitoy river, A7 | 2.4 | <0.50 | 27 | 20 | 37 | 7.8 | 40 |
| Bratsk, above dam, B3 | <0.5 | <0.50 | 6.4 | 9 | 12 | 2.3 | 22 |
| Bratsk, below dam, B15 | 3.4 | <0.50 | 31 | 210 | 39 | 6.1 | 50 |
| Above junction at Kulakovo, J15 | 3.2 | <0.50 | 14 | 13 | 18 | 4.5 | 31 |
| Junction, Strelka village, J9 | 5.1 | <0.51 | 47 | 35 | 37 | 14 | 100 |
| Below junction, below Lesosibirsk, J12 | 6.9 | <0.51 | 34 | 31 | 36 | 17 | 100 |
| Kazachinskoye, above junction, J7 | 5.8 | <0.51 | 54 | 43 | 390 | 18 | 150 |
| Below Kazachinskoye, above junction, J8 | 2.2 | <0.50 | 18 | 8.2 | 20 | 3.9 | 47 |
| ***River water samples (ppb)*** | | | | | | | |
| Irkutsk, below dam, I11 | 12 | <0.50 | <10 | <20 | <20 | <5.7 | <5.7 |
| Irkutsk, below dam, I15 | 6.3 | <0.50 | <10 | <20 | <20 | <5.7 | 150 |
| Irkutsk, below dam, A16 | 10 | <0.50 | <10 | <20 | <20 | <5.7 | <5.7 |
| Angarsk, A14 | 12 | <0.50 | <10 | <20 | <20 | <5.7 | <5.7 |
| Bratsk, above dam, B2 | 16 | <0.50 | <10 | <20 | <20 | <5.7 | <5.7 |
| Above junction, J14 | 12 | <0.50 | <10 | <20 | <20 | <5.7 | <5.7 |
| Below Lesosibirsk, below junction, J10 | 13 | <0.50 | <10 | <20 | <20 | <5.7 | <5.7 |
| Kazachinskoye, above junction, J2 | 8.6 | <0.50 | <10 | <20 | <20 | <5.7 | <5.7 |

**Table 4.15.** Comparison of results (ppm) of the laboratory analysis of the 1995 expedition materials on Angara river water quality (Analytical Services Center of Ecology and Environment, Inc., New York). Notation: nm, no measurements; na, beyond measurement accuracy (sensitivity of devices); $T_1 = [t_1, t_2]$; $T_2 = [t_2, t_4]$.

| *Type of data* | *Data of the Irkutsk Ecological Service on Angara water quality* | | | | | | *Results of water sample analysis* | |
|---|---|---|---|---|---|---|---|---|
| *Year* | *1978* | | *1979* | | *1980* | | *1996* | |
| *Season* | *T1* | *T2* | *T1* | *T2* | *T1* | *T2* | *August* | |
| ***The Ershov water collector (Irkutsk, above dam)*** | | | | | | | *Map ID I15* | *Map ID I11* |
| Mg | 3.2 | 3.0 | 3.3 | 3.1 | 3.4 | 3.3 | 3.5 | 4.9 |
| Fe | na | na | 0.01 | nm | na | na | <0.05 | <0.05 |
| As | na | na | na | na | na | na | 0.0063 | 0.012 |
| Zn | na | na | na | na | na | na | 0.15 | 0.12 |
| Cu | 0.002 | 0.002 | 0.003 | 0.002 | 0.001 | 0.003 | <0.02 | <0.02 |
| Mn | 0.001 | 0.001 | 0.003 | 0.005 | 0.003 | 0.006 | <0.01 | <0.01 |
| Al | 0.005 | 0.005 | 0.018 | 0.008 | 0.004 | 0.007 | <0.1 | <0.1 |
| Co | nm | na | nm | na | na | na | <0.02 | <0.02 |
| Ni | na | na | nm | 0.003 | 0.001 | 0.002 | <0.02 | <0.02 |
| V | na | na | na | na | na | na | <0.02 | <0.02 |
| Pb | na | na | na | na | na | na | <0.0057 | <0.057 |
| ***The water collector near the Sukhov station, above Angarsk*** | | | | | | | *Map ID A14* | |
| Mg | 3.4 | 3.4 | 3.7 | 3.5 | 3.3 | 3.4 | 3.2 | |
| Fe | 0.02 | 0.3 | 0.07 | 0.001 | na | na | <0.05 | |
| As | na | na | na | na | na | na | 0.012 | |
| Zn | na | na | 0.0053 | na | na | na | 0.24 | |
| Cu | 0.001 | 0.002 | 0.004 | 0.007 | 0.004 | 0.008 | <0.02 | |
| Mn | 0.011 | nm | 0.018 | 0.007 | 0.012 | 0.011 | 0.028 | |
| Al | na | 0.015 | 0.02 | 0.028 | 0.013 | 0.012 | <0.01 | |
| Co | nm | na | 0.004 | 0.002 | na | na | <0.02 | |
| Ni | nm | 0.001 | 0.003 | 0.002 | 0.002 | 0.0005 | <0.02 | |
| V | na | na | na | na | 0.001 | na | <0.02 | |
| Pb | na | na | na | na | na | na | <0.0057 | |

(*continued*)

**Table 4.15** (*cont.*)

| *Type of data* | *Data of the Irkutsk Ecological Service on Angara water quality* | | | | | | *Results of water sample analysis* |
|---|---|---|---|---|---|---|---|
| *Year* | *1978* | | *1979* | | *1980* | | *1996* |
| *Season* | *T1* | *T2* | *T1* | *T2* | *T1* | *T2* | *August* |
| ***The water collector, 0.5 km from Bratsk, below dam*** | | | | | | | *Map ID B2* |
| Mn | 4.4 | 3.8 | 4.4 | 3.9 | 4.8 | 5.2 | 4.9 |
| Fe | na | na | 0.02 | 0.13 | na | na | <0.05 |
| As | na | na | na | na | na | na | 0.016 |
| Zn | na | na | na | na | na | na | 0.087 |
| Cu | na | na | na | na | na | na | <0.02 |
| Mn | na | na | 0.004 | 0.001 | 0.002 | 0.001 | <0.01 |
| Al | nm | 0.002 | na | 0.002 | 0.003 | nm | <0.1 |
| Co | na | na | na | na | na | na | <0.02 |
| Ni | na | na | na | na | na | na | <0.02 |
| V | nm | na | na | 0.001 | 0.001 | 0.15 | <0.02 |
| Pb | na | na | na | na | na | na | <0.0057 |

to 27 Bq/kg. The latter value is significantly above the activity levels detected on the Angara river upstream of the junction.

In samples drawn from the Yenisey river at and upstream of the junction, the levels of neutron activation product $^{60}$Co (5.27-year half-life) and the fission products $^{137}$Cs and $^{152}$Eu (13.3-year half-life) were much higher than background levels: ranging from 9 to 240 Bq/kg for $^{60}$Co; from 14 to 400 Bq/kg for $^{137}$Cs; and from 7 to 150 Bq/kg for $^{152}$Eu. These are unmistakable indications of reactor products and are consistent with releases from the plutonium production reactors at Krasnoyarsk-26, which is approximately 270 km upstream on the Yenisey river. The large variations in measured activities were due to sampling location: the higher values came from a dead end backwater channel with thick sediment (J3) and from a flood plain area with thick mud deposits (J7) while the lower values came from sandy sediment (J2) where the current was swift and deposits were more likely to be washed away.

Above-background levels of $^{235}$U and $^{238}$U were observed in several samples taken just below the dam at Irkutsk. However, the measured concentrations for $^{238}$U have large uncertainties associated with them as a result of the relatively weak gamma ray emissions from this radionuclide. The ratios observed for $^{235}$U/

**Table 4.16.** Relative concentrations of $^{137}Cs$ in water ($\gamma_w$) and in bottom sediments ($\gamma_d$) normalized for values at $x = 0$.

| *Distance from Lake Baikal* (km) | *Time from the beginning of the simulation experiment* (days) | | | | | | | |
|---|---|---|---|---|---|---|---|---|
| | *30* | | *60* | | *90* | | *120* | |
| $x$ | $\gamma_w$ | $\gamma_d$ | $\gamma_w$ | $\gamma_d$ | $\gamma_w$ | $\gamma_d$ | $\gamma_w$ | $\gamma_d$ |
| 0 | 1.0 | 1.0 | 1.0 | 1.0 | 1.0 | 1.0 | 1.0 | 1.0 |
| 250 | 8.2 | 20.4 | 6.7 | 18.7 | 7.4 | 17.8 | 8.3 | 17.1 |
| 500 | 9.5 | 14.4 | 9.4 | 12.3 | 9.3 | 15.3 | 8.4 | 14.9 |
| 1,000 | 10.3 | 4.1 | 8.8 | 9.8 | 7.8 | 11.6 | 7.9 | 10.3 |
| 1,500 | 9.7 | 3.0 | 10.2 | 5.2 | 6.3 | 8.5 | 4.8 | 7.6 |
| 2,000 | 8.9 | 2.9 | 6.9 | 2.9 | 4.9 | 4.1 | 4.5 | 2.8 |
| 2,500 | 6.5 | 2.3 | 5.9 | 2.2 | 3.1 | 2.4 | 3.1 | 1.8 |
| 3,000 | 3.2 | 1.9 | 2.7 | 1.7 | 2.9 | 1.9 | 3.0 | 1.4 |
| 3,500 | 2.5 | 1.6 | 1.8 | 1.1 | 1.8 | 1.2 | 2.4 | 1.1 |
| 4,000 | 1.7 | 1.1 | 1.6 | 0.9 | 1.8 | 1.2 | 1.4 | 1.1 |

$^{238}U$ were larger than the natural abundance of 0.7% and were outside statistical uncertainties. However, these are counting errors only and do not include uncertainties in the interference between the overlying 186 keV gamma rays of $^{235}U$ and $^{226}Ra$ and possible errors from self-attenuation due to nonuniformity in the samples. These samples were taken in an area undergoing active landfill. As a result, uranium-containing fills could have been brought in from elsewhere.

*Heavy metals in river sediments.* Using the XRF unit as a screening instrument, a set of heavy metals were identified in all riverbank and river sediment samples collected during the expedition. After the expedition, 21 river sediment and 8 river water samples were sent for trace element analysis to an EPA-certified commercial laboratory. Some of the analytical results are shown in Tables 4.12 and 4.15. Five heavy metals (Cd, Cr, Cu, Ni, and Pb) were identified in all the samples at concentrations which were within the usual ranges of worldwide background levels. Two heavy metals, As and Zn, were identified in the river water samples at concentrations much higher than the median for natural freshwater (Lal and Stewart, 1994). In fact, the Zn concentration in four of the samples exceeded the normal range for natural freshwater. These water samples were collected near to or downstream of industrial complexes on both the Angara and Yenisey rivers.

Table 4.15 compares the 1995 U.S./Russian expedition results with archived 1978–1980 water quality data taken from the Angara at Irkutsk, Angarsk, and Bratsk by the Irkutsk Medical Inspection Service (personal communication by G.M. Svender and A.A. Krechetov, Chemical Department, Irkutsk State University, Russia). The measurements indicate a significantly higher Zn concentration in water collected above Angarsk than was measured previously.

#### *4.5.3.5 Experiments using the AYRSSM*

The AYRSSM database includes estimates of model coefficients, initial information for climatic and anthropogenic scenarios, and a set of identifiers $\{A_i\}$ describing the $\Omega$ area according to its boundaries, thus giving the structure of the land–water surface. The ICCE block provides an interface with the database and allows the user to modify its elements.

It is assumed that the vertical distribution of pollutants in the river water is homogeneous, the pollution of the soil and plants is negligible, and the cleaning process of the atmosphere has an exponential character with half-life $\tau_a$ ($\tau_a = 10$ days when $t \in [t_1, t_2]$, $\tau_a = 20$ days when $t \in [t_2, t_4]$). It is believed that the pollution sources located in the $\Omega$ territory support a constant level of distribution of pollutants. The distribution function is uniform on the intervals shown in Table 4.16. The values of $\xi$, $\mu$, and $\nu$ were determined by average estimates for 1992, 1993, and 1994 on the basis of published data (Rovinsky *et al.*, 1995) and data of the Irkutsk Scientific Center. The AYRS slope is taken equal to 0.21 m/km. Ecological elements of the AYRS are considered as having unique levels with biomass $\beta(t)$ a constant value at each of the intervals $[t_i, t_{i+1}]$ $(i = 1, 2, 3)$. The equation for $H_{5\nu}$ is written in the form: $H_{5\nu} = [g_1 + m_1 m_2 \beta] Q_{k\nu} \Delta$ where $g_1$ is the gravitation coefficient (0.05 day$^{-1}$), $m_1$ is the mortality coefficient (0.01 day$^{-1}$), and $m_2$ is the pollutant capture coefficient (0.03 kg$^{-1}$).

Table 4.16 shows the results of using the AYRSSM to estimate the distribution of $^{137}$Cs in the $\Omega$ area where the cells $\Omega_i$ are situated along the AYRS at discretization steps of $\Delta x = 10$ km. The $^{137}$Cs distribution along the river system has a single maximum that changes value and coordinates over time. A similar picture is observed for other radionuclides from Table 4.13. Such a result cannot be explained by the location of sources of radionuclear pollutants alone. It is possible that variations in the value and position of maximal $^{137}$Cs concentrations are caused by high turbulence in the river system and by the existence of reservoirs and eddies. The AYRSSM gives only averaged results. The fourth-level blocks in the AYRSSM structure are necessary to describe hydrological processes where $\Delta x < 100$ m.

Computer experiments show that the input of radionuclear pollution to the Kara Sea from the Yenisey river is stable and disperses at a rate equal to $\pm 32\%$. The role of the AYRS ecosystem in the process of transformation of radionuclear pollution is neglected as being small ($<3\%$). Vertical transport by organisms varies from 0.1% to 0.7%. Such calculations can be carried out for different scenarios.

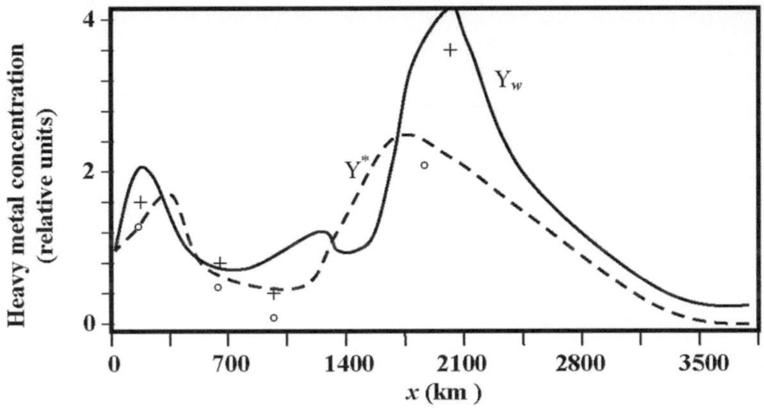


**Figure 4.19.** Distribution of the concentration of heavy metals in the water (dashed line) and in the sediments (solid line) as a function of distance $x$ from Lake Baikal. The signs ○ and + correspond to the measured concentrations of metals in the water and sediments, respectively.

Figure 4.19 shows a distribution of the concentration of heavy metals along the AYRS constructed by means of a computer experiment. We can see that there are three maxima of heavy metal concentration located at distances of 200, 1,200, and 2,000 km from Lake Baikal. This is the result of the distribution of pollution sources along the river system. The AYRS neutralizes pollutants over a distance of 600–1,000 km from the source. The locations of the maxima vary as a function of river flow rate. For the flow data of Figure 4.14 this variation is 150 km. The river system transforms the pollutant flow such that input into the Kara Sea is estimated at a pollution level less than 2% of the maximal concentration of pollutant in the AYRS.

The AYRSSM allows the dependence of pollution level in the AYRS estuary as a function of anthropogenic activity to be estimated. Let us suppose that the intensity of sources of heavy metals is such that their concentration in the water near Angarsk, Irkutsk, Krasnoyarsk, Bratsk, and Ust-Ilimsk is described by a stationary function, supporting heavy metal concentrations at the level $h$ at each of these cities. Computer experiments show that there is a stable correlation between $h$, the heavy metal concentration in the AYRS estuary, and the water flow rate $\mu$. An increase in $h$ of 10% equates to a rise of pollution input to the Kara Sea by 2.5%. An increase in $\mu$ of 1% equates to a rise of pollution input to the Kara Sea by 0.7%. These results are correct only when values $\mu$ and $h$ are varied near their average estimates. Near their critical values the estimates are unstable and more detailed models are required.

In conclusion, this section has demonstrated the possibility of using modeling technology to solve complex environmental problems demanding combined knowledge from different scientific fields. The AYRSSM is one such combination. It provides the typical elements needed to synthesize a simulation system to look into Arctic Basin pollution.

The results given in this section illustrate the functional features that lie at the very heart of a simulation experiment. It is obvious that the strategy of modeling technology is in the interplay between model calculations and on-site experiments. In the case considered, such a strategy secures economical benefit as it reconstructs the distribution of pollutants along the AYRS and provides estimates of environmental consequences should the chosen scenarios materialize.

A joint U.S./Russian expedition to Siberia's Angara and Yenisey rivers detected manmade contaminants in water and sediment samples from industrial regions and the wilds of Siberia. On-site analysis using sensitive instrumentation revealed radionuclides, heavy metals, and volatile organic compounds. Results indicate that the nuclear production facility near Krasnoyarsk on the Yenisey river has likely introduced radioactive contamination far downstream and is a probable source of previously detected radioactivity in the Yenisey estuary at its outlet into the Kara Sea.

The AYRSSM is a complex system that has a hierarchical structure made up of natural and anthropogenic elements. This study of the AYRS is the first time simulation results have been based on large-scale on-site measurements. The method proposed in this chapter can be used to investigate other Siberian river systems. Nevertheless, remote-sensing technology must be used if we want to expand the experimental base (Sellers *et al.*, 1995). Remote monitoring can give more precise data on the structure of identifiers $A_2$ and $A_3$, which describe the spatial distribution in the Arctic of pollutants and ice fields, respectively. The authors intend to continue their research, based on the database created, and plan to prepare a complete set of models and corresponding software to describe the process of transfer and transformation of pollution substances in natural Arctic ecosystems. This set will demand the synthesis of models for the kinetics of radionuclides and chemical compounds in the food chains of water and land ecosystems for the boreal zones, modeling the hydrological regime, and estimating pollutant flows in the Arctic Basin.

The ultimate aim of investigations is development of an environmental technology that is the main result of co-operative scientific work directed at evaluating the state of the Arctic ecosystem—a modeling system provided with as full a database as possible that has at its disposal a ramified informational measuring network and a complete set of computer models for the main biogeochemical, climatic, and biogeocenotic processes.

### 4.5.4 Dynamics of the carbon cycle in the Arctic

Studies of climate-forming processes in the Arctic, whose principal feature is determined by ice cover isolating the ocean from the atmosphere, have long been of particular concern. The most important impacts of sea ice on climate, as revealed through numerical modeling, are the following:

(1) maximum climate warming with increasing $CO_2$ concentration in the wintertime

Arctic as a result of increased heat input from the ocean through thinner ice as a consequence of warming;

(2) the effect of albedo on the atmospheric temperature of the more extended sea ice 18,000 years ago, which is compatible with the impact of continental glaciers; and
(3) possible reversal of the conventional relationship between the amplitude of annual change in temperature and depth of the oceanic mixed layer when sea ice dynamics is taken into account.

A thinner mixed layer favors wintertime sea ice increasing in strength, which causes a delay in springtime melting and produces a colder summer. The latest data of numerical modeling gathered by Krapivin and Varotsos (2008) by and large confirms these conclusions (Kondratyev and Johannessen, 1993).

When discussing the role played by polar regions in the formation of global changes, two aspects of global ecology of paramount importance should be kept in mind:

(1) anthropogenically induced redistribution of the heat balance components of the Earth (with emphasis on the greenhouse effect and its climatic impact); and
(2) anthropogenically induced breaking of global biogeochemical cycles (primarily, carbon, nitrogen, and sulfur).

These are the reasons polar regions are very special components of the global ecosystem. These aspects are not taken into account completely by existing global models of the nature–society system. A new approach to synthesizing geoinformation monitoring systems, proposed by Krapivin and Varotsos (2007), overcomes this shortfall. In this approach the interchange of $CO_2$ between high-latitude vegetation, the Arctic Ocean, and the atmosphere is considered part of a global biogeochemical cycle described by means of a GSM. Such a GSM comprises blocks parameterizing:

(1) the global hydrological balance;
(2) productivity of soil–plant formations, with 30 types defined;
(3) photosynthesis in ocean ecosystems taking into account the depth and surface inhomogeneity;
(4) demographic processes; and
(5) anthropogenic changes (Krapivin, 1993).

The GSM makes it possible to compute the dynamics of industrial $CO_2$ distribution between the oceans, terrestrial biota, and the atmosphere. The GSM describes the World Ocean as a spatial four-layer model with due regard for water chemistry. Spatial inhomogeneity of the World Ocean is represented by the structural distribution of surface temperature applied to upwelling and convergence zones. Sea ice in Arctic regions is considered by using CIESIN data (CIESIN, 2011; Edwards, 2010). The biogeochemical cycle of $CO_2$ is described by balance equations in accordance with Figures 4.20 and 4.21. Figure 4.20 develops Figure

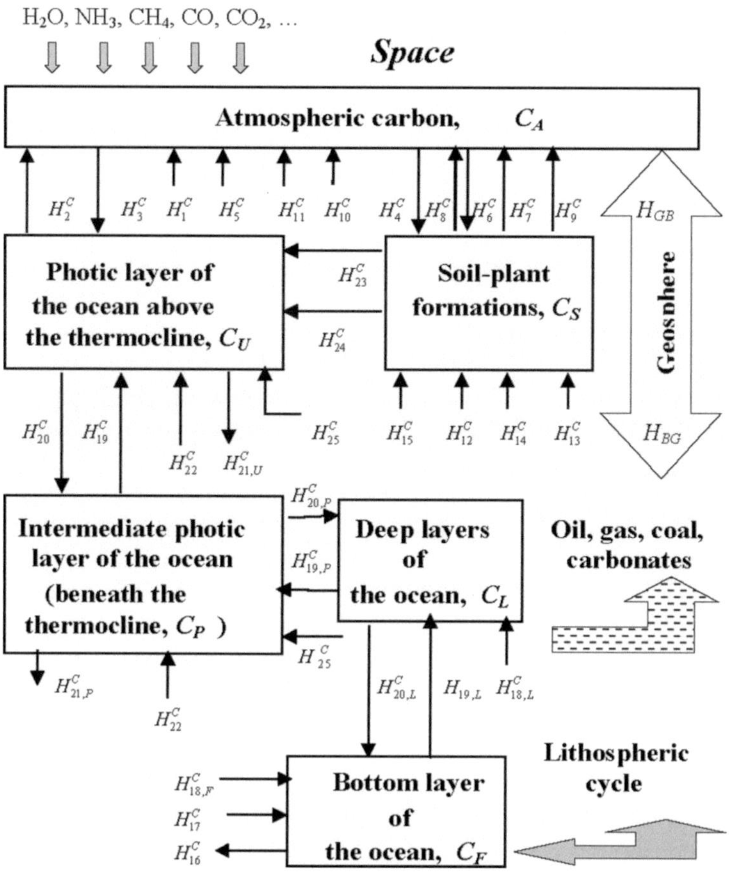


**Figure 4.20.** Modernized block diagram of the global biogeochemical cycle of $CO_2$ in the atmosphere–land–ocean system. $CO_2$ reservoirs and fluxes are described in Table 3.4 along with two additional fluxes $H_{BG}$ and $H_{GB}$, geospheric sink and source of carbon, respectively.

3.7 by considering geophysical components of the global carbon cycle. The GSM carbon dioxide block takes into consideration the dependence of flows $H_2$ and $H_3$ on water surface processes (wind–wave mixing, rough seas, foaming waves). Simulation experiments have shown that these flows vary between 16 and 1,250 mol m$^{-2}$ yr$^{-1}$. In the Arctic Ocean during the June–September period the partial pressure $p_a$ of $CO_2$ in the atmosphere exceeds the partial pressure of $CO_2$ in seawater by 20–110 ppm. These variations in the partial pressure of $CO_2$ have specific distributions for Arctic seas. For instance, the Norwegian Sea and Bering Sea have $CO_2$ deficits of 18–54 ppm and 33–69 ppm, respectively. The average $CO_2$ deficit reaches 450 g C $\cdot$ m$^{-2}$. The flow $H_3$ changes between 1.5 and 4.1 g C $\cdot$ m$^{-2}$ day$^{-1}$. In addition, linear correlations between the partial pressure of $CO_2$ and water temperature $T$ are observed, with a proportionality coefficient equal to 9.8 ppm/°C.

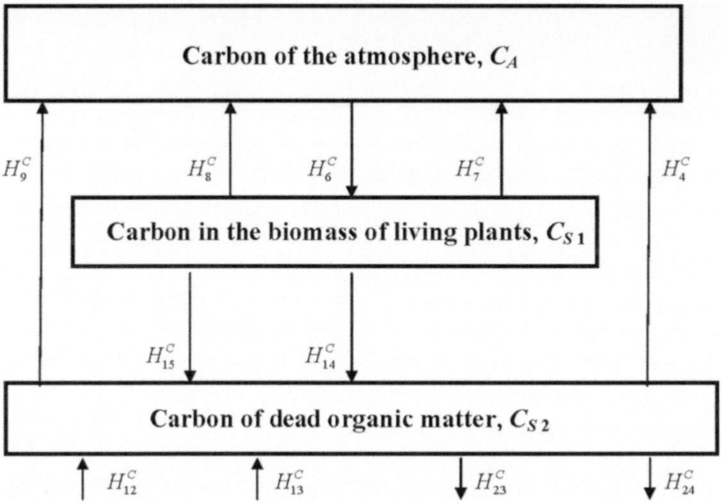


**Figure 4.21.** Block diagram of the carbon biogeochemical cycle in land ecosystems. See Table 3.4 for notation.

The flow $H_3$ is calculated by the formula $H_3 = \psi(T)p_a^{-1/2}(1 + 0.5S)$, where $S$ (‰) is water salinity and Weis's function $\psi$ describes the influence of water temperature $T$ (°K) on carbon dioxide solubility in seawater.

The greenhouse effect has been discussed by many authors within the confines of all manner of scenarios (Gorshkov *et al.*, 2000; Kondratyev, 1998b). The main conclusion is that global climate change brought about by $CO_2$ will be insignificant during the 21st century if:

(1) World Ocean pollution, especially by oil products, does not exceed the level of 1990 by 10%;
(2) agricultural land areas do not expand at the expense of forests;
(3) the rate of fossil fuel consumption has increased by 15% compared with 1990 levels;
(4) alternative energy sources (atomic, wind, etc.) are developed at a rate that does not hinder food production.

The role played by soil–plant formation in absorbing excess atmospheric $CO_2$ under the above scenario is estimated using data displayed in a geographic grid of 4° × 5° lat/long (see Table 4.17). The role played by the World Ocean is considered by taking water temperature in the surface layer into account. It has been shown that the atmospheric $CO_2$ concentration can reach a mixing level of 556.7 ppm during the 21st century. Industrial $CO_2$ distribution between the atmosphere, oceans, and vegetation will fluctuate less than 25%. With an increase in industrial $CO_2$ emission over the period from 1990 to 2090, the atmospheric portion will rise and absorption by the oceans will increase; whereas that of the

**Table 4.17.** Model estimation of surplus $CO_2$ absorption by vegetation in Russia. Anthropogenic emission of carbon is assumed to be 4.5 Gt C/yr.

| *Soil–plant formation* | *Carbon flow absorbed by vegetation* ($10^6$ t C/yr) |
|---|---|
| Arctic deserts and tundra, subarctic grassland, sand marshes | 2.2 |
| Tundra, mountain tundra, and forest tundra | 9.7 |
| North taiga forests | 10.8 |
| Mid-taiga forests | 31.2 |
| South taiga forests | 22.9 |

biota will stabilize after going through a small maximum. At the end of the 21st century and beginning of the 22nd century, during the highest level of human economic activity, the contribution of the oceans (especially the Arctic Basin) to industrial $CO_2$ absorption will be considerably higher than that of vegetation. This is because, with an increase in $CO_2$ concentration in the atmosphere, the ability of upper ocean layers to absorb industrial $CO_2$ will be supported by the transformation of biogeochemical processes in deep ocean layers. Ocean pollution and the restoration of plant cover are the problems that must be addressed today. For example, it was shown that if the natural/disturbed land relation changed from 2/3 in 1990 to 3/2 in 2050 then the atmospheric $CO_2$ concentration would reach no more than 497.3 ppm during the 21st century. This illustrates that the role played by the biospheric system in global change needs to be investigated more thoroughly (Gorshkov *et al.*, 2002; Krapivin, 2000a; Krapivin and Vilkova, 1990).

Simulation results show that, overall, the importance of global climate change caused by anthropogenic $CO_2$ emissions has been exaggerated. Thus, the existing arguments for climate change are not reliable. It is necessary to develop the GSM by incorporating newly discovered correlations between elements of the NSS and by taking biotic regulation processes into account.

This chapter has shown that the role played by the Arctic Ocean in global $CO_2$ balance has not been estimated accurately enough. Therefore, the Kyoto Protocol, which reflects climate warming as a result of greenhouse gases, suffers from a lack of objectivity. Moreover, existing climate models do not reliably reflect the correlation of global temperature with anthropogenic emissions of greenhouse gases. These models fail to take into consideration the role of biospheric processes in the land and oceans (biotic regulation of the environment and the functioning of the ocean chemical system). The Millennium Ecosystems Assessment Program (MEAP, 2007) is another means of solving the problem of global model synthesis to forecast the dynamics of the NSS.

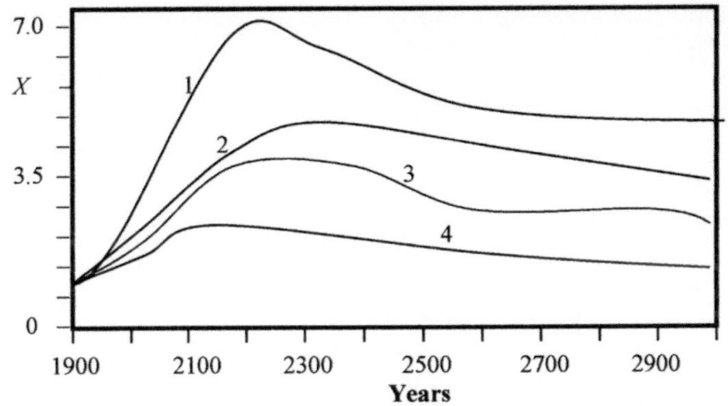


**Figure 4.22.** Forecast of $CO_2$ concentration in the atmosphere according to different scenarios of mineral resource expenditure: 1, pessimistic scenario (Bacastow, 1981); 2, optimistic scenario (Bjorkstrom, 1979); 3, IPCC scenario (Dore *et al.*, 2003); 4, realistic scenario (Demirchian and Kondratyev, 2004). $X = C_a(t)/C_a(1900)$.

There have been many long-term studies devoted to simulating the global $CO_2$ cycle. The main conclusion is that the exchange of carbon between the atmosphere and biosphere is "spotty" (i.e., of uneven quality). This spottiness is taken into account in models of the $CO_2$ cycle and produces serious errors. The conclusions of the Kyoto Protocol were unfortunately based on incomplete estimations. There are many carbon flows (natural and anthropogenic), mentioned in Figure 4.20, which have not been satisfactorily described parametrically. The principal key questions relating to the exchange of carbon between the atmosphere and the terrestrial pool of above-ground biomass, below-ground biomass, soils, and hydrospheric systems are discussed by Marchuk and Kondratyev (1992) and Krapivin and Varotsos (2007). The Arctic Basin is the least studied part of the biosphere when it comes to understanding regional global roles in $CO_2$ absorption processes. Figure 4.22 demonstrates the high variability in forecasting $CO_2$ dynamics under different anthropogenic scenarios:

(1) the pessimistic scenario of Keeling and Bacastow (1977) describes a situation where the ocean's role in carbon exchange with the atmosphere is restricted to physical processes only;
(2) the optimistic scenario of Bjorkstrom (1979) takes into account the ocean–carbonate system by parameterizing the $H_2$ and $H_3$ flows in Figure 4.20;
(3) the scenario of the Intergovernmental Panel on Climate Change (IPCC) (Watson *et al.*, 2000) is based on specific requirements concerning land use strategy (e.g., planting versus regeneration through silviculture); and
(4) the realistic scenario of Kondratyev (1999) foresees the development of existing tendencies in the world's energetic, demographic, and urbanization processes. This scenario is realized by the GSM (Krapivin, 1993; Kondratyev and Krapivin, 2001a).

In the third scenario a greenhouse effect via $CO_2$ is clearly problematic. To estimate the dynamics of atmospheric temperature dependence on $CO_2$ it is necessary to use a correlation describing the greenhouse effect. There are many empirical functional representations of warming effects (Krapivin, 2000a; Krapivin and Vilkova, 1990). The following simple correlation approximates the existing empirical dependence of atmospheric temperature deviation $\Delta T_{CO_2}$ on variations in the atmospheric $CO_2$ parameter $X = C_a(t)/C_a(1900)$:

$$\Delta T_{CO_2} = \begin{cases} 0.911 + 1.509 \ln X - 1.25 \exp\{-0.82(X-1)\} & \text{when } X \geq 1, \\ -2.63X^2 + 6.27X + 1.509 \ln X - 3.988 & \text{when } X < 1. \end{cases}$$

Finally, according to GSM calculations, future emissions of $CO_2$ will mean the Earth being up to 0.2°C to 1.3°C warmer than the year 1900. Variations in $\Delta T_{CO_2}$ depend on assumptions about urbanization processes and land use strategies. Results show that the dynamics of industrial $CO_2$ distribution between the atmosphere and oceans changes in accordance with the increasing preponderance of flows $H_3$ in the northern aquatories. At the end of the 21st century and beginning of the 22nd century, at a time when the pressure of human economic activity will likely have its maximum effect, the contribution of oceans to industrial $CO_2$ absorption will be considerably lower than that of vegetation. This is because an increase in $CO_2$ concentration in the atmosphere causes the partial pressure in the oceans to rise and its ability to absorb industrial $CO_2$ to decrease, whereas the productivity of vegetation does not fall. By the end of the 22nd century, with a projected decrease in human impact, the contribution of the oceans to absorption increases as a result of the growing role played by its deep layers. Atmospheric $CO_2$ will decrease at the same rate as the contribution to absorption by the oceans increases.

The entire area of the Arctic ocean is only 3.8% of the world's ocean surface but the role it plays in $CO_2$ absorption varies from 23% to 38% (16.7–28.9 Gt C/yr). This role is greatly influenced by seasonal variations in ice cover and ecosystem productivity. The results of this chapter are consequently preliminary in character and point to the need to further refine the parametrical descriptions of flows $H_i$ from Figure 4.20. The authors of this chapter understand that policy-oriented computer tools aimed at supporting decision-making processes related to global change require the design of a new methodology in global modeling—one based on simulation models that have detailed spatial descriptions of biospheric systems and operative informational services defining the parameters.

# 5

# GIMS-based study of inland seas

## 5.1 BRINGING GIMS TECHNOLOGY TO BEAR ON THE ARAL–CASPIAN AQUAGEOSYSTEM STUDY

### 5.1.1 Characteristics of the Aral–Caspian aquageosystem

An anthropogenic activity of interest is the influence it has on the biospheric water cycle (Krapivin *et al.*, 1982). The present scales of this influence are global and are composed of a hierarchy of regional changes especially in arid districts. The control of biospheric water systems is an element of climate system monitoring. The Aral–Caspian water system (ACWS) has been put under close scrutiny by scientists because it is deemed a significant subsystem of the biosphere. The catastrophic state of this system is well known (Bondur *et al.*, 2009; Chernenko, 1987; Hublaryan, 1995; Kondratyev *et al.*, 2006; Ratkovich, 1997; Salokhiddinnov and Khakimov, 2004) and clearly seen in Figure 5.1. In the opinion of many authors the causes of the anomalous increase in the Caspian Sea level and the decrease in that of the Aral Sea are global. According to the data of Bortnic and Chistyakova (1990) the fluctuations in the Aral Sea level during the last 200 years occurred as inter-annual deviations of up to 1 m. At that time the Aral Sea area covered an area of $(51\text{–}61) \cdot 10^3\ \text{km}^2$, about $55\ \text{km}^3/\text{yr}$ was evaporated from its surface, precipitation brought $9\text{–}10\ \text{km}^3/\text{yr}$, and river inflow ranged between 33 and $64\ \text{km}^3/\text{yr}$. Some estimates of the Aral Sea water regime and other parameters are given in Tables 5.1–5.4 (Bortnik and Chistyakova, 1990; Kadukin and Klige, 1991).

The total mean annual flow of all rivers into the Aral Sea Basin is estimated at about $116\ \text{km}^3$. This amount comprises the flow of the Amudar′ya at $79.4\ \text{km}^3/\text{yr}$ and the Syrdar′ya at $36.6\ \text{km}^3/\text{yr}$. In accordance with flow probabilities of 5% (very wet years) and 95% (dry years), the annual flow ranges from 109.9 to $58.6\ \text{km}^3$ for the Amudar′ya river, and from 51.1 to $23.6\ \text{km}^3$ for the Syrdar′ya river, respectively.

V.F. Krapivin and A.M. Shutko, *Information Technologies for Remote Monitoring of the Environment*, Springer Praxis Books, DOI 10.1007/978-3-642-20567-5_5, © Springer-Verlag Berlin Heidelberg 2012

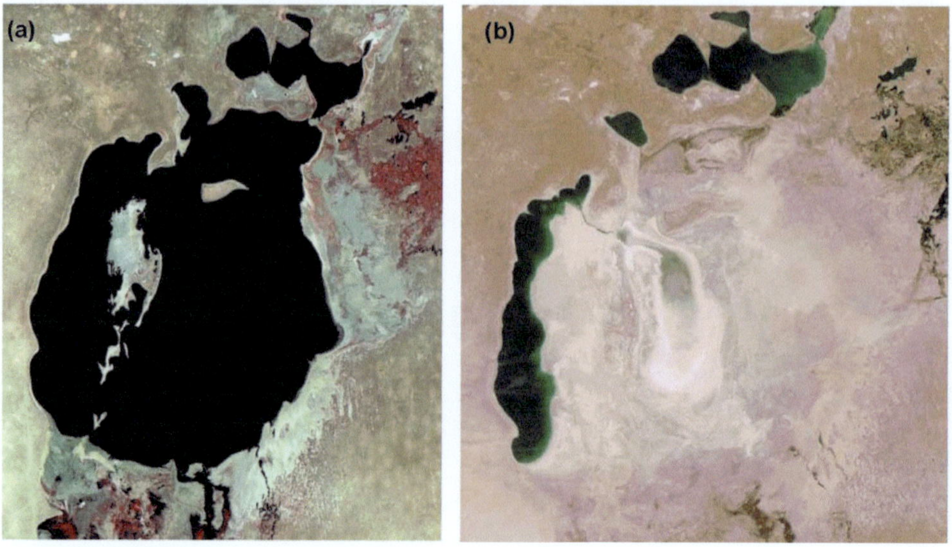


**Figure 5.1.** Satellite monitoring of the Aral Sea: (a) August 1989 (Resurs-O); (b) June 2009 (Aqua, MODIS).

At the time when anthropogenic interference with the water regime of Central Asia was negligible, the Caspian Sea level fluctuated at or around the absolute mark of $-28$ m, its aquatory was 375,000 $km^2$ in area, the evaporation volume from its surface was close to 380 $km^3$/year. Some data on the Caspian Sea water balance are given in Tables 5.5 and 5.6.

**Table 5.1.** Seasonal deviations of atmospheric temperatures (°C) from average values in the Aral Sea region (Kuksa, 1994).

| *Zone* | *Spring* | *Summer* | *Fall* | *Winter* |
|---|---|---|---|---|
| Aral Sea | 1.4 | 0.4 | −0.1 | 0.5 |
| Monsyr | 0.6 | 0.2 | 1.1 | 1.3 |
| Kazalinsk | 0.9 | 0.6 | 1.5 | −0.1 |
| Karak | 0.8 | 0.2 | 0.8 | 0.3 |
| Chabankazgan | 1.1 | 0.6 | 0.7 | −1.5 |
| Muinak | 0.9 | 0.5 | −0.3 | −0.6 |
| Chirik-Rabat | 0.9 | 0.5 | 1.0 | 0.4 |
| Kungrad | 1.4 | 1.4 | 1.1 | −0.1 |
| Kosbulak | 1.3 | 0.8 | 0.0 | 0.0 |

**Table 5.2.** The dynamics of water flow to the Aral Sea ($km^3\ yr^{-1}$) (Benduhn and Renard, 2004; Kuksa, 1994).

| *Year* | *Amudar'ya* | *Syrdar'ya* | *Total* | *Year* | *Amudar'ya* | *Syrdar'ya* | *Total* |
|---|---|---|---|---|---|---|---|
| 1959 | 40.0 | 18.3 | 58.3 | 1974 | 6.2 | 1.9 | 8.1 |
| 1960 | 37.8 | 21.0 | 58.8 | 1975 | 10.0 | 0.6 | 10.6 |
| 1961 | 29.2 | — | 29.2 | 1976 | 10.3 | 0.5 | 10.8 |
| 1962 | 29.1 | 5.7 | 34.8 | 1977 | 7.2 | 0.4 | 7.6 |
| 1963 | 29.9 | 10.6 | 40.5 | 1978 | 18.9 | — | 18.9 |
| 1964 | 36.5 | 14.9 | 51.4 | 1979 | 10.9 | 2.9 | 13.8 |
| 1965 | 25.2 | 4.6 | 29.8 | 1980 | 8.3 | — | 8.3 |
| 1966 | 33.1 | 9.5 | 42.6 | 1981 | 5.9 | — | 5.9 |
| 1967 | 28.6 | 8.6 | 37.2 | 1982 | 0.04 | — | 0.04 |
| 1968 | 28.9 | 7.2 | 36.1 | 1983 | 2.3 | — | 2.3 |
| 1969 | 55.1 | 17.5 | 72.6 | 1984 | 7.9 | — | 7.9 |
| 1970 | 28.7 | 9.8 | 38.5 | 1985 | 2.4 | — | 2.4 |
| 1971 | 15.3 | 8.1 | 23.4 | 1986 | 0.4 | — | 0.4 |
| 1972 | 15.5 | 6.9 | 22.4 | 1987 | 10.0 | — | 10.0 |
| 1973 | 33.4 | 8.9 | 42.3 | 1988 | 16.0 | 7.0 | 23.0 |

The constant fall in the level of the Aral Sea began in 1961. It averaged 21 cm/yr up to 1970, then it was 58 cm/yr until 1980, and after that it approached 80 cm/yr. The Aral Sea level dropped from the absolute mark of 53 m in 1960 to 41.4 m in 1985. By 1994 it had further reduced by 13 m. By 2007 it was 10% of its original size, splitting into four lakes: the North Aral Sea, the eastern and western basins of the once far larger South Aral Sea, and a smaller lake between the North and South Aral Seas. By 2009, the southeastern lake had disappeared and the southwestern lake retreated to a thin strip at the extreme west of the former southern sea. The maximum depth of the North Aral Sea is currently (2011) 40.4 m. The stable regime of the Caspian Sea began to change a short while later. The Caspian Sea level decreased to the absolute mark of −29.04 m in 1977, and then increased by 3 m by 1995. The area covered by the Caspian Sea increased to 440,000 $km^2$. The present increase in the Caspian Sea level began in 1978 at an average annual rate of 16–18 cm/yr. Between 1830 and 2004 the variations in the

**Table 5.3.** Long-standing average values of the water balance of the Aral Sea over separate periods. Components of the water balance are shown in the numerator as volumes ($km^3$) and in the denominator as levels (cm).

| *Time period* | *Input* | | *Evaporation* | *Water balance* | *Change in sea level* | *Balance discrepancy* |
|---|---|---|---|---|---|---|
| | *River flow* | *Precipitation* | | | | |
| 1911/1960 | 56.0/84.7 | 9.1/13.8 | 66.1/100.0 | −1.0/−1.5 | 0.1 | −1.6 |
| 1961/1970 | 43.3/68.5 | 8.0/12.7 | 65.4/103.5 | −14.1/−22.3 | −21.0 | −1.3 |
| 1971/1980 | 16.7/29.3 | 6.3/11.0 | 55.2/96.8 | −32.2/−56.5 | −57.6 | 1.1 |
| 1981/1985 | 2.0/4.1 | 7.1/14.7 | 45.9/96.2 | −36.8/−77.4 | −80.0 | 2.6 |
| 1986/1988 | 10.8/28.0 | 6.2/15.4 | 47.0/116.3 | −30.0/−72.9 | −65.6 | -7.3 |

**Table 5.4.** The repetitiveness of basic types of wind fields above the Aral Sea (%).

| *Wind type* | *Month* | | | | | |
|---|---|---|---|---|---|---|
| | *May* | *Jun* | *Jul* | *Aug* | *Sep* | *Oct* |
| Northern, northeastern, east-northeastern | 23 | 25 | 23 | 20 | 30 | 25 |
| Northwestern, northern, north-northeastern, west-southwestern | 19 | 19 | 17 | 17 | 22 | 28 |
| Southwestern, south-southwestern | 13 | 18 | 22 | 20 | 15 | 13 |
| Southwestern, southern, south-southeastern | 18 | 17 | 17 | 22 | 15 | 14 |
| Wind of 7 m/s and more | 9 | 6 | 6 | 7 | 10 | 15 |
| Wind of 3 m/s and less | 9 | 16 | 15 | 14 | 8 | 5 |

Caspian Sea level can be classified as having four stages (Karpychev, 2007; Lebedev and Kostianoy, 2005):

(1) 1830–1928 were characterized by sea level decreasing at the moderate speed of 11–12 cm/yr;
(2) 1929–1941 witnessed a sharp fall in sea level from the absolute mark of −29.96 m to −27.84 m;
(3) 1942–1977 were accompanied by sea level decreasing at the moderate speed of 11–12 cm/yr;
(4) 1978–2004 saw an increase in sea level from the absolute mark of −28.89 to −26.37 m.

**Table 5.5.** Water balance of the Caspian Sea.

| *Time period* | *Annual water inflow to the Caspian Sea* (km$^3$) | *Annual water outflow from the Caspian Sea to Kara-Bogaz-Gol gulf* (km$^3$) | *Evaporation* (cm) |
|---|---|---|---|
| 1880–1913 | 306 | 24.7 | 71.5 |
| 1914–1932 | 321 | 19.7 | 74.7 |
| 1933–1940 | 229 | 10.5 | 77.3 |
| 1941–1956 | 292 | 11.6 | 78.2 |
| 1957–1970 | 281 | 9.5 | 73.2 |
| 1971–1977 | 236 | 6.9 | 72.0 |
| 1978–1990 | 305 | 2.8 | 68.4 |
| 1990–1993 | 327 | 15.6 | 71.0 |

**Table 5.6.** More detailed water balance of the Caspian Sea (Ratkovich, 1997).

| *Years* | *Level* (m) | *River runoff* (km$^3$) | *Precipitation* (mm) | *Evaporation* (mm) | *Loss to Kara-Bogaz-Gol* (km$^3$) | *Balance* (km$^3$) | (cm) |
|---|---|---|---|---|---|---|---|
| 1900/1929 | −26.18 | 335.7 | 173 | 967 | 21.8 | −1.7 | −0.4 |
| 1930/1941 | −26.80 | 268.6 | 185 | 1004 | 12.4 | −61.7 | −15.7 |
| 1942/1969 | −28.18 | 285.4 | 200 | 964 | 10.6 | −3.4 | −0.9 |
| 1970/1977 | −28.64 | 240.5 | 243 | 1,039 | 7.1 | −49.9 | −13.8 |
| 1978/1985 | −28.35 | 304.5 | 222 | 928 | 1.8 | 48.1 | 13.1 |
| 1900/1995 | −27.35 | 298.2 | 194 | 974 | 13.6 | −11.8 | −3.1 |

The hydrometeorological situation that prevails at the current time (2011) in the ACWS demonstrates the urgency of finding new methods to hold up these catastrophic tendencies and to synthesize new algorithms to show how the ACWS can return to a controlled stable state. The natural water cycle of the Aral Sea was disturbed after 1960 as a result of the increased use of river water for irrigation. Water leakage from irrigation systems resulted in the formation of different lakes and other reservoirs. The Prearalie's deltaic lakes, which mainly depend for their existence from inflow from the Amudar′ya river, underwent a broad range of negative changes. At the same tame, with the completion of new large reservoirs, a new type of water body—so-called tail accumulators (sinks) for drainage water—occurred in Prearalie. The area covered by lakes was 70.2 $km^2$ in 1984 (considered a normal year), then increased up to 120 $km^2$ in the humid year 1997, and decreased to 26.0 $km^2$ in the dry year 2000.

This chapter proposes an approach to solution of this task. It is proposed to set up a monitoring system of the influence zone of the Aral and Caspian Seas. This system is based on the principle of GIMS technology. It involves remote monitoring and simulation blocks. These blocks function by a regime that alternates between measurement and prognosis procedures. The temporal parameters of such a regime are determined empirically.

The area $\Xi$ delimited by geographical coordinates $[41, 47]^{\circ}N \times [50, 70]^{\circ}E$ is chosen for analysis and study. Within the $\Xi$ the water regime is described by the ACWS theoretical information model (TIM). Outside the $\Xi$ it is described by the GMNSS.

### 5.1.2 Remote-monitoring data

The territory of the ACWS was studied by means of the flying laboratories of the Russian Academy of Sciences' Institute of Radio Engineering and Electronics during the last 25 years (Armand *et al.*, 1987; Borodin *et al.*, 1987, 1996). Figure 5.2 shows one of them. This laboratory was equipped by a remote system of trace radiometers of different frequencies, a side-viewing radar system, TV system, and GPS. The on-board system had an automated information-modeling device that made communicating data to land-monitoring services possible. The principal structure of this item is shown in Figures 5.3 and 5.4.

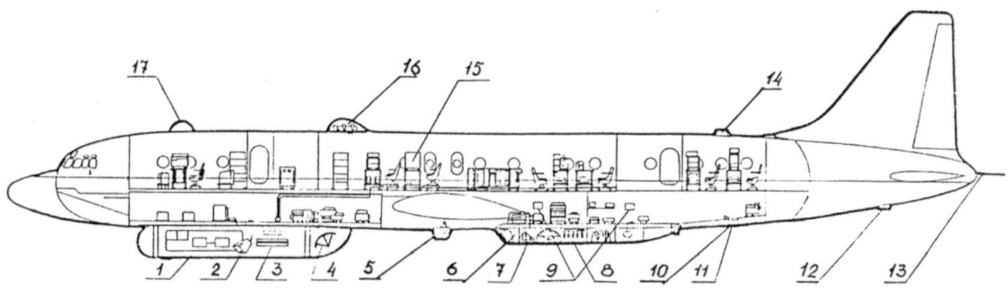


**Figure 5.2.** The IL-18 flying laboratory. Notation: 1 and 3, synthetic aperture radio locators of 2 m and 10 cm range, respectively; 2 and 6, trace polarimeters of 0.8 and 2.25 cm, respectively; 4, six-channel scanning polarimeter of 0.8, 1.35, and 2.25 cm range; 8, trace radiometers of 0.8, 1.35, 2.25, 10, 20, and 27 cm wavelengths; 7 and 9, side-viewing precision altimeter and interferometer of 2.2 cm wavelength; 13, sensors for subsurface sensing in the decimeter range; 11 and 14, radiometers of millimeter range; 16, trace radiometers of 0.8, 1.35, and 2.25 cm wavelengths; 15, gravimetric and inertial sensors; 17, aero-hatch; 5, 10, and 12, photo hatches.

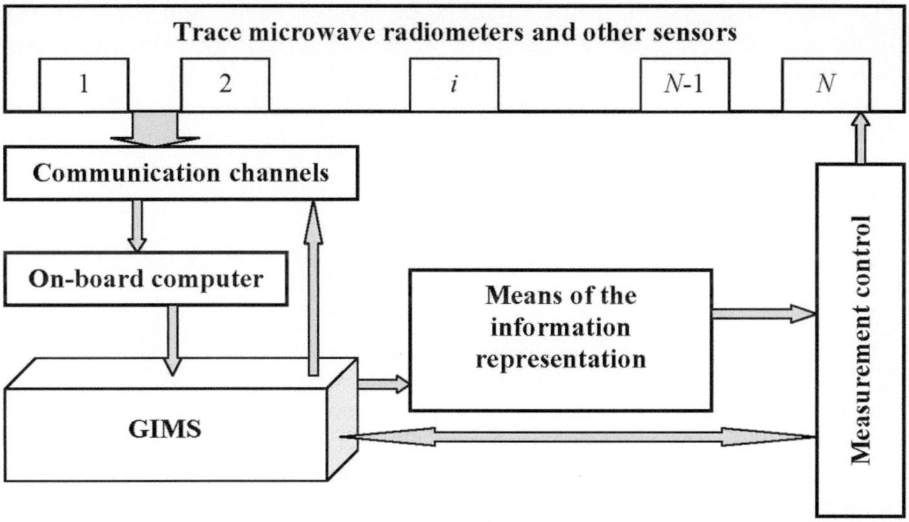


**Figure 5.3.** Typical structure of airborne information system on board the IL-18 flying laboratory (Borodin *et al.*, 1987).

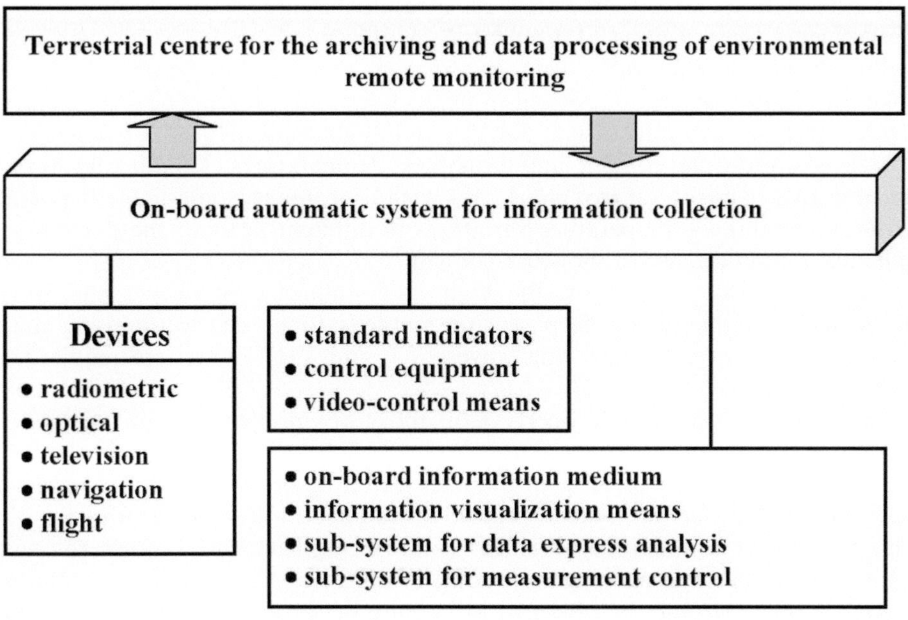


**Figure 5.4.** Exemple structure of devices located on board the multi-functional flying laboratory.

The on-board system forms archives of measurement data and attendant information. The complex of on-board sensors and GIMS software make it possible to measure remote data at a frequency of 0.01 s, including:

- radiation of the underlying surface measured by means of microwave radiometers;
- radiation of the atmosphere and cloudiness measured by means of radiometers using wavelengths of 0.8, 1.35, and 2.25 cm;
- radiation reflected by subsurface layers of the land cover;
- temperature of zones where receivers were located.

Measurements were made by microwave radiometry in the ranges of 0.8, 1.35, 2.25, 3.4, 10, 18, 20, 21, 27, and 30 cm. It was shown that SHF measurements ensure reliable grouping of land covers, indication of ground water levels, and estimation of water content in the atmosphere over the ACWS area. On the basis of measurements made in Central Asia the remote database was set up with a three-level structure:

(1) routing measurements;
(2) data processed by means of various algorithms (spectra, statistical characteristics, subject classification, correlation models),
(3) maps charted using spatiotemporal interpolation methods.

The database was set up to have algorithms at its disposal for the data processing of microwave monitoring. As a result, brightness temperature contrast measurements over regions of Central Asia at centimeter and decimeter wavelengths from the aircraft were performed. Essential negative brightness temperature contrasts were observed in flights over moist areas. The measurements carried out at wavelengths of 0.8, 3.4, 10, and 20 cm indicated a clear correlation between soil moisture content and brightness temperatures. Conducting aircraft measurements over the same regions over such a prolonged time made it possible to observe brightness temperature variations in different fields in the dynamics of the increase and decrease of moisture content.

The larger the wavelength, the higher the influence of deeper soil layers. This fact allows thick-layer deep sensing methods to be developed using multi-frequency radar systems. The long waves of P and VHF bands are required for subsurface sensing.

Information about soil property (soil moisture) profiles can be received by analyzing scattering measurements at different wavelengths. To ascertain the influence of the soil moisture profile on the backscattering cross section it is necessary to develop models that can reflect from layers situated at different depths. Solution of the inverse problem can be obtained by measuring backscatter at several wavelengths and in different polarization modes. Complete information about the soil moisture profile can only be obtained by solving the problem of image interpretation in the broad band including the meter band, where attenuation in soil and vegetation is comparably low.

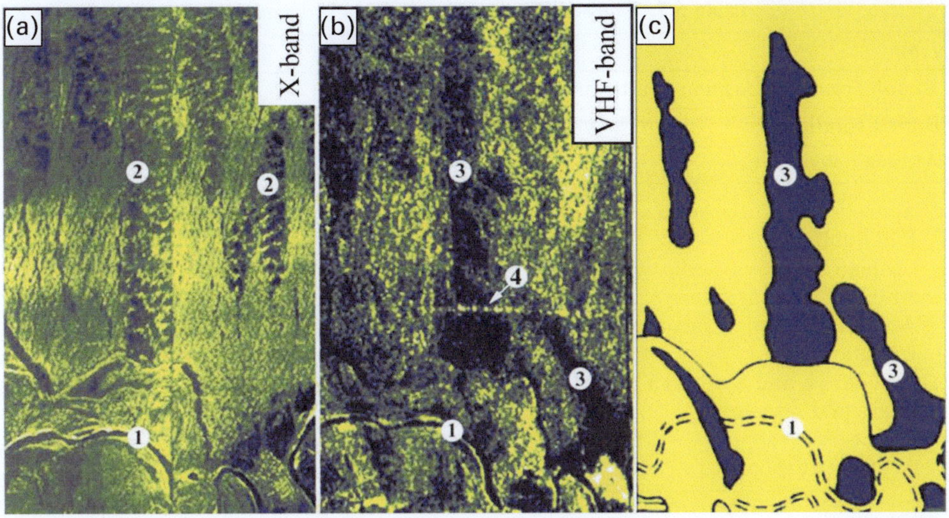


**Figure 5.5.** Radar images of the Kara-Kum desert region in the X-band ($\lambda = 4$ cm) (a) (near surface layer information); VHF-band ($\lambda = 2.5$ m) (b) (thick layer information), and geologic map of distribution of underground waters as a result of radar survey (c).

The results of multiband radar survey using the four-band airborne SAR IMARC of Vega (a Russian radioengineering corporation) illustrate the possibility of measuring hydrological soil regimes and water lens allocation in Kara-Kum desert (Figure 5.5). Lenses of underground water at a depth of 50–70 m were detected. The results were validated by the boring of a control well. Radar images show (1) the dry riverbed of the Uzboy, (2) sand dunes 6–15 m high, (3) underground water lenses, and (4) transmission facilities.

The data obtained convincingly indicate the possibility of spatiotemporal moisture content measurements by means of microwave radiometry (Reutov and Shutko, 1986; Shutko, 1986). The database of the Institute of Radio Engineering and Electronics includes data from radar, radiometry, optoelectronics, and optical measurements. The database has sets of parameters for salt marshes, water surfaces, watersheds, waterways of waste and drainage waters, irrigation nets and for processes of filtration, waterlogging, salting, bogging, and desertification. All data were compiled using a set of specific algorithms (Kondratyev *et al.*, 2002; Nitu *et al.*, 2004).

### 5.1.3 Theoretical information model of the Aral–Caspian aquageosystem

Basic elements of the water balance of the ACS are given in Figure 5.6. A computer realization of this scheme is based on $\Xi$ discretization by dividing $\Xi$ into areas $\Xi_{ij}$ of size $\sigma_{ij} = \Delta\varphi_i \cdot \Delta\lambda_j$. $\Xi_{ij} \subseteq \Xi$ is characterized by specific soil–plant formation, geophysical structure, and socioeconomic parameters. The area $\Xi_{ij}$ has a subset of flows $\{H_k\}$ marked by identifiers $\{A_s\}$ which are characterized in Table

**Table 5.7.** The structure of water flows in Figure 5.5.

| *Flow* | *Flow identifier* |
|---|---|
| Runoff from the Caspian Sea to the KBG gulf | $H_1$ |
| Buried solutions | $H_2$ |
| Artificial evaporation | $H_3$ |
| Evaporation from the surface | |
| KBG gulf | $H_4$ |
| Caspian Sea | $H_5$ |
| Aral Sea | $H_7$ |
| Accumulators of drainage waters | $H_{18}$ |
| Irrigation waters | $H_{19}$ |
| Amudar′ya and Syrdar′ya rivers | $H_{20}$ |
| Lakes and reservoirs | $H_{21}$ |
| Soils | $H_{23}$ |
| Artificial evaporators | $H_{37}$ |
| River runoff | |
| To the Caspian Sea | $H_6$ |
| To the Aral Sea | $H_8$ |
| Replenishment of Amudar′ya and Syrdar′ya from lakes and reservoirs | $H_9$ |
| Melting of glaciers and snowfields | $H_{10}$ |
| Water expenditure on irrigation | $H_{24}$ |
| Input of water to accumulators of drainage waters | $H_{12}$ |
| Leakage from irrigation systems | $H_{13}$ |
| Water accumulation in lenses | $H_{14}$ |
| Taking water from lenses for domestic needs | $H_{15}$ |
| Return waters | $H_{16}, H_{17}$ |
| Diversion flow to irrigation systems | $H_{11}$ |
| Surface runoff to rivers | $H_{22}$ |
| Surface runoff from irrigated territories | $H_{25}$ |
| Precipitation on | |
| Lakes and reservoirs | $H_{26}$ |
| Glaciers and snowfields | $H_{27}$ |
| Soil | $H_{28}$ |
| KBG gulf | $H_{29}$ |
| Caspian Sea | $H_{30}$ |
| Accumulators of drainage waters | $H_{31}$ |
| Amudar′ya and Syrdar′ya rivers | $H_{32}$ |
| Aral Sea | $H_{33}$ |
| Irrigation systems | $H_{34}$ |
| Artificial evaporators | $H_{38}$ |
| Replenishment of water bodies from ground waters | $H_{36}$ |
| Replenishment of rivers from ground waters | $H_{35}$ |

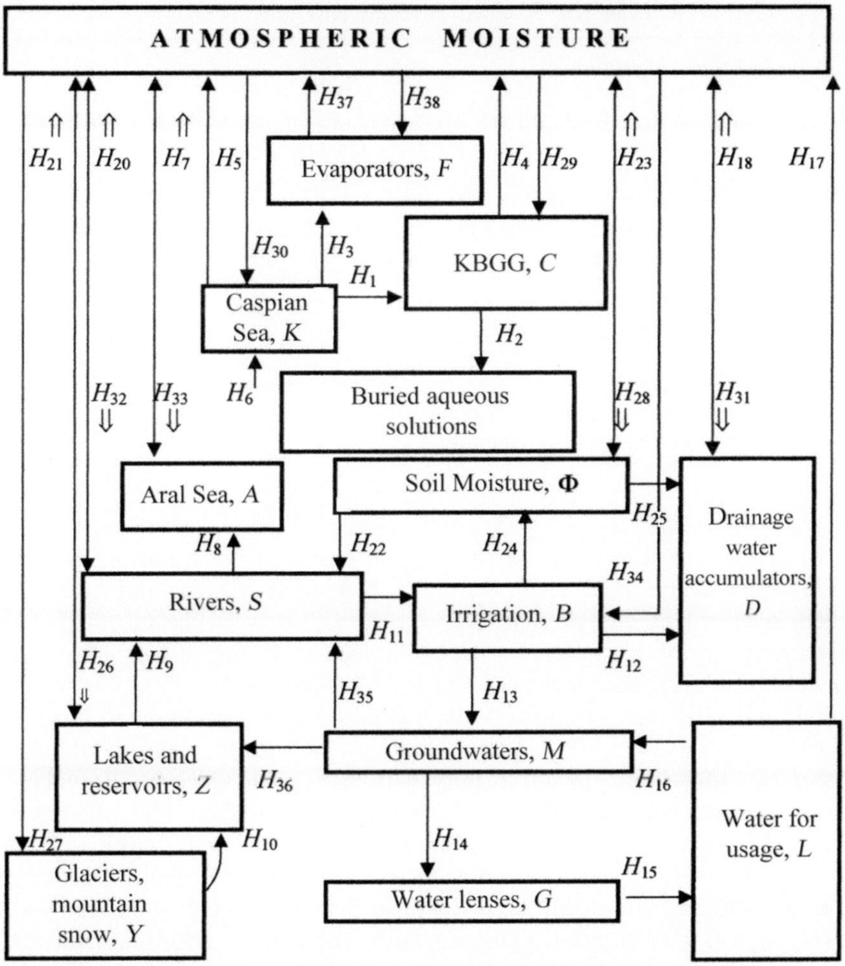


**Figure 5.6.** Block diagram of the ACWS water balance as represented in the framework of the TIM.

5.8. The set of $A_s$ reflects an information database structure and forms the elements of water balance equations:

$$\sigma_{ij}\left(\frac{\partial W_{ij}}{\partial t} + v_\varphi \frac{\partial W_{ij}}{\partial \varphi} + v_\lambda \frac{\partial W_{ij}}{\partial \lambda}\right) = \sum_{s \in I_{ij}} H_s^{ij} - \sum_{s \in J_{ij}} H_s^{ij} \tag{5.1}$$

$$\sigma_{ij}\frac{dE_{ij}}{dt} = \sum_{s=1}^{38} (\omega_s - \gamma_s) H_s^{ij} \tag{5.2}$$

$$\frac{dL_{ij}}{dt} = H_{15}^{ij} - H_{16}^{ij} - H_{17}^{ij}, \tag{5.3}$$

**Table 5.8.** A set of TIM database identifiers.

| *Identifier* | *Identifier description* |
|---|---|
| $A_1$ | Fixing $\Xi$; $A_1 = \|a^1_{ij}\|$, $a^1_{ij} = 0$ when $(\varphi, \lambda) \notin \Xi$, $a^1_{ij} = a_1$ when $(\varphi, \lambda) \in \Xi$; $a_1 = 1$ for the Aral Sea, $a_1 = 2$ for the Caspian Sea, $a_1 = 3$ for the Kara-Bogaz-Gol gulf, $a_1 = 4$ for saline land, $a_1 = 5$ for rivers, etc. |
| $A_k$ | Average statistical data of wind speed ($k = 2$), atmospheric temperature ($k = 3$), precipitation ($k = 4$), and wind direction ($k = 5$) |
| $A_6$ | Initial data for the all input parameters |
| $A_7$ | Relief of $\Xi$ |
| $A_8$ | Fixing $\Xi_{ij}$ as an artificial evaporator ($a^8_{ij} = 1$ for "yes", $a^8_{ij} = 0$ for "no") and as an area for forced precipitation ($a^8_{ij} = 2$ for "yes", $a^8_{ij} = 0$ for "no"). |

where $I_{ij}$ and $J_{ij}$ are identifiers of the evaporation and precipitation flows on the area $\Xi_{ij}$, respectively; $E = (K, C, A, \Phi, D, S, B, M, Z, Y, G)$; $\omega_s$ and $\gamma_s$ are binary identifiers reflected by the presence ($\omega_s = 1, \gamma_s = 1$) or absence ($\omega_s = 0, \gamma_s = 0$) of inflowing and outflowing processes for each element $E$, respectively. For example, if $E = A$, $\omega_s = 1$ only when $s = 8.33$ and $\gamma_s = 1$ when $s = 7$.

Initial and boundary conditions are needed to solve equations (5.1)–(5.3) over the whole territory $\Xi = \bigcup \Xi_{ij}$. Initial data can be found in the database or the TIM user can form them in conformity with their own scenario. Boundary conditions are formed by the GMNSS or are given by the user. The water flows $H_k$ ($k = 1, \dots, 38$) are described using analytical, table, and graphic functions. Evaporation from land and transpiration are described by empirical correlations from Bras (1990). Evaporation from the Aral Sea surface ($H_7$) and shallow reservoirs ($H_4, H_{18}, H_{20}, H_{37}$) are calculated with Goptarev's formula (Bortnik and Chistyakova, 1990). Evaporation from the surface of the Caspian Sea ($H_5$) is described by the Penman equation (Penman, 1948). Flow $H_6$, describing the river water influx to the Caspian Sea, has a normal distribution with an average value of 290 km$^3$/yr and a standard deviation of 30 km$^3$/yr.

The movement of atmospheric moisture over the territory $\Xi$ is described by using only the wind component of the standard system of finite difference balance equations. The parameterization of $H_k$ and control of the TIM are realized by means of the software mentioned in Table 5.9.

For the purposes of TIM validation, the dynamics of the ACWS was calculated from 1960 to 1985. The flows $H_5$, $H_6$, $H_7$, $H_8$, $H_{30}$, and $H_{33}$ were given as table functions with $\Delta t = 1$ year on the basis of published data (Bortnik and Chistyakova, 1990; Bortnik *et al.*, 1994; Golitsin, 1995). The other flows $H_i$ were calculated using the models in Table 5.9. An adaptive side to the TIM was realized within the framework of the simulation procedure (Kondratyev *et al.*, 2002). This was based on evolutionary modeling procedures that were described in detail in

**Table 5.9.** A list of TIM blocks.

| *Block* | *Block description* |
|---|---|
| CSII | Calibration and scaling of input information |
| IDF | Input data filtration |
| FTDM | Forming two-dimensional maps on the basis of trace measurements and fragmentary data using spatiotemporal interpolation methods |
| SEEM | Solution of equations (5.1)–(5.3) using the Euler method modified by Krapivin (1969) |
| IBU | Interface block that makes the database accessible to the user |
| MF | Modeling the flows $H_k$ $(k = 1, \ldots, 38)$ |
| FS | Forming the scenario |
| CRSR | Cartographic representation of simulation results |
| IBMI | Interface block that allows the set of identifiers $\{A_i\}$ to be modified |
| IBC | Interface block for control of the computer experiment regime and for the choice of schemes for result representation |
| ASWR | The Aral Sea water regime model |
| KBG | The Kara-Bogaz-Gol gulf water regime model |

Krapivin *et al.* (1991) and Nitu *et al.* (2004). During this adaptation procedure, TIM parameters are corrected to minimize the value of

$$\Delta h = \sum_{i=1960}^{1985} [|h_A(i) - h_{A,0}(i)| + |h_C(i) - h_{C,0}(i)|]/50,$$

where $h_A$ and $h_C$ are the water levels of the Aral and Caspian Seas, respectively, calculated by means of the TIM; and $h_{A,0}$ and $h_{C,0}$ are existing estimates of $h_A$ and $h_C$, respectively (Bortnik and Chistyakova, 1990). As a result of this procedure a precision level of 23% was achieved in approximating the dynamics of ACS water levels between 1960 and 1985.

### 5.1.4 Simulation experiments using a GIMS-adopted procedure

The main purpose of computer experiments is to find a scenario in the water regime control of the ACWS in which the Aral Sea level increases at the expense of the Caspian Sea level. Let $\Delta\varphi = \Delta\lambda = 10'$, $\Delta t = 10$ days. Let us consider the EP (evaporation/precipitation) scenario in which the Caspian Sea level decreases as a result of its water diverting to other reservoirs/evaporators. Such reservoirs include the saline lands and hollows situated on the East Caspian Sea coast. Their absolute levels are below the recent Caspian Sea level (−25.7 m). The following hollows are representative: lifeless Kultuk (−27 m), Kaidok (−31 m), Karagie (−132 m), Kaundy (−57 m), Karin Arik (−70 m), Chagala Sor (−30 m), and Kara-Bogaz-Gol gulf (−32 m).

Let us represent the water volume of the Caspian Sea streamed to these evaporators as $\zeta$. This water volume evaporates rapidly as a result of wind and temperature. Thus, the water content in the atmosphere is increased. As a result, atmospheric moisture over the Aral Sea increases as well. A procedure of forced precipitation over the Aral Sea watershed is realized. The flows $H_i$ $(i = 6, 8, 11, 15, 24)$ are fixed at 1985 levels.

Long-standing data indicate that the wind directions over the ACWS territory are such that northwest (NW), west (W), and southwest (SW) winds prevail for 130–160 days per year with a high degree of frequency. Hence, an atmospheric waterway towards the Aral Sea is realized by such a stable statistical regime (Figure 5.7).

The results of simulation experiments are given in Figures 5.8, 5.9, and 5.10 and in Tables 5.10 and 5.11. We can see the Caspian Sea level decreasing by 1.2 cm/yr as long as the following scenario conditions prevail: value $\zeta$ is uniformly distributed in the interval 50–60 km$^3$/yr, the atmospheric temperature is no lower than 15°C, the NW, W, and SW wind directions occur 50% of the time over $\tau \geq 80$ days. In this situation 0.6–1.3 km$^3$/day of atmospheric moisture reaches the Aral Sea hollow directly from the Caspian Sea aquatory (flow $H_5$) and 0.1–0.2 km$^3$/day is added at the expense of evaporators (flow $H_{38}$). As a result it is possible to have precipitation of 0.5–1.5 km$^3$/day in the Aral Sea hollow. This means the Aral Sea level reaching its absolute mark of −53 m in about 7–12 years.

Computer experiments have revealed the potential volume of precipitation in the Aral Sea hollow to be quasi-invariant of NW, W, and SW wind directions. This phenomenon means ACS water regime stabilization and its conservation through continuous remote monitoring is possible.

Figure 5.9 shows precipitation in the Aral Sea hollow raised by 12%. Moreover, the positive balance of water transport across the eastern boundary of the ACS is raised by 4% stimulating an increase in river flow to the Aral Sea of up to 40 km$^3$/yr ($34 \leq H_8 \leq 50$ km$^3$/yr). The east wind increases precipitation in the Aral Sea area by 4–7% by returning atmospheric moisture.

The dynamics of all water balance components of the ACS stabilizes 8–9 years after the beginning of the computer experiment. An increase in $M(t, \varphi, \lambda)$ at a distance of 100–170 km from the Aral Sea takes place. The value of $\Delta M$ equals 1.2 cm/yr during this time. Other important estimates include $H_7 = 51.1 \pm 9.2$ and $H_{33} = 9.7 \pm 1.8$ km$^3$/yr.

The results of this study inspire confidence to propose the following scheme for aero-cosmic monitoring of the water balance in the Central Asian arid zone under the influence of the Aral and Caspian Seas. The monitoring system registers water balance parameters in a regime of episodic measurements using remote-

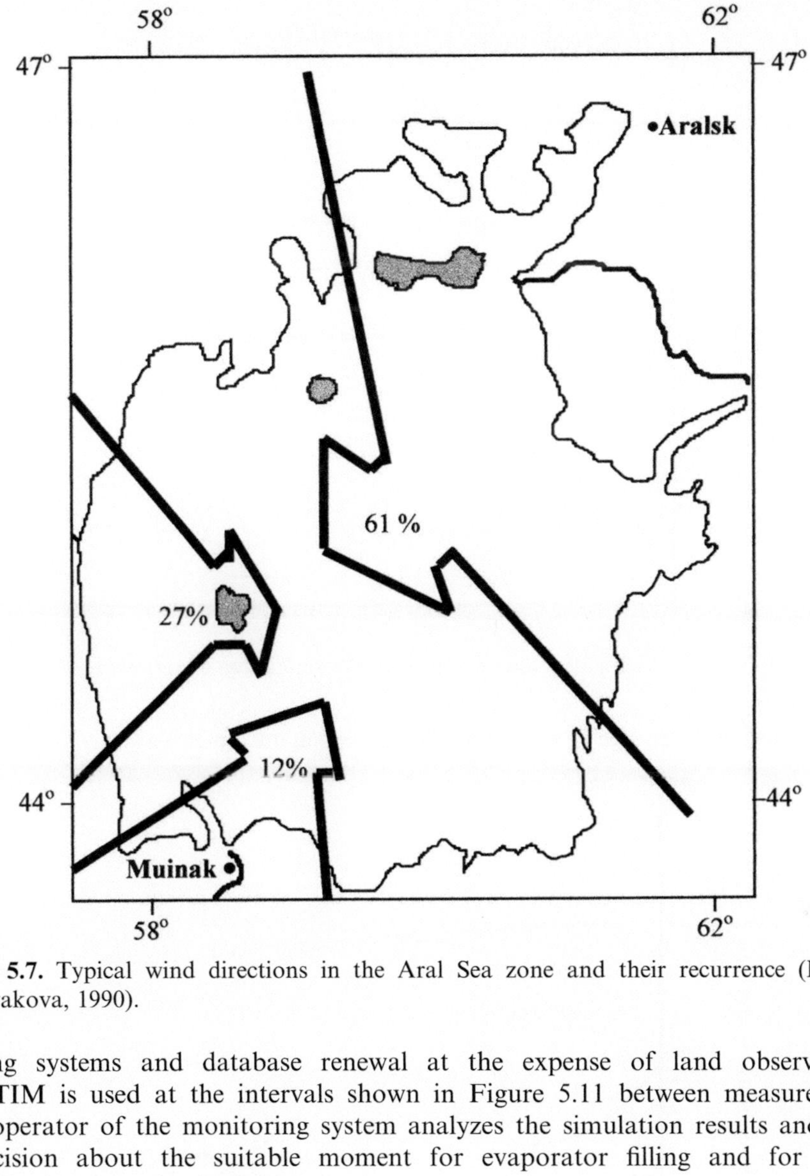


**Figure 5.7.** Typical wind directions in the Aral Sea zone and their recurrence (Bortnik, Chistyakova, 1990).

sensing systems and database renewal at the expense of land observations. The TIM is used at the intervals shown in Figure 5.11 between measurements. The operator of the monitoring system analyzes the simulation results and takes a decision about the suitable moment for evaporator filling and for forced precipitation over the Aral Sea area.

The TIM allows different hypothetical situations for the formation of atmospheric moisture flows to be considered and estimates its localization in the form of precipitation on different areas of the ACWS. This problem leaves a lot of room for future investigations. The water balance elements of the ACWS may also be obtained from a systematic approach based on evolutionary modeling, but the set of parameters and structures governing the accuracy of the monitoring system need to be further investigated, both experimentally and theoretically.

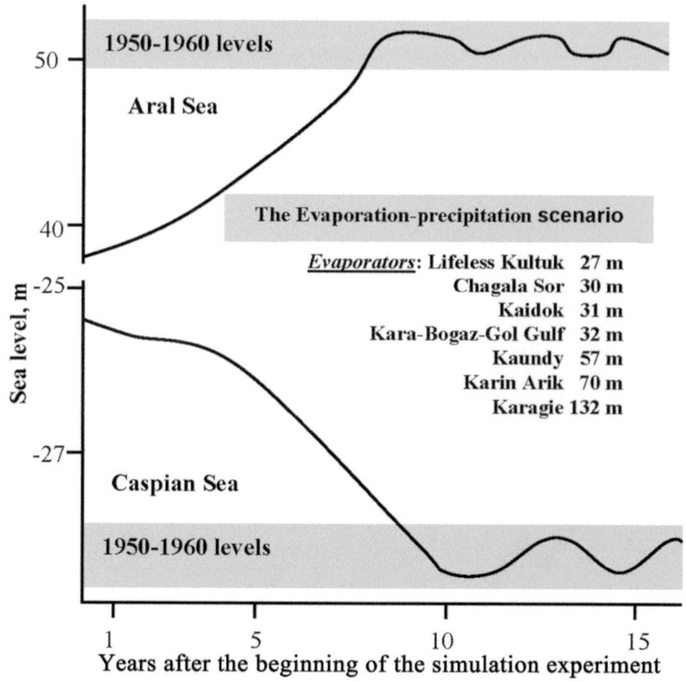


**Figure 5.8.** Results of the simulation experiment for the evaporation/precipitation (EP) scenario showing the change in water levels of the Aral and Caspian Seas.

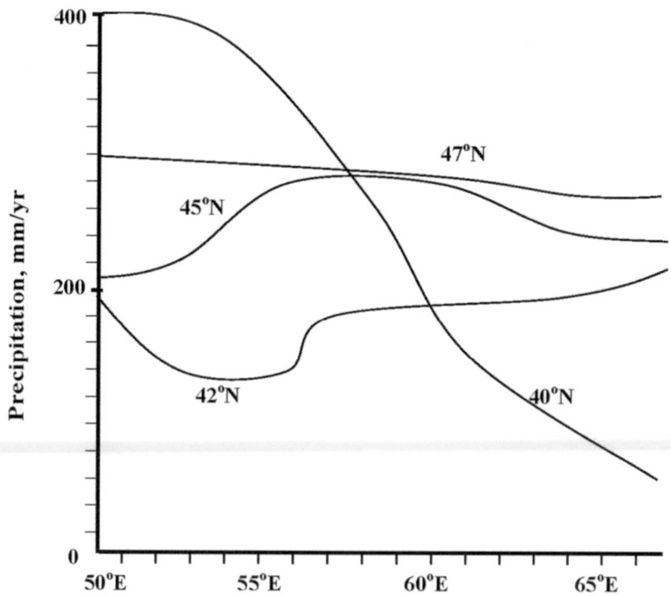


**Figure 5.9.** Spatial distribution of precipitation intensity on the studied territory after realization of the evaporation/precipitation scenario.

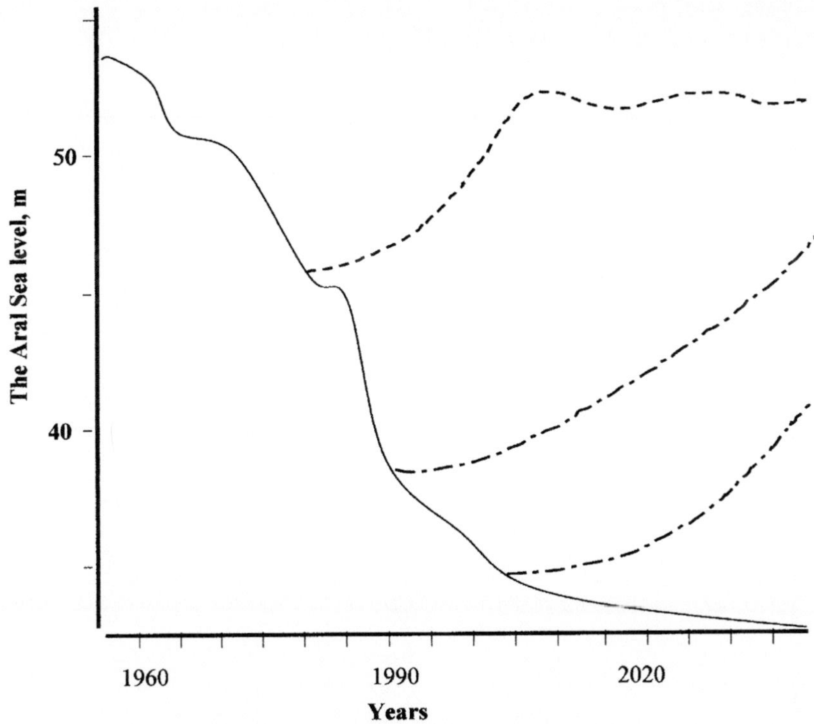


**Figure 5.10.** Possible dynamics of Aral Sea absolute levels (in meters with respect to the World Ocean level) as a result of changing the starting time of realizing the scenario in Figure 5.7 (dashed curves). The solid curve corresponds to the natural dynamics of the sea level under conditions in which the average indicators of natural/anthropogenic parameters are preserved.

When a model of the regional water balance such as that described in Section 5.1 is used, the GIMS allows the optimal distance between parallel flight paths to be estimated. Precision in soil moisture mapping depends on the distance between aircraft flight paths. Computer simulations make it possible to optimize the monitoring regime to give the precision required. An analogous procedure can be used in any region where operative knowledge of soil moisture content is the important factor for a decision about irrigation system functioning to be made.

## 5.2 SIMULATION MODEL OF THE OKHOTSK SEA ECOSYSTEM

### 5.2.1 The GIMS and cold seas

There are many effective marine models that can be used to study the dynamics of sea ecosystems. However, sea ecosystem modeling runs into difficulties whose removal requires the developent of new modeling methods. Namely, the sea ecosystem is complex and functions under unstable environmental conditions. These

**Table 5.10.** Comparison of on-site measurements of seawater temperature *T* (°C) using TIM simulation results.

| *Depth* (m) | *September, 1981* | | *September, 1989* | |
|---|---|---|---|---|
| | *Measurement* | *Model* | *Measurement* | *Model* |
| ***The west deep-water hollow*** | | | | |
| 0 | 19.6 | 20.0 | 16.3 | 16.5 |
| 5 | 19.5 | 19.5 | 17.5 | 17.8 |
| 10 | 19.4 | 19.5 | 17.4 | 17.8 |
| 15 | 19.3 | 19.0 | 17.5 | 17.6 |
| 20 | 19.2 | 19.0 | 17.5 | 17.6 |
| 25 | 14.3 | 17.0 | 17.5 | 17.2 |
| 30 | 6.7 | 9.0 | 17.5 | 17.1 |
| 40 | 4.5 | 6.0 | 15.7 | 15.3 |
| 50 | 1.5 | 4.0 | 15.7 | 15.3 |
| ***Central aquatory of the Aral Sea*** | | | | |
| 0 | 18.3 | 18.4 | 13.5 | 13.0 |
| 5 | 18.3 | 18.0 | 13.7 | 13.0 |
| 10 | 18.1 | 18.0 | 14.0 | 13.5 |
| 15 | 18.0 | 18.0 | 14.0 | 13.5 |

conditions can be defined by both measured and undefined parameters. Thus, sea ecosystems can be studied using expedition measurements, mathematical models, and space observations. A combination of these approaches gives useful results (Alfultis and Martin, 1987; Berdnikov *et al.*, 1989; Krapivin, 1996; Krapivin *et al.*, 1990; Terziev, 1992). The main difficulties of such approaches arise as a result of information incompleteness.

In the 1990s, several investigators (Kelley *et al.*, 1992a; Krapivin and Shutko, 1989; Sellers *et al.*, 1995) reported a variety of problems inhibiting the monitoring of complex systems for meaningful collection and synthesis of environmental information concerning the Okhotsk Sea. In response to these difficulties, a method was devised to integrate a GIS with models and field measurements. The developed GIMS = GIS + Model was focused on systematic observation and evaluation of the environment related to changes attributable to human impact on

**Table 5.11.** Model estimates of some Aral Sea water balance elements under different prevailing wind directions: $H_8$ is river flow (km$^3$/yr); $H_7$ is evaporation, and $H_{33}$ is precipitation (mm/yr).

| *Time since first simulation experiment* | *Prevailing wind direction* | | | | | | | | |
|---|---|---|---|---|---|---|---|---|---|
| | *NW* | | | *W* | | | *SW* | | |
| (years) | $H_8$ | $H_{33}$ | $H_7$ | $H_8$ | $H_{33}$ | $H_7$ | $H_8$ | $H_{33}$ | $H_7$ |
| 1 | 38 | 197 | 1,010 | 41 | 188 | 998 | 10 | 198 | 1,007 |
| 2 | 44 | 180 | 991 | 37 | 190 | 987 | 12 | 183 | 1,011 |
| 3 | 70 | 160 | 993 | 55 | 171 | 869 | 16 | 160 | 1,004 |
| 4 | 56 | 174 | 968 | 68 | 183 | 901 | 21 | 171 | 1,023 |
| 5 | 48 | 149 | 1,001 | 50 | 194 | 977 | 18 | 152 | 1,014 |
| 6 | 51 | 187 | 986 | 44 | 189 | 983 | 14 | 188 | 989 |
| 7 | 66 | 191 | 999 | 61 | 169 | 1,015 | 16 | 190 | 1,003 |
| 8 | 61 | 177 | 956 | 63 | 175 | 994 | 12 | 180 | 1,004 |
| 9 | 59 | 163 | 983 | 52 | 166 | 899 | 9 | 171 | 999 |
| 10 | 53 | 154 | 979 | 57 | 160 | 908 | 13 | 155 | 991 |
| 11 | 49 | 142 | 988 | 55 | 159 | 910 | 17 | 143 | 1,001 |
| 12 | 57 | 138 | 985 | 48 | 144 | 1,017 | 11 | 140 | 973 |
| 13 | 52 | 144 | 987 | 54 | 147 | 999 | 8 | 141 | 966 |
| 14 | 55 | 107 | 1,003 | 50 | 133 | 976 | 12 | 110 | 981 |
| 15 | 58 | 99 | 990 | 56 | 121 | 891 | 15 | 101 | 909 |

the marine system. An important functional aspect of integrated systems is the possibility of forecasting undesirable changes in the environment. Application of mathematical modeling to the monitoring effort greatly improves simulation of natural processes in the environmental complex under observation.

The development of models based on biogeochemical, biogeocenotic, demographic, and socioeconomic information, including consideration of the environmental dynamics of biospheric and climatic processes on the overall system, illustrates what is demanded of the GIMS structure and its database. According to guidelines proposed in the paper of Kelley *et al.* (1992a), the simulation of the dynamics of the biosphere is an important function of the GIMS. Accordingly, growing importance has been attached to the value of this new monitoring

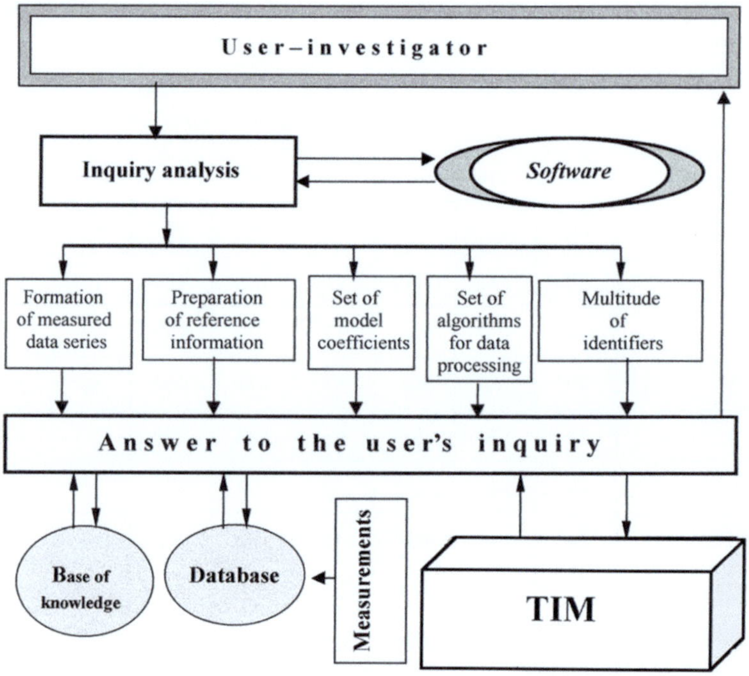


**Figure 5.11.** An approximate scheme of the dialog regime used for algorithmic provision of a hydrophysical experiment.

approach which integrates assessments of the status of the biosphere. The basic objective of all investigations in the development of GIMS technology is to describe the regional biogeosystem as a biospheric subsystem. The Okhotsk Sea comprises such a subsystem. Therefore, application of GIMS technology to the study of the Okhotsk Sea Environment (OSE) entails synthesizing a simulation model of the Okhotsk Sea environment (SMOSE) to describe the associated ecosystem dynamics (Figure 5.12). Preliminary results in this direction have been published by Aota *et al.* (1991a, b, 1992, 1993). These help to describe the spatial dynamics of OSE and to optimize the monitoring regime.

The OSE occupies a significant position in the global natural system. Currently (2011), it has a low level of pollution with fishery the main anthropogenic influence. A correlation between the OSE state and global changes is a problem discussed both within the framework of regional investigations and in global studies of environment. The OSE interacts with biospheric processes via its influence on global climate and on the Pacific Ocean. This influence is reciprocal. An approach to estimation of this influence is given below, wherein the common concept of complex system survivability is interpreted for the ecological system of Okhotsk Sea and criteria of survivability are defined.

For a complex system to survive it has to have the ability to resist the influence of external impacts and to preserve its structure and effectiveness in

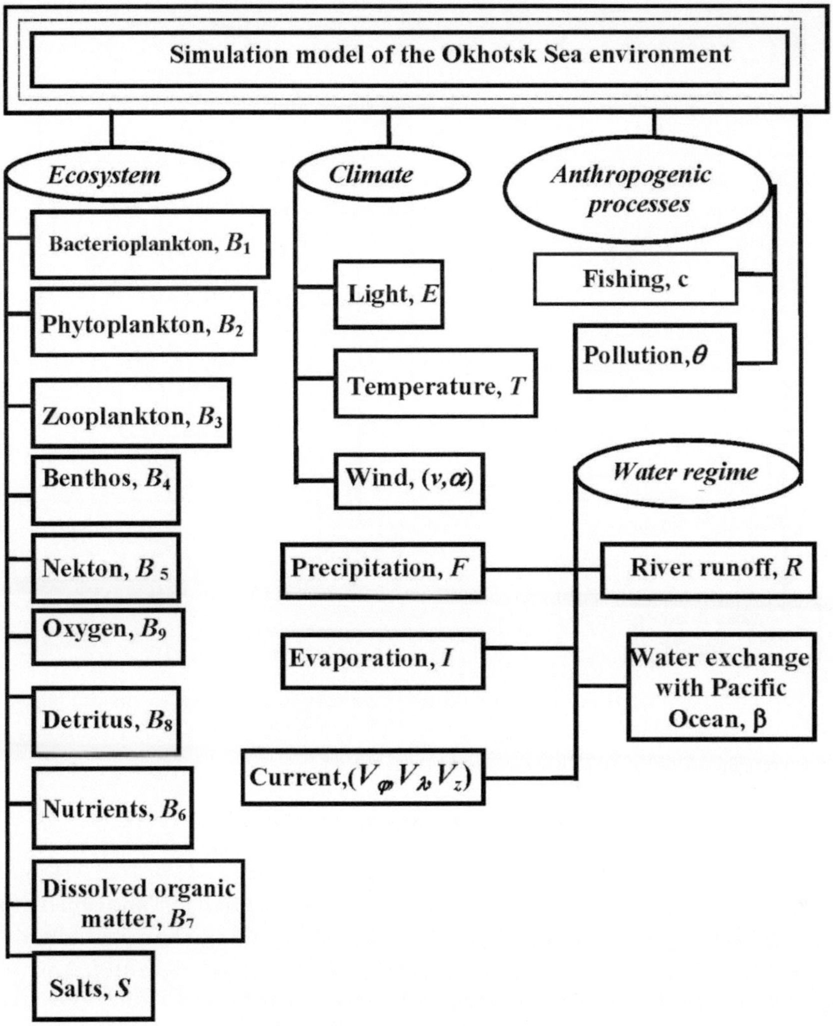


**Figure 5.12.** The SMOSE structure.

doing so. The OSE comprises more than energy and nutrient flows, trophic webs, and competition communities. The full interrelations among coexisting living organisms and nonliving elements are also components. In this study the definition problem between ecosystem *aim* and *behavior* are not discussed. These were discussed by Krapivin (1978). It is postulated that the OSE wants to maximize its biomass of living elements at high trophic levels and its behavior in the adaptive change of food chains. The model developed by Aota *et al.* (1992) is considered the OSE prototype. The final realization of all dynamic levels describing the interaction of ecological factors in the sea leads to the set of model units shown in Figure 5.13. This set is considered below.

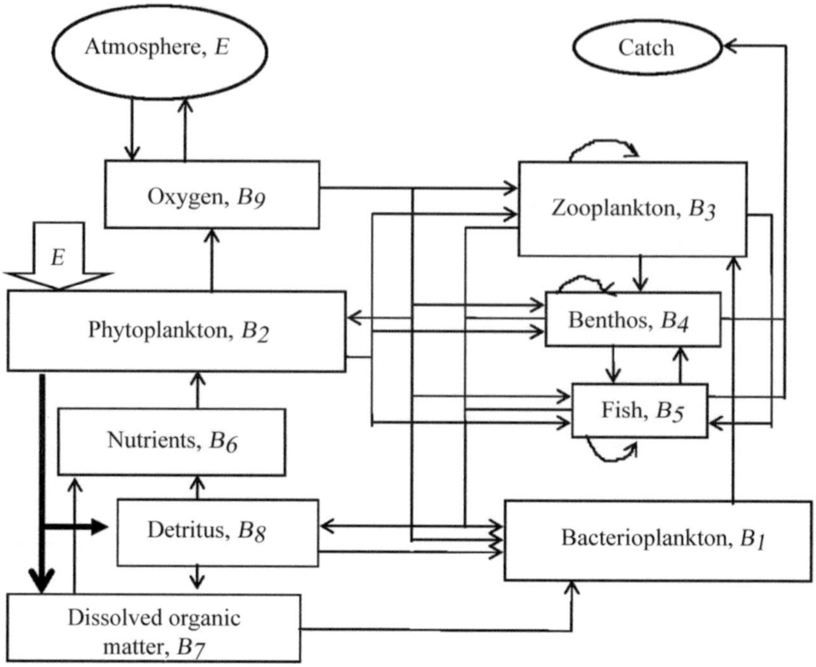


**Figure 5.13.** The trophic pyramid of the Okhotsk Sea ecosystem which forms the basis for the SMOSE.

### 5.2.2 Block diagram and principal structure of the SMOSE

Let us designate the Okhotsk Sea oceanic environment as $\Omega = \{(\varphi, \lambda)\}$ where $\varphi$ and $\lambda$ are latitude and longitude, respectively. The spatial inhomogeneity of the model of the Okhotsk Sea basin is provided by representing $\Omega$ as spatially discrete on the set of cells $\Omega_{ij}$ with latitude and longitude steps of $\Delta\varphi_i$ and $\Delta\lambda_j$, respectively. Each $\Omega_{ij}$ occupies a square $\sigma_{ij} = \Delta\varphi_i\Delta\lambda_j$. According to this, the Okhotsk Sea is considered to consist of $N = i_{\max} j_{\max}$ water bodies $\Xi_m$ $(m = j + (i-1)j_{\max})$. These cells are the basic spatial structure of $\Omega$ for the purpose of developing computer algorithms. Seawater flows between each $\Xi_m$ water body (Kawasaki and Kono, 1993). Each $\Xi_m$ has a vertical structure discretized in depth $z$ by steps $\Delta z_m$. Every now and then, step $\Delta t$ encounters a vertical structure fixed on the basis of remote information about the sea surface state and temperature. The vertical structure scheme developed by Legendre and Krapivin (1992) is taken into consideration.

Cells $\Omega_{ij}$ have heterogeneous parameters and functions. There is a set of cells that are adjacent to river mouths $(\Omega_R)$ and to ports $(\Omega_P)$, bordering on the land $(\Omega_\Gamma)$, in the Kuril–Kamchatka straits $(\Omega_B)$, and on the boundary with the Sea of Japan (Tartar and Soya straits, $\Omega_N$). The distribution of depths is given by the matrix $\|h_{ij}\|$ where $h_{ij} = h(\varphi_i, \lambda_j)$, $(\varphi_i, \lambda_j) \in \Omega_{ij}$. As a result, the full water volume

**Table 5.12.** Vertical structure of the OSE aquatic system.

| *Layer and its identifier* | *Parameters* | | | | | |
|---|---|---|---|---|---|---|
| | *Layer thickness* | *Temperature* | *Irradiance* | *Turbulence coefficient* | *Relaxation coefficient* | *Reflection coefficient* |
| Surface, $\Omega$ | | $T_0$ | $E_0$ | | | |
| Snow, $S$ | $g$ | $T_g$ | $E_g$ | 0 | $\alpha_g$ | $\beta_g$ |
| Floating ice, $I_1$ | $r$ | $T_r$ | $E_r$ | 0 | $\alpha_r$ | $\beta_r$ |
| Sinking ice, $I_2$ | $f$ | $T_f$ | $E_f$ | 0 | $\alpha_f$ | $\beta_f$ |
| Water, $W$ | $z-f$ | $T_w$ | $E_w$ | $\Delta$ | $\alpha_w$ | $\beta_w$ |

of $\Omega$ is divided into volumetric compartments $\Xi_{ijk} = \{(\varphi, \lambda, z)/\varphi_i \leq \varphi \leq \varphi_{i+1}; \lambda_j \leq \lambda \leq \lambda_{j+1}; z_k \leq z \leq z_{k+1})\}$ with volume equal to $\sigma_{ijk} = \Delta\varphi_i\Delta\lambda_j\Delta z_k$. Within compartment $\Xi_{ijk}$, the water body is considered as having a homogeneous structure. The water temperature, salinity, density, and biomass of $\Xi_{ijk}$ are described by points at a given location in the models. The four seasons are used to represent the effect of anthropogenic processes on the oceanic environment $\Omega$: $\tau_w$ winter, $\tau_s$ spring, $\tau_u$ summer, $\tau_a$ fall.

Simulation procedure realization is defined by the structure in which the OSE environment is discretized. When $\Delta z_{ij} = h_{ij}(i = 1, \ldots, i_{\max}; j = 1, \ldots, j_{\max})$, vertical mixing processes inside $\Xi_m$ are identified as having model parameters whose vertical distribution is uniform. In the case of $\Delta z_{ij} < h_{ij}$, compartment $\Xi_m$ is considered a vertical structure with such characteristics.

Figure 5.13 is a conceptual diagram identifying the components of the SMOSE. The dynamics of the SMOSE is supported in turn by a GMNSS. SMOSE input data are from pollutant sources indicated in the near-shore Okhotsk Sea and sea ice areas, as shown on current maps. The set of SMOSE components is divided into three types of information sources: mathematical models of natural, ecological, and hydrophysical processes, service software, and the scenario generator.

The OSE functions under climatic conditions when the greater part of the sea surface is covered with ice for several months. Therefore, the vertical structure is represented by a three-layer block diagram (Table 5.12). The trophic chains and energy fluxes of the OSE shown in Figure 5.13 are completed with specific channels (Figure 5.14).

### 5.2.3 Marine biota block

The source of energy for the entire system is solar radiation $E(t, \varphi, \lambda, z)$, the intensity of which depends on time $t$, latitude $\varphi$, longitude $\lambda$, and depth $z$. An

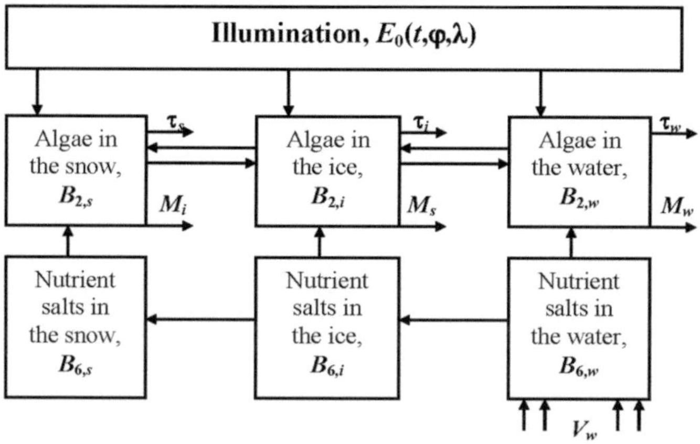


**Figure 5.14.** The conceptual structure of the model of phytoplankton biomass under the climatic conditions that prevail in the Okhotsk Sea region. The source of external energy is solar radiation $E(\varphi, \lambda, z, t)$, whose intensity changes with latitude $\varphi$, longitude $\lambda$, sea depth $z$, and time $t$.

equation that can describe the biomass dynamics of living elements is:

$$\frac{\partial B_i}{\partial t} + \xi_i\left[V_\varphi \frac{\partial B_i}{\partial \varphi} + V_\lambda \frac{\partial B_i}{\partial \lambda} + V_z \frac{\partial B_i}{\partial z}\right] = R_i - T_i - M_i - H_i - \sum_{j\in\Gamma_i} C_{ij}R_j$$
$$+ \xi_i\left[\frac{\partial}{\partial\varphi}\left(\Delta_\varphi \frac{\partial B_i}{\partial\varphi}\right) + \frac{\partial}{\partial\lambda}\left(\Delta_\lambda \frac{\partial B_i}{\partial\lambda}\right)\right.$$
$$\left. + \frac{\partial}{\partial z}\left(\Delta_z \frac{\partial B_i}{\partial z}\right) + \beta_V \frac{\partial B_i}{\partial z}\right]$$
$$(i = 1, \ldots, 5), \quad (5.4)$$

where $V(V_\varphi, V_\lambda, V_z)$ are components of the current velocity; $R_i$ is production; $T_i$ is the rate of exchange with the environment; $M_i$ is mortality; $H_i$ is non-assimilated food; and $\Gamma_i$ is the set of trophic subordinates to the $i$th component: $C_{ji} = k_{ji}F_i \Big/ \sum_{m\in S_i} k_{jm}F_m$; $S_i$ is the food spectrum of the $j$th component; $k_{jm}$ is the satisfaction index of nutritive requirements of the $j$th component at the expense of the biomass of the $m$th component biomass; $F_i = \max\{0, B_i - B_{i,\min}\}$; $B_{i,\min}$ is the minimal biomass of the $i$th component consumed by other trophic levels; $\Delta(\Delta_\varphi, \Delta_\lambda, \Delta_z)$ are components of the turbulent mixing coefficient (assuming vertical mixing isotrophism in the horizontal plane $\Delta_\varphi = \Delta_\lambda = v_H$); and $\beta_V$ is upwelling velocity. Functions $R_i$, $M_i$, $H_i$, and $T_i$ are parameterized according to the models of Krapivin (1996) and Legendre and Krapivin (1992). Equations describing the dynamics of abiotic elements are represented in conformity with Berdnikov *et al.* (1989). Functions $M_4$ and $M_5$ include biomass losses at the expense of fishing.

Parameter $\xi_i$ characterizes the way in which the $i$th component is subjected to the current. It is supposed that $\xi_i = 1$ for $i = 1, 2, 3$ and $\xi_i = 0$ for $i = 4, 5$.

Inert components are described by the following equations (Krapivin, 1996):

$$\frac{\partial B_6}{\partial t} + V_\varphi \frac{\partial B_6}{\partial \varphi} + V_\lambda \frac{\partial B_6}{\partial \lambda} + V_z \frac{\partial B_6}{\partial \lambda} = \delta_n \rho_* B_8 - \delta_1 R_2 + \beta_V \frac{\partial B_6}{\partial z} + \rho_1 \sum_{i=1}^{5} T_i + \frac{\partial}{\partial \varphi}\left(\Delta_\varphi \frac{\partial B_6}{\partial \varphi}\right) + \frac{\partial}{\partial \lambda}\left(\Delta_\lambda \frac{\partial B_6}{\partial \lambda}\right) + \frac{\partial}{\partial z}\left(\Delta_z \frac{\partial B_6}{\partial z}\right), \quad (5.5)$$

$$\frac{\partial B_7}{\partial t} + V_\varphi \frac{\partial B_7}{\partial \varphi} + V_\lambda \frac{\partial B_7}{\partial \lambda} + V_z \frac{\partial B_7}{\partial \lambda} = k_{7,0} R_2 + \beta_V \frac{\partial B_7}{\partial z} - k_{1,7} R_1 B_7 / P_1 + \frac{\partial}{\partial \varphi}\left(\Delta_\varphi \frac{\partial B_7}{\partial \varphi}\right) + \frac{\partial}{\partial \lambda}\left(\Delta_\lambda \frac{\partial B_7}{\partial \lambda}\right) + \frac{\partial}{\partial z}\left(\Delta_z \frac{\partial B_7}{\partial z}\right), \quad (5.6)$$

$$\begin{aligned} \frac{\partial B_8}{\partial t} + V_\varphi \frac{\partial B_8}{\partial \varphi} + V_\lambda \frac{\partial B_8}{\partial \lambda} + V_z \frac{\partial B_8}{\partial \lambda} &= \sum_{i=1}^{5}(M_i + H_i) - \rho_* B_8 - (v_* - \beta_V)\frac{\partial B_8}{\partial z} \\ &\quad - (k_{1,8} R_1/P_1 + k_{3,8} R_3/P_3 + k_{4,8} R_4/P_4 \\ &\quad + k_{5,8} R_5/P_5) B_{8,\min} + \frac{\partial}{\partial \varphi}\left(\Delta_\varphi \frac{\partial B_8}{\partial \varphi}\right) \\ &\quad + \frac{\partial}{\partial \lambda}\left(\Delta_\lambda \frac{\partial B_8}{\partial \lambda}\right) + \frac{\partial}{\partial z}\left(\Delta_z \frac{\partial B_8}{\partial z}\right), \end{aligned} \quad (5.7)$$

$$\begin{aligned} \frac{\partial B_9}{\partial t} + V_\varphi \frac{\partial B_9}{\partial \varphi} + V_\lambda \frac{\partial B_9}{\partial \lambda} + V_z \frac{\partial B_9}{\partial \lambda} &= \delta_2 R_2 - \sum_{i=1}^{5} \zeta_i T_i + \beta_V \frac{\partial B_9}{\partial z} \\ &\quad + \max\{0, (z_* - z)/|z - z_*|\} W_* - \delta_* \rho_* B_8 \\ &\quad + \frac{\partial}{\partial \varphi}\left(\Delta_\varphi \frac{\partial B_9}{\partial \varphi}\right) + \frac{\partial}{\partial \lambda}\left(\Delta_\lambda \frac{\partial B_9}{\partial \lambda}\right) \\ &\quad + \frac{\partial}{\partial z}\left(\Delta_z \frac{\partial B_9}{\partial z}\right), \end{aligned} \quad (5.8)$$

where $P_i = \sum_{j \in S_i} k_{i,j} B_{j,\min}$; $\rho_*$ is the speed at which detritus decomposes per unit biomass; $v_*$ is the velocity of gravity settling; $\rho_1$ are those exchange biomass losses that transform into nutrients (Prieur and Legendre 1988; Zenkevich, 1963); $\delta_1$ is nutrient assimilation velocity as a result of photosynthesis per unit phytoplankton production; $\delta_2$ is oxygen production per unit phytoplankton production; $\zeta_i$ $(i = 1, \dots, 5)$ and $\delta_*$ are oxygen losses at the expense of respiration and detritus decomposition, respectively; $z_*$ is the maximal depth of oxygen exchange of seawater with the atmosphere; $W_*$ is the difference between invasion and evasion processes; and $k_{ij}$ is the $i$th element's elective coefficient with respect to the $j$th element of the ecosystem.

Equations (5.4)–(5.8) are used for the complete volume of the OSE only when $(\varphi, \lambda, z) \in W$. In other cases (i.e., the layers of ice or snow) these equations are automatically reduced in accordance with the scheme presented in Figure 5.13.

The ice–water ergocline plays an important role in the biological productivity of the OSE. According to hypotheses of Legendre and Demers (1985) and Demers *et al.* (1986), energetic ergoclines are the preferential sites for biological production under Arctic conditions. Primary production in the food chains of Arctic ecosystems is determined by phytoplankton productivity. This is connected with complex variations in the meteorological, hydrodynamic, and energetic parameters of the sea environmment. Parameterizing phytoplankton production in northern seas was studied by Prieur and Legendre (1988). A vertical structure for the composition of seasonal conditions affecting primary production in $\Omega$ is given in Table 5.12.. This scheme is applied for each of the $\Omega_{ij} \subseteq \Omega$.

The climate block of the GMNSS calculates $T_w$ by averaging the temperatures of mixed water volumes. Moreover, the following correlations are realized (Kochetov, 1985):

$$T_g = T_r = Tf = \begin{cases} -0.024(g+r+f) + 0.76T_0 + 8.38 & \text{when } g+r+f \leq 50\text{ cm}, \\ -0.042(g+r+f) + 0.391T_0 - 0.549 & \text{when } g+r+f > 50\text{ cm}. \end{cases}$$

Energy input is provided by solar radiation $E_0(t, \varphi, \lambda)$ and the upflow of nutrients from deep-sea layers. The concentration of nutrients $B_{8,A}(t, \varphi, \lambda, z)$ at depth $z$ is determined by photosynthesis $R_{2,A}$, advection, and decomposition of suspended dead organic matter $B_{7,A}$. As a basic scheme for water circulation, that proposed by Vinogradov *et al.* (1972) is accepted by Demers *et al.* (1986) as correct under OSE conditions. The role played by hydrodynamic conditions is in ensuring the necessary supply of nutrients for photosynthesis by transporting them from other layers or aquatories of the sea where the concentration of chemical elements is sufficiently high. Taking into account the designations of Table 5.12 we have (Legendre and Krapivin, 1992):

$$E(t, \varphi, \lambda, z) = \begin{cases} E_0 & \text{when } z \leq -(g+r); \\ E_g = (1-\beta_g)E_0 \exp(-\alpha_g z) & \text{when } z \in [-(g+r), -r]; \\ E_r = (1-\beta_r)E_g(t, \varphi, \lambda, -r)\exp(-\alpha_r z) & \text{when } z \in [-r, 0]; \\ E_f = (1-\beta_f)E_r(t, \varphi, \lambda, 0)\exp(-\alpha_f z) & \text{when } z \in [0, f]; \\ E_w = (1-\beta_w)E_f(t, \varphi, \lambda, f)\exp(-\alpha_w z) & \text{when } z > f, \end{cases} \tag{5.9}$$

where the values of $\alpha_A$ $(A = g, r, f, w)$ depend on the optical properties of the $A$th medium. Irradiance $E_0$ is that which arrives at the surface of $\Omega$. $E_0$ can be estimated within the framework of the monitoring regime or calculated by the climatic model . The flow of $E_0$ is attenuated by snow, ice, and water according to the scheme of Table 5.12. In each cell $\Omega_{ij}$ the structure of these layers is changed to correspond with the time of year. Within each of the layers, the attenuation of irradiance with depth is described by an exponential model (5.9). Parameters $\alpha_A$

and $\beta_A$ are functions of salinity, turbidity, temperature, and biomass. The form of this dependence is given by standard functions or a scenario. Phytoplankton production $R_{2,A}$ is a function of solar radiation $E_A$, the concentration of nutrients $B_{6,A}$, temperature $T_A$, phytoplankton biomass $B_{2,A}$, and concentration of pollutants $\eta_A$. There are many models that can be used to describe photosynthesis (Legendre and Krapivin, 1992; Prieur and Legendre, 1988; Vinogradov *et al.*, 1972). For its description in the present study, an equation of Michaelis–Menten type is used:

$$R_{2,A} = a_A k_{1,A} B_{2,A,\max}/(E_A + k_{1,A}), \tag{5.10}$$

where $k_{1,A}$ is the irradiance level at which $R_{2,A} = R_{2,A,\max}$; and $B_{2,A,\max}$ is the maximum quantum yield (Prieur and Legendre, 1988). A coefficient $\alpha_A$ reflects the dependence of phytoplankton production on environmental temperature $T_A$ and nutrient concentration $B_6$. The SMOSE realizes the following equation to calculate $a_A$:

$$a_A = a_1 K_0(T_A, t)/[1 + B_{2,A}/(a_2 B_{6,A})], \tag{5.11}$$

where $a_1$ is the maximal velocity at which nutrients are absorbed by phytoplankton (day$^{-1}$); $a_2$ is the index of the velocity of saturation of photosynthesis; and

$$K_0(T_A, t) = a_3 \max\left\{0, \frac{T_c - T_A}{T_c - T_{\text{opt}}} \exp\left[1 - \frac{T_c - T_A}{T_c - T_{\text{opt}}}\right]\right\}, \tag{5.12}$$

where $a_3$ is the weight coefficient; and $T_c$ and $T_{\text{opt}}$ are the critical and optimal temperatures for photosynthesis (°C).

Kiefer and Mitchel (1983) showed that equation (5.10) adequately fits laboratory data. Relations (5.11) and (5.12) make the description of the phytoplankton production more accurate under critical environmental conditions when nutrient concentration and temperature have high fluctuations. The coefficients of these relations are defined on the basis of estimates given by Demers *et al.* (1986).

### 5.2.4 Hydrological block

Water circulation in the OSE is a complex system of cycles and currents that function on different scales (Belkin and Carnillon, 2004; Moroshkin, 1964; Wakatsuchi and Martin, 1991). Water dynamics in the $\Omega$ can be represented by flows between compartments $\Xi_{ijk}$. The directions of water exchanges at every level $z_k = z_0 + (k-1)\Delta z_k$ follow Aota *et al.* (1992). The external boundary of $\Omega$ is determined by the coastline, the sea bottom, the Kuril straits, the boundary with the Sea of Japan and the water–atmosphere boundary.

Hydrological regime data can be synthesized as a four-level structure functioning on a seasonal time scale. The velocity of currents in the Kuril straits is estimated by the following binary function:

$$V(t) = \begin{cases} V_1 & \text{for } t \in \tau_u \cup \tau_a, \\ V_2 & \text{for } t \in \tau_w \cup \tau_s. \end{cases} \tag{5.13}$$

Water exchange through the border with the Sea of Japan is $V_3$. Water temperature $T^W_{ijk}$ in $\Xi_{ijk}$ is a function of evaporation, precipitation, river flow, and inflows of water from the Pacific Ocean (Kawasaki and Kono, 1993). How it changes with time in $\Xi_{ijk}$ is described by the heat balance equation:

$$\zeta C\sigma_{ijk}\frac{\partial\Gamma^W_{ijk}}{\partial t} = \sum_{s,l,m}(W^{ijk}_{slm} + f^{ijk}_{slm}) - W_{ijk}, \tag{5.14}$$

where $\zeta$ is seawater density (g/cm$^3$); $C$ is the thermic capacity of water (cal $\cdot$ g$^{-1}$ grad$^{-1}$); $\sigma_{ijk}$ is the volume of $\Xi_{ijk}$; $W^{ijk}_{slm}$ is heat inflow to $\Xi_{ijk}$ from $\Xi_{slm}$; $f^{ijk}_{slm}$ is heat exchange between $\Xi_{slm}$ and $\Xi_{ijk}$ caused by turbulent mixing; and $W_{ijk}$ is total heat outflow from $\Xi_{ijk}$. Equation (5.14) can be used for layers $k > 1$. Temperature distribution in the $k = 1$ layer is calculated from monitoring data (Itoh *et al.*, 2003; Shinohara and Shikama, 1988).

The dissipation of moving kinetic energy, geothermic flow on the ocean bed, the heat effects of chemical processes in the ocean ecosystem, and freezing and melting of ice are not considered global determinants of water temperature fields. The SMOSE does not consider these effects.

The dynamics of water salinity $s(t, \varphi, \lambda, z)$ can be described by the balance equation of Berdnikov *et al.* (1989), which supposes that $s(t, \varphi, \lambda, z) = s_0$ for $z > 100$ m. Scenarios of river flow, ice fields, and synoptic situations can be set up by the user of the SMOSE. Snow layer thickness $g(t, \varphi, \lambda)$ can be described via statistical data with given dispersion characteristics: $g = g_1 + g_0$ where the value $g_1$ is defined as the mean characteristic for the chosen time interval and the function $g_0(t, \varphi, \lambda)$ shows how $g$ varies for a given time interval.

Alternatively, the dynamics of the snow layer process can be parameterized within the framework of an atmospheric process simulation algorithm that relates the thickness of the growth and melting of the snow layer to temperature and precipitation (Sellers, 1983):

$$g(t + \Delta t, \varphi, \lambda) = g(t, \varphi, \lambda) + S_F - S_M,$$

where $S_F$ is the proportion of snow precipitated at temperatures close to freezing ($265\,\text{K} \le T_0 \le 275\,\text{K}$); and $S_M$ is snow ablation (i.e., evaporation + melting).

Initially, the water space occupies the semi-space $z \ge f$. If we designate the density of snow, ice, and water by $\rho_g$, $\rho_\gamma$, and $\rho_w$, respectively, we obtain:

$$f = (g\rho_g + rg_r)/\rho_w.$$

The process of ice formation at the water surface is described by the Stephan equation:

$$\frac{\partial T(t, \varphi, \lambda, z)}{\partial t} = a^2(z)\frac{\partial^2 T(t, \varphi, \lambda, z)}{\partial z^2} \tag{5.15}$$

$$\lambda_1 \frac{\partial T(t, \varphi, \lambda, z)}{\partial z}\bigg|_{z=f} = q\rho_r \frac{\partial r(t, \varphi, \lambda)}{\partial t} + \lambda_2 \frac{\partial T(t, \varphi, \lambda, z)}{\partial z}\bigg|_{z=f} \tag{5.16}$$

with initial and boundary conditions:

$$r(0,\varphi,\lambda) = \psi_1(\varphi,\lambda), \quad T(0,\varphi,\lambda,z) = \psi_0(\varphi,\lambda,z),$$
$$T(t,\varphi,\lambda,-r) = f_1(t,\varphi,\lambda) \leq 0^\circ\text{C}, \quad T(t,\varphi,\lambda,f) = 0^\circ\text{C},$$
$$T(t,\varphi,\lambda,h) = Y(t,\varphi,\lambda),$$

where $f_1$ is a function representing how near-land air temperature changes in the absence of snow or snow temperature specified in a scenario or calculated from a climatic model; $Y$ is the temperature of deep-water layers; and $a^2(z)$ is the thermal conductivity coefficient of the medium:

$$a^2(z) = \begin{cases} a_1^2(z) & \text{when } -r \leq z \leq f, \\ a_2^2(z) & \text{when } z > f. \end{cases}$$

As alternative process for ice thickness formation is given by the empirical formula proposed by Truskov *et al.* (1992):

$$r + f = 2.1(t_\Delta)^{0.47} \quad \text{(cm)},$$

where $t_\Delta$ is the accumulated number of degree days of frost ($T_0 \geq -7^\circ$C).

Many models describe the process of ice and snow thickness formation in an efficient way. For example, the model developed by Ivanov and Makshtas (1990) describes various situations that pervade all meteorological and thermodynamic correlations. Nevertheless, the use of such a model is impossible in the SMOSE structure. A more detailed parameterization of climate processes is necessary before this model can be used as a SMOSE block.

The most complex element to reconstruct in the SMOSE is mathematical description of fish migration. Lyapunov (1971) proposed identifying the migration process with inverse turbulent diffusion by coefficient $\Delta^* > \Delta$. Krapivin (1978) incorporated this in an algorithm, but found estimation of $\Delta^*$ to be fraught with difficulties. Radakov (1972) noticed fish migration is accompanied by the external appearance of purposeful behavior. Therefore, let us formulate a law of migration that follows from the general biological principle of adaptation and long-term adaptability: fish migration is subordinate to the principle of complex maximization of effective nutritive ration $P_5$, given the preservation of favorable temperature, salinity, and pollution conditions. In other words, migrating species travel at characteristic speed toward the maximum gradient of effective food, subject to environmental restrictions (Brandhorst, 1966).

### 5.2.5 Simulation procedure and experiments

A complete system of equations of ecosystem dynamics can be formed from the set of traditional balance, hydrodynamical, and biogeochemical equations (5.4)–(5.16). These equations have a set of coefficients $X = X_1 \cup X_2 \cup X_3$, where subset $X_1$ contains coefficients whose values are determined with high precision, subset $X_2$ consists of coefficients determined by low precision, and subset $X_3 = X \setminus (X_2 \cup X_1)$ has coefficients that have not been estimated. The simulation procedure foresees a two-tier process of model adaptation. The first tier of the model adaptation process is realized at the expense of determining coefficients of subset $X_3$. This process consists in varying coefficient values within ranges fixed by the SMOSE user. The quality criterion is also formulated by the user; this proceeds from existing experimental data or from other considerations. The service block of the SMOSE has a set of such possibilities. The user can demand any divergence between his data about some parameter of the OSE and its model estimation to be minimized. A spectrum of such parameters spreads throughout many individual and integral elements of the OSE.

The second tier of the model adaptation process involves making subset $X_2$ more precise. The quality criterion of this process is based on improving the model quality reached during the first stage of the adaptation process. Thus, the simulation procedure of the SMOSE is a continuous process of model adaptation via coefficient changes or trophic graph modification. A set of models $\{A_i\}$ is formed by sorting out model parameters. Every model $A_i$ is characterized by a quality level $Q_i$. A complete model $A^*$ is formed as a limit to the set of these models. The adaptation procedure ends when the user's definition of quality criterion $Q^*$ is met, which depends on the values $|Q^* - Q_i|$. Model $A^*$ ensures the $\min_i|Q^* - Q_i|$. In this chapter it is believed that

$$Q_i = \frac{1}{h_1 \sigma} \int_{(\varphi,\lambda,z) \in \Omega} B_2(t, \varphi, \lambda, z) \, d\varphi \, d\lambda \, dz.$$

The value of $Q^* (= 20 \text{ g/m}^3)$ is estimated at time $t^*$ (middle of summer) as the average concentration of the biomass of phytoplankton in the upper layer, which has a depth of $h_1 (= 100 \text{ m})$.

Simulations of the OSE can be used to study various aspects of it (e.g., how it functions). Some simulation results from the values of OSE parameters given in Table 5.13 are given in what follows. Specifically, the present status of the OSE is of great interest in understanding its susceptibility to anthropogenic influence. According to the common theory of complex systems, a dynamic system is in a living state at the time interval $[t_0, t_1]$ if the biomass of its elements is within limits $B_i \geq B_{i,\min}$. Correlations between trophic levels and nonliving elements give summary conditions for survivability criteria: $B_{\min} \leq \sum B_i$. Finally, survivability criteria for the OSE can be written in the form:

$$J(t) \geq \alpha J(t_0),$$

where $J = U(t)/U(t_0)$; $U(t) = \sum_{k=1}^{9} \int_{(\varphi,\lambda,z)} B_k(t, \varphi, \lambda, x) \, dx$; $t_0$ is the instant of time when the value of function $A(t)$ is considered to be known; and $\alpha < 1$ is the level of survivability. We consider the OSE to be in a living state if the condition $J(t) > \alpha J(t_0)$ applies to $t > t_0$. Calculations of $J(t)$ for $t > t_0$ demonstrate how the OSE reacts to various environmental conditions. For example, fluctuations in the oxygen saturation of water may be of anthropogenic origin (oil pollution, large discharge of sewage waters, etc.). $J(t)$ shows that the OSE displays a high degree

**Table 5.13.** Values of some parameters in the framework of simulation experiments using the SMOSE.

| *Parameter* | *Value* |
|---|---|
| Spatial digitization by<br>latitude, $\Delta\varphi$<br>longitude, $\Delta\lambda$<br>depth, $\Delta z$<br>$z \leq 100$ m<br>$z > 100$ m | <br>2°<br>2°<br><br>10 m<br>$h - 100$ m |
| Ice thermal conductivity coefficient, $\lambda_1$ | 2.21 W m$^{-1}$ deg$^{-1}$ |
| Water thermal conductivity coefficient, $\lambda_2$ | 0.551 W m$^{-1}$ deg$^{-1}$ |
| Coefficient of vertical turbulence, $\Delta_z$ | $10^2$ m$^2$ s$^{-1}$ |
| Ice-thawing specific heat, $q$ | 334 kJ · kg$^{-1}$ |
| Velocity coefficient of detritus decomposition, $\delta_*$<br>In the snow and ice layers<br>In the water layer | <br>0<br>0.1 |
| Velocity of the water current in the Kuril–Kamchatka straits, $V_i$<br>$i = 1$<br>$i = 2$ | <br>0.25 m · s$^{-1}$<br>0.15 m · s$^{-1}$ |
| Velocity of water current between the Okhotsk Sea and the Sea of Japan, $V_3$ | 0.3 m · s$^{-1}$ |
| Water thermal capacity, $C$ | 4.18 kJ · kg$^{-1}$ K$^{-1}$ |
| Water salinity for $z > 100$ m, $s_0$ | 34.95‰ |
| Area of the Okhotsk Sea, $\sigma$ | 1,583,000 km$^2$ |
| Volume of the Okhotsk Sea waters, $W$ | 1,365,000 km$^2$ |
| Average depth, $h_*$ | 859 m |
| Critical temperature for photosynthesis, $T_c$ | −0.5°C |

of stability with respect to the initial saturation of water with oxygen, it then shows that the OSE proceeds rapidly to a quasi-stationary regime of functioning at $B_9(t_0, \varphi, \lambda, z) \geq 1.8$ mL/L, and finally the length of time it takes the OSE to overcome the initial shortage of oxygen in the case of $B_9(t_0, \varphi, \lambda, z) = 1.1$ mL/L. For $B_9(t_0, \varphi, \lambda, z) \leq 0.8$ mL/L, the OSE is unable to proceed to a stable regime of functioning. The OSE is observed to be more sensitive to dynamic effects when water is saturated with oxygen. A reduction in oxygen production by 12% does not influence OSE dynamics. However, the OSE does not survive when oxygen production is decreased by 20%.

Calculating $J(t)$ makes it possible to detect zones whose survivability in $\Omega$ is at risk. For instance, a particularly dangerous anthropogenic influence is change in nutrient concentration. Simulation experiments enable us to determine variations in the vertical velocity of water within a given range.

The turbulent escape of nutrients into layers overlying the maximal depth of photosynthetic layer ($z \leq h_*$) can be estimated by the SMOSE by assuming the vertical velocity of water equal to $10^{-3}$ cm/s. Data point to the fact that, on average, the integrated pattern of the distribution of community elements is not subject to any significant variations within the velocity range from $3.5 \cdot 10^{-4}$ to $10^{-2}$ and even $10^{-1}$ cm/s, but is observed to be drastically distorted under a higher and, what is more important, under a lower ($< 10^{-4}$ cm/s) vertical advection of water.

Table 5.14 gives results from varying the concentration of nutrients at the initial moment of time $t_0$. Comparison with the results in this chapter reveals that variation of the nutrient concentration within a wide range at moment $t = t_0$ has

**Table 5.14.** Simulation results in which various nutrient concentrations are used. Spatiotemporal mean results are given. Initial data for the phytoplankton biomass is $B_2(t_0, \varphi, \lambda, z) = 2\ \text{mg} \cdot \text{m}^{-3}$.

| *Depth* (m) | *Phytoplankton biomass* $(\text{g} \cdot \text{m}^{-3})$ | | | | | |
|---|---|---|---|---|---|---|
| | $B_6(t_0, \varphi, \lambda, z) = 0.05\ \text{mg} \cdot \text{m}^{-3}$ | | | $B_6(t_0, \varphi, \lambda, z) = 3\ \text{mg} \cdot \text{m}^{-3}$ | | |
| | $t - t_0$ (days) | | | | | |
| | *10* | *30* | *50* | *10* | *30* | *50* |
| 0 | 1.1 | 5.4 | 10.7 | 6.3 | 11.3 | 10.9 |
| 10 | 1.9 | 5.3 | 12.3 | 5.4 | 11.9 | 15.8 |
| 20 | 2.4 | 10.2 | 21.9 | 7.1 | 19.2 | 21.1 |
| 30 | 1.3 | 6.6 | 18.4 | 7.2 | 14.6 | 14.9 |
| 40 | 0.6 | 3.1 | 9.2 | 5.5 | 10.1 | 9.7 |
| 50 | 0.2 | 1.2 | 8.7 | 4.2 | 7.7 | 8.4 |
| 70 | 0.1 | 0.9 | 5.3 | 1.7 | 4.8 | 5.1 |
| 100 | 0 | 0.4 | 2.1 | 1.1 | 1.9 | 2.2 |
| 150 | 0 | 0.1 | 0.9 | 0.8 | 1.1 | 1.3 |
| 200 | 0 | 0.1 | 0.1 | 0.2 | 0.3 | 0.2 |

**Table 5.15.** Vertical distribution of some OSE elements as a result of a simulation experiment with the SMOSE. Initial data are given for $t_0$ = May 15; the OSE dynamics are calculated for $t_1$ = September 15. The symbol "B" means "bottom".

| *Depth* (mg) | *Biomass* | | | | | | | | | |
|---|---|---|---|---|---|---|---|---|---|---|
| | $B_1$ ($mg \cdot m^{-3}$) | | $B_2$ ($mg \cdot m^{-3}$) | | $B_3$ ($mg \cdot m^{-3}$) | | $B_4$ ($mg \cdot m^{-3}$) | | $B_9$ ($mg \cdot m^{-3}$) | |
| | $t_0$ | $t_1$ | $t_0$ | $t_1$ | $t_0$ | $t_1$ | $t_0$ | $t_1$ | $t_0$ | $t_1$ |
| 0 | 15 | 21 | 500 | 672 | 20 | 19 | 200 | 181 | 8 | 9.3 |
| 10 | 15 | 19 | 500 | 935 | 20 | 21 | 200 | 181 | 8 | 10.1 |
| 30 | 15 | 12 | 500 | 59 | 20 | 31 | 200 | 181 | 8 | 11.4 |
| 50 | 0 | 3 | 0 | 10 | 0 | 15 | 200 | 181 | 8 | 9.9 |
| 100 | 0 | 2 | 0 | 1 | 0 | 2 | 200 | 181 | 8 | 8.6 |
| 500 | 0 | 0 | 0 | 0 | 0 | 0 | 200 | 181 | 8 | 8.4 |
| B | 0 | 0 | 0 | 0 | 0 | 0 | 200 | 181 | 8 | 7.5 |

little effect on the behavior of the OSE at moments $t \gg t_0$. The system, so to speak, *heals* with time the *blows* it has suffered and continues to function at the same level. When initial nutrient conditions are limited, the OSE reaches a stable state within just 50 days. If the nutrient concentration at time $t_0$ is unlimited, the OSE reaches a stable state on the 30th day of the simulation experiment.

The SMOSE can be used for many other simulation experiments. For example, Table 5.15 considers the vertical distribution of OSE elements as functions of initial values. The concentrations of all OSE elements have maximums in the area with the highest density gradient; these magnitudes have different values that change with time. Such results can be received for each point $(\varphi, \lambda) \in \Omega$. Comparison of these results with data of Terziev *et al.* (1993) shows that the SMOSE can be a useful instrument for OSE investigation. Other examples demonstrating model possibilities are given in Figures 5.15 and 5.16. Comparison of curves and estimations of benthos biomass by Zenkevich (1963) in Figure 5.15 indicate that the SMOSE unit describing the dynamics of the benthic comminity of the OSE does not reflect the benthic community's real distribution in space. There are differences between the spatial variations of the biomass of benthic animals estimated with the model and by Zenkevich (1963). These variations are explained by innacurate initial data and inaccurate values of SMOSE coefficients. Certainly, significant seasonal variations in mean estimations of SMOSE parameters do not approximate sufficiently accurately the time scale adopted here of four seasonal periods. Figure 5.16 shows a numerical result of the simulation procedure for ice cover reconstruction. This map shows that the SMOSE unit describing the ice formation process is accurate enough to be used to simulate the

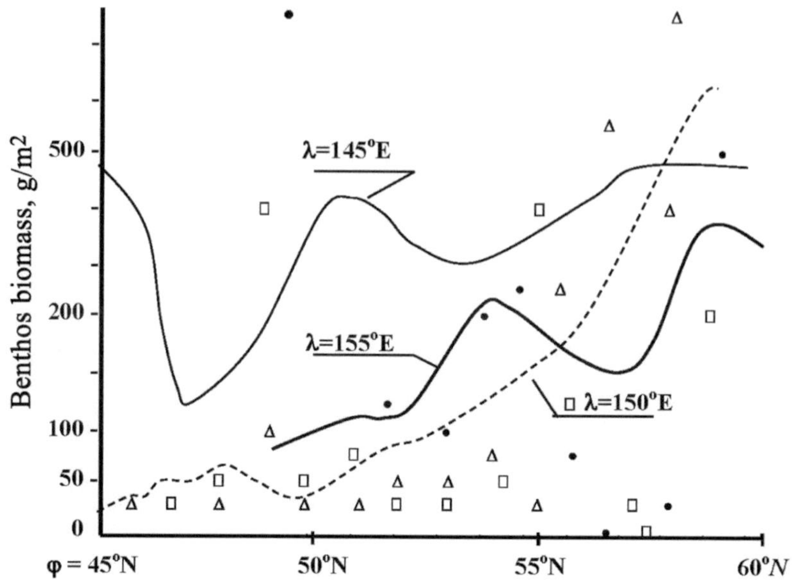


**Figure 5.15.** The spatial distribution of the biomass of benthos $D$ ($\mathrm{g\,m^{-2}}$) in the Okhotsk Sea. The symbol △ marks experimental or theoretical estimates published elsewhere over the last 5 years.

general behavior of the Okhotsk Sea ice fields observed by satellite measurements (Alfultis and Martin, 1987).

To demonstrate the possibility of using the SMOSE to study pollution processes, Figure 5.17 gives an example of oil carbohydrate dynamics as a function of different processes. The simulation experiment shows that 47% of oil carbohydrates ($Q_O$) are transformed to other forms in environments $g, r, f$ and 68% in water. The distribution of oil carbohydrates stabilizes 3.2 years after $t_0$. The hierarchy of flows $H_O^i$ ($i = 1, \ldots, 7$) can be estimated by the set: $H_O^2 > H_O^4 > H_O^1 > H_O^5 > H_O^6 > H_O^7$. This set changes for each OSE aquatory. The preponderance of oil carbohydrate neutralization processes is defined by seasonal conditions. Refining at the expense of oil carbohydrate evaporation ($H_O^2$) prevails over the other processes in summer. Oil carbohydrates that evaporate from the OSE surface partly return to the Ω with atmospheric precipitation. Oil carbohydrates are neutralized at the maximum rate of 0.0026 g/m²/day. This study indicates that the vegetative period affects the ecosystem's contribution to self-cleaning. In the case of the Ω, the OSE neutralizes about 25% of oil carbohydrates during the vegetative period. At other times this value oscillates near 3%.

### 5.2.6 Concluding remarks

Marine environmental research methods include mathematical models as the main unit because they make it possible to reconstruct spatial images of sea ecosystems

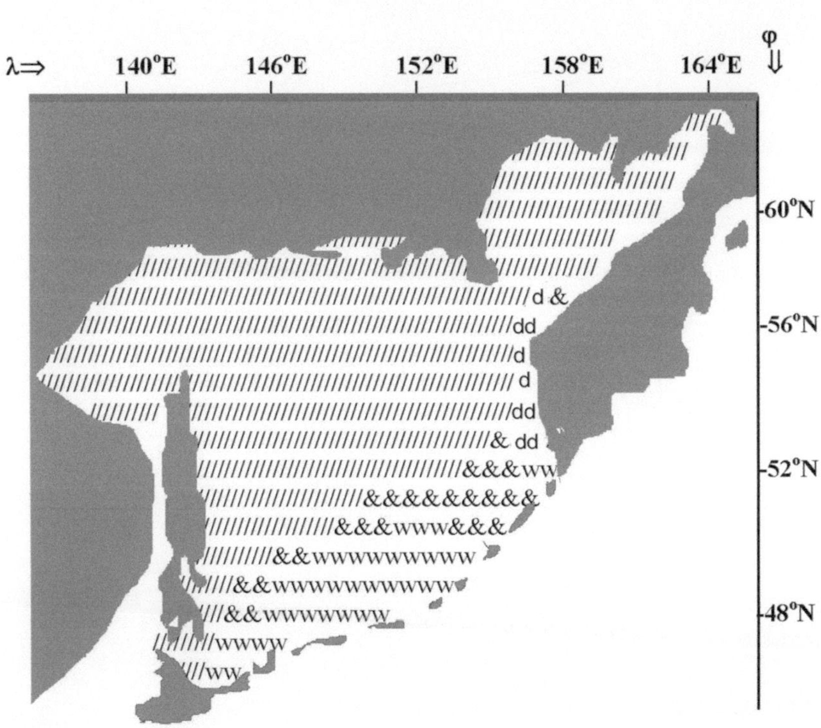


**Figure 5.16.** Forecast of the ice situation in the Okhotsk Sea for February 2004 from initial data available to the Japan Meteorological Service in November 2003. The difference between forecast and satellite monitoring data is marked with the symbol "d". Notation: / = stable ice; & = unstable ice (ice cover with polynyas); w = water surface.

on the basis of fragmentary information. The findings from this chapter show that the SMOSE can be considered such a unit to synthesize a monitoring system of the Okhotsk Sea environment. Clearly, the SMOSE is a simple bringing together of a set of parametrical descriptions of OSE functions. There are many problems for study in the future. We mention the following three which American, Japanese, and Russian scientists are close to solving (Aota *et al.*, 1991a, b, 1992, 1993; Kelley *et al.*, 1999a, b; Krapivin and Mkrtchyan, 2009).

The first problem relates to future modernization of the SMOSE by extending parametrical descriptions of the energy and heat exchange between the OSE and atmosphere to increase model validity and establish the dependence between model parameters and satellite measurements (Sellers *et al.*, 1995). The second problem of the SMOSE relates to greater accuracy by elaborating OSE elements in greater detail with a view to extending its significance for fishery. Better description of the biological balance of the OSE is a priority for future investigation, particularly with regard to fish farming. Finally, the third problem relates to improving the SMOSE by synthesizing an expert system at the center of which the SMOSE will be the main unit.

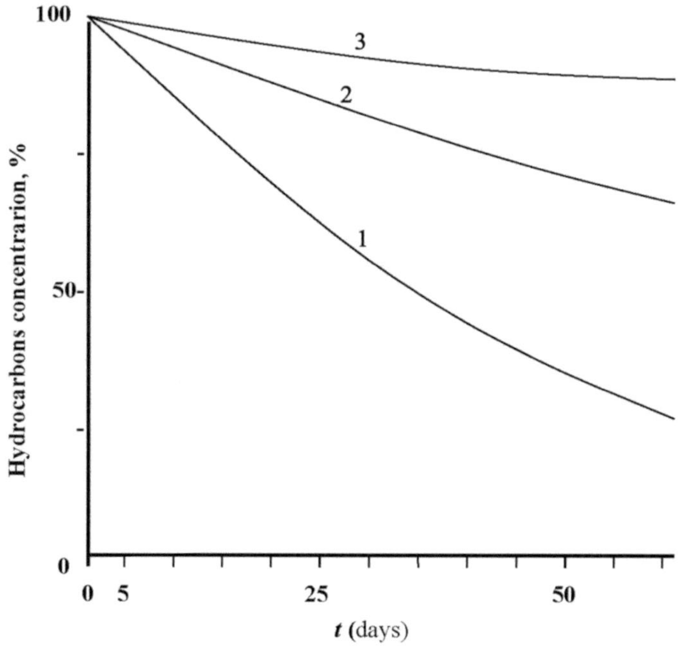


**Figure 5.17.** The OSE's self-cleaning process regarding oil hydrocarbons. It is supposed that $O(t_0, \varphi, \lambda, z) = 0.05\,\text{mg/m}^2$ when $z = 0$ and $O(t_0, \varphi, \lambda, z) = 0$ for $z > 0$; $H_O^1 = 0.1\,\text{mg} \cdot \text{m}^{-3}\,\text{day}^{-1}$; $H_O^2 = 0$; $H_O^3 = 0.01\,\text{mg} \cdot \text{m}^{-3}\,\text{day}^{-1}$; $H_O^4 = 0.02\,\text{mg} \cdot \text{m}^{-3}\,\text{day}^{-1}$; $H_O^5 = k_D(B_8)^{1/3}$ ($k_D$ is an absorption coefficient equal to zero for $(\varphi, \lambda, z) \in S \cup I_1 \cup I_2$ and to $0.005\,\text{day}^{-1}$ when $(\varphi, \lambda, z) \in W$); $H_O^6 = k_Z(B_3)^{1/4}$ ($k_Z$ is a bio-sedimentation coefficient equal to zero for $(\varphi, \lambda, z) \in S \cup I_1 \cup I_2$ and to $0.004\,\text{day}^{-1}$ when $(\varphi, \lambda, z) \in W$); $H_O^7 = k_B B_1$ ($k_B$ is a bacterial decomposition coefficient equal to $0.001\,\text{mg} \cdot \text{m}^{-3}\,\text{day}^{-1}$ for $(\varphi, \lambda, z) \in S \cup I_1 \cup I_2$ and to $0.005\,\text{mg} \cdot \text{m}^{-3}\,\text{day}^{-1}$ when $(\varphi, \lambda, z) \in W$). Three causes are considered: full spread of OSE self-cleaning ($t_0$ = May 15); 2, accumulation by ice only with porosity equal to $80\,\text{cm}^3/\text{kg}$ ($t_0$ = November 15); 3, decomposition by bacterioplankton only ($t_0$ = May 15).

# 6

# Decision-making procedures in the GIMS

## 6.1 BASIC DEFINITIONS

Environmental monitoring foresees decision-making situations in real time based on data accumulated up to the moment of decision making or as a result of prior data analysis without correlation to the current time. The statistical analysis of events involved in how the monitoring system works can be realized by many methods whose applicability at each event is determined by the combination of parameters that characterize the process studied. However, non-steady state and parametrical uncertainty call for new methods to be found capable of facilitating the decision using fragmentary time and space data. Such methods include sequential analysis and evolutionary technology (Bukatova *et al.*, 1991; Krapivin *et al.*, 1997a, b; Nitu *et al.*, 2004; Lai, 2001; Schaum *et al.*, 1990; Soldatov, 2010, 2011; Soldatov *et al.*, 2010).

### 6.1.1 Classical and sequential procedures

The classical decision-making procedure is based on the given volume, $n$, of measurements. The size of $n$ is determined by *a priori* information about the probability density function $f_a(x_1, \dots, x_n)$ where the random variables $\{x_i\}$ are the observation data. The hypotheses $H_0$ and $H_1$ are that a 0 or a 1 is prescribed, respectively. The distinction between these hypotheses is based on synthesizing the boundary of the optimal critical area, $E_1$, in the hyper-surface as:

$$L_n = L_n(x_1, \dots, x_n) = f_{a_1}(x_1, \dots, x_n)/f_{a_0}(x_1, \dots, x_n) = C, \tag{6.1}$$

where

$$f_a(x_1, \dots, x_n) = \prod_{i=1}^{n} f_a(x_i),$$

V.F. Krapivin and A.M. Shutko, *Information Technologies for Remote Monitoring of the Environment*, Springer Praxis Books, DOI 10.1007/978-3-642-20567-5_6, © Springer-Verlag Berlin Heidelberg 2012

$f_a(x)$ is a probability density function of variable $x$ with an unknown parameter $a$; $C$ is a constant determined by $E_1$ with a given level of error of the first kind $\alpha$.

The ratio of conditional probabilities in equation (6.1), called the likelihood ratio, provides the final choice between the above hypotheses:

(1) if $L_n \leq C$ then hypothesis $H_0$ is accepted;
(2) if $L_n > C$ then hypothesis $H_1$ is chosen.

In general, many criteria exist such as Bayesian minimum error, minimax, etc. Dalton and Dougherty (2011) derived a closed-form analytic representation of the Bayesian minimum mean square error estimator for linear classification by assuming Gaussian models. This is presented in a general framework whose structure includes covariance matrices and a very flexible class of prior parameter distributions with four free parameters. Closed-form solutions are provided for known, scaled identity, and arbitrary covariance matrices. Minimax theory developed from game theory (Krapivin, 1972; Krapivin and Klimov, 1995, 1997; Myerson, 1997).

The content of hypotheses $H_0$ and $H_1$ depends on the specific conditions of the task in hand. In reality, there are two steps to be followed when using these criteria. The first step involves synthesizing the empirical function of the probability distribution of the observed variable $x$. The second step is transforming it to $f(x)$ using, for example, the Neyman–Pearson criterion. The sequential procedure of the decision-making process does not differentiate between these steps.

Let us consider the case of a uniform dataset where the selected values $x_i$ $(i = 1, \ldots, n)$ are independent realizations of the same causal value $\xi$ having density $f_{a_0}(x)$ when hypothesis $H_0$ is false and density $f_{a_1}(x)$ when hypothesis $H_1$ is true. For this case, parameter $a$ of the real density $f_a$ cannot be equal to $a_0$ or $a_1$.

Errors of the first kind, $\alpha$, and the second kind, $\beta$, satisfy the following formulas:

$$\alpha \approx \exp[-0.5\{(E_{a_1}\xi - E_{a_0}\xi)(D_{a_0}\xi)^{-1/2}\}^2 n], \tag{6.2}$$

$$\beta \approx \exp[-0.5\{(E_{a_1}\xi - E_{a_0}\xi)(D_{a_1}\xi)^{-1/2}\}^2 n], \tag{6.3}$$

where

$$E_a\xi = \int_{-\infty}^{\infty} \ln[f_{a_1}(x)/f_{a_0}(x)] f_a(x)\, dx, \tag{6.4}$$

$$D_a\xi = \int_{-\infty}^{\infty} \{\ln[f_{a_1}(x)/f_{a_0}(x)]\}^2 f_a(x)\, dx - (E_a\xi)^2. \tag{6.5}$$

The sequential procedure has the following basic characteristic:

$$L(a) \approx [A^{h(a)} - 1]/[A^{h(a)} - B^{h(a)}], \tag{6.6}$$

where $A$ and $B$ are the boundaries of $L_n(x)$; and $h(a)$ is a solution to the equation:

$$\int_{-\infty}^{\infty} [f_{a_1}(x)/f_{a_0}(x)]^{h(a)} f_a(x)\, dx = 1. \tag{6.7}$$

The values of $A$ and $B$ have estimates:

$$B \approx \beta/(1-\alpha), \qquad A \approx (1-\beta)/\alpha. \tag{6.8}$$

Accordingly, $L(a_0) = 1 - \alpha$ and $L(a_1) = \beta$. From this it follows that the average number of observations in a sequential procedure equals:

$$\left.\begin{aligned} E_a\nu &= [(1-\alpha)\ln[\beta/(1-\alpha)] + \alpha\ln[(1-\beta)/\alpha]/E_{a_0}\xi, \quad \text{when } a = a_0, \\ E_a\nu &= [\beta\ln[\beta/(1-\alpha)] + (1-\beta)\ln[(1-\beta)/\alpha]/E_{a_1}\xi, \quad \text{when } a = a_1. \end{aligned}\right\} \tag{6.9}$$

For the value $a = a^*$ and when $E_{a^*}\xi = 0$ and $E_{a^*}\xi^2 > 0$ we have:

$$E_{a^*}\nu \approx [-\ln[\beta/(1-\alpha)]\ln[(1-\beta)/\alpha]/E_{a^*}\xi^2. \tag{6.10}$$

### 6.1.2 Universality of the sequential procedure

According to equations (6.9) and (6.10), the number of observations of a sequential procedure is a random variable, $\nu$, the average value of which ($E_a\nu$) can be smaller or larger than $n$. To judge the possible values of $\nu$, it is necessary to know the distribution $P(\nu = n) = P_a(n)$:

$$E_a\nu P_a(n) = W_c(y) = c^{1/2}y^{-3/2}(2\pi)^{-1/2}\exp[-0.5c(y + y^{-1} - 2)], \tag{6.11}$$

where

$$0 \le y \le n|E_a\xi| < \infty, \quad c = K|E_a\xi|/D_a\xi = (E_a\nu)^2/D_a\nu > 0,$$
$$D_a\nu - KD_a\zeta/(E_a\zeta)^3, \quad E_a\nu = K/E_a\xi,$$
$$K = \ln A \text{ for } E_a\xi > 0 \quad \text{and} \quad K = \ln B \text{ for } E_a\xi < 0.$$

The distribution function $W_c(y)$, according to (6.11), has the form:

$$W_c(x) = \int_0^x w_x(z)\,dz, \tag{6.12}$$

where $w_c(z) = (c/2\pi)^{1/2}z^{-3/2}\exp[-0.5c(z + z^{-1} - 2)]$, $c = [E(\nu)]^2[D(\nu)]^{-1}$.

The universality of the distribution (6.12) follows from its duality to the Gaussian distribution. As far back as 1960 Wald showed that if $|E_a\xi|$ and $D_a\xi$ are sufficiently small in comparison with $\ln A$ and $\ln B$ then the distribution of $\nu/E_a\nu$ defined by expression (6.11) will be a close approximation to reality even when $\xi$ is not distributed by the Gaussian law.

The theoretical aspects of the universality of the distribution $W_c(x)$ are important for full estimation of sequential procedure efficiency. However, these are not the principal aspects for experimental applications. This is the reason synthesizing the decision-making system is generally realized without taking these considerations into account. In reality, measurement volume is generally small and the central limit theorem does not work. The statistical difficulties that arise as a result can be overcome by evolutionary modeling (Bukatova *et al.*, 1991), intelligent technology (Nitu *et al.*, 2000b), and the use of other algorithms.

## 6.2 SCHEME OF THE DECISION-MAKING PROCEDURE USING SEQUENTIAL ANALYSIS

In contradiction to the Neyman–Pearson criterion (Yan and Blum, 2001), the sequential procedure does not separate the stages of measurement and data processing but rather alternates them. This is represented schematically in Figure 6.1. For comparison, the classical procedure is given in Figure 6.2. These figures show that the algorithmic load of the sequential procedure changes dynamically while at the same time the classical procedure realizes the data-processing stage only after the last step of the experiment is completed. Thus, synthesizing an efficient decision-making system (DMS) poses the following problems:

i. selecting the criterion necessary for parameter estimation;
ii. revealing the probabilistic characteristics of the process studied;

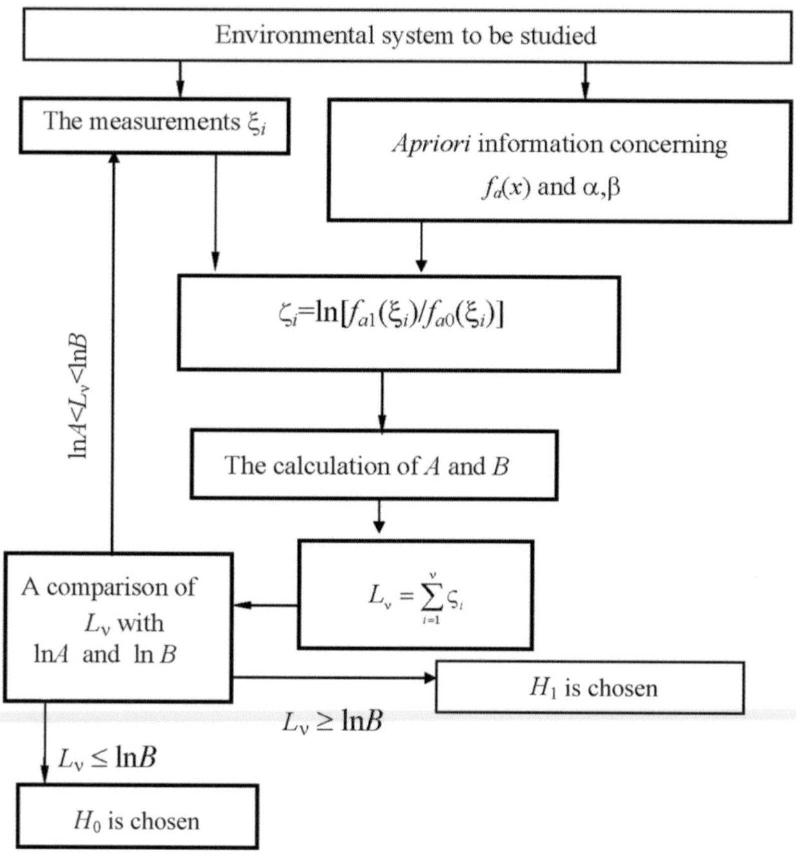


**Figure 6.1.** Scheme of the sequential analysis procedure using the sequential analysis procedure to decide between hypotheses $H_0$ and $H_1$.

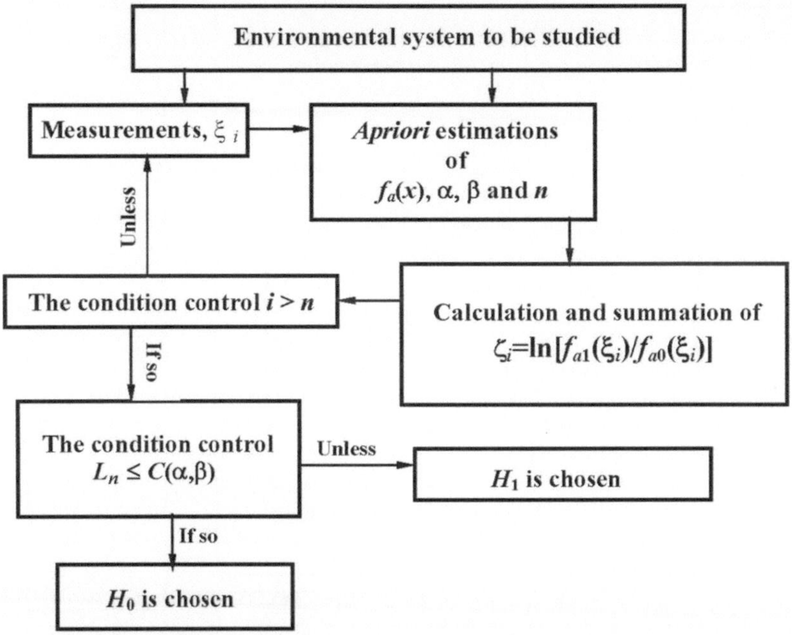


**Figure 6.2.** Classical Neyman–Pearson decision-making procedure to choose between the two hypotheses $H_0$ and $H_1$.

iii. *a priori* estimating possible losses in precision of the decisions taken;
iv. prognosis of the dynamic stability of experimental results.

A DMS should encompass a wide spectrum of functions:

i. visualizing measurement data as soft copy on a computer, discrete distribution, and statistical parameters;
ii. calculating statistical characteristics (mean, central second-order and third-order moments, asymmetry and excess coefficients, entropy, etc.);
iii. synthesizing empirical and theoretical distribution functions;
iv. valuing the parameters to be used in the Neyman–Pearson criterion and the sequential procedures for hypothesis decision;
v. ensuring user access to all the functions of the decision-making system.

According to the scheme represented in Figure 6.3 the decision-making system should have an expert control level. Namely, unit DMSP-I controls through its inputs and outputs the decision-making procedure. According to the functions of the subunits described in Table 6.1 the user can intervene promptly at any arbitrary stage of the computer experiment, correcting parameters of the decision-making procedure or ending the procedure completely. Subunit CTT manages the

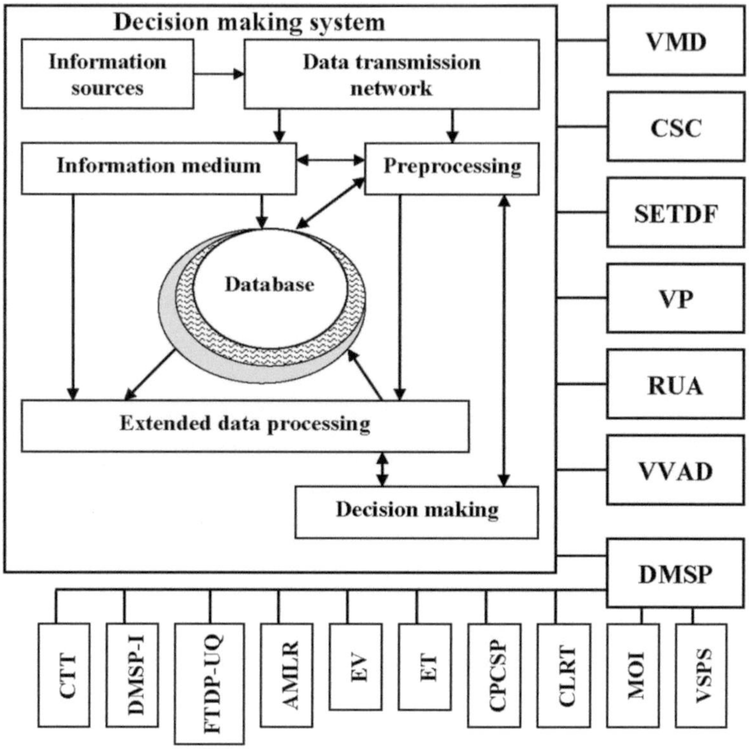


**Figure 6.3.** Principal scheme of the decision-making system based on the sequential analysis procedure. A description of the subunits is given in Table 6.1.

calculation process by considering the character of the task in hand. It forms variants that correspond to concrete combination of errors of the first and second kind, $\alpha$ and $\beta$. Based on this combination, subunit CTT produces a set of parameters that manages the other subunits. Depending on $\alpha$ and $\beta$, simplified procedures are possible. For example, two variants for the asymmetric thresholds $A$ and $B$ are: (1) $B = \beta/(1-\alpha) \to 0$, $A = (1-\beta)/\alpha \to \text{const}$; and (2) $B = \beta/(1-\alpha) = \text{const}$, $A = (1-\beta)/\alpha \to \infty$. In other words, errors $\alpha$ and $\beta$ are unequal in value, namely: (1) $\beta \to 0$, $\alpha = \text{const}$; (2) $\beta = \text{const}$, $\alpha \to 0$. In these cases, the sequential procedure will end with the probability equal to 1 if the following conditions are in place: (1) $E_a\zeta > 0$, $\zeta = \ln[f_{a_1}(\xi)/f_a(\xi)]$; (2) $E_a\zeta_0 < 0$. The probability of completion of the procedure is small when one of the following conditions is not realized: (1) $\beta \gg \alpha$, $E_a\zeta > 0$; or (2) $\alpha \gg \beta$, $E_a\zeta < 0$. The user can visualize the state of the procedure in Figure 6.4.

Unit VMD coordinates system input with separate channels of measured data. As each random process $\{x_i\}$ is realized, analyses are carried out to eliminate errors and to represent input signals in a form that is acceptable for the other units. Unit CSC calculates signal characteristics:

$$M_1 = \frac{1}{n}\sum_{i=1}^{n} x_i, \quad M_2 = \frac{1}{n}\sum_{i=1}^{n}(x_i - M_1)^2, \quad r_1 = M_3 M_2^{-3/2}, \quad V = \frac{\sqrt{M_2}}{M_1}\cdot 100\%,$$

$$M_3 = \frac{n}{(n-1)(n-2)}\sum_{i=1}^{n}(x_i - M_1)^3, \quad r_2 = \frac{M_4}{M_2^4} - \frac{3(n-1)^2}{(n-2)(n-3)},$$

$$M_4 = \frac{n(n+1)}{(n-1)(n-2)(n-3)}\sum_{i=1}^{n}(x_i - M_1)^4, \quad H = -\int_{-\infty}^{\infty} f_a(x)\ln[f_a(x)]\,dx,$$

$$R_i = x_{i,\max} - x_{i,\min}, \quad M_6 = \frac{M_4}{M_2^2} - 3.$$

**Table 6.1.** Description of the decision-making system (DMS) and subunits of the decision-making sequent procedure (DMSP) in Figure 6.3.

| *Subunit* | *Description* |
|---|---|
| VMD | Visualizing measurement data as computer soft copy, discrete distribution, and statistical parameters |
| CSC | Calculating the statistical characteristics (mean, central second-order and third-order moments, asymmetry and excess coefficients, entropy, etc.) |
| SETDF | Synthesizing empirical and theoretical distribution functions |
| VP | Valuing parameters used in the Neyman–Pearson criterion and the sequential procedures of hypothesis decision |
| RUA | Ensuring user access to all functions of the decision-making system |
| VVAD | Visualizing the accepted decision |
| DMSP-I | DMSP input: $\alpha, \beta, a, \{x_i\}$ |
| AMLR | Accumulating measurements as likelihood ratios |
| ET | Evaluating thresholds $A$ and $B$ |
| CLRT | Comparing the likelihood ratio with thresholds $A$ and $B$ |
| VSPS | Visualizing the state of the sequential procedure |
| FTDP-UQ | The FTDP unit query with respect to the form of $f_a(x)$ and activating the appropriate knowledge base level |
| EV | Evaluating $E_v, D_v, c$ |
| CPCSP | Computing the probability of completing the sequential procedure |
| CTT | Controlling the task type for the choice between $H_0$ and $H_1$ by taking errors of the first and second kind into account |
| MOI | Managing user intervention in how the DMSP unit functions |

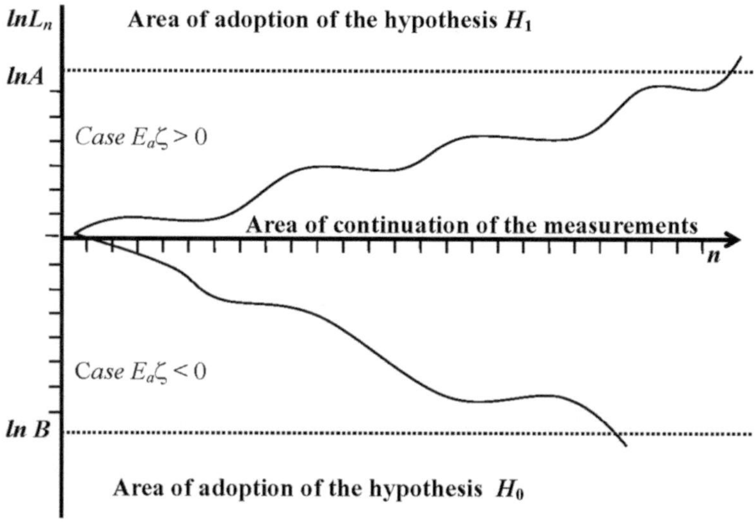


**Figure 6.4.** An example of the dynamics of the accumulated sum of the likelihood function logarithm visualized by the subunit VSPS.

These characteristics and other standard parameters are used to reconstruct $f_a(x)$. For instance, if $M_3 \approx 0$ and $r_1 \approx 0$ then $f_a$ should be looked for in symmetric distributions. Unit RUA realizes this search by selecting continuous distributions from the knowledge base and estimating their closeness to the empirical distribution using the criterion:

$$X^2 = \sum_{i=1}^{L} \frac{[m_i - f_a(x_i)\Delta x_i]^2}{f_a(x_i)\Delta x_i}.$$

The continuous distribution $f^*$ is chosen from the following condition:

$$\min_{f_a} X^2(f_a) = X^2(f^*).$$

The decision-making system is synthesized according to the principal scheme of Figure 6.3. How it functions is characterized in Figure 6.5.

## 6.3 PARAMETRICAL ESTIMATES FOR THE SEQUENTIAL ANALYSIS PROCEDURE

Investigation of the sum $x_n$ of independent random values having the same distribution imposes a double task on the distribution function, $P(x_n < x) = F_1(x)$, both for fixed $n$ and for the variable case. Comparison between $x_n$ and some level $C$ arises in both cases. However, in the second case this task is transformed into studying the distribution function, $P(\nu < n) = F_2(n)$, of the chance number $\nu$ of

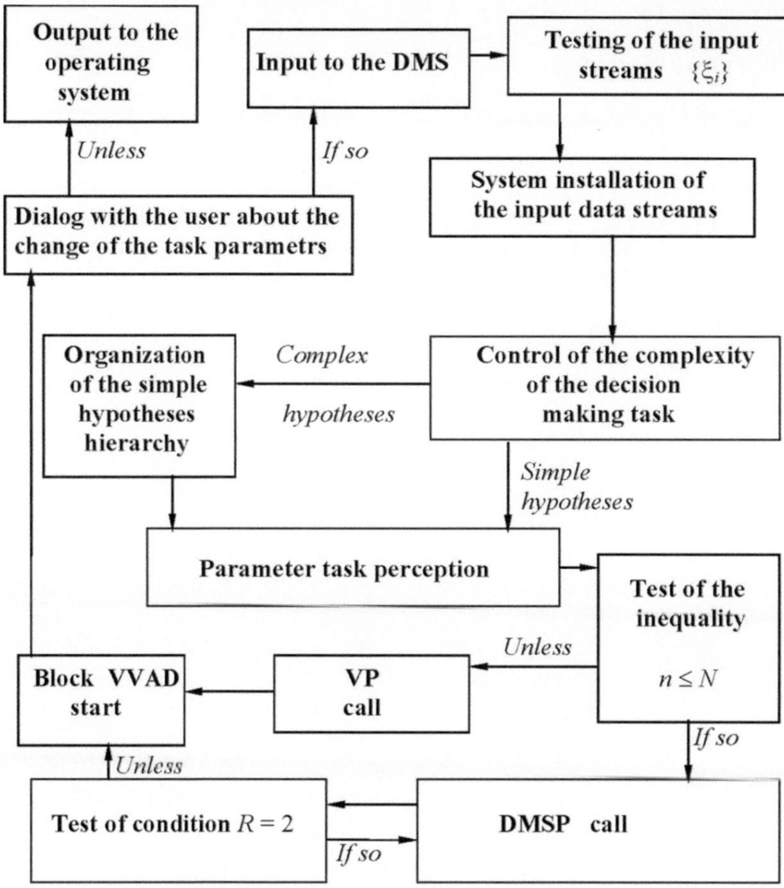


**Figure 6.5.** Block scheme of data processing in the sequential analysis procedure.

components by which $x_\nu$ first exceeds the level $C = C(\alpha, \beta)$: $x_i < C$ $(i = 1, \ldots, \nu - 1)$, $x_\nu \geq C$.

In conformity with the central limit theorem, in the first case the distribution $F_1(x)$ under $n \to \infty$ approaches the normal distribution. In the second case we have the distribution represented by expression (6.12). The following correlation between these distributions exists:

$$W_c(x) = \Phi[(x-1)(c/x)^{1/2}] + \Phi[-(x+1)(c/x)^{1/2}] \exp\{2c\}, \qquad (6.13)$$

where $\Phi$ is the normal distribution function.

Expression (6.13) makes it possible to study the sequential analysis distribution using the characteristics of $\Phi$. As can be seen from (6.11) the distribution function $W_c(x)$ is defined on the half-space $[0, \infty]$ and has one maximum at the point $x = m_c$. In fact, we have:

$$dW_c(y)/dy = (d/dy)\{c^{1/2}/[y^{3/2}(2\pi)^{1/2}]\exp[-0.5c(y + y^{-1} - 2)]\} = 0.$$

After the set of transformations, this equation can be solved:

$$y = m_c = [(9 + 4c^2)^{1/2} - 3]/(2c). \tag{6.14}$$

The position of the $W_c(y)$ maximum changes if and only if parameter $c$ remains less than one ($m_c < 1$). Moreover, $m_c \to 0$ when $c \to 0$ and $m_c \to 1$ when $c \to \infty$. Comparing equations (6.1) and (6.14), we find:

$$W_c(m_c) = (3/(2\pi)^{1/2})(1/[m_c(1 - m_c^2)^{1/2}]) \times \exp\{-3(1 - m_c)/[2(1 + m_c)]\}. \tag{6.15}$$

Hence, the function $W_c(y)$ degenerates under $c \to 0$ or $c \to \infty$ into the delta functions $\delta(y)$ or $\delta(y - 1)$, respectively.

Since $W_c(x) \to 1$ when $x \to \infty$, from equation (6.12) we have:

$$W_c(x) = \int_0^x W_c(y)\,dy = \int_{1/x}^{\infty} zW_c(z)\,dz = 1 - \int_0^{1/x} zW_c(z)\,dz. \tag{6.16}$$

The important parameters of distributions $\Phi$ and $W_c$ deal with solution of the following equations:

$$\Phi(u_p) = p, \qquad W_c(x_p) = p. \tag{6.17}$$

When $c > u_p^2/4$ we clearly find:

$$1 + (u_p - \varepsilon)c^{-1/2} + (u_p^2 - \varepsilon)c^{-1}/2 + (u_p^3 - \varepsilon)c^{-3/2}/8 - (u_p^5 - \varepsilon)c^{-5/2}/128$$
$$< x_p(c) < 1 + u_p c^{-1/2} + u_p^2 c^{-1}/2 + u_p^3 c^{-3/2}/8, \tag{6.18}$$

where

$$\varepsilon = \Phi(-2c^{1/2})\exp\{2c\} < 0.5(2\pi c)^{1/2}. \tag{6.19}$$

For $c < u_p^2/16$ we have:

$$c/u_a^2 < x_p(c) < c/u_b^2, \tag{6.20}$$

where $a = 0.5p\exp(-2c)$; and $b = p/2$.

When $c \to \infty$, we have:

$$x_p(c) = 1 + u_p c^{-1/2} + O(1/c). \tag{6.21}$$

From equations (6.13) and (6.21) we obtain:

$$W_c(x) = \Phi[(x - 1)c^{1/2}] + O(1/c). \tag{6.22}$$

Thus, the random value $\nu/E_\nu$ is asymptotically normal with an average value and dispersion equal to 1 and $1/c$, respectively, when $c \to \infty$. Expressions (6.13) and (6.22) can receive various analytical approximations of $W_c$. For example, let us represent $\Phi(y)$ by the following:

$$\Phi(y) = \Phi(y_0) + \varphi(y_0)(y - y_0)[1 - 0.5y_0(y - y_0)], \tag{6.23}$$

where $y = (x-1)(c/x)^{1/2}$; $y_0 = (x_0-1)(c/x_0)^{1/2}$; and $x_0$ is the point at which the value of $\Phi$ was estimated.

The following formula can be easily derived using equations (6.12), (6.13), and (6.23):

$$\begin{aligned} W_c(x) = W_c(x_0) + \varphi(y_0)(y-y_0)[1-0.5y_0(y-y_0)] \\ + \varphi(v_0)(v-v_0)\exp(2c)[1-0.5v_0(v-v_0)], \end{aligned} \tag{6.24}$$

where

$$v = -(x+1)(c/x)^{1/2}, \quad v_0 = -(x_0+1)(c/x_0)^{1/2},$$
$$\text{and} \quad \varphi(z) = \varphi(z_0)[1 - z_0(z-z_0) + 0.5(z_0^2-1)(z-z_0)^2].$$

Let us designate

$$H(x) = \frac{2}{\sqrt{\pi}}\int_0^x \exp(-t^2)\,dt. \tag{6.25}$$

We have

$$H(x) = \begin{cases} \dfrac{2}{\sqrt{\pi}}\displaystyle\sum_{k=1}^{\infty}(-1)^{k+1}\frac{x^{2k-1}}{(2k-1)(k-1)!} & \text{when } |x| \le x_1, \\ \dfrac{2}{\sqrt{\pi}}\exp(-x^2)\displaystyle\sum_{k=0}^{\infty}\frac{2^k x^{2k+1}}{(2k+1)!!} & \text{when } x_1 < |x| < x_2, \\ 1 - \dfrac{1}{\pi}\exp(-x)\displaystyle\sum_{k=0}^{\infty}(-1)^k x^{-(k+1/2)}\Gamma(k+1/2) & \text{when } |x| \ge x_2. \end{cases}$$

The functions $\Phi$ and $H$ are connected by the obvious correlation:

$$\Phi(x) = [1 + \sin(x)H(2^{-1/2}|x|)/2, \tag{6.26}$$

from which we have:

$$\Phi(x) = \begin{cases} g(x) & \text{for } x \le 0, \\ 1-g(x) & \text{for } x > 0, \end{cases}$$

where

$$g(x) = \begin{cases} 1/2 - \dfrac{1}{\sqrt{2\pi}}\displaystyle\sum_{k=1}^{\infty}(-1)^{k+1}\frac{|x|^{2k-1}}{2^{k-1}(2k-1)(k-1)!} & \text{for } |x| \le x_1, \\ 1/2 - \dfrac{1}{\sqrt{2\pi}}\exp(-x^2/2)\displaystyle\sum_{k=0}^{\infty}\frac{|x|^{2k+1}}{(2k+1)!!} & \text{for } x_1 < |x| < x_2, \\ \dfrac{1}{\sqrt{2\pi}}\exp(-x^2/2)\displaystyle\sum_{k=0}^{\infty}(-1)^k\frac{(2k-1)!!}{|x|^{2k+1}} & \text{for } |x| \ge x_2. \end{cases}$$

Thus, formula (6.26) allows us to calculate $W_c(x)$ for various values of $c$ and $x$. The free parameters $x_1$ and $x_2$ influence calculation error. Values $x_1 = 2.2$ and $x_2 = 7.5$ are acceptable in practice. Moreover, there is a convergence problem in the above rows when parameter $c$ is increasing. Clearly, $W_c(x) \approx \Phi[(x-1)c^{1/2}]$ when $c \gg 1$ and $W_c(x) \approx 1 - \exp(-cx/2)$ when $x \to \infty$.

Formula (6.13) is the basic expression used to calculate $W_c(x)$. The large factor, $\exp(2c)$, can be neutralized by the following expression:

$$\exp(2c)\Phi[-(x+1)\sqrt{c/x}] = \frac{1}{\sqrt{2\pi}}\exp(2c - x^2/2)\sum_{k=0}^{\infty}(-1)^k\frac{(2k-1)!!}{x^{2k}}.$$

The calculation of $W_c(x)$ can also be realized using the Bessel ($J$) and Whittaker ($Y$) functions:

$$W_c(x) = \sqrt{\frac{c}{2\pi}}\exp\left[c\left(1-\frac{1}{2x}\right)\right]\sum_{k=-\infty}^{\infty}\frac{x^l}{c^m}J_k(-c)\Psi_{-l,-s}(-c/x),$$

where $l = (2k+3)/4$; $m = (6k+9)/4$; and $s = (1+2k)/4$. The following correlations are also very useful here:

$$\exp[(1/z - z)c/2] = \sum_{k=-\infty}^{\infty} J_k(-c)z^k, \qquad \int_u^{\infty} e^{-x}x^{-\nu}\,dx = u^{-\nu/2}e^{-u/2}\Psi_{-\nu/2,(1-\nu)/2}(u).$$

Finally, let us use the following algorithm to evaluate the distribution function $W_c(x)$:

$$W_c(x) = A_0\varphi_0(x) + A_1\varphi_1(x) + \cdots + A_m\varphi_m(x) + \cdots,$$

where

$$A_i = \frac{(-1)^i}{i!}\int_{-\infty}^{\infty} w_c(z)R_i(z)\,dz,$$

and $R_i(z)$ is the Chebyshev–Hermite polynomial approximation $R_i(z) = (-1)^i\varphi_i(z)/\varphi_0(z)$, which is here designated by $\varphi_0(z) = (2\pi)^{-1}\exp(-t^2/2)$, $\varphi_i(z) = d^i\varphi_0(z)/dz^i$ ($i = 1, 2, \ldots$).

The normalization requirement is also accepted:

$$\int_{-\infty}^{\infty}\varphi_0(z)R_i(z)R_j(z)\,dz = \begin{cases} 0, & i \neq j, \\ 1, & i = j. \end{cases}$$

Thus, the sequential analysis distribution can be written in the form:

$$W_c(x) = \varphi_0[\sqrt{c}(x-1)] + \sum_{k=1}^{\infty}(-1)^k c^{-k/2}\frac{(2k+1)!!}{(k+2)!}\varphi_{k+1}[\sqrt{c}(x-1)].$$

## 6.4 AN ALGORITHM FOR MULTI-CHANNEL DATA PROCESSING

### 6.4.1 Introduction

A schematic diagram illustrating how a monitoring system detects anomalies on the Earth's surface of necessity involves many levels. The hierarchy of the structure of a search organization may be much deeper and include the processing of information from satellites, airborne and floating laboratories, and stationary systems. Despite the feasibility of bringing about such a structure and an applicable computer technique, there exists, nevertheless, one general problem whose solution is vital to detection efficiency. This task involves organizing in-line data processing at all levels of the detection system under real-time conditions, ensuring all cybernetic devices of the system operate in unison, taking into account any limits in the way they work.

Statistical analysis of the information at each level of the monitoring system is completed by making two subsequent decisions:

(1) storing features of landscape elements in such a way that additional information about them can be accumulated;
(2) completing analysis of the presence of an anomaly and transmitting its characteristics to the next data-processing level.

These two stages determine the efficacy of the detection system. At the first stage the selection and storage of parameters of landscape elements suspected of being anomalous depends on the algorithm being used to single out two-dimensional signals against background noise that are associated with the gradual filling of the main storage. The second stage determines the magnitude of detection probability depending on how landscape elements selected at the first stage for further analysis have been registered. As a result the problem arises of matching the flow of information between processing stages under monitoring conditions and when losses are minimal. Solution of this problem involves two procedures of parallel component processing: intermediate information delay either for a constant computer memory buffer size or for a fixed storage time.

### 6.4.2 Statistical analyzer

Let the on-board computer memory have $M$ cells. The statistical analyzer simultaneously and in parallel selects the components of the suspected variant of the vectorial parameter $b = b^*$. This is achieved by means of $K$ consecutive devices or by algorithms that select component values $b_i$ $(i = 1, \dots, K)$ corresponding to the space of anomalous features. The process of selecting the anomaly variant can be stabilized by creating constant delays of two types between the selection devices of the anomalous components: constant delay with the number of candidates and constant delay with time.

The two delay modes allow greater flexibility in the organization of the distribution of $M$ memory elements for storing intermediate values of suspected components. Because of the random character of the whole process, the statistical analyzer is apt to make transient errors as a result of the overflow of computer memory capacity intended for such delays. Therefore, it is necessary to assess the probability of errors and to find memory capacities for delays $M_1 + \dots + M_k < M$ and for the distribution of time delay intervals $t_1, \dots, t_K$.

### 6.4.3 Error probability assessment of the system and requisite delay memory capacity with constant expectation time

Let selection $b_i^*$ be realized in the volume sample $n_i$, basing each selection on $N_i$ reasonable values $b_i$. In this case the non-anomalous value $b_i$ is taken as an anomalous value with probability $\alpha_i$ and is rejected with probability $1-\alpha_i$. The probability of appearance of candidates during time $t_i = r_i\alpha_i S_i$ equals:

$$P\{\mu_i = S_i\} = C_{r_i}^{S_i}\alpha_i^{S_i}(1-\alpha_i)^{r_i-S_i} = \nu_i(S_i).$$

When $\mu_i \leq r_i\alpha_i$, candidates arrive intermittently and have time to be processed without delay. If $\mu_i > r_i\alpha_i$, candidates $\{b_i^*\}$ arrive often and do not have time to be analyzed before the arrival of $\{b_{i+1}^*\}$. Therefore, variants are delayed at $F_i$. The probability that the number of candidates will not exceed the mean values $r_i\alpha_i$ by more than $\varepsilon_i$ equals:

$$P\{\mu_i \leq r_i\alpha_i + \varepsilon_i\} = \sum_{s=0}^{m(i)} \nu_i(s),$$

where $m(i) = r\alpha_i + \varepsilon_i$. If $r_i$ are sufficiently large, we obtain according to Laplace's limit theorem:

$$P\{\mu_i = s_i\} \approx \phi[(s_i - r_i\alpha_i)\{r_i\alpha_i(1-\alpha_i)\}^{-1/2}],$$

where $\phi(u) = (2\pi)^{-1/2}\exp(-u^2/2)$.

Denoting memory capacity intended for delay of the $i$th component of $b$ by $M_i$ and fixing the condition that emerges from the limitation, we obtain: $r_i\alpha_i + \varepsilon_i \leq M_i$ $(i = 1, \ldots, K)$. Then the probability of non-overflow of memory $M_i$ will equal: $P\{\mu_i \leq M_i\} = \Phi(u_i)$ $(i = 1, \ldots, K)$, where

$$u_i = \varepsilon_i\{r_i\alpha_i(1-\alpha_i)\}^{-1/2} = (M_i - r_i\alpha_i)\{r_i\alpha_i(1-\alpha_i)\}^{-1/2}. \tag{6.27}$$

Utilizing the Boolean formula, let us calculate the probability of non-overflow of memory $M_i$ on the $i$th component during delay $t_i$ and uninterrupted transfer of candidates from $F_i$ to $F_{i+1}$ without delay. Let us denote this probability by $P(1, 2, \ldots, N_i$:

$$P(1, 2, \ldots, N_i) \geq 1 - N_i[1 - \Phi(u_i)], \quad i = 1, \ldots, K;$$

where $N_i$ is the number of possible variants of values $b_i$. Let us now demand that this probability differ from unity by no more than $\delta_i$. Then we obtain an equation for determining the delay value $t_i$ $(i = 1, \ldots, K)$:

$$\Phi(u_i) = 1 - \delta_i/N_i. \tag{6.28}$$

From equations (6.27) and (6.28) we obtain:

$$\Phi[(M_i - r_i\alpha_i)\{r_i\alpha_i(1-\alpha_i)\}^{-1/2}] = 1 - \delta_i/N_i. \tag{6.29}$$

With $u \gg 1$ we may write approximately: $\Phi(u) \approx 1 - \exp(-u^2/2)$. Then equation (6.29) assumes the form:

$$r_i^2 - 2r_i[\alpha_i/M_i + (1-\alpha_i)\ln(N_i/\delta_i)] + (M_i/\alpha_i)^2 = 0, \quad i = 1, \ldots, K. \tag{6.30}$$

For unambiguous determination of the numerical value of $r_i$ let us find the probability of "non-emptying" the delay (i.e., the probability that during the exhaustive search for $N_i$ values of the $i$th component at least one candidate will be under delay) and then demand that this probability be within permissible limits. Thus,

$$P\{\mu_i > r_i\alpha_i - v_i[r_i\alpha_i(1-\alpha_i)]^{1/2} = 0\} = 1 - \Phi(-v_i) = \Phi(v_i). \tag{6.31}$$

The probability $Q(1,2,\ldots,N_i)$ of fulfilling the inequality in equation (6.31), during an exhaustive search for $b_i$ according to the Boolean formula, can be estimated by the inequality:

$$Q(1,2,\ldots,N_i) \geq 1 - N_i[1 - \Phi(v_i)]. \tag{6.32}$$

Thus, by solving equation (6.30) and inserting its root into equation (6.32), we obtain two probable values for emptying the delay: $q_i = N_i[1 - \Phi(v_i)]$.

When the binomial distribution $\nu_i(s)$ cannot be approximated precisely enough by the normal distribution, we can use the Poisson distribution: $\nu(s) \approx p(s) = (r\alpha)^s \exp(-r\alpha)/s!$. We obtain:

$$p_i(s_i) \approx (2\pi s_i)^{-1/2} \exp\{-r_i\alpha_i - s_i\{1 - \ln(r_i\alpha_i s_i)\}\}.$$

Further, all the arguments that can be inferred from this equation are similar to those of equation (6.29); for determining $t_i$ they remain the same.

Generally, the calculation procedure for $t_i$ can be realized for any expression $\nu_i(s)$ on a computer. But since $N_i\alpha_i \gg 1$ is almost always the same, the agreements presented are very real and simple.

Thus, the probability of error of the whole anomaly detection system with a given time delay can be specified by the expression:

$$P_t = 1 - \sum_{i=1}^{K} Q(1,2,\ldots,N_i).$$

### 6.4.4 Evaluating system error probability and requisite memory capacity delay with a constant number of computer storage registers

Let us consider another variant of constant delay: delay in a constant number of candidates. The expectation time for this delay to be completely filled is a random value with a Pascal distribution:

$$P\{\tau_i = t_i^*\} = C_{R_i-1}^{m_i-1}\alpha_i^{m_i}(1-\alpha_i)^{R_i-m_i},$$

where $R_i = t_i^*/n_i$ is the number of $b_i$ variants surveyed for time $t_i^*$; and $n_i$ is sample capacity.

When $\tau_i \geq m_i n_i/\alpha_i$, variants from $F_i$ enter $F_{i+1}$ more rarely than normal. According to the matching condition of the whole flow, they have time to be processed without delay. At $\tau_i < m_i n_i/\alpha_i$, the $b_i^*$ variants do not have time to be processed at $F_{i+1}$ and are delayed at $F_i$. The numerical values of $m_i$ $(i = 1, \ldots, K)$ should be determined such that $m_i \leq M_i$ and in the process of an exhaustive search for the $i$th component there should be a definite probability of no overflow and no delay emptying.

Let us take advantage of the approximate expression of Pascal's distribution via distribution (6.12). At $R_i\alpha_i \gg 1$, the following expression is known

$$P\{\tau_i = t_i^*\} = \alpha_i m_i^{-1} w_{c_i}(y_i).$$

For predetermined probability of non-overflow of $M_i$ memory, it is necessary to choose $m_i$ such that the time between the arrivals of candidates should be as normal as possible. Let us calculate the probability that $\tau_i$ exceeds the value $m_i n_i/\alpha_i$ by more than some constant $d_i$

$$P\{\tau_i \geq m_i n_i/\alpha_i - d_i\} = 1 - W_{c_i}[(m_i n_i - d_i\alpha_i)m_i^{-1}n_i^{-1}]. \tag{6.33}$$

The $Q(1, 2, \ldots, N_i)$ probability of fulfilling inequality (6.33) during the exhaustive search for values of the $i$th component can be evaluated as:

$$Q(1, 2, \ldots, N_i) \geq 1 - N_i W_{c_i}[(m_i n_i - d_i\alpha_i)m_i^{-1}n_i^{-1}].$$

Let us try to make this probability differ from unity by no less than the value of $\varepsilon_i$. Then the equation for determining the delay value $m_i$ will take the following form:

$$W_{c_i}[(m_i n_i - d_i\alpha_i)m_i^{-1}n_i^{-1}] = \varepsilon_i/N_i. \tag{6.34}$$

When $c_i > 0$ regarding the designated approximation of the normal distribution function, we have $m_i \approx d_i\alpha_i[2c_i \ln(N_i/\varepsilon_i)]^{-1/2}/n_i$ $(i = 1, \ldots, K)$. Let us further demand that the probability of appearance during time $m_i n_i/\alpha_i - d_i$ of more than $M_i$ candidates be close to zero. Then we have:

$$P\{\mu_i \leq M_i\} \approx \Phi[(M_i - R_i\alpha_i)/\sqrt{R_i\alpha_i(1 - \alpha_i)}]. \tag{6.35}$$

The probability of realizing $\mu_i \leq M_i$ during an exhaustive search for values of the $i$th component according to the Boolean formula is estimated by the inequality:

$$\Omega(1, 2, \ldots, N) \geq 1 - N_i[1 - \Phi\{(M_i - R_i\alpha_i)/\sqrt{R_i\alpha_i(1 - \alpha_i)}\}] = 1 - \Delta_i.$$

Given that a certain small value is expressed by $\Delta_i$, we obtain—in addition to condition (6.34)—a second condition for determining $m_i$:

$$\Phi[(M_i - q_i\alpha_i)/\sqrt{q_i\alpha_i(1 - \alpha_i)}] = 1 - \Delta_i/N_i, \tag{6.36}$$

where $q_i = m_i/\alpha_i - d_i/n_i$.

Thus, by calculating the delay capacities from equation (6.36), the probability of error for the whole system will equal:

$$P_m = 1 - \prod_{i=1}^{K} \Omega(1, 2, \ldots, N_i)$$

because of memory overflow.

This evaluation of the probability of on-board computer memory overflow when monitoring processing information through $K$ channels allows in given situations the parameters of the system to be calculated, system efficacy to be estimated, and one of two processing methods (indicated above) to be chosen. Both delay variants discussed above are equivalent in requisite additional memory capacity and operation time. Indeed, if $\Delta_i = \sigma_i$, then $r_i = R_i$ (i.e., the delay in the first variant equals the number of variants considered in the second). Similarly, if $q_i = \varepsilon_i$, we obtain the equality $r_i = R_i$. Therefore, the choice of delay type should be determined by the technical considerations necessary for its realization.

## 6.5 APPLICATIONS OF THE SEQUENTIAL DECISION-MAKING PROCEDURE

The sequential decision-making procedure can be used to estimate land surface parameters based on remote data received from the IL-18 flying laboratory. Measurements are obtained by radiometers working in the 0.4, 0.8, 2.25, 11, 21, and 27 cm wavelengths. Sets of radiobrightness temperatures $\{T_j\}$ are registered and $T_{j,\min}$, $T_{j,\max}$, and $T_{j,\text{mean}}$ are calculated. As a result they can be transformed to normalized sets $\{T_j/T_{j,\max}\}$ or $\{\Delta T_j = T_j - T_{j,\text{mean}}\}$. Figure 6.6 gives examples

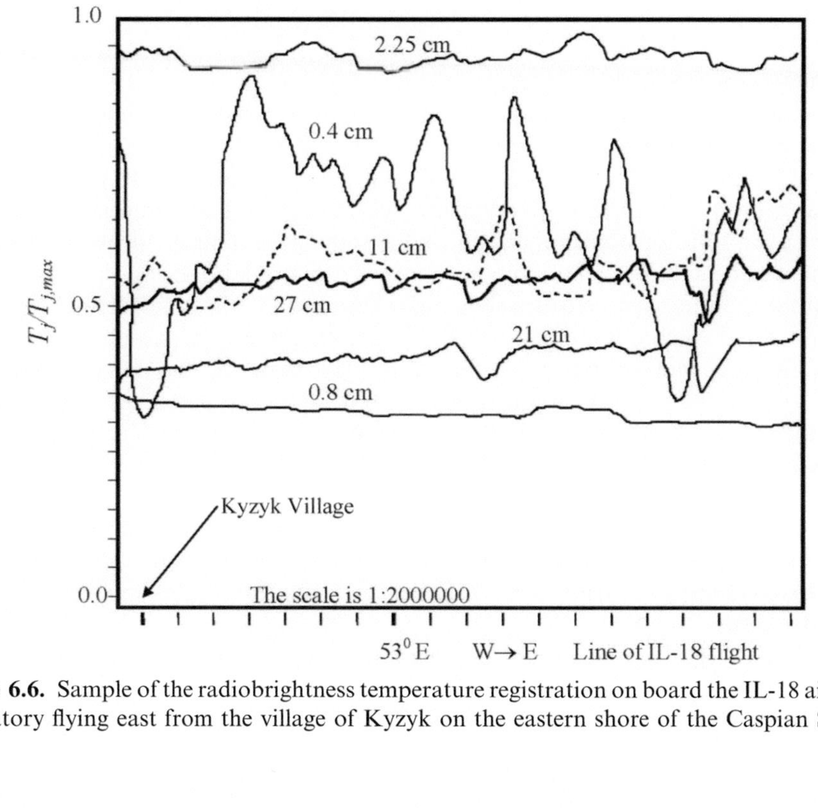


**Figure 6.6.** Sample of the radiobrightness temperature registration on board the IL-18 airborne laboratory flying east from the village of Kyzyk on the eastern shore of the Caspian Sea.

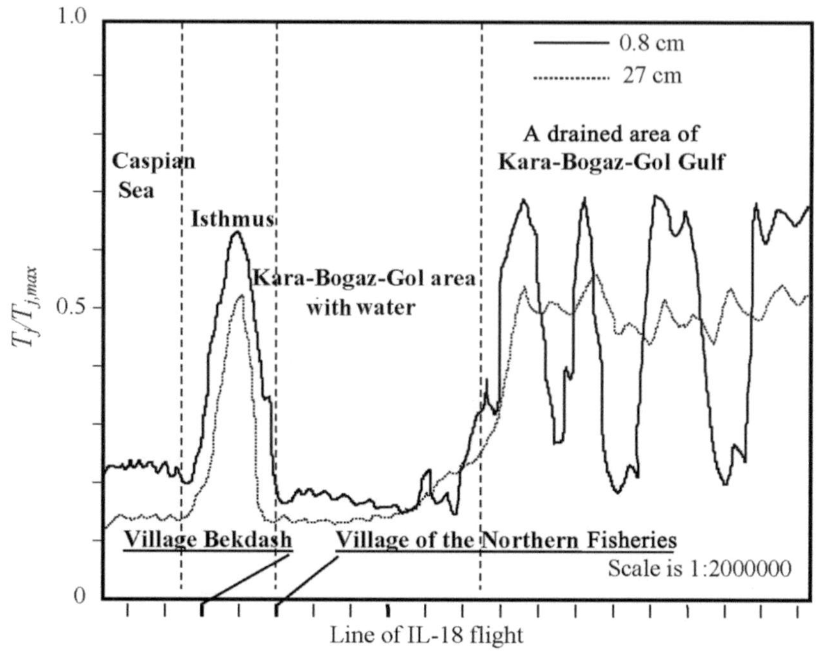


**Figure 6.7.** Radiobrightness contrasts in the area of the Kara-Bogaz-Gol gulf on the eastern shore of the Caspian Sea as registered on board the IL-18 flying laboratory.

of such transformations. The radiobrightness temperatures registered are typical of the land cover of the eastern Caspian Sea coast. This is a region of saline land, dry sand, saline water, dense bush, and populated landscapes. Figure 6.7 shows datasets of multi-channel measurements which facilitate differentiating between surface covers. Moreover, application of the sequential analysis procedure ensures this task can be solved in real time on board the flying laboratory.

Radiobrightness temperature sets, made discrete by quasi-uniform subsets, can be transformed to combinations of distribution characteristics which make it possible to form a multi-dimensional image of the state of the land surface. An example of cluster analysis application is given in Figure 6.8. The real brightness temperatures of this example are given in Figure 6.9. The statistical parameters of distributions are given in Table 6.2. The brightness temperatures of multi-channel measurements allow land surfaces to be reliably recognized irrespective of the climatic zone. This is confirmed by Figure 6.10 where varied distributions do not correspond to the land surface structure. However, expanding the parametrical space by including distribution characteristics increases the efficiency of the decision system.

The best way for the decision system to work is joint use of classical and sequential algorithms for data processing (see Figure 6.5). This procedure saves time and supplies operative parameters of distributions that can be used within

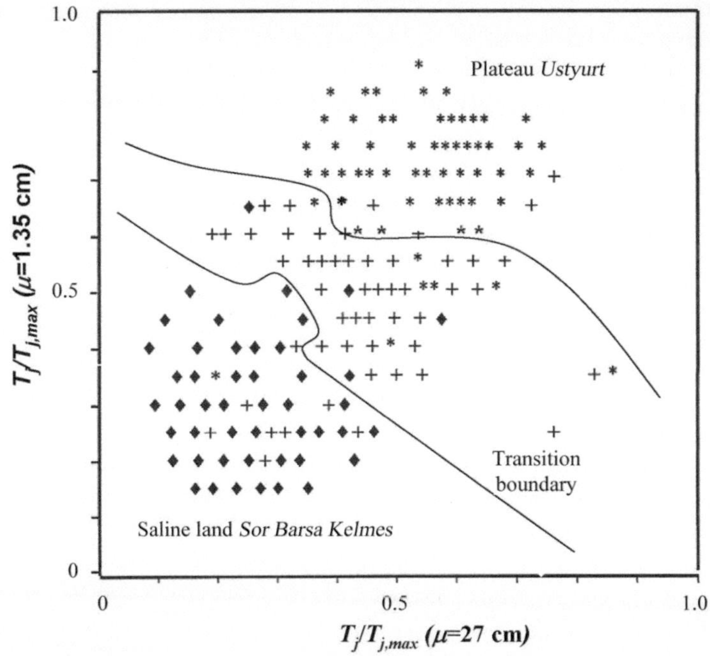


**Figure 6.8.** The state of the system for the Sor Barsa Kelmes saline land and Ustyurt plateau from IL-18 in-flight laboratory measurements. Notation: –, saline land; *, plateau; +, the boundary between them.

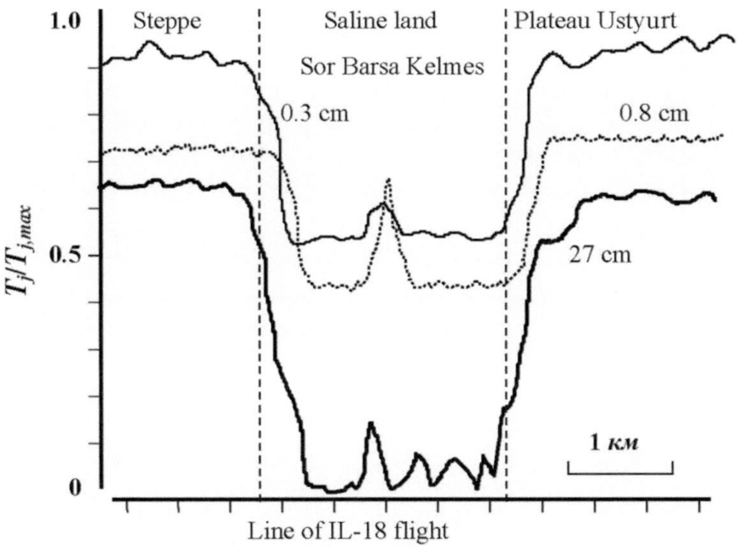


**Figure 6.9.** Fragment of the record from the IL-18 flying laboratory at the boundary of the Ustyurt plateau.

**Table 6.2.** Results of processing multi-channel radiometric measurements taken on board the IL-18 flying laboratory in Sor Barsa Kelmes, a tract of saline land bordering the Aral Sea. Notation: *R* is the Rayleigh distribution, *E* is the exponential distribution, *L* is the lognormal distribution, *G* is the Gaussian distribution, *V* is the Weibull distribution, *W* is the sequential procedure, *N–P* is the Neyman–Pearson procedure.

| *Channel* (cm) | *Distribution* | | | *Procedure used for decision making* | | |
|---|---|---|---|---|---|---|
| | *Steppe* | *Saline land* | *Boundary* | *Steppe* | *Saline land* | *Boundary* |
| 0.8 | *R* | *G* | *E* | *W* | *W* | *W* |
| 1.35 | *R* | *L* | *G* | *W* | *N–P* | *W* |
| 2.25 | *R* | *L* | *G* | *W* | *N–P* | *W* |
| 21 | *R* | *L* | *E* | *W* | *W* | *W* |

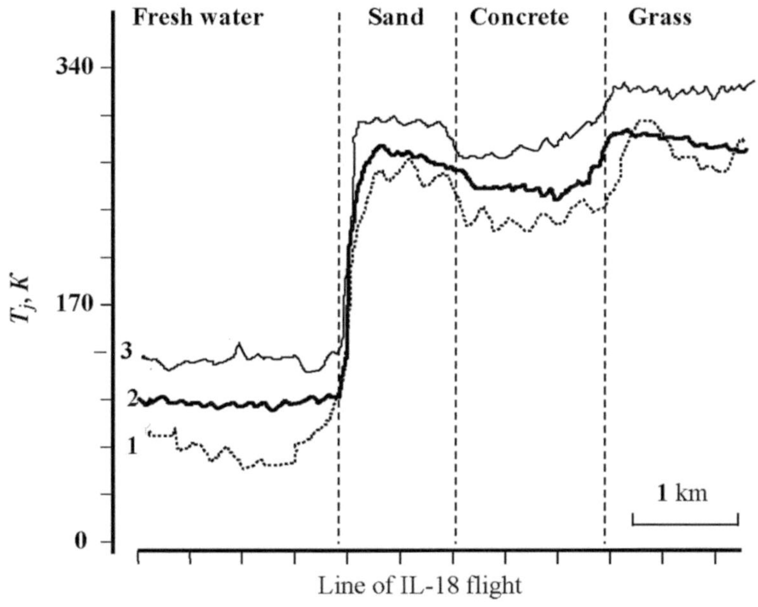


**Figure 6.10.** A registrogram received from the IL-18 flying laboratory near Yeniseysk City airport using radiometers of 2.25 cm (1), 10 cm (2), and 30 cm (3) wavelengths.

other procedures. Table 6.3 shows the advantage of such an approach. Figure 6.11 shows the results of reconstructing the empirical distribution of precipitation in three regions of the Russian Federation and Kustanay in northern Kazakhstan. Parameters for these distributions are given in Table 6.4.

**Table 6.3.** Results of processing temperature anomalies for the Arctic atmosphere (during the last 10 years). Notation: $M_1$ is first-order moment (mean), $M_2$ is central second-order moment (dispersion), $M_5$ is the asymmetry coefficient, and $M_6$ is the excess coefficient, $NW$ is the number of measurements made in the sequential procedure, $\Delta$ is the advantage the sequential procedure has over the classical procedure in the fixed level of 100 measurements, $G$ is the Gaussian distribution, $K$ is the Cauchy distribution, $Si$ is the Simpson distribution, and $St$ is the Student distribution.

| *Season* | $M_1$ | $M_2$ | $M_5$ | $M_6$ | $NW$ | $\Delta$ (%) | *Distribution* |
|---|---|---|---|---|---|---|---|
| Winter | 0.02 | 0.29 | 0.16 | 0.41 | 44 | 56 | *K* |
| Spring | 0.08 | 0.87 | −0.52 | 0.05 | 108 | 0 | *St* |
| Summer | 0.00 | 0.39 | −0.01 | −0.65 | 51 | 49 | *Si* |
| Fall | 0.09 | 0.64 | 0.11 | 4.74 | 73 | 27 | *G* |
| Year | 0.05 | 0.55 | 0.26 | −0.75 | 59 | 41 | *G* |

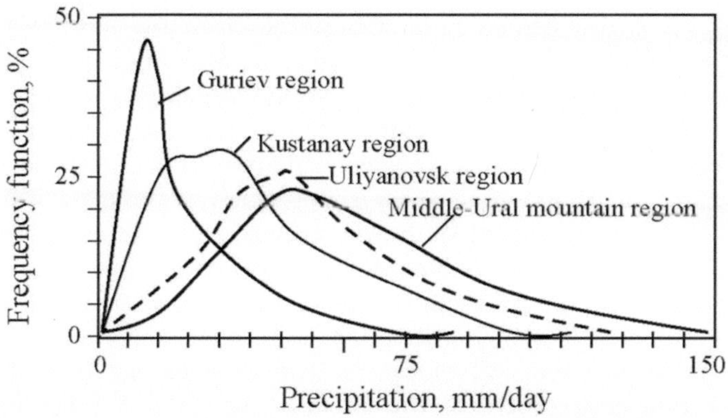


**Figure 6.11.** Empirical distributions of precipitation in four regions of Russia and Kazakhstan estimated by means of the sequential procedure. The distribution parameters are given in Table 6.4.

## 6.6 EXPERT SYSTEM FOR WATER QUALITY CONTROL IN THE ESTUARY ZONE OF THE SOUTH CHINA SEA COAST IN VIETNAM

The development of industry and agriculture in Vietnam poses the problem of coastal water pollution control and synthesizing monitoring systems for water quality assessment in active regimes. Vietnam's coastline is 3,260 km long. There are many estuaries and lagoons on the seashore where the development of anthropogenic systems depends on seawater quality. At the same time these systems generate pollutants and contaminate the sea. Pollutant sources in Vietnam can be grouped in five classes (Bui, 1999):

(1) agricultural sewage;
(2) domestic, municipal, and industrial waste;
(3) shipping;
(4) aquaculture;
(5) coastal mining.

These sources are compounded in territories bordering the South China Sea. This is the reason an environment-monitoring system synthesized to assess the water quality in the estuary zone requires a systematic approach. We now give a short description of the expert system for ecological control of estuary zones (ESECEZ) which was developed as part of an agreement between the Russian Academy of Sciences and the Vietnamese Academy of Sciences and Technology to collaborate on ecoinformatics and remote sensing (Cao *et al.*, 2009a, b; Klimov *et al.*, 1999; Nguyen *et al.*, 2011a, b; Mkrtchyan *et al.*, 2004).

The expert system is based on combining modeling technology, algorithms that reconstruct measurement data spatiotemporally, and computer visualization software. A block diagram of the expert system is given in Figure 6.12 and Table 6.5.

The ESECEZ is oriented at finding processes to study how best to regulate and control the release of potential hazardous chemicals into the typical Vietnam estuary environment. The main functions of the expert system can be realized automatically, but not the human components. Human activity can be displayed when it is needed to explain the ecological situation or to make decisions about how to protect the environment.

The ESECEZ is constructed of blocks connected by means of parametrical inputs and outputs. There are blocks with the same functional peculiarity. Selection between these blocks is automatically realized in concordance with the database. Only blocks for which the database has the necessary parameters for block input can be inserted. All blocks activated within the ESECEZ are compatible. Realization and control of this process are tasks that belong to the SC block. If these tasks cannot be carried out the system requests the necessary data

**Table 6.4.** Estimates of the distribution parameters for precipitation ($mm \cdot month^{-1}$) in three regions of the Russian Federation and in Kustanay, Kazakhstan.

| *Region* | *Variation coefficient* | *Distribution* | *Mean* | *Dispersion* |
|---|---|---|---|---|
| Uliyanovsk | 16 | Pearson | 310 | 51 |
| Middle Ural mountains | 15 | Pearson | 435 | 67 |
| Guriev | 31 | Pearson | 109 | 33 |
| Kustanay | 31 | Pearson | 211 | 61 |

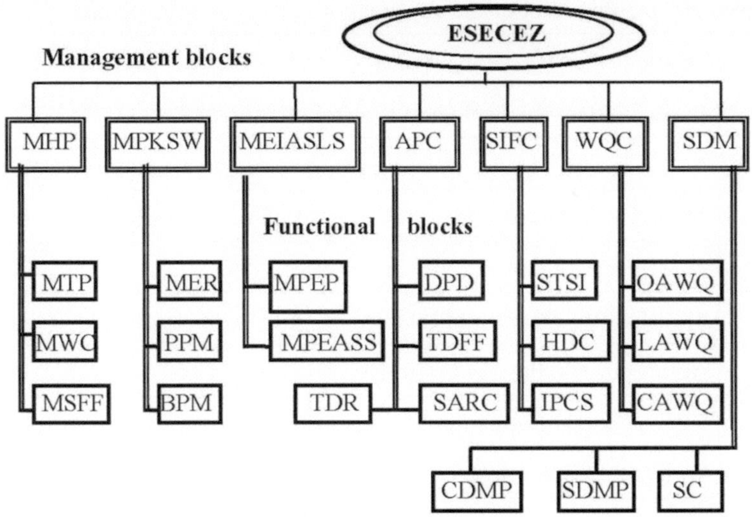


**Figure 6.12.** Block structure of the ESECEZ. See Table 6.5 for an interpretation of the abbreviations.

from the database. The working regime of the ESECEZ is such that the user can form his or her own system structure.

The main ESECEZ function is risk assessment with the purpose of informing groups of user about the water quality in estuary zones. There are various components to the risk assessment software system, each of which reflects a different approach to using the information contained in the system. The SDM block realizes such functions as hazard identification, dose–response assessment, human exposure evaluation, and risk characterization (Rochon *et al.*, 1996). These functions conform with Vietnamese water standards. The ESECEZ generates signals that reflect the water quality in a site defined by the system user. The CDMP and SDMP blocks realize a risk assessment approach that improves regulatory decisions. The user can pinpoint his or her own specific features and decide how to use the CDMP and SDMP blocks (Krapivin *et al.*, 1997b; Ward *et al.*, 1990). In this way the ESECEZ produces risk assessment of the highest possible quality. Moreover, it ensures each assessment is sensitive to site-specific conditions.

The WQC blocks are based on sampling, laboratory analysis, data handling, data analysis, reporting, and information utilization. The SIFC blocks regulate information flows between other blocks and provides the necessary data for analysis and interpretation.

Specific features in the design of the ESECEZ were developed to assess water quality in real estuary zones including those subject to anthropogenic activity. One of the most important and difficult tasks here is simulating pollutant kinetics in seawater. The MPKSW and MEIASLS blocks resolve this by taking typical processes in Vietnam's estuaries into consideration. The DPD and TDR blocks realize adaptation procedures to fit the ESECEZ to individual estuary zones.

**Table 6.5.** Interpretation of the abbreviations in Figure 6.12 (Soldatov, 2011).

| *Identifier* | *Description of the block function* |
|---|---|
| ESECEZ | Expert System for Ecological Control of Estuary Zones |
| | ***First-level blocks*** |
| MHP | Modeling Hydrodynamics Processes (Monin and Krasitskyi, 1985) |
| MPKSW | Modeling Pollutant Kinetics in Seawater |
| MEIASLS | Modeling Environmental Interactions in the Atmosphere–Sea–Land System |
| APC | Anthropogenic Process Control |
| SIFC | Subsystem of Information Flow Control |
| WQC | Water Quality Control |
| SDM | Subsystem of Decision Making |
| | ***Second-level blocks*** |
| MTP | Model of Tide Processes |
| MWC | Model of Waterflow from the Coast |
| MSFF | Model of Sewage Flow Formation |
| MER | Model of the Ecosystem Role in processes of pollutant transformation |
| PPM | Physical Process Model of interactions between pollutants and seawater |
| BPM | Biological Process Model of pollutant transformation in seawater |
| MPEP | Model of Pollutant Exchange Processes between estuary zones and other territories (Krapivin, 1995) |
| MPEASS | Model of Pollutant Exchange in the Atmosphere–Sea System |
| DPD | Description of Passport Data for an industrial and agricultural system of defined type, specific parameters of production, and activity scale. On the basis of these data in the ESECEZ, control etalons can be formed to describe the possible specter of pollutants generated |
| TDFF | Technology of Database Fragment Forming based on official information about the real volume of generated pollutants entering the atmosphere and water area |
| TDR | Technology for Database Renewal |
| SARC | Set of Algorithms to Reveal Correlations in the functioning of the anthropogenic system |

| *Identifier* | *Description of the block function* |
|---|---|
| | ***Second-level blocks*** (*cont.*) |
| SC | System Control under realization of its functions |
| CDMP | Classical Decision Making Procedure |
| SDMP | Sequential Decision Making Procedure |
| STSI | Spatial Tie between Surface Identifiers |
| HDC | Heterogeneous Documents for Control of the environment and their coordination |
| IPCS | Information Procedure of Control Signal formation about disturbances to the ecological condition |
| AOWQ | Algorithms for Operative Water Quality control |
| LAWQ | Laboratory Analysis of Water Quality |
| CAWQ | Complex Assessment of Water Quality based on a set of criteria |

How contaminants diffuse in seawater depends on their state. For example, the dissolved proportion of heavy metals ($\psi$) takes a more intensive part in biogeochemical processes than suspended particles ($e$). However, suspended particles like heavy metals fall out more rapidly to the sediment. A description of the entire spectrum of these processes in the framework of this study is impossible. Therefore, the PPM block describes processes that can be estimated. The transport of heavy metals in seawater includes absorption of the dissolved proportion $\psi$ by plankton ($H_Z^\psi$) and by nekton ($H_F^\psi$), sedimentation of the solid fraction ($H_1^e$), deposition with detritus ($H_D^\psi$), adsorption by detritophages from bottom sediments ($H_L^{e\psi}$), and release from bottom sediments owing to diffusion ($H_a^{e\psi}$). As a result, the dynamic equations for heavy metals can be given as:

$$\frac{\partial e_W}{\partial t} + v_\varphi^W \frac{\partial e_W}{\partial \varphi} + v_\lambda^W \frac{\partial e_W}{\partial \lambda} + v_z^W \frac{\partial e_W}{\partial z} = \sum_{i=1}^{3} \alpha_2^i Q_{e\psi}^i - H_1^e + \alpha_1 H_a^{e\psi},$$

$$\frac{\partial \psi_W}{\partial t} + v_\varphi^W \frac{\partial \psi_W}{\partial \varphi} + v_\lambda^W \frac{\partial \psi_W}{\partial \lambda} + v_z^W \frac{\partial \psi_W}{\partial z} = (1-\alpha_1) H_a^{e\psi} + k_2^W \frac{\partial^2 \psi_W}{\partial z^2} - H_Z^\psi - H_F^\psi - H_D^\psi - H_a^\psi,$$

$$\frac{\partial e^*}{\partial t} = H_1^e - \alpha_1 (H_L^{e\psi} + H_a^{e\psi}),$$

$$\frac{\partial \psi^*}{\partial t} = H_D^\psi - (1-\alpha_1)(H_L^{e\psi} + H_a^{e\psi}),$$

where $e^*(e_w)$ and $\psi^*(\psi_w)$ are concentrations of heavy metals in bottom sediments (water) in their solid and dissolved phases, respectively; $H_a^\psi$ is output of heavy metals from the sea to the atmosphere by evaporation and spray; $Q_{e\psi}^i$ is input of heavy metals to the sea by river water ($i = 1$), atmospheric deposition ($i = 2$), and ship waste ($i = 3$); $\alpha_2^i$ is the proportion of suspended particles in the $i$th flow of heavy metals; and $\alpha_1$ is the proportion of heavy metals in bottom sediments.

ESECEZ need measurements of a series of environmental characteristics, which can be realized by national hydrometeorological services and derived from satellite data. Islam *et al.* (2005) demostrated a means of receiving monthly time series of SST data from NOAA satellites in the South China Sea region.

## 6.7 DECISION-MAKING WHEN MONITORING NATURAL DISASTERS

### 6.7.1 Natural catastrophes and monitoring problems

As civilization relentlessly develops the problem of forecasting future environmental change and the linked change in people's living conditions becomes all the more pressing. The origin and spread of dangerous natural phenomena not only ruin the lives of living beings they also cause serious economic damage (Gardner, 2002; Kondratyev and Krapivin, 2004; Milne, 2004). Such phenomena are called natural catastrophes. Historically, natural anomalies on different spatio-temporal scales played important roles in the evolution of nature, causing and actuating the mechanisms of natural system regulation. Wildfires can serve as a special example of such anomalies (Kondratyev *et al.*, 2006). With the development of industry and growth in population density these mechanisms have suffered significant change and the threat to life is thereby intensified (Nazaretyan, 2004; Podlazov, 2001; Shahramantyan, 2003). This has to do with the growth and spread of anthropogenic changes in the environment. Numerous studies of the problems that have cropped up in recent years have shown the frequency of occurrence of disastrous natural phenomena and their power to be growing and posing increased risk of greater losses of human lives, economic losses, and breaches to the social infrastructure. In 2001, for example, about 650 natural catastrophes occurred, taking the lives of more than 25,000 people and causing more than U.S.$35 billion of economic damage. Subsequent years witnessed natural disasters that resulted in big losses of human lives. The overall picture in these years was dominated by a series of severe earthquakes seldom experienced in recent decades. Several major catastrophes in 2010 resulted in substantial losses and an exceptionally high number of fatalities. Altogether, 950 natural catastrophes were recorded, nine tenths of which were weather-related events like storms and floods. This total makes 2010 the year with the second-highest number of natural catastrophes since 1980, markedly exceeding the annual average for the last 10 years (785 events per year). Overall losses amounted to around U.S.$130 billion, of which approximately U.S.$37 billion was insured. This puts 2010 among the six most loss-intensive years for the insurance industry since 1980. The level of overall losses was slightly above the high average of the past 10 years. The current year (2011) continues this tendency. A series of very severe natural catastrophes makes 2011 the highest-ever loss year on record, and we are only six months into the year. Unfortunately, no reliable prognosis of the future level of global natural disasters exists.

Such factors are functions of the readiness of the territory to reduce the risk of losses and change over time. In 2002, for example, there were 700 significant catastrophes in which 11,000 people perished and economic damage reached U.S.$55 billion. The greatest loss was that caused by floods. The spatial distribution of the catastrophes showed the typical strong spottiness (i.e., unevenness). A list of human victims worldwide in 2002 can be characterized as: Africa suffered 51 catastrophes (661 people perished), America 181 (825), Asia 261 (8570), Australia and Oceania 69 (61), and Europe 136 (459) (Bondur *et al.*, 2009). Subsequent years were characterized by numerous human victims and economic losses. For example, in 2008 there were 354 natural disasters in which 236,000 people died and 211,628,186 people were affected. This compares with the average for 2000–2007 of 397 natural disasters, with 66,812 deaths and a total of 231,588,104 people affected.

Natural catastrophes can be divided into different categories. Environmental phenomena that kill and displace thousands of people who are essential to the economy of a given region are classified as large natural catastrophes. It thus follows that the scale of natural catastrophes depends on the level of economic development of the region and determines the degree of protection needed by the population from such disastrous natural phenomena. So, the study of phenomena in natural catastrophes must be accompanied by analyzing the poverty level of the population of a given region. The results of studies over the last 25 years show that in developing countries losses from natural catastrophes are much greater than in economically developed regions. Being mindful of the fact that the number and power of natural catastrophes in recent decades has increased by about five times and the danger they pose by about nine times, the threat to the population of these countries in the near future becomes abundantly clear. This is the reason that forecasting and warning of crisis phenomena globally must be a subject of worry for all countries, regardless of their economic development.

At present, the theory of catastrophes and the analysis of risk are well developed (Brebbia, 2004). Their application to events and processes in the real environment requires synthesizing a global model of the NSS that employs the technical facilities of satellite monitoring. Solution of the problems that arise here is the subject of ecoinformatics, which provides a link between analytically simple, semi-empirical, and complex nonlinear models of ecosystems and an updated global database. Many international and national programs of environmental study, directed at problem solving and spatial orientation, have recently raised the level of thematic coordination to reach the necessary level of efficiency. An example of such coordination is the Global Carbon Project and Earth Observing System, which between them contain information and technical facilities of the highest quality for estimation and forecasting of NSS dynamic features.

The development of constructive ways to forecast natural catastrophes requires decisions regarding a set of problems:

- Adapting ecoinformatics methods to the problem of diagnosing and forecasting natural catastrophes in all their variety.
- Ascertaining the statistical characteristics of natural catastrophes historically, selecting categories, and defining the spatiotemporal scales at which human habitation has occurred. Analyzing the history of natural catastrophes is important to understand modern dependences between nature and society at times of crisis. Statistical characteristics of the dynamics of natural catastrophes allow formulating the basics of a mathematical theory of catastrophes and determining the directions studies should take.
- Developing a concept of and synthesizing a survivability model to estimate the influence of natural catastrophes on human habitation.
- Studying objective laws that apply to the interaction between different components and processes in the global NSS in light of such a capacious notion as the biological complexity of the ecosystem (i.e., biocomplexity), considering it as a function of biological, physical, chemical, social, and behavioral interactions of environmental subsystems, including living organisms and their community. The notion of biocomplexity is linked to objective laws that govern the way the biosphere functions by uniting its ecosystems and natural/economic systems on different scales, from local to global. A joint formalized description of the biological, geochemical, geophysical, and anthropogenic factors and processes occurring at a given level of the spatiotemporal hierarchy of scales is called for here. It is also important to evaluate the possibility of using different indicators, such as biocomplexity, to show the approach of a natural catastrophe.
- Investigating correlations between survivability, biocomplexity, and NSS evolution using global-modeling technologies. Creating global model blocks, describing regularities and trends in the environment, to ascertain stressful situations initiated by economic or political activity.
- Considering demographic premises for natural catastrophes to occur and discovering the mechanisms that regulate the natural environment and prevent such catastrophes from happening.
- Synthesis of the NSS model and the development of a computer method to carry out calculations in order to assess the consequences of a realization of various anthropogenic activity scenarios.

### 6.7.2 Natural catastrophes as dynamic phenomena

Walker (2003) observed quite reasonably that the notion of natural catastrophe was vague and its determination depended on many factors. Grigoryev and Kondratyev (2001) define a natural catastrophe as "exceeding and dangerous situation in vital activity of the population, caused by essential disadvantage changes in the environment," or "as uneven changes in the system, appearing in the manner of its sudden answer to fluent changes of the external environment." The number of critical situations in the environment is increasing. At present, the following phenomena are considered components of natural catastrophes: floods, droughts, hurricanes, storms, tornadoes, tsunamis, earthquakes, landslides, severe cold, volcanic eruptions, wildfires, windstorms, sandstorms, avalanches, severe frosts, heat, epidemics, plagues of locusts, and many other natural phenomena. In the future, this list may well increase to account for new types of natural catastrophes: collisions with cosmic bodies, bioterrorism, nuclear catastrophes, magnetic field reversal, plague, cultural decline, etc. This is the reason it is important to develop efficient quantitative technologies and criteria so that warnings can be issued with high reliability about the appearance of dangerous disastrous natural phenomena.

Many authors consider the notion of natural catastrophe to be associated with the notion of ecological safety, which first appeared when the need arose to estimate risks facing the population in a given area to get insurance to cover health, buildings, or other property as a result of changing environmental parameters. These changes can be triggered naturally and anthropogenically. In the first case, dangerous situations appear as fluctuations in natural processes connected with change in the ecological situation, in the origin of epidemics, or in types of natural disasters. In the second case, danger appears as nature reacts to anthropogenic activity. For instance, Gardner (2002) investigated change in the natural environment of the Himalayas in India and came to the conclusion that such factors as reforestation and change in vegetation cover induce and amplify instability in this region. This results in land resource degradation and increased destruction of the natural environment as a result of floods. Abrahamson (1989) and Field and Raupach (2004) associate change in the regularity of appearance of natural catastrophes with increased instability in the carbon–climate–society system. Field *et al.* (2002) believe instability will increase considerably in the next two decades as a result of changes in many features of the World Ocean ecosystem. Milne (2004), analyzing the history of different large catastrophes, makes pessimistic forecasts for the fate of humankind, using terms like "doomsday".

Ecological danger anywhere is generally the result of environmental parameters deviating from limits, where longstanding living organisms begin to behave in a way that does not correspond with the natural process to evolution. The notions "ecological danger" or "ecological safety" are connected with notions of stability, vitality, and wholeness of the biosphere and its elements. Moreover, the NSS, being a self-organizing and self-structuring system that develops according to the laws of evolution, creates collections of ecological niches, the acceptability of which for the population of given territory is defined, as a rule, in accordance with national criteria (a set of environmental quality criteria, religious dogmas, national traditions, etc.).

When considering the viability conditions for life on Earth it is necessary to move away from human-oriented standards to assess the level of environmental degradation, since in the course of time local and regional changes in the environment become global. The amplitudes of these changes are defined by the mechanisms involved in NSS functioning, which provide optimality for any changes in its elements. Humankind has increasingly deviated from this optimality in its interaction with environmental abiotic and biotic formations. However, human society, as just another NSS element, will in time understand its large-scale relationship with nature, as a result of many sciences and studying the causal relationship in this system (Kondratyev *et al.*, 2003b).

The ecological environment in which people are living is a complex dynamic system. How stable it is in time depends on the regularity of its structure, material composition, energy balance, and stability of its reactions to the same external influences. System stability can be disturbed as a result of influences from both passive and active external impacts. In other words, under modern conditions nature $N$ and human society $H$, a united planetary system with hierarchical structures $(|N|, |H|)$, interact as a result of pursuing their own goals $(\underline{N}, \underline{H})$. From a formal standpoint this interaction can be considered a causal process $\eta(x,t)$ with an unknown distribution law, representing the tension level between subsystems $N$ and $H$ or evaluating the state of one of them. In this process $x = \{x_1, \ldots, x_n\}$ is a set of features identifying subsystems $N$ and $H$ as components of a possible indicator of a natural catastrophe (i.e., deflections $\eta(x,t)$ from limits), where the state of subsystem $N$ becomes threatening for $H$. It follows from this that the purposes and behaviors of subsystems $N$ and $H$ are functions of the factor $\eta$, depending on which their behaviors can be antagonistic, indifferent, or cooperative. The main goal of subsystem $H$ consists in achieving high living standards in comfortable conditions and guaranteed permanent survival. The goal and behavior of subsystem $N$ are defined by the objective laws of co-evolution (in other words, biotic regulation of the environment; Kondratyev *et al.*, 2003b, 2004b). In this sense the division between $N$ and $H$ is conditional and can be interpreted as the division of all natural processes into controlled and non-controlled. As world population increases, natural catastrophes will intensify the sensation of unconformity reflected in the social and cultural conditions of many regions.

From a theory of systems viewpoint, $H$ and $N$ are open systems. Without going into the philosophical and methodical aspects of this division, we assume that systems are symmetrical from the viewpoint of their simulation. In addition, system $H$ disposes technology, science, economic potential, industry and agriculture, sociological structure, size of population, etc. Interaction between systems $N$ and $H$ brings about change in $\eta$, the level of which influences the structures of $\underline{N}$ and $\underline{H}$. There exists a threshold $\eta_{max}$ beyond which humankind ceases to exist but nature survives. Any asymmetry between subsystems $N$ and $H$ in this sense causes the goal and strategy of system $H$ to change. Probably, under today's conditions the value of $\eta$ is rapidly approaching $\eta_{max}$, but then separate components of $\underline{H}$ may possibly refer to the cooperative class. Since the modern socioeconomic structure of the world is represented by countries, it is reasonable to consider them as functional system elements of $H$. The function $\eta(x,t)$ reflects the result of the interaction between countries and nature. It is possible to describe the results of these interactions by matrix $B = \|b_{ij}\|$, each element of which carries the symbolic semantic load:

$$b_{ij} = \begin{cases} + & \text{for cooperative behavior,} \\ - & \text{for antagonistic interaction,} \\ 0 & \text{for indifferent behavior.} \end{cases}$$

Many scientific theories study the different ways in which complex systems interact. In the asymmetrical case considered here the question is about the survivability of system $H$ and finding a way to estimate the future dynamics of system $H$. According to Podlazov (2001), the reflexive behavior of $H$ will help humankind one day to find a behavioral technology capable of comparing advantages and danger, of understanding limits to possibilities, and of identifying new threats at an early stage.

### 6.7.3 Method for forecasting natural catastrophes

As a result of the many aspects of environmental change in recent decades many authors have come up with multiple concepts to describe the NSS on a global basis and models of different complexity to parameterize the dynamics of biospheric characteristics. The availability of a global database containing highly detailed information allows the consequences of possible realization of different scenarios for the development of subsystem $H$ to be considered and evaluated. Traditional approaches to synthesizing global models are founded on considering collections of balance equations, in which parameters $\{x_i\}$ take the form of functions, arguments, factors, and conditions of the transition between parametric descriptions of environmental processes. Moreover, other approaches based on evolutionary and neural network algorithms have been used (Nitu *et al.*, 2004). How the functioning of the $N \cup H$ global model is organized can be presented as a conceptual scheme like that described by Kondratyev *et al.* (2004a). This scheme can be realized by introducing a geographical cell $\{\varphi_i, \lambda_j\}$ with spatial steps $\Delta\varphi_i$ and $\Delta\lambda_j$ for the land surface and World Ocean by latitude and longitude, respectively. All processes and NSS elements are considered as uniform and parameterized by point models within the pixel $\Omega_{ij} = \{(\varphi, \lambda) : \varphi_i \leq \varphi \leq \varphi_i + \Delta\varphi_i, \lambda_j \leq \lambda \leq \lambda_j + \Delta\lambda_j\}$. The choice of pixel size is defined by a set of conditions that depend on the spatial resolution of satellite measurements and the availability of a necessary global database. In the case of the water surface, the water body of pixel $\Omega_{ij}$ is divided by depth $z$ into layers (i.e., three-dimensional volumes $\Omega_{ijk} = \{(\varphi, \lambda, z) : (\varphi, \lambda) \in \Omega_{ij}, z_k \leq z \leq z_k + \Delta z_k\}$ are formed). All elements of $\Omega_{ijk}$ are considered uniform. Finally, the atmosphere above pixel $\Omega_{ij}$ is digitized by height in accordance with atmospheric pressure levels or by height on typical layers.

Creation of a global model is clearly only possible if knowledge and data are available on a multidisciplinary level. Among the ensemble of global models we consider the best to be that described by Kondratyev *et al.* (2004a). A block diagram of this model is submitted in Figure 3.2. Kondratyev *et al.* (2002) describe an adaptive procedure for fitting a global model into a geoinformation monitoring system.

### 6.7.4 Search and discovery of the imminence of natural catastrophes

The time leading up to the occurrence of a natural catastrophe is characterized by hits of vector $\{x_i\}$ appearing in a certain cluster of multi-dimensional space $X_c$. In other words, for quantitative determination of this process, we use a generalized feature $I(t)$ of the natural catastrophe and identify it with a graduated scale $\Xi$, for which the presence of relations of type $\Xi_1 < \Xi_2$, $\Xi_1 > \Xi_2$, or $\Xi_1 \equiv \Xi_2$ is postulated. This means that a value of $I(t) = \rho$ always exists that defines the imminence of a natural catastrophe of given type: $\Xi \to \rho = f(\Xi)$, where $f$ transforms the notion of natural catastrophe into a number. As a result, value $\theta = |I(t) - \rho|$ defines the expected interval of time before the catastrophe occurs.

Let us try to find a satisfactory model to portray a natural catastrophe by means of notions and signs that comply with a formalized description and transformation. For this purpose, we select $m$ element subsystems of the bottommost level in system $N \cup H$, whose interactions are defined by the matrix function $A = \|a_{ij}\|$, where $a_{ij}$ is a factor of the dependence level of relations between subsystems $i$ and $j$. Then characteristic $I(t)$ can be defined as a sum:

$$I(t) = \sum_{i=1}^{m} \sum_{j>i}^{m} a_{ij}.$$

In the general case we clearly have $I = I(\varphi, \lambda, t)$. For limited territory $\Omega$ with area $\sigma$ the indicator $I$ can be defined as an average value:

$$I_\Omega(t) = (1/\sigma) \int_{(\varphi,\lambda) \in \Omega} I(\varphi, \lambda, t)\, d\varphi\, d\lambda.$$

Entering indicator $I$ brings a scheme to monitor and forecast natural catastrophes that much closer. We now present the possible structure of a monitoring system that has functions to search for, forecast, and plot the course of a natural catastrophe. Three levels are selected in the system: holder, resolver, and searcher. Blocks representing these have the following functions:

(1) periodic examination of Earth's surface elements;
(2) memorizing suspicious elements;
(3) looking for traces of the anomaly moving as a result of suspect elements;
(4) accumulating data on fixed elements of the Earth's surface over time to carry out a statistical analysis to decide whether fixed suspicious elements are background noise or signals;
(5) stage-by-stage localization of the procedure in the search for an anomaly.

The efficiency of such a monitoring system depends on the parameters used by technical facilities to take measurements and on algorithms for data processing.

An important role is played here by the environmental model, used in parallel with the monitoring system to set up and statistically test row $\{I_\Omega(t)\}$ and adapted to the monitoring mode favored by a GIMS-based adaptation scheme.

Employing these criteria when studying natural catastrophes reveals that form and behavior $I_\Omega(t)$ have a typical type for each process in the environment. A problematic task consists in determining these forms and categorizing them. For instance, dangerous natural phenomena such as landslides and mudflows often appear and show typical signs of preliminary change in the relief and landscape, which can be successfully registered by satellites in the optical range. Moreover, aerial images and *in situ* measurements of relief slopes, hillsides, and hydrologic networks permit forecasting such events several days before. However, the optical range is limited under cloudy conditions or thick vegetation cover. Nevertheless, this can be overcome by introducing remote-sensing systems that operate in the microwave range of the electromagnetic spectrum. Then, in addition to the above-mentioned indicators of landslides and mudflows it is possible to add such informative parameters as soil moisture and biomass, since increase in soil moisture leads to the appearance of landslides, while increase in the biomass is indicative of strengthening the restraining role played by vegetation cover in preventing soil and rocks moving. This is particularly important when monitoring snow–stone avalanches or purely snow avalanches. Creating a catalog of such indicators for all possible natural catastrophes and adding them to the knowledge base of a monitoring system is a necessary stage in increasing its efficiency.

Knowledge of the set of informative indicators $\{x_i^j\}$ that characterize a natural catastrophe of $j$th type and *a priori* determination of the cluster $X^j$ in space of these indicators allows calculating the velocity $v_j$ of approach of point $\{x_i^j\}$ to the center of $X^j$ during satellite observation and, thereby, to evaluate the time of arrival of the catastrophe. Other algorithms can be used to forecast natural catastrophes. For instance, forecasting forest fire is possible by considering the dependences of forest radiothermal radiation of different wavelengths on the degree of wetness/dryness of combustible timber. Knowledge of such dependences facilitates assessing the risk of forest fire by taking into account the water content in plant cover and in the upper layer of soil.

Many studies have shown that there is a real possibility to assess the risk of fire in waterlogged and marsh forest by considering the water content in the plant cover and upper soil layer, using microwave measurements in the 0.8–30 cm range. Multi-channel observations together with cluster analysis algorithms facilitate forest classification according to the category of fire. The efficiency of such methods depends on detailed descriptions of the forest structure in the model which reflect the state of the canopy and density of trees. The most dangerous and difficult to detect are undergrowth fires. The three-layer model of the soil–stem–canopy system is efficient in this case when it is coordinated with the fire risk indicator $I(\lambda_1, \lambda_2) = [T_b(\lambda_1) - T_b(\lambda_2)]/[T_b(\lambda_1) + T_b(\lambda_2)]$. For instance, under $\lambda_1 = 0.8$ cm and $\lambda_2 = 3.2$ cm indicator $I$ changes approximately from $-0.25$ in areas where there is no forest fire to 0.54 where there is. At the first sign of undergrowth fires $I \approx 0.23$. In addition, the value $I$ depends weakly on the distribution of combustible layers in the forest such as lichens, mosses, grass, dead pine-needles or leaves.

Realization of such a three-level mode for decision making about the imminence of a natural catastrophe depends on the spatiotemporal scales of the monitoring system corresponding with characteristics of the natural phenomena. The most complex for decision making are delayed action natural catastrophes, realization of which can take decades. Ozone holes, global warming, desertification, biodiversity reduction, overpopulation of the Earth, etc. belong to expected catastrophes. The solution to the basic task of reliably forecasting such undesirable natural phenomena can be realized via the GMNSS—using global databases, current satellites, and *in situ* measurements.

## 6.8 DECISION-MAKING PROCEDURES IN MONITORING THE OCEAN–ATMOSPHERE SYSTEM

### 6.8.1 Phases of the ocean–atmosphere system

Understanding the interaction between the ocean and atmosphere is critical for predicting changes in global temperature patterns and climate variability as a result of global warming and for predicting tropical cyclone onset. Ocean–atmosphere coupling is a concept of climate dynamics essential for understanding many climate phenomena including the El Niño–Southern Oscillation (ENSO) and climate variability on multi-decadal timescales. Heat exchange between the ocean and the atmosphere (heat fluxes), atmospheric circulation and hence the effect of winds on the ocean surface (wind stress), and large-scale anomalies of sea surface temperature are principal areas of investigations based on numerous observations and models.

Ocean–atmosphere coupling at different latitudes remains a subject of intense research and even heated debate. Climate variability is determined to a large extent by random atmospheric variability, also called atmospheric noise, and passive oceanic response to atmospheric forcing. Numerous efforts to parameterize interactions between the ocean and atmosphere by simulation modeling have not been successful. The main difficulty is caused by instability of the ocean–atmosphere system (OAS). This is the reason Soldatov (2009a, b) proposed a new approach to the study of ocean–atmosphere interactions: one that considers the state of this system as an aggregate of phases.

There are scales that can be used to classify OAS states. The best known are the Saffir–Simpson hurricane scale, the Beaufort scale, and the Fujita–Pearson scale.

The Saffir–Simpson scale was developed in 1971 by civil engineer Herbert Saffir and meteorologist Bob Simpson, who at the time was director of the U.S. National Hurricane Center (NHC). This scale is the classification used for some

**Table 6.6.** The Saffir–Simpson scale for measurement of hurricane magnitude (Webster *et al.*, 2005).

| *Hurricane category* | *Maximum wind speed* | | *Minimum atmospheric pressure* | *Storm surge height* |
|---|---|---|---|---|
| | m/s | km/h | (hPa) | (m) |
| Tropical depression | 0–17 | 0–62 | | |
| Tropical storm | 18–32 | 63–118 | | |
| One | 33–43 | 119–153 | ≥980 | 1.0–1.7 |
| Two | 43–50 | 154–177 | 979–965 | 1.8–2.6 |
| Three | 50–56 | 178–209 | 964–945 | 2.7–3.8 |
| Four | 56–67 | 210–249 | 944–920 | 3.9–5.6 |
| Five | >67 | >250 | <920 | ≥5.7 |

western hemisphere tropical cyclones (i.e., hurrricanes) that exceed the intensities of tropical depressions and tropical storms. The scale divides hurricanes into five categories distinguished by the intensities of sustained winds. To be classified as a hurricane, a tropical cyclone must have maximum sustained winds of at least 74 mph (33 m/s, 64 kn/h, 119 km/h). The highest classification in the scale, Category 5, is reserved for storms with winds exceeding 155 mph (69 m/s, 135 kn/h, 249 km/h). There are two additional categories that correspond to storms and depressions (Table 6.6).

The Beaufort scale (Table 6.7) is an empirical measure for describing wind speed based on observed sea conditions. The scale was created in 1806 by Sir Francis Beaufort, an Irish-born British admiral and hydrographer. The Beaufort scale was extended in 1946, when Forces 13 to 17 were added. However, Forces 13 to 17 were intended to apply only to special cases, such as tropical cyclones. Nowadays, the extended scale is only used in Taiwan and mainland China, which are often affected by typhoons. Wind speed on the extended Beaufort scale is based on the empirical formula:

$$v = 0.837B^{3/2}\ \text{m/s},$$

where $v$ is the equivalent wind speed at 10 m above the sea surface; and $B$ is the Beaufort scale number. For example, $B = 9.5$ equates to 24.5 m/s, equal to the lower limit of "10 on the Beaufort scale". Using this formula the highest winds in hurricanes would be 23 on the Beaufort scale.

Today, hurricane force winds are sometimes described as 12–16 on the Beaufort scale, which very roughly relate to the respective category speeds of the Saffir–Simpson hurricane scale, by which actual hurricanes are measured, where

**Table 6.7.** The modern (extended) Beaufort scale.

| *Beaufort number* | *Description* | *Wind speed* (m/s) | *Wave height* (m) | *Sea conditions* |
|---|---|---|---|---|
| 0 | Calm | <0.3 | 0 | Flat |
| 1 | Light air | 0.3-1.5 | 0–0.2 | Ripples without crests |
| 2 | Light breeze | 1.6–3.4 | 0.2–0.5 | Small wavelets |
| 3 | Gentle breeze | 3.5-5.4 | 0.5–1 | Large wavelets |
| 4 | Moderate breeze | 5.5–7.9 | 1–2 | Small waves with breaking crests |
| 5 | Fresh breeze | 8.0–10.7 | 2–3 | Moderate waves of some length |
| 6 | Strong breeze | 10.8–13.8 | 3–4 | Long waves begin to form |
| 7 | High wind, moderate gale, near gale | 13.9–17.1 | 4–5.5 | Heavy swells |
| 8 | Gale, fresh gale | 17.2–20.7 | 5.5–7.5 | Moderately high waves with breaking crests and spindrift |
| 9 | Strong gale | 20.8–24.4 | 7.5–9 | High waves whose crests sometimes roll over |
| 10 | Storm, whole gale | 24.5–28.4 | 9–11.5 | Very high waves with overhanging crests |
| 11 | Violent storm | 28.5–32.6 | 11.5–14 | Exceptionally high waves |
| 12 | Hurricane force | ≥32.7 | ≥14 | Huge waves |

Category 1 is equivalent to 12 on the Beaufort scale. However, the extended Beaufort numbers above 13 do not match the Saffir–Simpson scale. Category 1 tornadoes on the Fujita–Pearson scale and TORRO (Tornado and Storm Research Organization) scales also begin roughly at 12 on the Beaufort scale, but are indeed independent scales. The TORRO scale (or T-scale) measures tornado intensity between T0 and T10. It was developed by Meaden (1985), who worked for TORRO (a meteorological organization in the U.K.), as an extension of the Beaufort scale. The T-scale (Table 6.8) sets T0 as equivalent to 8 on the Beaufort scale and is related to the Beaufort scale ($B$) by the formula $B = 2(TI + 4)$ and, conversely, $TI = (B/2 - 4)$. The T-scale formula for wind speed is $v = 2.365(T + 4)^{3/2}$ m/s or $v = 8.511(T + 4)^{3/2}$ km/h.

In most cases, tornado intensity is measured by the Fujita–Pearson scale (Table 6.9) which really reflects the risk level. It is important to know when and how a tornado forms. It is known that many tornadoes have their origin as

Table 6.8. TORRO scale parameters.

| *TORRO intensity* | *Wind speed* (m/s) | *Tornado description* | *Intensity description* | *Damage description* |
|---|---|---|---|---|
| FC | | Funnel cloud aloft (not a tornado) | Weak | No damage to structures (apart from tops of tallest towers) or to radiosondes, balloons, and aircraft. No damage in the countryside (apart from agitation of the highest tree tops and effect on birds and smoke). Recorded as FC when known not to have reached ground level. A whistling or rushing sound may be noticed overhead |
| T0 | 17–24 | Light | | • Loose litter raises from the ground as spirals<br>• Tents, marquees, awnings seriously disturbed<br>• Exposed tiles and slates dislodged. Twigs snap; trails visible through crops<br>• Wheelie bins topple over and roll<br>• Garden furniture and pots disturbed |
| T1 | 25–32 | Mild | | • Deck chairs, small plants, heavy litter become airborne<br>• Minor damage to sheds<br>• More serious dislodging of tiles, slates<br>• Chimney pots dislodged. Wooden fences flattened<br>• Slight damage to hedges and trees<br>• Windows already ajar blown open breaking latches |
| T2 | 33–41 | Moderate | | • Mobile homes displaced. Caravans blow over<br>• Garden sheds destroyed. Garage roofs torn away and doors implode<br>• Much damage to tiled roofs and chimneys. Ridge tiles missing<br>• General damage to trees, large branches twist or snap off, small trees uprooted<br>• Bonnets blown open on cars<br>• Weak or old brick walls topple<br>• Windows blown open or glazing sucked out of frames |

(*continued*)

**Table 6.8.** TORRO scale parameters (*cont.*)

| *TORRO intensity* | *Wind speed* (m/s) | *Tornado description* | *Intensity description* | *Damage description* |
|---|---|---|---|---|
| T3 | 42–51 | Strong | Weak | • Mobile homes overturned/badly damaged. Caravans destroyed. Garages and weak outbuildings destroyed<br>• House roof timbers exposed. Large trees snapped or uprooted<br>• Heavier debris becomes airborne causing secondary damage breaking windows and impaling softer objects<br>• Debris carried considerable distances. Garden walls blown over<br>• Eyewitness reports of buildings physically shaking<br>• Mud sprayed up the side of buildings |
| T4 | 52–61 | Severe | Strong | • Automobiles lifted from the ground. Mobile homes airborne/destroyed<br>• Sheds airborne and carried considerable distances. Entire roofs removed from houses<br>• Roof timbers of strong brick or stone houses completely exposed. Gable ends torn away<br>• Trees uprooted or snapped. Traffic signs folded or twisted<br>• Large trees uprooted and carried several yards<br>• Debris carried up to 2 km leaving an obvious trail |
| T5 | 62–72 | Intense | | • Heavier motor vehicles (4 × 4s, 4 t trucks) lifted from the ground<br>• Wall plates, entire roofs, and several rows of bricks on top floors removed<br>• Items sucked out of houses including partition walls and furniture<br>• Older weaker buildings collapse completely<br>• Support poles snapped |

| *TORRO intensity* | *Wind speed* (m/s) | *Tornado description* | *Intensity description* | *Damage description* |
|---|---|---|---|---|
| T6 | 73–83 | Moderately devastating | Strong | • Strongly built houses suffer major damage or are demolished completely<br>• Bricks and blocks, etc. become dangerous airborne debris<br>• National grid pylons damaged or twisted<br>• Exceptional or unusual damage found (e.g., objects embedded in walls or small structures lift up and land with no obvious damage) |
| T7 | 84–95 | Strongly devastating | | • Brick and wooden-frame houses wholly demolished<br>• Steel-framed warehouse-type constructions destroyed or seriously damaged<br>• Trains disrailed<br>• Trees de-barked by flying debris |
| T8 | 96–107 | Severely devastating | Violent | • Automobiles carried great distances<br>• Some steel-framed factory units severely damaged or destroyed<br>• Steel and other heavy debris strewn over great distances<br>• Much damage within the confines of the damage path |
| T9 | 108–120 | Intensely devastating | | • Many steel-framed buildings demolished<br>• Trains hurled some distances<br>• Complete de-barking of standing tree trunks.<br>• Inhabitant survival reliant on shelter below ground level |
| T10 | 121–134 | Super | | • Entire houses and similar buildings lifted bodily from foundations and carried some distance<br>• Severe destruction, leaving a broad linear track largely devoid of vegetation, trees, or man-made structures |

**Table 6.9.** Characteristics of the Fujita–Pearson scale (Marshall, 2001).

| *Level of scale* | *Intensity* | *Wind speed* (km/h) | *Damage scale* | *Type of danger* |
|---|---|---|---|---|
| F0 | Storm | 40–72 | Light | Some damage to chimneys, broken tree branches, uprooted trees, signs torn off buildings |
| F1 | Moderate tornado | 73–112 | Moderate | Lowest wind speed of early-stage hurricanes. Blows off roofs, pulls houses from foundations or blows them over, moves cars, destroys detached or attached garages |
| F2 | Substantial tornado | 113–157 | Considerable | High level of danger. Blows off roofs, destroys mobile homes, overturns box garages, breaks or uproots big trees, carries light-weight objects off |
| F3 | Strong tornado | 158–206 | Severe | Roofs are ripped off and some walls (even of well-constructed houses) collapse, the majority of forest trees are uprooted, skyscrapers are twisted and deformed suffering massive damage to their exteriors, automobiles (even large ones) are lifted off the ground and thrown some distance |
| F4 | Devastating tornado | 207–260 | Devastating | Destroys solid buildings, removes houses with poor foundations, cars, and large objects. |
| F5 | Incredible tornado | 261–318 | Incredible | Lifts houses from foundations and carries them away, moves cars as much as 100 m, snaps trees, damages steel-fastened and concrete-fastened structures |
| F6 | Inconceivable tornado | 319–379 | Maximal | Such winds are improbable. The damage they would do is much the same as tornadoes of type F4 and F5. The vortex would apply massive power over a small territory, lifting up large volumes of water and transporting them great distances |

tropical hurricane products. Therefore, information about conditions for tropical hurricane onset and the track it follows are principal stages to predict tornado origin. Resolving these tasks depends on data of different origin. The scale allows a tornado's wind speed to be determined by various means:

- visiting the tornado damage site to make non-engineering assessments;
- obtaining an engineering assessment of the damage;
- using Doppler radar;
- applying photogrammetric analysis;
- directly measuring observed tornadoes.

### 6.8.2 Monitoring data

The existing global Earth-observing system includes several levels of subsystems. Their interactions are presented symbolically in Figure 6.13. The Global Observation Earth System of Systems (GEOSS) is based on present monitoring of the ocean–atmosphere system (Charvat *et al.*, 2010). GEOSS is being developed by the Group on Earth Observations (GEO) on the basis of a 10-year implementation plan running from 2005 to 2015 (Klein, 2008). Earth observations include measurements and monitoring of the Earth under water, on and beneath the land surface, air and water quality, atmospheric conditions, and measures of the health of humans, plants, and animals. Measurements can be made directly or by sensors *in situ* or remotely, and all these observations support modeling or other tools that create information for environmental decision making.

**Figure 6.13.** Schematic representation of the existing global observing system (*www.meted.ucar.edu/tropical/textbook-2nd-edit*).

The Global Change Observation Mission (GCOM) is concerned with retrieval algorithms for geophysical products, product validation, and data application. GCOM seeks to establish and demonstrate a global long-term satellite observing system to measure essential geophysical parameters for understanding global climate change and the water cycle mechanism and, thereby, contribute to improving future climate projections through a collaborative framework with climate model institutions. Demonstrating the capabilities of operational applications by providing continuous data to operational agencies is another important purpose. GCOM will take over the Advanced Earth Observing Satellite-II (ADEOS-II) mission and transition into long-term monitoring of the Earth. To achieve global, comprehensive, long-term, and homogeneous observation, GCOM will consist of two satellite types and three consecutive generations with a one-year overlap, resulting in more than a 13-year observation period. The two satellites are GCOM-W (water) and GCOM-C (climate). The GCOM-W1 satellite will carry the Advanced Microwave Scanning Radiometer-2 (AMSR2) to contribute to understanding the water and energy cycle. The GCOM-C1 satellite will be equipped with the Second-Generation Global Imager (SGLI) to observe the Earth's atmosphere and surface and thereby contribute to understanding of the carbon cycle and radiation budget (GCOM, 2008).

By various means, numerous data sources about atmospheric and oceanic processes are available for decision making concerning the state of the OSE. The basic purposes of most global Earth observing systems are to predict climate change. For realistic prediction of climate change, these data and oceanic and atmospheric models must be coupled. It is evident that oceanic processes exert a major influence on climate. Moreover, sea surface temperature determines the heat exchange between the atmosphere and oceans. In order to assess the OSE state, results from different Earth observing systems and general circulation models are processed. According to Cracknell *et al.* (2009), the OSE state is characterized by the unstable dynamics of its components and, therefore, a lot of monitoring data cannot be used as a result of their temporal incompleteness. In this sense, data of numerous meteorological stations such as the TAO/TRITON&PIRATA/RAMA (Tropical Atmosphere Ocean/Triangle Trans-Ocean Buoy Network & Pilot Research Moored Array in the Tropical Atlantic/Research Moored Array for Monsoon Analysis) are more acceptable.

The Global Climate Observing System (GCOS) and Global Ocean Observing System (GOOS) produce large amounts of archive and operative data about different environmental characteristics on a global basis. TAO/TRITON/PIRATA/RAMA is an efficient subsystem of these systems: the data of this subsystem are the most precise and informative.

Table 6.10 describes the data employed in this chapter. More precisely, this dataset is based on archived data of the TAO/TRITON&PIRATA system for the months in which hurricane activity is greatest. The basic goal of data processing is to search for a procedure and a criterion to determine with high probability the time at which hurricanes arise and the conditions that precede the event. The tropical moored buoys that deliver numerous data about different meteorological and geophysical characteristics of the ocean–atmosphere system are located in the Pacific and Atlantic Oceans. The most densely spaced meteorological stations are in the tropical sector of the Atlantic Ocean off the coast of Florida. Assessment of the algorithm proposed in this chapter can be made by choosing meteorological stations that are located close to the equator, where severe hurricanes are registered by existing monitoring systems. To this end, the standard meteorological data of meteorological stations of the U.S. National Data Buoy Center have also been used.

### 6.8.3 Percolation procedure

Let us consider the cubic grid $\Xi_d = \{\Xi_{i_1,\dots,i_d}\}$ of dimension $d$ with side $h$ in the phase space of the meteorological parameters enumerated in Table 6.10. The data flow from TAO/TRITON&PIRATA is not restricted in time and is characterized by data entry $\bar{x} = \{x_i; i = 1, \dots, 10\}$ regularly at hourly intervals. A limited time interval, $T$, is chosen with the successive times $\{t_j; j = 1, \dots, N\}$ when data on the meteorological situation in the previously surveyed site of the ocean–atmosphere system are delivered. In the course of time, the cells $\Xi_{i_1,\dots,i_d}$ in which a point $\{x_1^*(t_j), \dots, x_d^*(t_j)\}$ is hit are colored. The value $x_i^*(t_j)$ here is the observed value of the $i$th parameter in time $t_j \in T$. Following the percolation procedure (Soldatov, 2006, 2007; Sukov and Soldatov, 2007) we form clusters from the interfacing colored cells. Parameter $\lambda$ denotes the proportion of the total number of cells that are colored and $\mu$ is the relative volume of the cluster space; the distribution of clusters is assessed according to size.

The structure of the cluster space depends on the interval $T$, the choice of which is determined by the specific character of the given task. It follows from data processing that parameters $\lambda$ and $\mu$ can achieve their asymptotic levels $\lambda^*$ and $\mu^*$ when observations are gathered over a long time. Under this consideration the character of such a transition between different phase states of the ocean–atmosphere system reflects the objective laws for the onset of a hurricane. Consequently, calculation of percolation cluster parameters for different environmental conditions in which the ocean–atmosphere system functions makes it possible to construct a decision-making rule to assess the moment at which phase transition in the dynamics of the ocean -atmosphere system occurs.

An average cluster size $S$ (number of colored cells) and correlation length $\xi$ (linear size of the cluster) can be approximated by the following functions

$$S(\lambda) \approx |\lambda - \lambda^*|^{-\gamma}, \qquad \xi(\lambda) \approx |\lambda - \lambda^*|^{-\nu}, \tag{6.37}$$

where parameters $\gamma$ and $\nu$ depend on the cluster space dimension $d$ and, in a particular case, can be determined empirically. Functions $S$ and $\xi$ defined in formulas (6.37) increase sharply at hurricane onset. By definition (Essam, 1980) an average number $N_t$ of colored sites in the percolation cluster after numerous steps $t$ of the random walk of point $A$ and under the absence of a critical transition

**Table 6.10.** List of environmental characteristics used to detect tropical hurricane onset.

| *Measurement of meteorological parameters using the TAO/TRITON&PIRATA system* | *Parameter identifier* | *Parameter characteristic* |
|---|---|---|
| Ocean surface temperature (°C) | $x_1$ | Temperature is measured at depths 1.0 m or 1.5 m depending on buoy station type with precision ±0.003°C–±0.02°C |
| Wind speed (m/s) and direction (in degrees clockwise from the north) | $x_2, x_3$ | Wind parameters are measured at altitude 3.5 m. Wind speed is measured with precision 3%. Its direction is determined with precision 5.0°–7.8° |
| Precipitation (mm/h) | $x_4$ | Precipitation is registered every 10 minutes at altitude 3.5 m with precision ±0.4 mm/h |
| Seawater density (kg m$^3$) | $x_5$ | Water density is calculated as a function of its salinity |
| Depth of the isotherm (20°C) | | Depth of isotherm is calculated on the basis of vertical temperature distribution by linear interpolation at 20 m steps in depth |
| Water temperature of fixed depths $\{h_i\}$ (°C) | $x_7(h_i)$ | Water temperature is measured up to 750 m on fixed horizons with precision ±0.02°C |
| Relative humidity of atmospheric air (%) | $x_8$ | Air humidity is measured at altitude 2.2 m above sea level with precision ±2.7% |
| Atmospheric pressure (atm) | $x_9$ | Atmospheric pressure is fixed with precision ±0.01% at altitude 3 m above sea level |
| Atmospheric temperature (°C) | $x_{10}$ | Atmospheric temperature is measured at latitude 2.2 m with precision ±0.2°C |
| Seawater salinity (‰) | $x_{11}$ | Salinity is assessed by data on water temperature and conductivity with precision ±0.02‰ |
| Significant wave height (m) | $x_{12}$ | Calculated as the average of the highest one-third of all wave heights during a 20 min sampling period |
| Dominant wave period (s) | $x_{13}$ | Period with maximum wave energy |
| Average wave period (s) | $x_{14}$ | Period (s) of all waves during a 20 min period |

| *Measurement of meteorological parameters using the TAO/TRITON&PIRATA system* | *Parameter identifier* | *Parameter characteristic* |
|---|---|---|
| Direction from which waves at the dominant period are coming (deg) | $x_{15}$ | The units are degrees from true north, increasing clockwise, with north as 0 (zero) degrees and east as 90 degrees |
| Dewpoint temperature (°C) | $x_{16}$ | Dewpoint temperature taken at the same height as the air temperature measurement |
| Station visibility (nmi) | $x_{17}$ | Buoy stations are limited to reports from 0 to 1.6 nmi |

equals:

$$\langle N_t \rangle = t^{\frac{-\theta}{2}},$$

where parameter $\theta \in [1.29 - 2.84]$ is evaluated in each case by means of a modeling procedure.

Using TAO/TRITON&PIRATA/RAMA data in the monitoring regime by employing the percolation model is equivalent to the scheme of point $\{x_i\}$ randomly walking in the sites of $\Xi_d$. In the case $d = 2$ we consider wind speed, $x_2$, and atmospheric pressure, $x_9$, which are two meteorological parameters that can be expected to vary very strongly in the presence of a tropical cyclone. In Figure 6.14 we plot the hourly values of $x_2$ and $x_9$ for the period in August 2005 when Hurricane Katrina formed, developed, and dissipated. The random walk of point $\{x_i\}$ in $\Xi_2(x_9, x_2)$ is a function of a changing meteorological situation that is characterized by the transition from a background state to the state in which a tropical hurricane occurs. In this diagram the existence of the tropical hurricane within the time period $T$ is characterized by a double percolation transition of point $\{x_9, x_2\}$ through zone $\Xi_2^2$ between zones $\Xi_2^1$ and $\Xi_2^3$. Analysis of the motion of point $\{x_9, x_2\}$ agrees with the assessment of probability characteristics for the drift of its projection toward $\eta$ at the boundary of zone $\Xi_2^2$. The drift $\nu$ of this projection from some initial position $A$ represents the sum $\sum_v$ of a random number of components.

The distribution structure of zones in the plane $\Xi_2$ for the Bahamas in August 2005 (Figure 6.14) has specific elements bounded by the zone $\Xi_2^2$ in which the event leading to Hurricane Katrina was arising but not detected. When $(x_9, x_2) \in \Xi_2^1$, the meteorological situation was characterized as background up to August 20, 2005. Hurricane Katrina was recorded in the morning of August 23, 2005 by Landsat and other meteorological satellites above the southeastern sector of the Bahamas and after that it was classified in as Category 1 by the Saffir–Simpson scale. Moving to the Gulf of Mexico towards New Orleans, Louisiana, it

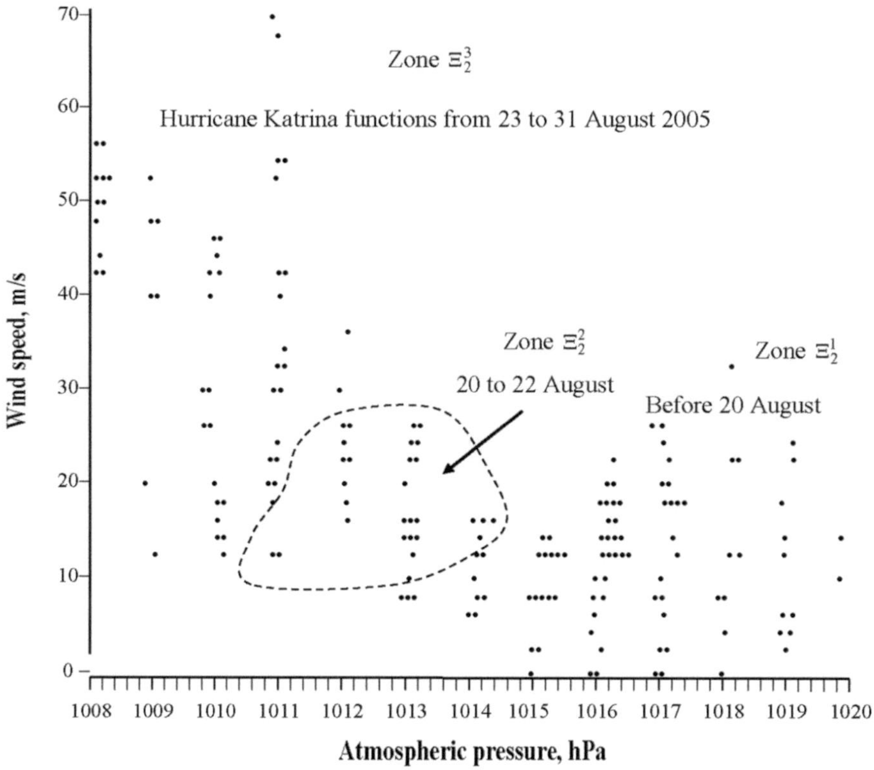


**Figure 6.14.** Cluster structure of the meteorological situation that took place during August 2005 in the region of the Bahamas from data of Meteorological Station No. SPGF1 (26°42′16″N, 78°59′40″W). Hurricane Katrina formed on August 23, hit New Orleans on August 28 and dissipated on August 30, 2005.

reached Category 5 on the morning of August 28 and then its strength decreased slowly until August 31. Hurricane Katrina passed New Orleans on August 29, 2005 and by August 31 New Orleans was flooded as a result. Analysis of the meteorological data shows that the first signs of environmental instability began to manifest themselves on August 20. Consequently, the interval between August 20 and 23 can be considered as the uncertainty zone (see Figure 6.14). Thus, it is necessary to find the indicators of environmental instability within this zone. Undoubtedly, the value of $\Delta$, the width of zone $\Xi_2^2$, and the structure of zone $\Xi_2^2$ depend on the dynamics of hurricane category. Searching for such a dependence is important and finding it would permit us to forecast hurricane origin and prognosis of its category.

Figure 6.15 represents the percolation cluster that formed in the region of the Bahamas in August of the 2005 hurricane season. The process of percolation cluster formation shows that there exists a moment when the ocean–atmosphere system changes its phase state. This moment is determined by the distances

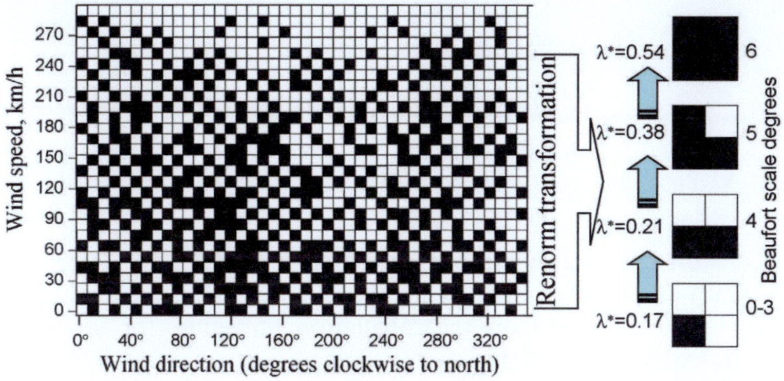


**Figure 6.15.** Principal scheme of the percolation transition in OAS phase situations between background state and hurricane state during August/September 2005. Data from Meteorological Station No. SMKF1 (24°37′36″N, 81°06′36″W).

between $\lambda$ and $\lambda^*$, $\mu$ and $\mu^*$. Simulation of this process has shown that the moment at which this transition occurs corresponds to the boundary between a storm and a Category 1 hurricane. This is not a very acceptable result. Therefore, it is necessary to introduce an additional indicator.

It can be seen that both percolation cluster sections have transition zones to the infinite cluster but are distinguished by the structural characteristics of the percolation cluster. Section $(x_2, x_9)$ has the following parameters: $\lambda^* = 0.2$, $\mu^* = 0.87$, and $S = 7.08$. Section $(x_{10}, x_8)$ has parameters $\lambda^* = 0.5$, $\mu^* = 0.94$, and $S = 22.5$. The phase transition zone can be defined in each case by geometrically reducing the route between the two zones of the basic cluster, each of which correspond to one state of the environment. It is evident that the moment this reduction takes place corresponds to hurricane onset. The calculation of hurricane origin probability depends on the structural dynamics of the cluster space.

To formalize the detection procedure for hurricane onset we introduce an algorithm to assess ocean-atmosphere system instability $I_m(t)$ (Krapivin *et al.*, 2007; Soldatov *et al.*, 2008; Sukov *et al.*, 2008):

$$I_m(k) = \frac{1}{s(m-N)} \sum_{j=1}^{m-N} \sum_{i=1}^{s} \beta_i \alpha_i(t_j) \quad (k = 1, 2, 3, \ldots, K), \tag{6.38}$$

where $N$ is the length of the quasi-stationary period of observations taken to calculate average values of the ocean–atmosphere system characteristics; $m$ is the length of the time interval in which indicator $I_m(k)$ is assessed; $s$ is the number of registered parameters of the ocean–atmosphere system; $\beta_i$ is the weight coefficient for the $i$th characteristic; $k$ is the number of time intervals taken for $I_m(k)$ assessment:

$$\alpha_i(j) = \begin{cases} 1 & \text{when } \Delta x_i(j) \cdot \Delta x_i(j+1) \le 0, \\ 0 & \text{when } \Delta x_i(j) \cdot \Delta x_i(j+1) > 0, \end{cases} \tag{6.39}$$

where

$$\Delta x_i(j) = \frac{1}{N} \sum_{l=m(k-1)+j}^{m(k-1)+j+N-1} x_i(l) - x_i(m(k-1)+j+N),$$

$$\Delta x_i(j+1) = \frac{1}{N} \sum_{l=m(k-1)+j+1}^{m(k-1)+j+N} x_i(l) - x_i(m(k-1)+j+N+1).$$

An indicator of ocean–atmosphere system instability $I_m(t)$ defined by equations (6.38) and (6.39) points to changeability of the current average value of vector $\{x_i\}$. Figure 6.16 explains the calculation procedure for this indicator. As traditional statistics are used to process meteorological data it is necessary to assess the quasi-stationary time interval. The analysis of one-dimensional distributions for selected meteorological parameters measured by TAO/TRITON&PIRATA shows this interval to be 7 hours. Figure 6.17 gives an example of the calculation of $I_m(t)$ for Hurricane Katrina. The approach of hurricane origin can be seen to be accompanied by $I_m(t)$ increasing. Three regions of $I_m(t)$ are involved here:

(1) the background region where variations in ocean–atmosphere system parameters during a restricted time interval change within the limits of statistical stability;

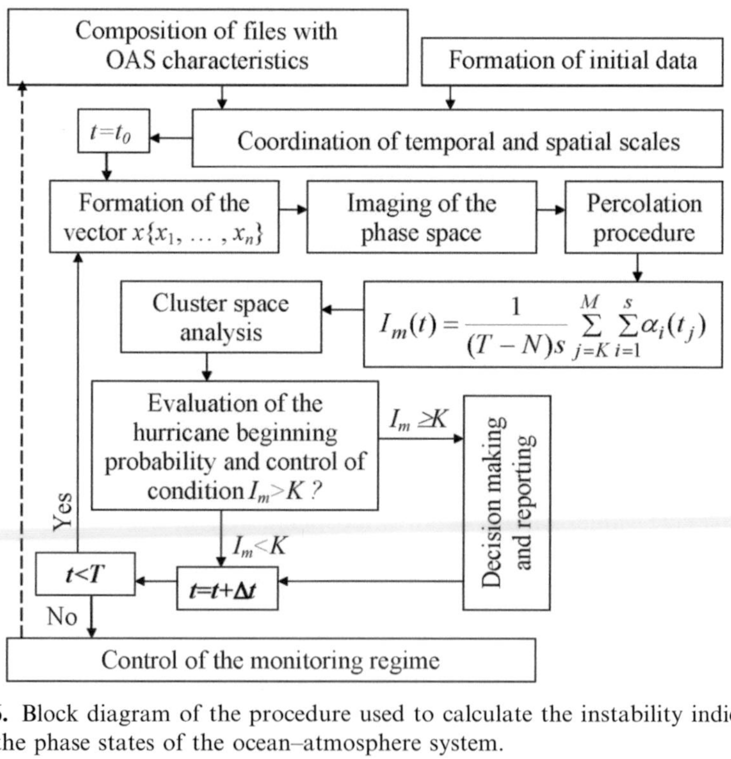


**Figure 6.16.** Block diagram of the procedure used to calculate the instability indicator characterizing the phase states of the ocean–atmosphere system.

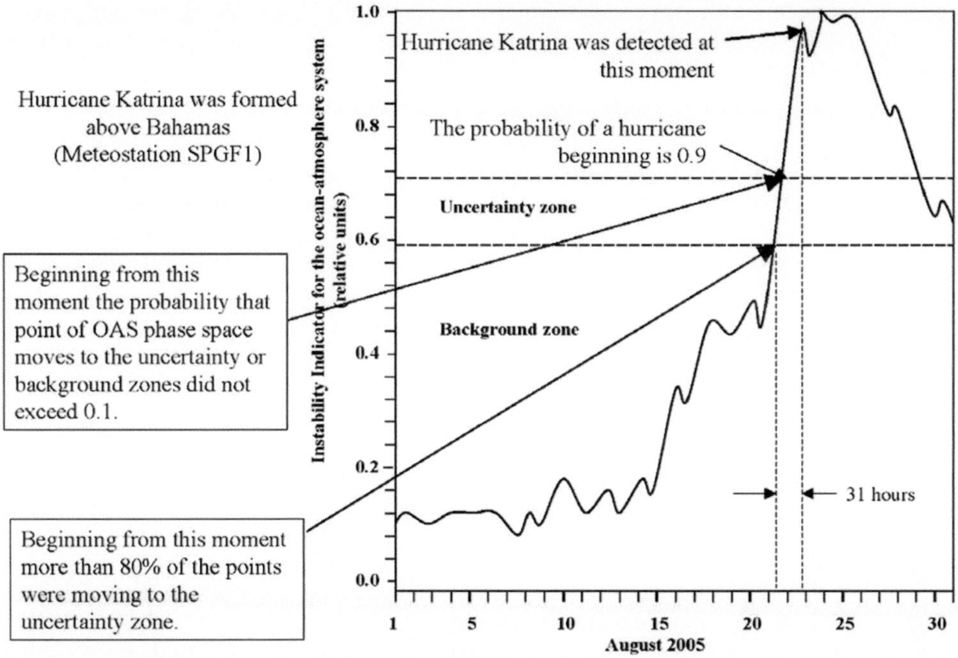


**Figure 6.17.** Dynamics of the instability indicator of the ocean–atmosphere system assessed using meteorological data from Meteorological Stations FWYF1 (25°35′25″N, 80°05′48″W); MLRF1 (25°00′36″N, 80°22′48″W), and SPGF1 (26°42′16″N, 78°59′40″W). The indicator was calculated using the following parameters: $s = 17$ (see Table 6.10 where all environmental parameters are taken into consideration), $\beta_i = 1$ ($i = 1, \ldots, 17$; the weight coefficients of environmental parameters are assumed equal), $m = 24$ hours (time period used to calculate the instability indicator equals one day), $N = 7$ (quasi-stationary time interval for environmental data), $K = 31$ days (another time period used to calculate the instability indicator). All parameters were estimated by numerical experiments to have high reliability within the calculation results.

(2) the uncertainty region where the probability of hurricane onset exceeds the level 0.5; and
(3) the hurricane proper region where the velocity of changes of all parameters $\{x_i\}$ achieve maximal values.

Using sequential analysis to control the dynamics of $I_m(t)$ in Figure 6.14 shows that $E(\nu) = 17$, $D(\nu) \to 0$, and $W_c(x) \to 0$ in zone $\Xi_2^1$. At the same time $E(\nu) = 9$, $D(\nu) \to 5$, $x = \nu/D(\nu) \to 1.35$, $c \to 16$, and $W_{16}(1.35) \to 0.912$ in zone $\Xi_2^3$. The latter explains the result represented in Figure 6.17. In other words, proceeding from verbal reasoning to quantitative estimation of the dynamics of $I_m(t)$, we introduce a generalized calibrated scale $\Xi$ for which we postulate the presence of relationships of the type $\Sigma_1 < \Xi_2$, $\Xi_1 > \Xi_2$, or $\Xi_1 = \Xi_2$. This means there always exists a value of $I_m(t) = \rho$ that can determine when phase transition of a given type in the ocean–atmosphere system can be expected. As a result, the magnitude $\theta = |I_m(t) - \rho|$ determines the expected time interval before phase transition occurs.

The region showing a sharp increase in the derivative of $I_m(t)$ corresponds to phase transition of the ocean–atmosphere system. Consequently, tracking the meteorological situation and synchronous calculation of the instability indicator by comparing it with the structure of the cluster space makes it possible to detect the moment the ocean–atmosphere system transitions from background state to the hurricane origin state. Figure 6.17 shows that this instability indicator would have been able to detect the start of Hurricane Katrina at least 31 hours before its detection by satellite. Figure 6.18 shows $I_m(t)$ within the 2005 hurricane season.

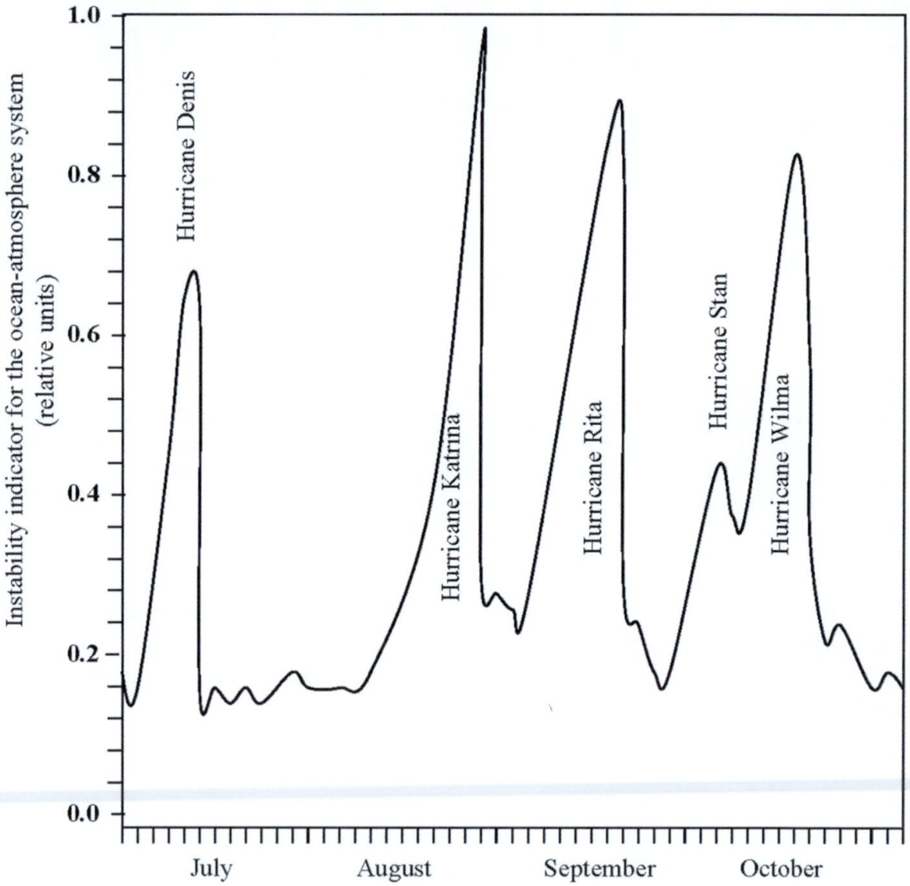


**Figure 6.18.** Dynamics of the instability indicator of the ocean–atmosphere system calculated for the 2005 tropical hurricane season based on data from Meteorological Station No. SMKF1 (24°37′36″N, 81°06′36″W). The parameters of $I_m(k)$ are $s = 17$, $m = 24$ hours, $N = 7$, $K = 123$ days.

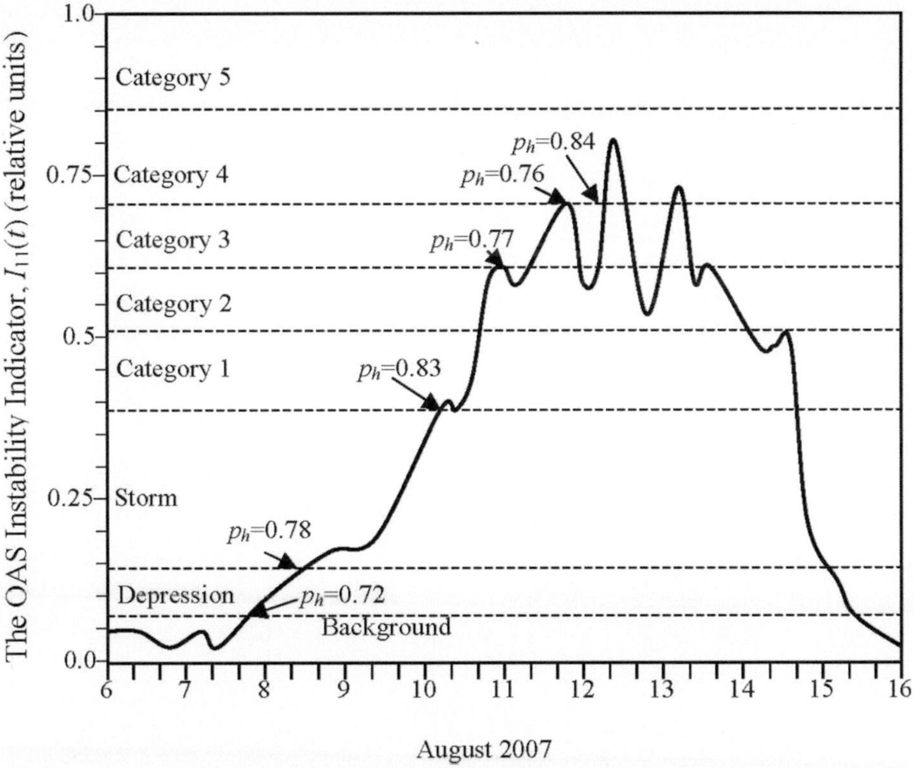


**Figure 6.19.** Dynamics of the instability indicator of the ocean–atmosphere system for Tropical Cyclone Flossie which developed over August 8–15, 2007. Data from Meteorological Station No. 51003 (19°22′N, 160°82′) were used to assess $I_m(t)$ ($m = 17$) as well as $s = 17, m = 24$ hours, $N = 7$, $K = 30$ days.

The behavior of $I_m(t)$ is identical to others in the registered hurricane series. This is confirmed by the results given in Figure 6.19 for the case of Tropical Cyclone Flossie that formed in the eastern sector of the Pacific Ocean on August 8 and reached Hawaii on August 15, 2007. The probability of the ocean–atmosphere system transitioning between its phase positions can be seen in the figure to be more than 0.7. Figures 6.20 and 6.21 further corroborate the indicator's information capability. Therefore, indicator $I_m(t)$ reflects the phase structure of the ocean–atmosphere system with high reliability and can be used to predict hurricane onset.

### 6.8.4 Predicting the path of a tropical cyclone

Storms, hurricanes, and typhoons can be defined as natural unstable events that occur in the ocean–atmosphere system with the potential of becoming catastrophic events leading to loss of life and livelihoods. Many authors have tried and failed

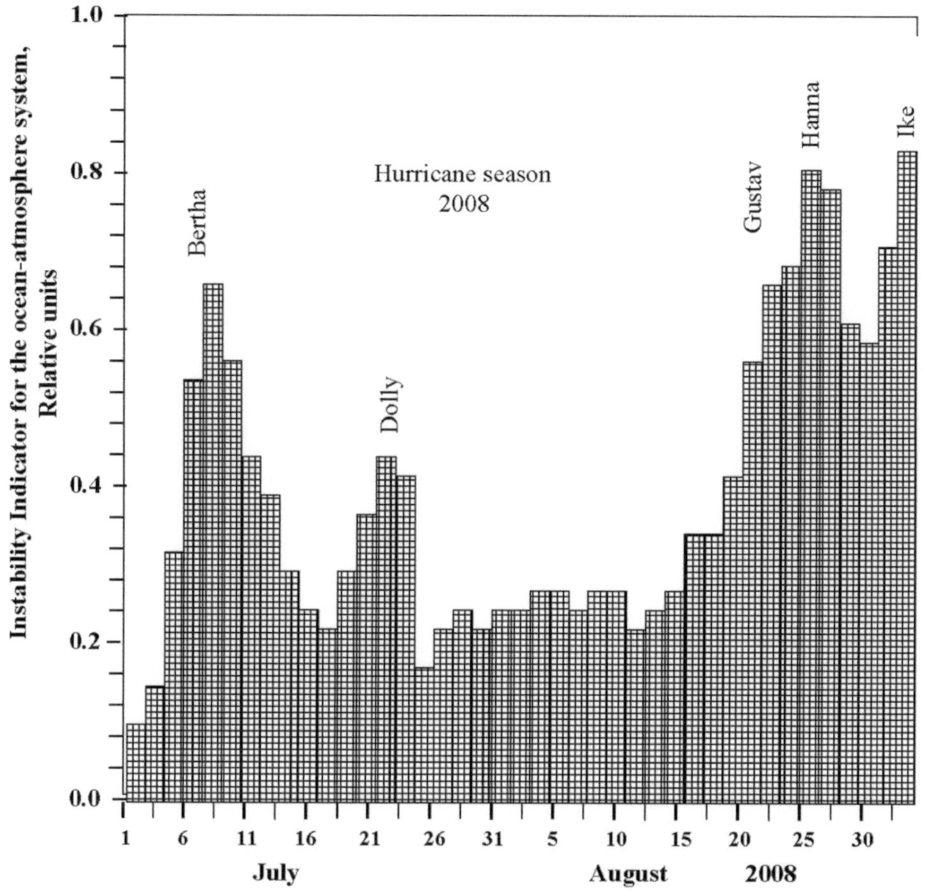


**Figure 6.20.** Dynamics of the instability indicator for the ocean–atmosphere system during the 2008 tropical hurricane season calculated using data from meteorological stations located in the Caribbean Basin and the Gulf of Mexico. The parameters of $I_m(k)$ are $s = 17$, $m = 24$ hours, $N = 7$, $K = 62$ days.

to forecast extreme processes in the ocean–atmosphere system. To solve this problem, a network of meteorological stations has been built up including the TAO/TRITON/PIRATA system of buoy stations. However, forecasting the development of a tropical cyclone is a complex problem that includes several elements: track forecasting, intensity forecasting, rainfall forecasting, and storm surge forecasting. Another possibly even more difficult task is detection of the origin of a tropical cyclone.

The methods used to forecast tropical cyclones have changed over the years in response to the availability of different information sources including satellites. These methods are based on tropical cyclone forecast models that can be broken down into three main types: statistical, dynamic, or combined statistical/dynamic.

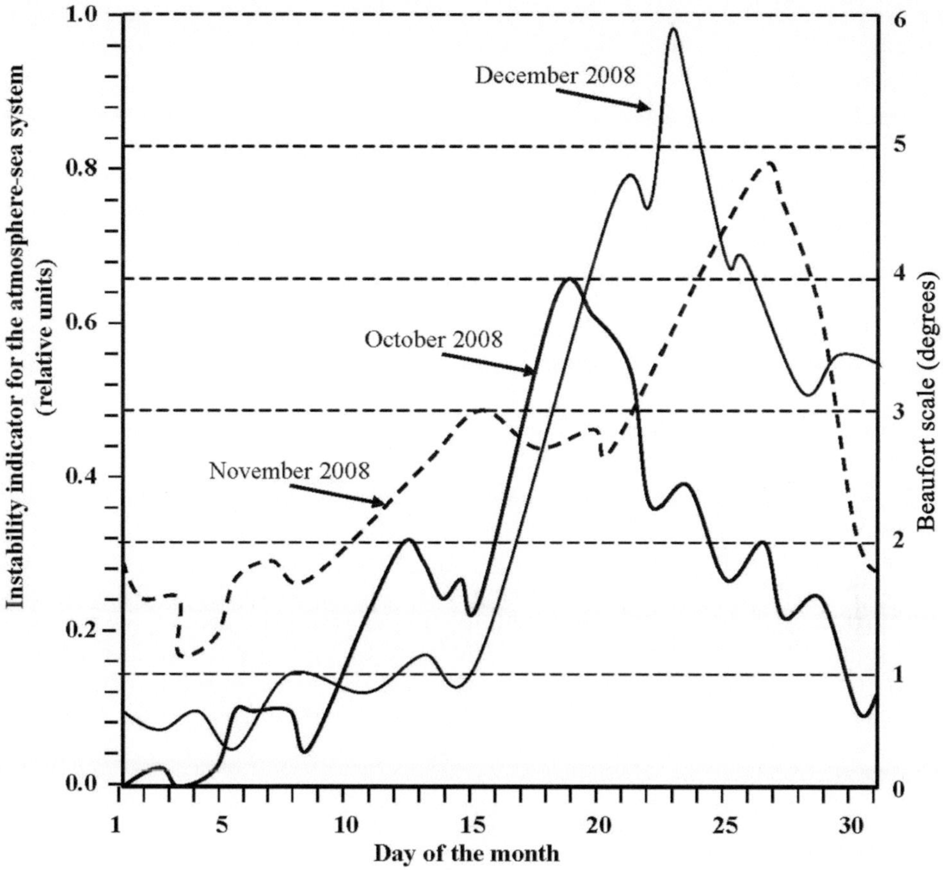


**Figure 6.21.** Instability indicator for the Baltic Sea. Calculations were made by parameters $s = 2$, $m = 24$ hours, $N = 7$, $K = 31$ days. Only wind speed and direction were considered.

Such models only utilize meteorological data to calculate paths and intensities. New forecasting techniques have been developed in recent years by Soldatov (2009b) as a result of the study of global ecodynamics. Natural disasters, generally, and tropical cyclones, in particular, are considered interactive components of global ecodynamics. The approach we have described could be considered as making it possible to forecast tropical cyclone onset. This approach differs drastically from traditional concepts. However, to produce a fully operational system requires routine processing, in real time, of vast amounts of data and determination of reliable thresholds to prevent false alarms being issued. It may then come down to a question of cost/benefit. In practical terms, given the time that passes between detection of a cyclone and its landfall would there be any practical advantage in having an extra 12–24 hours warning?

Tropical cyclones are, of course, but one of many major natural disasters that affect humankind; the others include earthquakes, floods, fires, volcanic eruptions, and many more. It is, of course, desirable, where possible, to give advance warning of an impending disaster so that people can take evasive action. An earthquake strikes suddenly and, as we know, often without specific warning. A lot of scientific research has been devoted to attempting to predict earthquakes. In a sense tropical cyclones are similar inasmuch as they strike suddenly; they come in from the ocean and damage is (mostly) done when and where they make landfall. But before they make landfall they have been tracked for a while and they have gathered strength as they travel across the ocean. The problem is not so much making an initial prediction but in following cyclones over a matter of a few days before they make landfall and predicting exactly where they will strike the land. The issue is not so much detecting the start of the typhoon but in trying to predict its path accurately in order to make precise predictions of where and when it will strike the coastline. Initially, tropical cyclones are little more than severe weather conditions that can be revealed by conventional weather forecasting and observation; navigators at sea or in the air can take evasive action. If the study presented in this chapter is able ultimately to contribute to detecting a cyclone earlier than it would be detected by conventional methods then that of course would be very useful. But, we are a long way from being in that situation. Remote sensing has its role to play—indeed, it already plays a large part in tracking hurricanes and in providing input to models trying to predict the paths of hurricanes. From the point of view of humankind the issue is not to predict or detect the beginning of a hurricane but to predict its development and the path it will take. Extolling the virtues of some aspects of remote sensing that cannot be proven to benefit humankind simply discredit the subject.

### 6.8.5 Concluding remarks and future tasks

Analysis of the observational data that characterize the dynamics of the ocean–atmosphere system in the tropical sector of the World Ocean shows that differentiating the phases of the ocean–atmosphere system by means of transitions into the meteorological percolation cluster allows zones with different instability levels in the ocean–atmosphere system to be identified. For more reliable agreement between the state of the ocean–atmosphere system and the parametrical space of the meteorological percolation cluster it is necessary to establish an algorithm to calculate the critical level of the renormalization probability $p^*$ taking into consideration the procedure of renormalization group transformations in more than two dimensions. It would only be then that the real space renormalization process could make the moment $I_m(t)$ transits from the background zone $\Xi_2^1$ to the signal zone $\Xi_2^3$ more precise (Figure 6.14).

Thus, the following conclusions can be drawn:

(1) Application of the percolation procedure makes it possible to detect the moment the ocean–atmosphere system switches its phase state from background to a state with a high level of environmental instability.
(2) Introduction of an indicator of ocean–atmosphere system instability allows

hurricane onset to be detected before existing monitoring systems are able to do so.

(3) Increase in the precision of the proposed algorithm is possible by optimizing the parametrical structure of indicator $I_m(t)$.

Clearly, to predict hurricane onset reliably requires the development of an efficient information technology for environmental monitoring systems to use. This technology should encompass the planning of measurements, development of algorithms for complex processing of data from different sources, and relevant risk assessment. The spatiotemporal heterogeneity of existing information obtained from different observational systems is the main difficulty here. Overcoming this heterogeneity can be accomplished by using spatiotemporal interpolation algorithms (Kondratayev *et al.*, 2006; Krapivin and Varotsos, 2007). Ultimately, however, the proof of the usefulness of a prediction algorithm such as this can only be established by using it in a live situation (i.e., in real time) to forecast hurricanes in their initial stages. As with all attempts to predict environmental disasters, a major difficulty lies in striking a balance between issuing warnings of imminent disasters and, at the same time, not wishing to issue false alarms. To a considerable extent it seems to be a matter of successful definition of thresholds.

Figure 6.22 presents a structural scheme of a multi-channel monitoring system set up to search and detect hurricane onsets. This system can be realized on the basis of existing meteorological stations located in areas where hurricanes or typhoons are known to begin. In the diagram $T_i$ and $\xi_i$ are the deterministic and stochastic components of $I_m$ or $\lambda^*$, respectively.

Future development of this approach to detect hurricane onset needs, first, to solve the following problems:

- Studying the correlations between hurricane origin and solar activity and understanding the interdependence between the state of the ocean–atmosphere system and large-scale meteorological processes.
- Assessing the role that global degassing processes play in the formation of tropical cyclones and how the state of the ozone layer correlates with events leading to hurricanes.
- Searching for technologies to determine hurricane onset.
- Perfecting climate models so that they are effective at detecting hurricane origins.

## 6.9 EXPERT SYSTEM TO IDENTIFY POLLUTANTS ON THE WATER SURFACE

The detection and identification of patches of pollutants on water surfaces, especially of oil spills, is being addressed by many scientists. The recently developed technology of adaptive identification of environmental elements from measurements in the visible spectral region makes it possible to synthesize an expert system for adaptive identification of environmental parameters (ESAIEP).

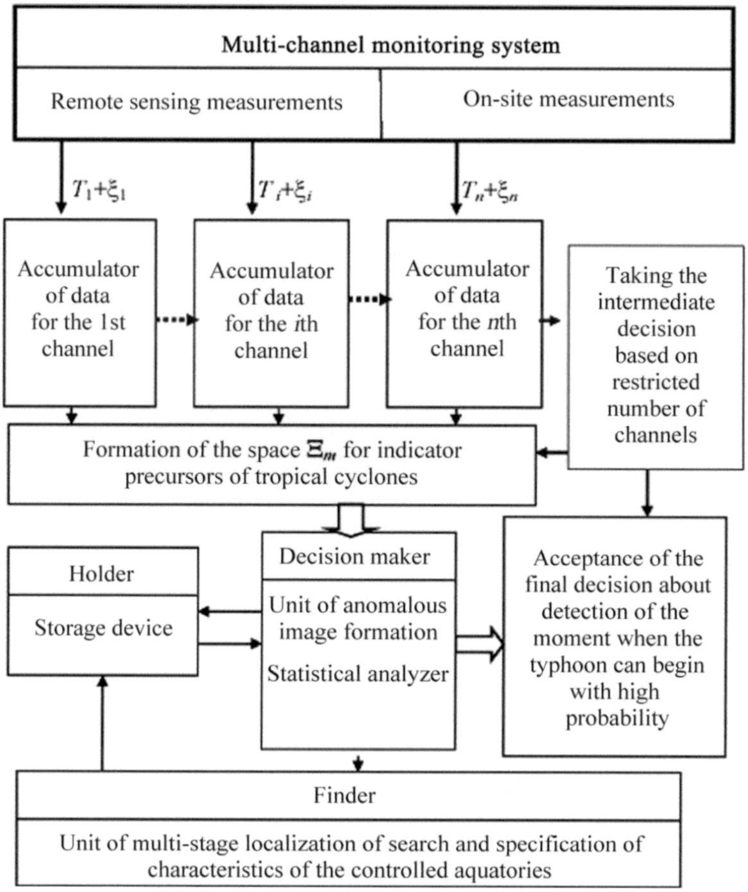


**Figure 6.22.** The structural scheme of a monitoring system to search and detect anomalies in the ocean–atmosphere system.

The system's structure includes a compact multi-channel spectropolarimeter (MSP) (Figure 6.23), information interface with computer (IIC), computer software (STW), and extendable database (EDB). The STW realizes a number of algorithms to process data fluxes from the MSP and provides the service functions of visualization and control of measurements. The EDB consists of sets of standard spectral images of patches of pollutants represented by points in the multi-dimensional vector space of indicators, pre-calculated on the basis of learning samples.

The principle behind the ESAIEP is based on fixing changes in light flux from the MSP and digitizing them. Further processing of these data to make them more efficient is determined by STW with algorithms capable of recognizing two-dimensional objects. The degree of adaptability of the recognition procedure is determined by the level of accumulated knowledge about intensity fluctuations

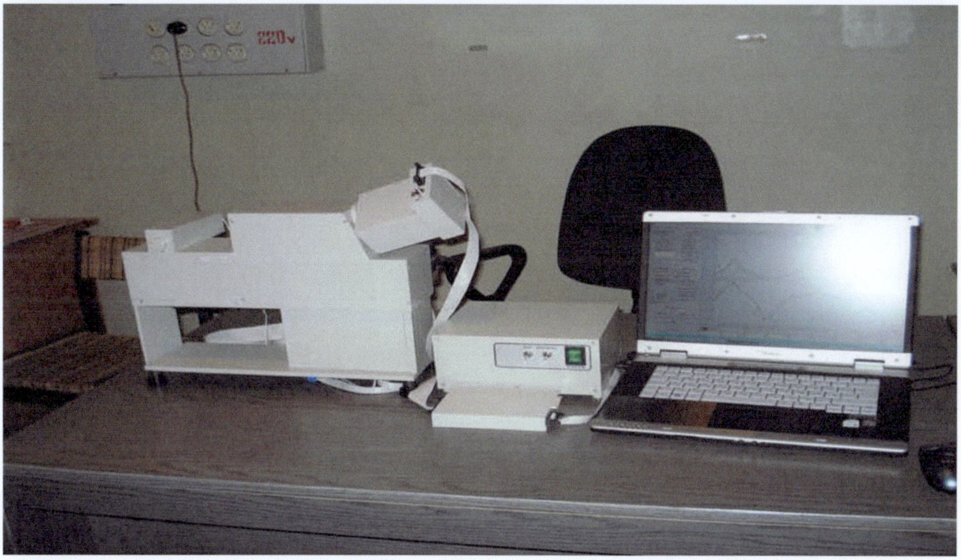

**Figure 6.23.** General view of the multi-channel spectropolarimetric system (Soldatov, 2011).

and the polarizing properties of light reflected from the water surface. The STW enables, in case of uncertain identification of pollutant patches, to make an expert decision based on visual analysis of a spectral image. This procedure is realized in dialog mode with the ESAIEP and, if a decision is made, the operator can fix it in the database in the form of a standard for subsequent appearances of similar patches.

The principal scheme of the STW unit providing the identification procedure involves transforming $\Phi$ (Figure 6.24, Table 6.11). The light intensity $\xi_j^i$ recorded at time $t_i$ in channel $\lambda_j$ enters the algorithm $\Phi$ where two hypotheses $H_0$ and $H_1$ are identified (Figure 6.25). The ESAIEP operator determines initial data $\mathbf{v}_i$, $\alpha$, and $\beta$ and decides which parameters $\mathbf{u}_i = (u_1, \dots, u_r)$ will be calculated from measurements of $\{\xi_i^j\}$. The service unit IIC makes it possible to form vector $\mathbf{u}_i$ from statistical characteristics of the series $\{\xi_i^j\}$ or to use direct measurements. *A priori* information characterizes the type of distribution $f_a(\mathbf{u}_i)$. The function

$$L_i = \sum_{i=1}^{\phi} \psi_j = f_{a_1}(\mathbf{u}_i^j)/f_{a_0}(\mathbf{u}_i^j)$$

is compared with its ultimate values $L_{i,\min}$ and $L_{i,\max}$. At the first stage, these values are chosen arbitrarily, but then they change until they recognize hypotheses $H_0$ and $H_1$ at the highest level of accuracy. We have $L_{i,\min} \to L_{i,\min}^*$ and $L_{i,\max} \to L_{i,\max}$. The values $L_{i,\min}^*$ and $L_{i,\max}$ are memorized by the EDB.

After the learning procedure, the expert system is limited only by the volume of measurements fixed by the operator and proceeds from statistical reliability and the real-time regime. The operator has two possibilities to regulate this regime:

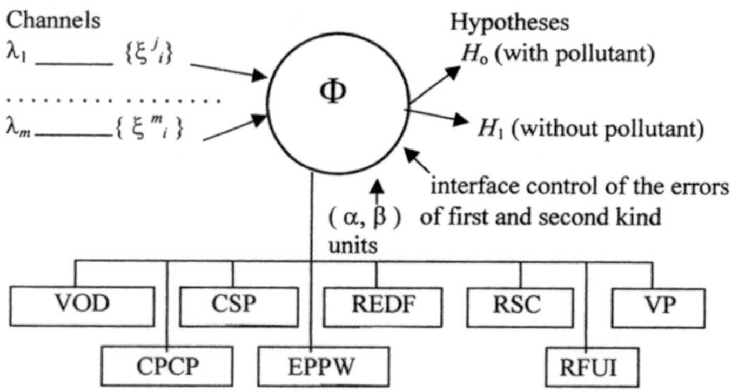


**Figure 6.24.** A conceptual block diagram of the procedure used to identify pollutant patches on the water surface. Notation is given in Table 6.11.

**Table 6.11.** Description of the ESAIEP units schematically shown in Figure 6.24.

| *Unit* | *Characteristic of the unit* |
|---|---|
| VOD | Visualization of observed data (spectra, correlations, statistical parameters, empirical and theoretical functions of distribution) |
| CPCP | Calculation of parameters for the classical procedure of decision making when using the Neuman–Pearson criterion |
| CSP | Calculation of statistical parameters and other characteristics, formation of the vector space $\{u_i\}$ |
| EPPW | Estimation of parameters for the Wald successive analysis procedure in decision making |
| REDF | Reconstruction of empirical distribution functions |
| RFUI | Reconstruction of the functions of a user's interface equipped with software to intervene in the working regime of an expert system at any stage |
| RSC | Realization of statistical criterion to estimate the theoretical function of distribution |
| VP | Visualization procedures |

establishing the volume of series $\{\xi^j_i\}$ or fixing the time of their accumulation. Usually the latter characteristic is chosen equal to 1 s. Figure 6.26 explains this procedure. The operator is combined with ESAIEP units by means of the man/machine interface IIC, which provides selective control of operations in all units.

In the event of an oil slick on the water surface the system analyzes its thickness, age, source, and geometry. The most informative measurements here

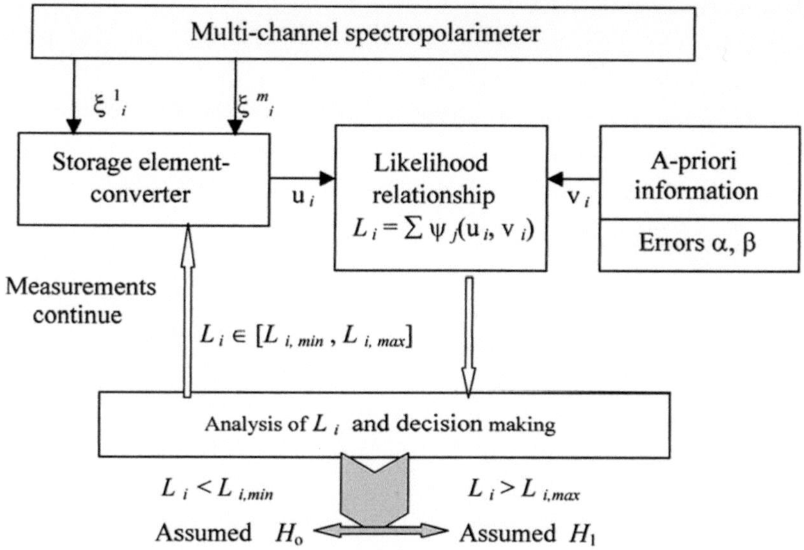


**Figure 6.25.** Sequential analysis procedure used to identify hypotheses $H_0$ and $H_1$ realized in the ESAIEP.

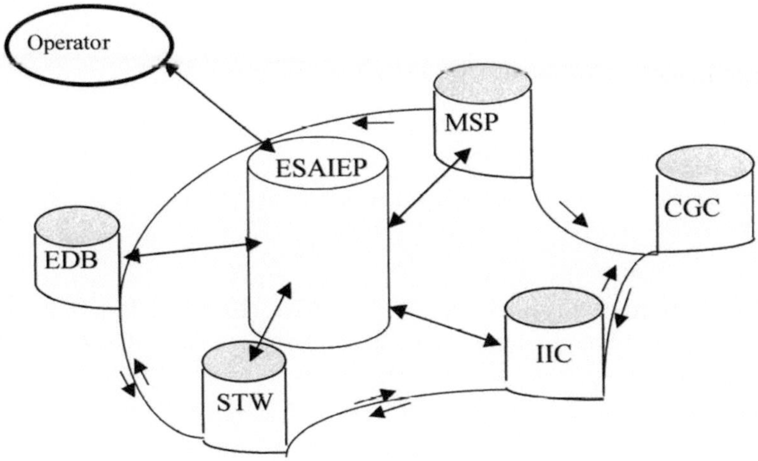


**Figure 6.26.** The ESAIEP model.

are those made at the 398 nm, 439 nm, and 480 nm wavelengths. In the event of dissolved or suspended components in the water the system estimates their concentration and, from data in the EDB on the hydrodynamic parameters of the water body, it calculates their spatial distribution using methods that process two-dimensional images.

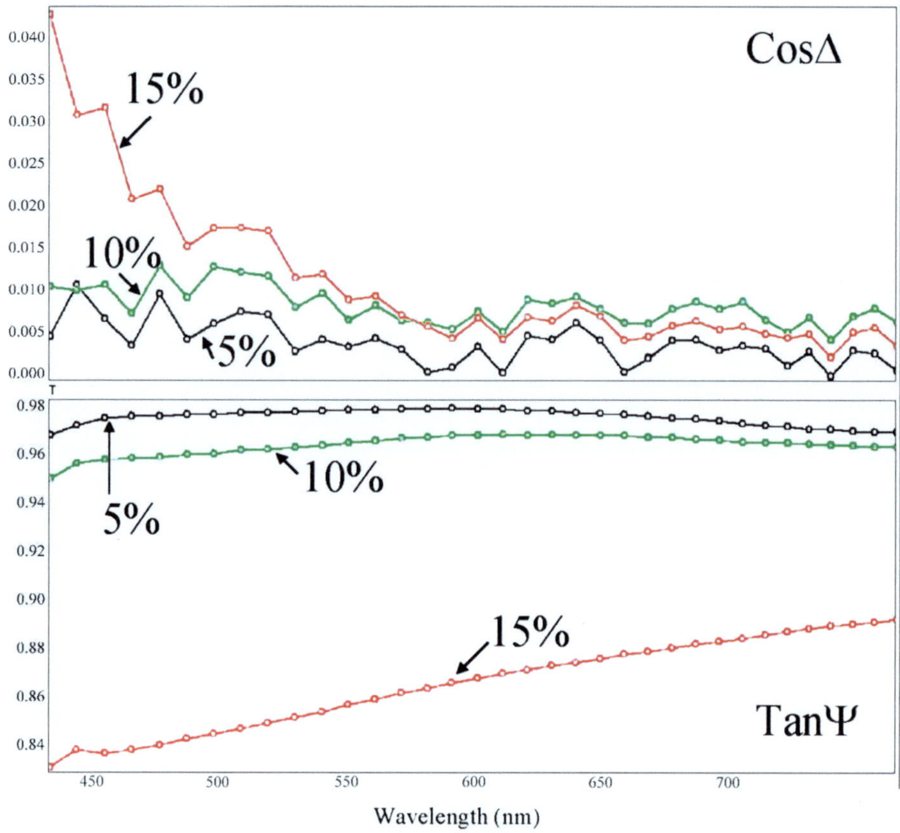


**Figure 6.27.** Real results of spectral measurements of $ZnSO_4$ dissolved in water as retrieved by the ESAIEP. Concentrations of zinc sulfate are shown on the curves.

Figure 6.27 shows visual results of spectral measurements carried out at the Institute of Applied Mechanics and Informatics, Hô Chi Minh City, Vietnam, during Russian–Vietnamese hydrophysical experiments in 2002–2011 (Nguyen *et al.*, 2011b).

# 7

# Typical remote-sensing technologies and data-processing algorithms

## 7.1 MICROWAVE METHODS

Remote sensing of land covers, the atmosphere, and World Ocean aquatories is based on registering background or reflected and scattered electromagnetic radiation. Such knowledge is necessary to get data about the properties of environmental elements as it explains the dependence of thermal emission on its own physical and geothermal parameters. Moreover, scattering mechanisms and active radiation reflection are functions of these parameters (Golovachev *et al.*, 2004; Krapivin and Shutko, 2002; Petty, 1995; Schimel, 1995; Shutko *et al.*, 1994, 2010).

The waves (or frequencies) of electromagnetic emission used for remote sensing in environmental monitoring systems occupy a wide spectrum from 0.3 μm to 1.3 m and cover the subranges: near ultraviolet (0.3–0.4 μm), visible (0.4–0.76 μm), near infrared (IR: 0.76–1.5 μm), middle and far IR (1.5 μm–1 mm), super high frequency (SHF: 1 mm–1.3 m). The SHF range is divided into three basic subranges: millimeter (1–10 mm), centimeter (1–10 cm), and decimeter (10–130 cm). Two subranges—the L-band (15.8–63 cm) and P-band (63–100 cm)—are normally used by many authors.

The real range chosen for objective study of the environment has to take on board many conditions such as the absorption and scattering of electromagnetic waves by the Earth's atmosphere and its interactions with the Earth's land covers and water surfaces. The atmosphere is an extremely limiting factor when choosing a working range for remote monitoring. For instance, the P-band can be used for abyssal sensing from flying laboratories, but its application in satellite monitoring systems is beset with problems because the ionosphere acts as a screen. A similar situation is found with ultraviolet radiation which is intensively absorbed by atmospheric gases. In other words, there is a question about atmosphere clarity for specific wavelengths. For example, some wavelengths in the IR and submillimeter ranges are intensively absorbed by water vapor which practically precludes

V.F. Krapivin and A.M. Shutko, *Information Technologies for Remote Monitoring of the Environment*, Springer Praxis Books, DOI 10.1007/978-3-642-20567-5_7, © Springer-Verlag Berlin Heidelberg 2012

them from consideration as remote-sensing tools. Clouds have a powerful effect on light radiation and create obstacles to land surface observation from space over many territories (Grankov, 1994; Justice *et al.*, 1995; Klyuev, 2000).

Knowledge of what the atmosphere will let through under given synoptic and geographical conditions is the key task when synthesizing a remote monitoring system. As a rule this is solved as measurements are realized. For this, one or several channels use waves that are relatively powerfully absorbed or are dispersed by different atmospheric components; these are added to the basic channels. The content of relative components is determined by the effect relaxation of these waves has on various latitudes (or the integrated content, depending on the task). Measurement result correction is introduced by basic channels using these additional data.

Certain physical features of environmental remote sensing are inherent to super high frequency and ultra short wavelengths, which fundamental to radiometric geoinformation monitoring systems. These systems allowed the many tasks involved in operative identification of natural phenomena to be solved and new methods and technologies for remote diagnosis of natural and natural/technogenic systems to be developed.

Microwave methods can be conditionally divided into two classes: active and passive. The methods that study the reflection, scattering, and absorption of waves emitted by a source with a known spectral density are defined as active. The optical range includes such sources: the Sun, lasers, and other light emitters. Active methods used widely in radiolocation systems are in the radio range. The power of reflected and dispersed radiation, its spectral composition and polarization characteristics, phase and propogation time are the study objects in this case.

Passive methods are based on analyzing the heat emission of natural formations. The fact that thermal emission is determined by substance temperature and its physical parameters is taken into account here. This is the reason passive methods are used for temperature measurement and for the determination of different environmental parameters. These measurements are effective particularly under multi-channel sensing coordinated with data-processing algorothms (Bondur *et al.*, 2010; Krapivin and Mkrtchyan, 1991; Krapivin *et al.*, 2010; Sarkisyan, 2000; Sellers *et al.*, 1996).

Let us now move on and consider the specific features of radiophysical methods (i.e., those in which radio waves are used). Basic attention is here concentrated on the SHF range since it is this range that is applied in the remote-sensing systems located on flying laboratories and satellites (Kondratyev, 2000a; Krapivin and Phillips, 2001; Shutko *et al.*, 2010). Microwave radiometer systems are used intensively on numerous satellites. These systems secure the operative control of more then 20 basic geophysical parameters that are involved in weather and climatic processes (Bondur *et al.*, 2009).

The principal features of radiometric methods have to do with the high radio clarity of the atmosphere. This is an advantage they have over optical and infrared methods. The latter are greatly restricted by the absorbing and dispersing qualities of the atmosphere. The main problem here is cloud which is opaque in these wavelengths and, hence, the receipt of necessary data about the environmental state is prevented. This makes it difficult to obtain operative data when such tasks as the diagnosis of extraordinary and disastrous natural or anthropogenic anomalies need to be solved. Optical and infrared range devices can only be used in regions with little cloud. Hence, the main advantage of radiophysical methods for remote monitoring is that they can be used irrespective of the weather. Nevertheless, there are still problems here, which scientists from many countries are addressing. For example, the absorption line of water vapor only exists at $\lambda = 1.35$ cm and the oxygen absorption zone only at $\lambda = 0.5$ cm.

Radio waves can penetrate beneath the canopy of the vegetation cover and deep into the soil layer. This is the reason radiophysical methods are used to assess the vegetation and soil state and to define many of its properties (Burkov and Krapivin, 2009; Krapivin, 2000b; Krapivin *et al.*, 1998b; Metternicht 1998).

Remote-sensing topics cover many theoretical and experimental investigations in geoscience, ecology, radiophysics, biochemistry, oceanology, applied mathematics, cybernetics, climatology, and other sciences (Grankov and Mil′shin, 2004; Haarbrink *et al.*, 2011; Klyuev, 2000). This chapter mainly considers the role played by the microwave range in the measurement of environmental parameters.

## 7.2 PHYSICAL, THEORETICAL, AND EXPERIMENTAL BACKGROUND OF MICROWAVE MONITORING

Microwave radiometry, or passive microwave remote sensing, is a radiophysical method used for remote observations of the environment. It is based on measurements of the natural electromagnetic radiation of objects in the millimeter to decimeter range of wavelengths. The theoretical aspects of radiophysical monitoring are determined by organizing a measurement procedure in the active or passive regimes of electromagnetic radiation propagation. An active sensing technology is based on radiolocation methods while a passive one is based on background radiation registration. In both cases, the theoretical tasks of microwave monitoring are connected with the study of electromagnetic wave propagation in the environment and, primarily of course, in the atmosphere and near ionosphere (Chukhlantsev, 2006; Gulyaev *et al.*, 2009; Kondratyev and Fedchenko, 1992; Yakovlev, 2001; Yakovlev *et al.*, 2009).

In the active regime the talk is about the propagation of electromagnetic waves of various ranges from the emitter to the receiver. Two cases are possible here. In the first case, the transmitter and receiver occupy different positions in the environment and environmental parameters are assessed on the basis of distortions registered in the transmitted signal. In the second case, the transmitter and receiver are combined so that estimation of environmental parameters is made by analyzing emitted signal absorption and any distortion received from its reflection. Clearly, the time it takes the signal to propagate in the environment is the information parameter. The active form of microwave monitoring is typical for the atmospheric radio-translucence method.

The passive methods of microwave monitoring are based on SHF radiometry. The existence of resonance absorption fields in the SHF range allows remote reconstruction of meteorological atmosphere parameters such as the vertical profiles of temperature, humidity, total water vapor mass, water storage in clouds, and precipitation intensity, etc. The possibility to receive data not only about the properties of water and land surfaces but also about deep characteristics depends on the choice of electromagnetic range. In the infrared range, total emission is concentrated into a very thin surface layer. Electromagnetic waves of the microwave range are absorbed powerfully by land and water surfaces. The depth of its spread into the water environment is measured in hundredths of a millimeter; whereas in dry soils, continental ices, and dry snow this value can reach several tens of wavelengths. Therefore, remote investigations of soil, ice, and snow covers can be made at considerable depths.

The penetrating ability of radio waves is particularly advantageous when sensing land covers. Vegetation without a dense canopy (grass, cereals, etc.) absorbs radio waves weakly and this is the reason radio-observation of soil cover through this vegetation is possible. Radio waves can spread through the soil to a depth of one meter.

The main defect of microwave radiometric observations consists in its comparative low spatial resolution compared with the optical range. In the radio wave range high spatial resolution is achieved by multi-channel application and specific data-processing methods. It demands large economic investments.

Let us consider some aspects of active location. Let us suppose that irradiation at the nadir of a smooth plot is realized. In this case the signal power reflected from the land surface equals $W = pGA|\kappa_0|^2(16\pi H^2)^{-1}$, where $p$ is emitted power, $G$ is the coefficient of directed action of the antenna, $A$ is efficient antenna area, $H$ is antenna height above the ground surface, and $\kappa_0$ is the coefficient of surface reflection. If all instrumental parameters and antenna height are known the ratio of power accepted to that of the one received determines the value of the reflection coefficient. When the soil is uniform in depth the reflection coefficient equals $\kappa_0 = (\varepsilon^{1/2} - 1)/(\varepsilon^{1/2} + 1)$, where $\varepsilon$ is dielectric ground permittivity. To estimate $\varepsilon$ the following approximate formula exists

$$\varepsilon = (\varepsilon_w^{1/2} p_w + \varepsilon_s^{1/2}[1 - p_w])^2, \tag{7.1}$$

where $\varepsilon_w$ is dielectric water permittivity; $p_w$ is relative volume concentration of free moisture in the soil; $\varepsilon_s \approx (1 + 0.5\rho_s)^2$ is the dielectric permittivity of dry ground; and $\rho_s$ is dry ground density (1–2 g $\cdot$ cm$^{-3}$).

From (7.1) it can be seen that reflected signal power makes it possible to determine the reflection coefficient and, consequently, dielectric ground permeability. The value of $\varepsilon$ defines ground density when it is dry or water content when it is wet.

The dielectric properties of land covers can be diagnosed under different situations by applying radiometric methods. Specifically, knowledge of the spectral coefficient of emission in the millimeter range allows snow or ice layer properties to be estimated.

The reflection coefficient can be determined using measurements of radiothermal emission intensity as well as by passive methods. The intensity of background radiation in the microwave range according to Kirhhoff's law is characterized by the brightness temperature

$$T_j = \kappa T_e, \tag{7.2}$$

where $\kappa$ is the emission coefficient (or absorbing surface ability, or blackness level); and $T_e$ is effective surface temperature.

Expression (7.2) characterizes the thermal emission of the surface but does not take into account the emission falling on the surface or reflected by it. Coefficient $\kappa$ is described by Fresnel's reflective formula. Thermodynamic and brightness temperatures are measured in kelvins: $T(\mathrm{K}) = t(^\circ\mathrm{C}) + 273$. Emissivity is a function of the dielectric permittivity of the object/surface being observed. For a land surface, dielectric permittivity is mainly a function of soil moisture: the higher a soil moisture content, the higher the permittivity of soil and the lower the emissivity/intensity of radiation/brightness temperature.

The dielectric permittivity of a water surface is mainly a function of the electrical conductivity of a water solution that is dependent on the concentration of salts, acids, oil films, and many other chemical substances: the higher the salinity of water, the higher the dielectric permittivity of water solution and the lower the emissivity/intensity of the radiation/brightness temperature of this water body. The radiation characteristics of several surface types within the 2 to 30 cm band, for $t = 10$–$30^\circ$C, are shown in Table 7.1. Table 7.2 shows the sensitivity of radiation in the X-band (2–3 cm) and the L-band (18–30 cm) to changes in the free water content in bare soil, soil density, and salinity and temperature of the soil surface. These data show that the main parameter affecting the intensity of bare soil radiation, practically independent of spectral band, is soil moisture. Based on this sensitivity it is feasible to estimate the value of soil moisture without *a priori* data on soil parameters.

The power registered by a combined receiver equals $W = \kappa T_j \Delta f$ where $\Delta f$ is the branch of the emission received. When a soil model with a flat surface is considered the emission and reflection coefficients are correlated by the formula $\kappa = |\kappa_0|^2$. Consequently, knowledge of the emission coefficient allows the reflection coefficient and electrophysical properties of the ground to be estimated. For this to happen the soil temperature $T$ has to be known or the radiometer has to be calibrated using land covers whose emission coefficients are known. The theoretical bases of remote methods, as a rule, are approximate in character. This has to do with the need for simplified prepositions. For example, among the simplifications often used are uniformity of environmental parameter distribution by depth or height, absolute transparency of the area, and land cover smoothness. Models synthesized under these symplifications reflect the restricted spectrum of properties of the phenomenon studied. This is the reason estimation of model adequacy is necessary.

**Table 7.1.** Basic microwave radiation characteristics of some typical surface types.

| *Surface* | $T_j$ (K) | $\kappa$ |
|---|---|---|
| Metal | 0 | 0 |
| Water surface | 90–110 | 0.3–0.4 |
| Very wet soil | 160–180 | 0.55–0.65 |
| Very dry soil | 250–270 | 0.85–0.93 |

**Table 7.2.** Sensitivity of bare soil microwave radiation to variations in soil moisture ($W$), soil density ($D$), salinity ($S$), and surface temperature ($T$).

| *Wavelength* (cm) | *Spectral band* | $\Delta T_j/\Delta W$ (K/g/cm$^3$) | $\Delta T_j/\Delta D$ (K/g/cm$^3$) | $\Delta T_j/\Delta S$ (K/ppt) | $\Delta T_j/\Delta T$ (K/°C) |
|---|---|---|---|---|---|
| 2–3 | X | −200 | −15 | 0.05 | 0.5 |
| 18–30 | L | −(200 to 300) | −10 | −0.5 | 0.1 |

The problem of wavelength choice and the combination of its ranges with classes of solved tasks is now within sight of many investigators. The theoretical foundation of this choice lies in the area of thermal radiation transmission theory. Specifically, for the monochromatic case the transmission equation has

$$dI(z)/dz = J(z) - \alpha I(z), \tag{7.3}$$

where $\alpha$ is the absorption coefficient; $J(z)$ is the emission source distributed by $z$; and $I(z)$ is emission intensity at point $z$. Under these conditions the following equation is valid for an unscattering atmosphere: $J(z) = \varepsilon_1(z)B(z)$, where $\varepsilon$ is atmospheric emissivity,

$$B(z) = 2hf^3c^{-2}[\exp\{hf(kT)^{-1}\}]^{-1};$$

$h$, $c$, and $k$ are fundamental constants; and $f = \nu c$ is frequency. For the microwave branch where the observed frequency is in the region of a few hundred gigahertz or less there are conditions under which $hf \ll kT$ and the Rayleigh–Jeans approximation is valid: $B(z) = 2kT\lambda^{-2}$, where $k = 1.38 \cdot 10^{-20}\ \mathrm{mW \cdot Hz^{-1}\,K^{-1}}$ is the Boltzmann constant, and $\lambda$ is the wavelength.

Equation (7.3) can be solved by

$$T_j = \kappa T_e \tau(0, \infty) + \int_0^\infty T(u)Q(u)\,du, \tag{7.4}$$

**Table 7.3.** Brightness temperatures typical of some land covers.

| *Land cover* | *Wavelength,* $\lambda$ (cm) | | | |
|---|---|---|---|---|
| | 0.8 | | 1.35 | |
| | $T_{j,\min}$ | $T_{j,\max}$ | $T_{j,\min}$ | $T_{j,\max}$ |
| Desert | 251 | 261 | 245 | 270 |
| Cultivated steppe | 255 | 268 | 257 | 277 |
| Steppe covered by snow | 246 | 261 | 241 | 260 |
| Ice covered by snow | 247 | 263 | 231 | 253 |
| Ice | 239 | 247 | 235 | 254 |
| Sea surface | 136 | 171 | 110 | 168 |

where

$$\tau(a,b) = \exp\left[-\int_a^b \alpha(z)\,dz\right].$$

The weight function $Q(u) = \partial\tau(u,\infty)/\partial u$ describes the influence of various atmospheric layers on the receiving signal. Formula (7.4) converts expression (7.2) into a whole. Examples of specific values $T_j$ are given in Table 7.3.

Microwave measurement calibration is important in remote monitoring. The calibration procedure can be defined as a process that relates signals that can be extracted from a measurement radiometric system to the physical characteristics of a target such as its brightness temperature. The main goal of a calibration process is in getting all measurements to accord with a unique scale. For this to happen in the passive regime objects having known brightness temperatures are used (see Table 7.4). This usually means forest or freshwater reservoirs. In this case the radiobrightness temperature of forest is equal to the thermodynamical temperature of the tree canopy, and the radiobrightness temperatures of reservoirs are calculated using known algorithms (Burkov and Krapivin, 2009; Ferm and Hultberg., 1999; Yakimov, 1996).

## 7.3 REMOTE-SENSING TECHNOLOGIES IN THE INFRARED AND OPTICAL BANDS

There exist no fewer than 10 technologies for remote sensing of the environment. Among them are some that are not widely known or not widely applicable to our case: ultrasound methods, radar sensing beyond the horizon, solar radiation sensing, and a few others.

**Table 7.4.** Emission coefficients for some natural and anthropogenic microwave standards (Shutko, 1987).

| *Land cover* | *Microwave range* | |
|---|---|---|
| | *Centimeter* | *Decimeter* |
| Dry sand | $0.93 \pm 0.03$ | $0.93 \pm 0.03$ |
| Concrete | $0.84 \pm 0.02$ | $0.82 \pm 0.02$ |
| Asphalt | $0.85 \pm 0.03$ | $0.83 \pm 0.02$ |
| Dense forest | 0.95–0.98 | 0.92–0.96 |
| Fresh water | $0.38 \pm 0.01$ | $0.36 \pm 0.01$ |
| Waterlogged soils | $0.58 \pm 0.03$ | $0.56 \pm 0.03$ |

For our case—that is, using the GIMS as the key instrument to solve environmental problems, for decision-making procedures in the GIMS, for geoscience and remote sensing, and for oceanic engineering—and for the very typical example of diagnosing and preventing at an earlier stage the buildup of excessive moisture in the soil, water barrier failure, and flood development the most acceptable, useful and effective methods, instruments, and technologies that have been proven in a number of domestic and international campaigns are the following:

(1) optical (digital photo and/or video cameras, spectrometers);
(2) lidar (laser 3-D scanners);
(3) thermal IR radiometers;
(4) microwave radiometers;
(5) radar (synthetic aperture radar, SAR);
(6) short-pulse nadir-viewing subsurface dielectric irregularity finders;
(7) georadar (subsurface dielectric irregularity finders located on the ground and nadir viewing).

Table 7.5 presents the effectiveness estimates of the first five of the seven widely used instruments regarding their application to remote sensing of water and land areas. The plus signs in Table 7.5 indicate the degree of capability (and thus of usefulness) of different sensors at registering different parameter variations or changes under different situations in typical/characteristic environments. Three pluses indicate the highest possible rank of effectiveness whereas one plus represents the minimum.

Why are five of the seven, rather than seven of the seven, presented types of sensors discussed? The answers are as follows:

(a) sensors 6 and 7 are still not as widely used as the first five sensors;
(b) these two sensors were not used in the development of the GIMS; and
(c) special conditions are needed to organize a flight campaign and have it include these sensors and radar, including synthetic aperture radar which, though effective and useful, are heavy, unwieldy, and expensive tools for aircraft and spacecraft remote sensing.

These technologies are currently under development—primarily at JPL and the Vega Radio Corporation in Russia—and currently available for potential users.

Table 7.5 gives rough but effective assessments of how basic remote sensors, operating at both passive and active microwaves in the infrared and optical bands, behave under the basic parameters and conditions inherent in the following scenarios:

- oceanic, sea, or fresh water surfaces;
- plane and rough water surfaces;
- foam-covered and oil spill-covered water surfaces;
- water–vegetative canopy system;
- bare land surfaces;
- vegetation-covered soil;
- atmosphere–tree crown system (when sensing upward through the canopy).

As mentioned above, the pluses in Table 7.5 characterize the information content (usefulness) of different sensors, beginning with a single cross (the lowest rate of usefulness) and moving to three crosses (highest usefulness). This table can be used to get preliminary assessments of the effectiveness of expected results of experimental campaigns as they are developed.

Of course, Table 7.5 reflects the personal opinion of this book's authors regarding the usefulness of different sensors. This opinion is the culmination of multi-year studies of the spectral properties of different environments. This personal experience was used to develop, test, and put into practice truly multispectral remote-sensing technologies providing state-of-the-art information by means of an optimum set of sensors, such as those installed on board of the Dutch/U.S./Russian Miramap aircraft equipped with the abovementioned multispectral passive/radiometric microwave, thermal infrared, color digital orthophoto cameras, and lidar (laser 3-D scanners working in the near-infrared band).

## 7.4 MONITORING SOIL–PLANT FORMATIONS

Land cover change is an important environmental process. Studying and monitoring this change are significant when assessing the variations and dynamics of the global biogeochemical cycles of greenhouse gases. The Moderate Resolution Imaging Spectroradiometer (MODIS) of NASA's Earth Observing System (EOS)

**Table 7.5.** Typical remote-sensing technologies, their information content, and their effectiveness. Notation: *O*, optical (digital photo and/or video cameras, spectrometers); *L*, lidar (laser 3-D scanners); *T*, thermal IR radiometers; *M*, microwave radiometers; and *R*, radar (SAR).

| *Sensors and parameters they determine* | *O* | *L* | *T* | *M* | *R* |
|---|---|---|---|---|---|
| Sea/water surface temperature | | | ++ or +++ | ++ or +++ | |
| Sea/water surface natural salinity | | | | ++ or +++ | |
| Oil pollution on water surface | + | | + | ++ | +++ |
| Water pollution with dissoluble chemicals | + | | | ++ | + |
| Water pollution with waste outflow | + | | | ++ | + |
| Soil moisture | + | | + or ++ | +++ | + or ++ |
| Subsurface moisture (depth to water table) | | | | +++ | + or ++ |
| Zones subject to water seepage through water barriers | + | + | + | +++ | + |
| Flooding | + or ++ | + or ++ | + | +++ | +++ |
| General water regime of the terrain | + or ++ | + | + | +++ | ++ |
| Terrain relief | + | ++ or +++ | + | + or ++ | ++ or +++ |
| Water areas and/or wetlands overgrown with vegetation | + | + | + | ++ | +++ |
| Vegetation classification | ++ | + | | + | + or ++ |
| Vegetation biomass, dynamics, seasonal changes | + | + | | + | ++ |
| Monitoring crowns of free-standing trees | | | | ++ or +++ | |
| Monitoring diseases and pests in the vegetation | + or ++ | | + | | + |

| | | | | | |
|---|---|---|---|---|---|
| Deforestation | + or ++ | | + | | + |
| Monitoring timber/forest clearance | + or ++ | | + | | + |
| Risk of forest or peat fire | + or ++ | | + or ++ | ++ or +++ | + |
| Revealing epicenter and/or border of fire under foggy and smoky conditions | + or ++ | | ++ or +++ | ++ or +++ | |
| Monitoring the terrain's traffic ability | + or ++ | + or ++ | ++ or +++ | ++ or +++ | ++ or +++ |

is an effective satellite instrument that facilitates detecting disturbances in the globe's land surfaces.

Understanding Earth system processes and their interdependences through observation, theory, and modeling is the objective of many international programs such as the International Geosphere–Biosphere Program (IGBP), the International Human Dimensions Program (IHDP), the World Climate Research Program (WCRP), and the Global Carbon Project (GCP). However, this is only possible if reliable and operative information about land covers is available. Undoubtedly, for this information to be used effectively there need to be algorithms and models to improve the quality of environmental monitoring.

Remote monitoring of soil–plant formations is aimed at assessing biological productivity, understanding interactions inside the soil–vegetation–atmosphere system, evaluating biome dynamics, modeling biogeochemical cycles paying particular attention to the role played by vegetation, and controlling vegetation resources.

Land covers are characterized by a wide range of landscapes with various soil and plant types; water, salt, and temperature regimes; a surface layer that is cultivated, and macrorelief. The soil includes solid particles, water (soil mixture), and air. Solid particles consist of mineral and organic matter. The basic mass of most soils comprises mineral substances (mineral soils): particles more than 3 mm in size form the stony part, particles from 0.01 to 1 mm are called physical sand, and particles less than 0.01 mm are physical clay. The minerals that make up the soil consist basically of two oxides: $SiO_2$ and $Al_2O_3$. The content of organic substances (mainly humus) fluctuates in such soils between 1 and 10% by weight. All of this needs to be taken into account when designing a land cover monitoring system, one in which the choice of microwave channel needs to be informative. The many lengthy investigations in this field facilitate choosing the most effective wave range for monitoring many types of vegetation and soils. Studies of agricultural systems, forest, and arid zones are inevitably very detailed. Such areas as control of the hydrological soil regime under the forest canopy and the study of reforestation dynamics in regions where forest clearing and forest fires are prospective scientific fields urgently in need of study by microwave monitoring. This can be done by using the decimeter range in which the forest canopy is transparent in the microwave range and, consequently, soil up to the ground water level can be sensed. Experimental characteristics of natural microwave L-band emissions from forests are described by Chukhlantsev (2006) and Chukhlantsev and Shutko (1988).

Building up an inventory of forest–scrub vegetation by means of radiolocation sensing is possible by correlating the coefficients of SHF wave backscattering with the structure and dielectric properties of vegetation. Using radiolocation systems to control forests is stipulated by the possibilities of additional data being received together with remote measurements in the visible and IR ranges (Savin and Bondur, 2000).

Global mapping of vegetation on a scale determined by the economic needs of a given region is the main task of satellite-based radiolocators in forest observations. Specifically, protecting the forest from fire is an objective when developing a

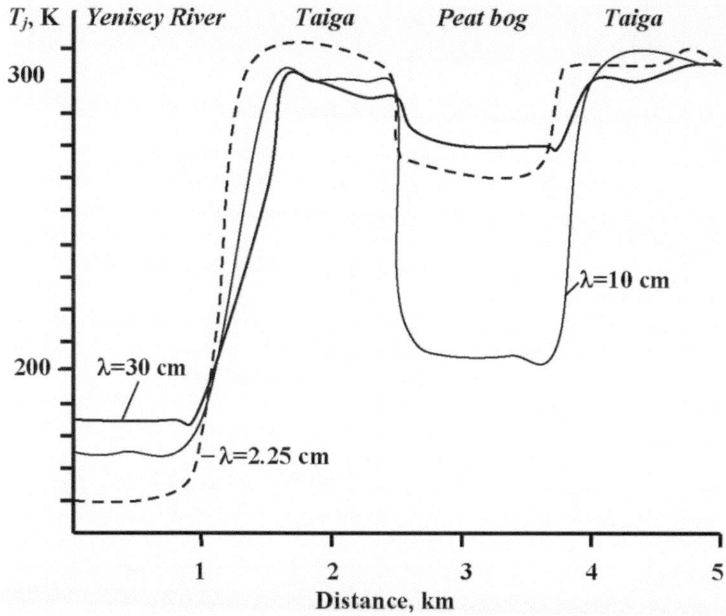


**Figure 7.1.** Registrograms received by means of the IL-18 flying laboratory near the city of Yeniseysk using radiometers of 2.25 cm (1), 10 cm (2), and 30 cm (3) wavelengths.

remote monitoring system to assess the vegetation state. Current satellite-based and aviation-based monitoring systems are capable of solving many significant problems:

- operative assessment of a forest's fire risk;
- mapping natural levels of forest water that can help in extinguishing fire in the forest;
- mapping the contours of forest fires through the smoke layer and tree canopy;
- determining the energetic paramaters of fire;
- forest state determination after fire.

Investigations of radiothermal emission from the natural layers of a forest that are combustible such as lichen, mosses, grasses, dead pine needles, and scrub have brought about methodologies to identify forest fire parameters based on measurements of radiothermal emission in the microwave range of the electromagnetic spectrum. Examples of radiobrightness temperatures from forest fires are given in Figures 7.1 and 7.2. The leading and rear edges of fire and the burnt forest in between can clearly be distinguished.

Comparing radiobrightness temperatures registered in different wavelengths shows that forest fire emissivity increases as wavelength decreases and depends on the fire type. Figure 7.3 shows the dependences of radiobrightness temperatures of

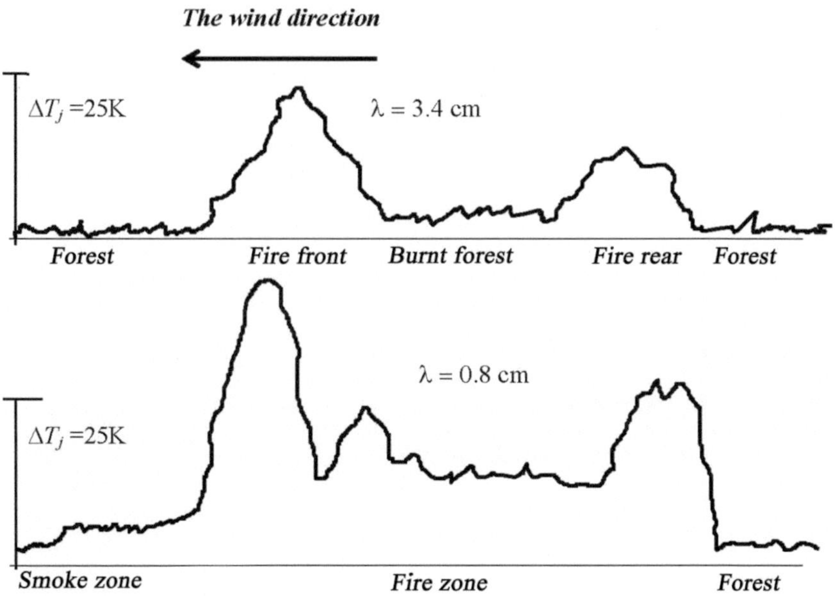


**Figure 7.2.** Registrograms in the wildfire zone near the city of Krasnoyarsk received by means of the IL-18 flying laboratory.

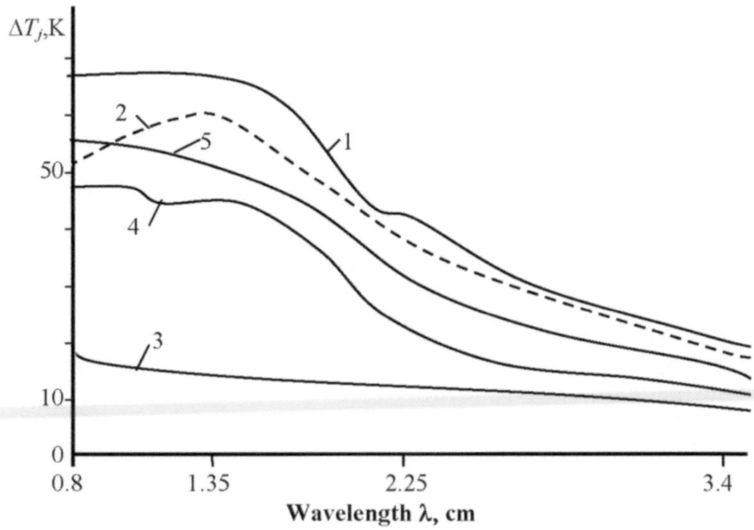


**Figure 7.3.** Averaged spectra of radiobrightness contrasts for specific elements of the forest fire. Notation: 1, smoke zone; 2, leading edge; 3, burnt zone; 4, trailing edge; 5, boundary between the leading edge and burnt zones.

various components of forest fire on different wavelengths and, as a result, these components can be classified with high precision.

For the GIMS to synthesize the functions of forest fire control over large areas, applied models describing the processes that lead to forest fires need to be created. Such models are founded on understanding the intricacies of how the intensity of radiothermal emission of forest landscape elements depends on water content. Many observations show that the statistical characteristics of forest microwave emissions change in accordance with the growth of layers that are combustible. Specifically, such a dependence can be represented in the following form (Yakimov, 1996):

$$T_j = \kappa(\lambda, w)T_S(1 - \exp\{-\alpha(\lambda, w)h\} + T_{jS}\exp\{-\alpha(\lambda, w)h\},$$

where $\kappa$ and $\alpha$ are the emission and absorption coefficients, respectively; $w$ is water content in timber; $h$ is thickness of the emitting layer; $T_S$ is soil temperature; and $T_{jS}$ is soil brightness temperature. The following deciding factor, based on calculating the first-order ($M$) and second-order ($\sigma$) moment for the set of $T_j$ corresponding to the forest territory and registered on the wavelength $\lambda = 2.25$ cm, was proposed by Yakimov (1996). The value $M$ (mean) is compared with a threshold equal to $d = 7.251\sigma + 249.876$. If $M \geq d$ then a forest plot has an average or higher risk of fire. If $M < d$ then the fire risk of the given forest plot is low. This deciding factor can then be extended by a forest fire model. An example of such a model is one describing forest fire as a running wave or as a self-supporting process of local energy release into the active environment.

The millimeter range (1–30 mm and 30–300 GHz bands) is best for land cover diagnosis. As numerous experimental investigations show, adding millimeter range channels to the monitoring system enhances the accuracy of soil–plant formation classification by separating out roughness and small formations. It is especially important for anthropogenic landscape monitoring.

McCloy and Lucht (2004) developed a method for comparing the complex spatiotemporal patterns present in two long-time series of data of the seasonal cycles of vegetation for a large part of the global land surface. It is based on the use of two datasets derived from global satellite observations (AVHRR) and the Lund–Potsdam–Jena dynamic global vegetation model (LPJ-DGVM). McCloy and Lucht (2004) showed that the LPJ model generally corresponds well with natural vegetation whereas the former is primarily dependent on climate. Because agriculture was missing from the model, its correspondence with the actual vegetative status was poor.

## 7.5 MICROWAVE MONITORING OF SOIL MOISTURE

The fundamental principles of soil moisture retrieval using passive microwave sensors and microwave radiometric techniques to measure the properties of soil moisture were established by Shutko (1986) and his colleagues. During a series of experiments, microwave radiometers were put on satellites, aircraft, and ground-based vehicles. L-band passive microwave measurements were the most informative in these experiments.

Interrelations between the microwave emission field of wet soil and the liquid water content of soil, soil density, temperature, and mineralization level of liquid water are properties that are studied both theoretically and by field measurements. Both research and field studies show that microwave radiometric measurements recognize seven to ten moisture levels in the top 0.1 to 1.0 wavelengths of soil and three to seven levels in subsurface water between 0 and 1.5–3.5 m.

Soil moisture is divided into solidly united, loosely united, and free. United moisture is the water adsorbed by ground particles at the surface and takes the form of a film of thickness equal to no more than six to eight molecular layers. The volume of united moisture in the soil layer is determined by the soil type and fluctuates widely from 2–3% for sandy soil to 30–40% for clay and loess soils. United water is unattainable by plants and does not influence the salt regime of soil. This is the reason a monitoring system needs to classify the types of moisture held in the soil.

Soil moisture is expressed as a percentage of dry soil weight. Radiation models of different types of soil moisture that consider soil density, temperature, and salinity have been studied in great detail. The data required to ascertain moisture levels can be obtained by means of brightness temperature contrasts, degree of polarization, and spectral characteristics at centimeter and decimeter wavelengths. Wavelengths 2.25, 18, and 30 cm have proven to be the most informative at carrying out this task. The microwave radiation model set up to describe land cover emittively under conditions of a heterogeneously moist layer is based on the existence of vertical heterogeneity in the dielectric permeability coefficient (see formula 7.1): $\varepsilon \cong (1 + 0.5\rho_s)^2$. The value of $\varepsilon$ is the indicator of soil moisture variation. As soil moisture begins to grow, $\varepsilon$ increases at first slowly if the soil is dry, then it increases more rapidly. The weak dependence of $\varepsilon$ on the initial moisture stage, explained by moisture tied up in the soil, can be characterized as having small dielectric permeability. A variety of experimental dependences of $\varepsilon$ on soil types and moisture levels has been given by many authors (Borodin and Krapivin, 1998; Engman and Chauhan, 1995; Schmugge, 1990; Shutko, 1986; Tsegaye *et al.*, 2005). Such a knowledge base combined with application of GIMS technology allows ascertaining water content, $W_s(z)$, diagnostics in the soil layer. The basic types of moisture profiles $W_s(z)$ of real soils and corresponding profiles, $\varepsilon(z)$, of the dielectric constant were described by Reutov and Shutko (1986). Four major types of vertical moisture profiles have been identified by analyzing the wetness characteristics of various climatic zones in the former Soviet Union. Such an identification is necessary in other climatic zones. As a matter of fact, the task can be reduced to reconstructing function $W_s(z)$ when the values $W_{si} = W_s(z_i)$ are known and the following conditions are satisfied:

$$\partial W_s/\partial z|_{z=0} \leq C, \qquad 0 \leq W_s(z) \leq K_s(z), \tag{7.5}$$

where $K_s(z)$ is the given function.

Practice has shown that, irrespective of the climatic zone, microwave radiometers need to be used to achieve acceptable precision when reconstructing the vertical profile of moisture in a soil layer to a depth of one meter. There is *a priori* information about average soil moisture for depths 50 cm ($W_{s2}$) and 100 cm ($W_{s3}$). $W_{s1}$ can be estimated using microwave radiometers with wavelengths $\lambda = 10$ cm and 30 cm. The following approximation $W_s^*(z) = \alpha z^3 + bz^2 + cz + d$ can then be considered, where unknown coefficients are calculated from the smallest deviations between $W_s^*(z_i)$ and $W_{si}$ ($i = 1, 2, 3$). For this to be done conditions (7.5) have to be realized. This methodology allows soil moisure to be determined by remote monitoring over large areas with an error no greater than $0.3\,\mathrm{g \cdot cm^{-3}}$ for a biomass of vegetable cover no smaller than $2\,\mathrm{kg \cdot m^{-2}}$ and with an error up to $0.07\,\mathrm{g \cdot cm^{-3}}$ for a biomass greater than $2\,\mathrm{kg \cdot m^{-2}}$.

Knowledge of function $W_s(z)$ allows the water balance model of a territory to be used to reconstruct the dynamics of soil properties and other water balance elements such as the functions of geophysical and ecological parameters. Experimental evaluation of the thickness of the upper soil layer can also be reconstructed; at the lower boundary of the upper soil layer capillary connections break every six days after rain or watering. The typical thickness of this dried-up layer oscilates between 3 and 5 cm. Coefficient estimations of the microwave range reflected from this layer are given in Table 7.6.

Optimal choice of the electromagnetic radiation range and model synthesis based on methodologies to measure the moisture content in the soil demands further investigation of the conditions that influence emission fields in the atmosphere–soil–vegetation system (Engman and Chauhan, 1995; Grankov and Mil′shin, 1994). These investigations can be correlated by means of remote sensing with agricultural crop forecasting and with water resource management.

Figure 7.4 shows the application of microwave radiometers and the use of spatial interpolation methods to monitor agricultural fields and map the soil moisture. This can be done by radiometers with 18 and 27 cm wavelengths located on a flying laboratory. Such experiments have taken place under the climatic conditions of Moldavia, the Ukraine, Bulgaria, Cuba, Vietnam, Russia, and the U.S.A. (Shutko *et al.*, 2010). These experiments showed GIMS technology allows specific geophysical conditions to have practically valid results from monitoring. To reconstruct the spatial distribution of soil moisture with a precision of 20%, the flying laboratory in most cases needs to fly parallel routes at height $H \approx 200$ m with a distance between them of no more than 500 m. The interval between measurements, when the precision of forecast for synoptical parameters is 10% and for precipitation is 15%, can be monthly if the precision of soil moisture estimation is planned to be less than 50%.

Guha *et al.* (2003) considered a series of experiments using the Electronically Scanned Thin Array Radiometer (ESTAR). The Southern Great Plains Experiment (SGPE-99), within the framework of which ESTAR-derived soil moisture fields were studied, took place in a region of the U.S.A. that is characterized by a relatively homogeneous geography, variable surface flux properties, and large seasonal variation in temperature and specific humidity. In addition to ESTAR, four different microwave radiometers operating in the L-band (1–2 GHz), S-band

**Table 7.6.** The reflection coefficients of microwaves for soils with different moisture contents.

| *Moisture content in the soil* ($g\,cm^{-3}$) | *Thickness of the soil layer* (cm) | | | | |
|---|---|---|---|---|---|
| | *0.5* | *1.0* | *2.0* | *3.0* | *5.0* |
| | | | ***λ = 18 cm*** | | |
| 0.001 | 0.98 | 0.94 | 0.95 | 0.96 | 0.97 |
| 0.02 | 0.91 | 0.92 | 0.93 | 0.95 | 0.97 |
| 0.05 | 0.88 | 0.89 | 0.92 | 0.95 | 0.96 |
| 0.1 | 0.82 | 0.84 | 0.89 | 0.95 | 0.96 |
| 0.2 | 0.73 | 0.76 | 0.85 | 0.94 | 0.95 |
| 0.3 | 0.65 | 0.69 | 0.83 | 0.94 | 0.93 |
| 0.4 | 0.59 | 0.64 | 0.82 | 0.94 | 0.93 |
| 0.5 | 0.53 | 0.59 | 0.82 | 0.92 | 0.92 |
| | | | ***λ = 30 cm*** | | |
| 0.001 | 0.93 | 0.93 | 0.94 | 0.95 | 0.96 |
| 0.02 | 0.91 | 0.91 | 0.92 | 0.93 | 0.96 |
| 0.05 | 0.88 | 0.88 | 0.89 | 0.92 | 0.96 |
| 0.1 | 0.82 | 0.83 | 0.85 | 0.89 | 0.96 |
| 0.2 | 0.72 | 0.73 | 0.78 | 0.85 | 0.96 |
| 0.3 | 0.64 | 0.66 | 0.73 | 0.83 | 0.95 |
| 0.4 | 0.57 | 0.59 | 0.69 | 0.82 | 0.94 |
| 0.5 | 0.52 | 0.55 | 0.66 | 0.81 | 0.92 |

(2–4 GHz), and C-band (4–8 GHz) were flown on the aircraft. The instruments included the Polarimetric Scanning Radiometer (PSR/C), the Step Frequency Microwave Radiometer (Step-C), and the Airborne C-band Microwave Radiometer (ACMR) operating in the C-band and the Passive and Active L- and S-Band System (PALS) in the L- and S-bands. Passive microwave data were also collected by the Tropical Rainfall Measuring Mission (TRMM) satellite and the Special Sensor Microwave Imager (SSM/I) sensor.

As a result of these measurements, the problem of distinguishing between the water content of soil and that of vegetation cover was solved. An empirically

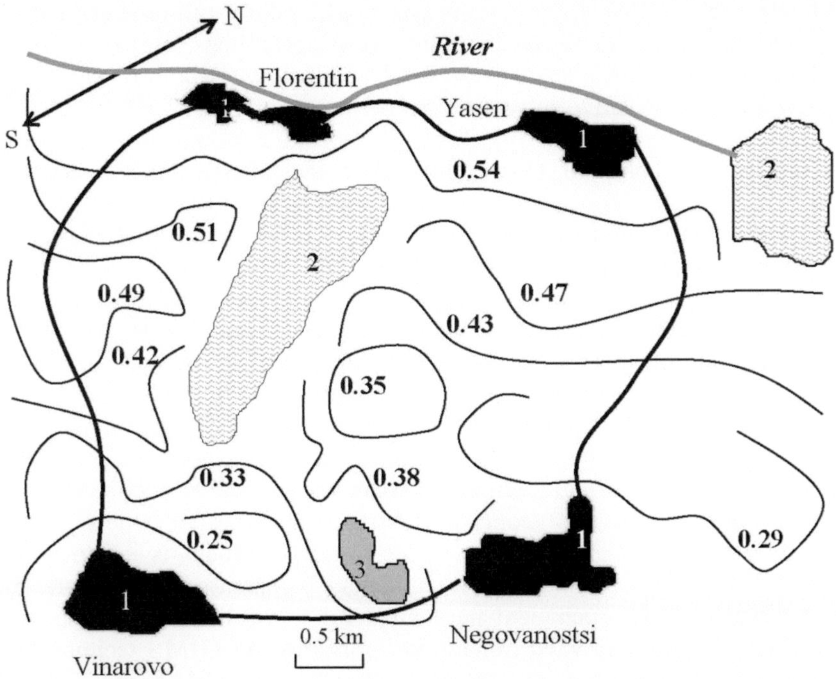


**Figure 7.4.** The application of GIMS technology to microwave mapping of soil moisture in Bulgaria at the beginning of May. The moisture values are shown on the curves in $g \cdot cm^{-3}$. Notation: 1, grassland; 2, reservoirs. The directions the flying laboratory moved are shown by arrows.

derived piecewise function was used to solve this problem. The vegetation water content (VWC) was calculated with the following correlation:

$$\mathrm{VWC} = \begin{cases} 1.9134(\mathrm{NDVI})^2 - 0.3215(\mathrm{NDVI}) & \text{when NDVI} \leq 0.5; \\ 4.2857(\mathrm{NDVI})^2 - 1.5429 & \text{when NDVI} > 0.5; \end{cases}$$

where VWC is measured in $kg \cdot m^{-2}$.

## 7.6 MICROWAVE RADIOMETRIC OBSERVATIONS OF TEMPERATURE ANOMALIES

The microwave range is an informative component of the multi-channel systems that monitor temperature anomalies on the Earth surface such as forest fires, peat bog fires, areas of geothermal activity, etc. Measurements of radiation contrasts make it possible to localize areas that have anomalously high and low brightness temperature values as a result of temperature variations with respect to a mean value of effective background radiation. Numerous experimental investigations have shown that radiobrightness contrasts registered by radiometers with wavelengths 0.8 and 3.4 cm in those areas of forests, peat bogs, etc. that are on fire can achieve values of $\Delta T_j \in [200, 300]$ K. The most important result is that the microwave range is sensitive to surface temperature variations and, hence, possible outbreaks can be prevented. Specifically, microwave monitoring of peat bogs means the interior zones of temperature anomalies can be fixed when external signs are absent.

Thermal radiation measurements taken by radiometers with wavelengths $\lambda_1$ and $\lambda_2$ commonly result in $\Delta T_j \approx \kappa \, \Delta z \, dT_S/dz$, where $\Delta z = z_2 - z_1$ is the difference between the depths of emitting layers ($z_2 > z_1$), $dT_S/dz = -Q/k_T$, $Q$ is thermal flux intensity, and $k_T$ is the thermal conductivity coefficient for the layer $\Delta z$.

The sensitivity of microwave radiometers equals $dT_S/dz \approx 0.1$–$1 \deg \cdot \mathrm{m}^{-1}$. The remote discovery of warmth escape zones of plants and buildings is now possible as a result of microwave radiometers operating in the centimeter range.

Another important area for the application of microwave radiometry is remote control of geothermal sources. Microwave monitoring facilitates analyzing the activity and classifying geophysical structures in areas where geothermal sources are active. In such areas the water content of rocks is insignificant and, hence, the depth of electromagnetic wave spread is increased to equal the inverse value of the absorption coefficient per unit depth $z$. Multi-channel measurements based on use of the microwave range and application of GIMS technology allow soils to be reliably classified and thermal situations mapped. Figure 7.5 shows a map of the Tolbachik volcano zone (Ostry Tolbachik, 55°57′N, 160°25′E; Plosky Tolbachik, 55°49′N, 160°24′E). This map was put together on the basis of the radiobrightness temperature spectra in Figure 7.6.

Looked at from the climate change perspective, the monitoring of forest fires has played an important role in global ecoinformatics. The forest state and its influence on greenhouse gas concentrations is too dynamic for reliable assessment of present and future tendencies in environmental change. This is the reason a global geoinformation monitoring system has to have built into it a function to effect such an assessment. For such a function to be brought about there needs to be a model that can identify the forest state from radiothermal radiation. For such a model to be created there needs to be knowledge of the dependences of forest radiothermal radiation on different wavelengths to reconstruct the moisture content of forest timber at various layers and its combustibility. The investigations of many authors have shown that application of microwave sensing in the range 0.8–20 cm allows the fire risk of waterlogged forests to be estimated. Multi-channel sensing allows using cluster analysis algorithms to distinguish and classify fire risk classes. The accuracy of this methodology depends on detailed description of the forest structure reflecting the canopy state and tree density. The most difficult to identify are forest edge fires. In this case the three-layer model of the flame–smoke–canopy system is effective:

$$\alpha_P = (1 - \exp\{-\kappa_1\}) \exp(-\kappa_2) \exp(-\kappa_3),$$

where $\alpha_P$ is the emissive ability of forest edge fires; $\kappa_i$ is the emission relaxation

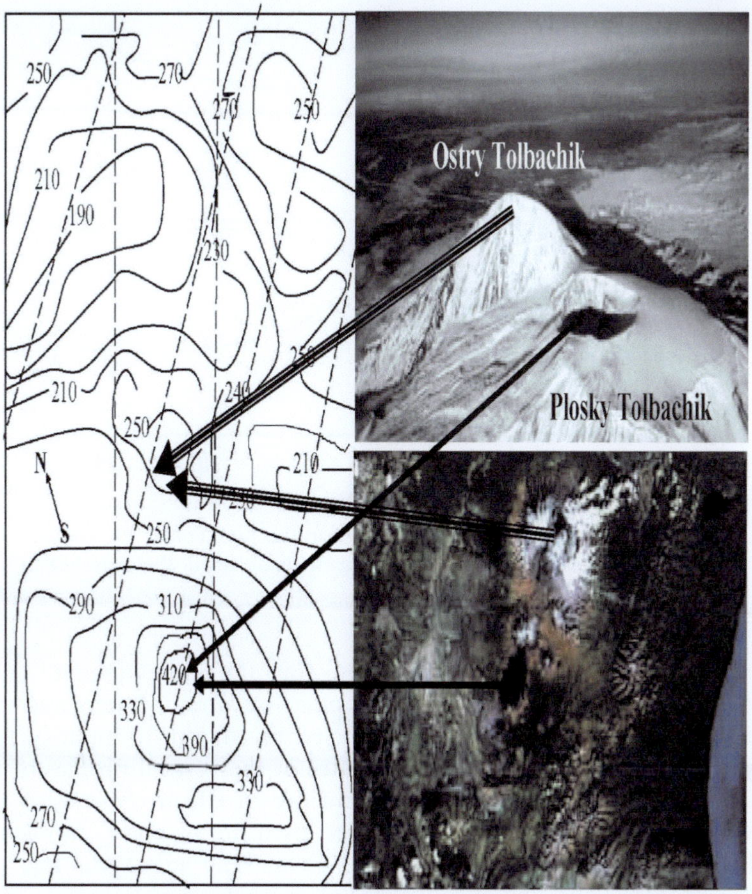


**Figure 7.5.** A map of the distribution of radiobrightness temperature $T_j$ near the Gt. Tolbachik volcano in Kamchatka by means of the IL-18 flying laboratory at wavelength $\lambda = 27$ cm. This map was reconstructed by means of GIMS technology based on trace data.

coefficient of flames ($i = 1$), of smoke ($i = 2$), and of the canopy ($i = 3$). Emission relaxation coefficients are functions of the type of tree. It has also been established that absorption in the canopy increases rapidly as the wavelength decreases. Consequently, the receipt of emissions and the discovery of forest edge fires become difficult when trees have a compact structure. Nevertheless, wavelengths 0.8–1.5 cm enable forest fires to be estimated with high reliability in most cases. This is because of the contribution to microwave emission of flame particles of sizes up to 100–300 μm. The fire zone results in a radiobrightness temperature increase at the 0.8 cm wavelength equal to about 200 K. This exceeds the sensitivity limit of radiometers. Therefore, microwave monitoring of forest fires in contradistinction to optical methods does not depend on the weather conditions.

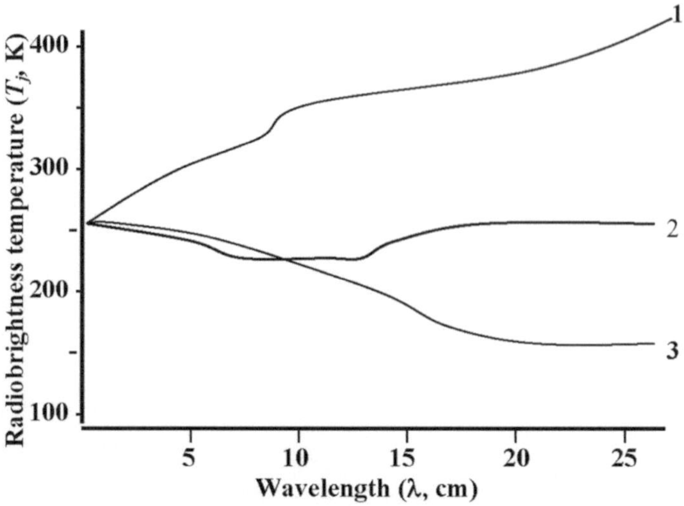


**Figure 7.6.** Radiobrightness temperature spectra for the Gt. Tolbachik volcano. Notation: 1, fresh lava flow; 2, isothermic cold lava; 3, slag fields showing the presence of subsurface moisture (Borodin and Krapivin, 1998).

Temperature anomaly diagnosis calls for the development of specific algorithms for data processing. Joint use of microwave radiometers and models of the atmosphere–soil–vegetation system have led to efficient technology to resolve this. Microwave observations provide estimates of the vertical profiles of temperatures in the ground as a result of the dependence of effectively emitted layers on wavelength. Two wavelengths are normally used in practice. Applied models describing the temperature profile are based on making an assumption about the form of the profile. Thus, the effectiveness of microwave radiometry for thermal source detection and intensity measurement depends on just how adequate this assumption is. The minimum temperature gradient determined in such a way from the radiation data at two wavelengths is approximately $0.3\text{–}1^{\circ}\mathrm{C}\cdot\mathrm{m}^{-1}$. Note that the effectiveness of the microwave range is much higher than infrared radiometry because of the ability of electromagnetic microwaves to penetrate soils.

## 7.7 MICROWAVE MONITORING OF THE ATMOSPHERE

The present state of industrial technologies, energy supply, and transportation worldwide coupled with the growing number of massive industrial complexes are in continuous conflict with ecological standards for natural environment quality, leading to greater risks of accidents, anthropogenic and natural catastrophes, and natural disasters in many parts of the world. All of which highlights the necessity for uninterrupted control of the most vulnerable environment subsystem—the atmosphere. The role microwave methods play in this consists in expanding the capabilities of atmospheric monitoring systems based on optical control principles. Since the atmosphere is almost limpid for many radio waves its effectiveness when measured by atmospheric parameters turns out to be insignificant especially on local scales. Nevertheless, numerous experimental investigations using a flying laboratory and satellites show that microwave sensing of the atmosphere is not as good as optical methods on short traces but is indispensable for atmospheric control over large areas. The choice of wave range for atmospheric sensing is determined both by the special features of the task in hand and by spatial scales.

Assumed ecological loading on the atmosphere is a concept that essentially depends on the extent to which the preceding processes in the interaction between the natural environment and anthropogenic factors can be considered spatio-temporally. Four levels corresponding to ecological monitoring systems can be shown.

The global level (planetary) predicts the impact of anthropogenic influences on the atmosphere during intervals from several months to tens of years. The macro level (continental) considers processes connected with the trans-boundary movement of atmospheric pollutants between countries or continents. International legislation determines how monitoring systems are regulated. Spatiotemporal scales cover thousands of kilometers and several months to years. The meso level (regional) considers atmospheric pollution processes over smaller territories and locates pollution sources and their chemical components. The spatial structure of such processes ranges from hundreds of meters to 100 km. Temporal scales range from several hours to a few days. Such monitoring systems control the atmospheric state in large industrial cities. The micro level (local) considers atmospheric pollution processes over areas having linear sizes from several meters to hundreds of meters with timescales from a few minutes to tens of minutes. Such monitoring systems consider a single phenomenon by studying the spatial distributions of atmospheric parameters in the local area in great detail. Such systems are normally used to control atmospheric contamination where hazardous industries are located.

The sources of atmospheric pollution can be divided into stationary and non-stationary categories. Each source is characterized by the quantity of emitted contaminants with specific properties per unit time, by the temperature of aerial mixtures emitted, by the altitude above the Earth's surface, by the speed at which aerial mixtures are emitted, and by geographical coordinates. Furthermore, source types are distinguished by geometric form (point, linear, areal) and by the emission regime (continuous action, periodic, bursts). Moreover, atmospheric pollution sources can be classified by their hazard levels. This involves large chemical industries, surface and underground oil and gas reservoirs, oil and gas pipes, industrial reservoirs of liquid ammonia, chlorine, concentrated sulfuric acid, and industries whose by-products are emitted as sulfur gas, nitrogen oxide, and hydrogen sulfide (Liou *et al.*, 2010; Straub, 1989).

The way in which the scale and physicochemical state of atmospheric pollution sources is classified helps to simplify the design of a monitoring system. Although pollution source categories can be precisely classified, adequate monitoring systems often do not exist. Nevertheless, in each concrete case such correlations are inevitably established when an atmospheric pollutant dynamics model is created.

Atmospheric remote-sensing methods are based on the study of electromagnetic and acoustic wave propagation. Optical radiation scattering on aerosol and atmospheric gas molecules is used in laser sensing. The effects of turbulent pulsation is taken into account in radiolocation sensing. Laser sensing is certainly effective at determining the chemical composition of the atmosphere, wind speed, temperature, cloud parameters, and dust formations (Klyuev, 2000). The water content in clouds as well as the compounds in atmospheric gases and aerosols can be ascertained by thermal radiation in the microwave range based on spectral measurements (Strelkov and Soldatova, 1989).

Many authors use the millimeter wavelength band to diagnose atmospheric gases. Strelkov (1995, 1996) used estimates of the optical depths of atmospheric tracks between two satellites to ascertain the absorption frequency of resonance of the CO and $N_2O$ admixture in the millimeter and submillimeter bands by taking data on their atmospheric concentrations into account. It was shown that optical depths can be fully defined by admixture gases from observational distances of $\approx 15$ km and can be measured, which indicates the value of using the active remote-sensing method of the atmosphere to ascertain the content of CO and $N_2O$. The intrusion of these gases to the atmosphere is overwhelmingly anthropogenic and a problem for many biogeochemical processes such as the ozone cycle. This is the reason that the possibility of measuring the concentration of these gases by satellite systems is a significant achievement.

Strelkov (1996) developed a method for measuring the concentration of water vapor in the mesopause by active sensing at resonance wavelengths of the rotational spectrum. He showed that for two-frequency sensing, simultaneous measurement of the height profiles of the concentration of water vapor and of the temperature within the mesopause is possible. This method is based on the fact that within the mesopause water vapor completely determines the optical depth of the sensing path in the centers of its spectral lines using wavelengths $\lambda \geq 0.5$ mm. Moreover, the absorption coefficient in the center of the line that has resonance frequency $f_i$ is described by the expression

$$\alpha_i = \frac{8\pi^2 N}{3hcG(T)} f_i |\mu_i| \left\{ \exp\left[-\frac{hcf_{i1}}{kT}\right] - \exp\left[\frac{hcf_{i2}}{kT}\right] \right\} \frac{c}{\Delta f_i(P,T)},$$

where $N$ is the concentration of water vapor molecules; $h$ is the Planck constant ($= 6.55 \cdot 10^{-34}$ joule-s); $c$ is the speed of light ($= 3 \cdot 10^8$ m $\cdot$ s$^{-1}$); $k$ is the Boltzmann constant ($= 1.3 \cdot 10^{-23}$ joules $\cdot$ K$^{-1}$); $G(T)$ is the rotational statistical sum; $f_{i1}$ and $f_{i2}$ are the lower and upper energy levels of transition with frequency $f_i = f_{i2} - f_{i1}$; $\mu_i$ is the dipole moment matrix element for transition with frequency $f_i$; $\Delta f_i$ is the half-width of the line which depends on pressure $P$ and air temperature $T$; $z$ is height.

Scientific and technical achievements in the satellite monitoring of atmospheric gases are important from the perspective of the greenhouse effect and ozone problems (Kondratyev and Varotsos, 2000; Zuev, 2000). Active and passive sensing of some atmospheric gases from satellite radiometers working in the millimeter and submillimeter wavelengths allow the content of $O_2$, $O_3$, $N_2O$, $H_2O$, CO, BrO, ClO, and HCl to be measured. For example, the possibility of sensing additional gas actively is equivalent to reliably measuring the intersatellite trace optical depth, $\tau$, stipulated by the gas:

$$\tau(f, z_m) = 2\int_{z_m}^{H} \frac{\gamma(t, z)(R+z)\,dz}{\sqrt{(R+z)^2 - (R+z_m)^2}},$$

where $\gamma$ is the gas absorption coefficient; $f$ is the sensing frequency; $z$ is height; $z_m$ is minimal distance of the trace from the Earth's surface; $R$ is the Earth's radius; $H$ is height of satellite orbits. The possibility to sense ClO passively, for instance, is defined by the expression for limb brightness temperature:

$$T_j(f, z_m) = 2\int_{z_m}^{H} \frac{T(z)\gamma(f, z)(R+z)}{\sqrt{(R+z)^2 - (R+z_m)^2}} \times [Q_1(z) + Q_2(z)]\,dz,$$

where

$$Q_1(z) = 0.5\exp\left(-\int_{z}^{H} \frac{\gamma(f, \xi)(R+\xi)\,d\xi}{\sqrt{(R+\xi)^2 - (R+z_m)^2}}\right),$$

$$Q_2(z) = 0.5\exp\left(-\tau(f, z_m) + \int_{z}^{H} \frac{\gamma(f, \xi)\,d\xi}{\sqrt{(R+\xi)^2 - (R+z_m)^2}}\right).$$

The radio-translucence method, which needs two satellites with strong gas absorption line frequencies, has proven to be an effective approach to measuring the characteristics of the ozone layer (Strelkov, 1995; Yakovlev, 2001; Yakovlev *et al.*, 2009). This method is further enhanced by Yakovlev *et al.* (2009), where Earth sensing is proposed to be carried out by the radio-eclipsing method. The principal scheme of this method is characterized in Figure 7.7. This method is based on the impact caused by variations in the intensity of the electromagnetic field and in the phase and frequency of radio waves under transmission through the atmosphere and ionosphere. Eclipse monitoring of the atmosphere and ionosphere is dependent on the inverse problem being solved so that it can define the vertical profiles of electron concentrations, refraction coefficients, and atmospheric temperatures. Subsequent development of this method would involve creation of algorithms that might make it possible to identify atmospheric and ionospheric characteristics based on regular measurements.

The greatest role played by microwave radiometry according to many authors is in the monitoring of precipitation and clouds. These possibilities were first demonstrated by experiments on board Cosmos-243 in 1968. Later, many space microwave measurements were taken by means of SMMR (Scanning Multichannel Microwave Radiometer), SSM/I, TRMM, and multi-channel systems on board the Mir space station, which fell out of orbit in 2001.

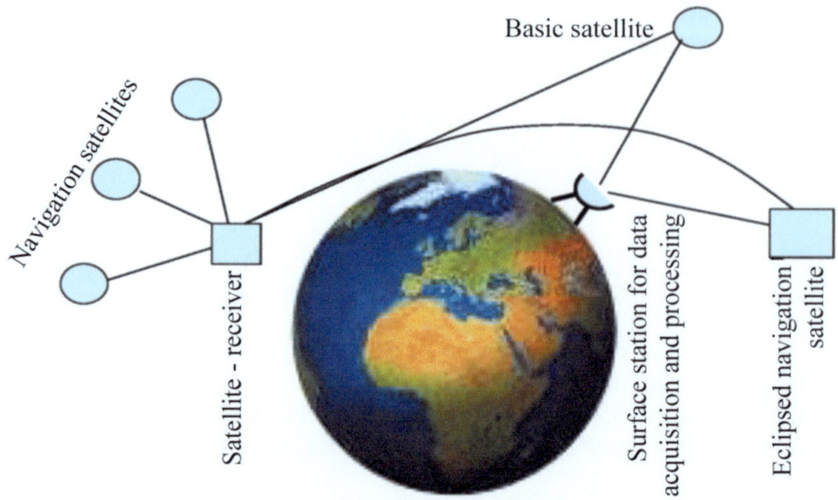


**Figure 7.7.** Schematic diagram of global monitoring system based on the radio-translucence method.

Local measurements of the amount of water vapor in the atmosphere can reliably be registered using wavelength 1.35 cm. Under this the dependence of atmosphere radiobrightness temperature on the integrated content of water vapor can be well described by a linear function with a slope equal to $\approx 19\,\mathrm{K}\cdot\mathrm{g}^{-1}\,\mathrm{cm}^{-1}$.

The application of radio waves from the centimeter and millimeter ranges to diagnose atmospheric properties has been pretty successful because of the weak, unlike the optical and IR ranges, interaction with cloud particles, dust, and atmospheric gas molecules. Radio wave interaction with rain drops, on the contrary, is resonant and expressed by the intensive absorption and scattering of wave energy. Therefore, the active and passive sensing microwave devices of today are unique in being able to receive data about precipitation parameters on the global scale from space.

Microwave observations of precipitation are based on the measurements and calculations of four Stock's parameters (Kutuza *et al.*, 1998, 2000; Zagorin and Kutuza, 1998): $S = \{S_1, S_2, S_3, S_4\}$, where $S_1$ is the total intensiveness of radiation, $S_2$ is the difference between the radiation intensiveness of horizontal and vertical signal polarizations, $S_3$ is the difference in radiation intensiveness of linear orthogonal polarizations with respect to the coordinate system turned by 45°, $S_4$ is the difference in radiation intensiveness of right and left circular polarizations. The Rayleigh–Jeans approximation gives $S_1 = 2k_B T_j \lambda^{-2}$, where $k_B$ is the Boltzmann constant ($=1.38\cdot 10^{-20}\,\mathrm{mW}\cdot\mathrm{Hz}^{-1}\,\mathrm{K}^{-1}$), and $\lambda$ is the wavelength. These parameters are functions of the radiobrightness temperatures $T_j$ of radiation received by vertical $T_V$, horizontal $T_h$, and under the angle ±45° channels $T_{\pm 45}$, as well as circular (right $T_r$ and left $T_l$) polarizations. They are also expressed as functions of complex amplitude for the vertical $E_V$ and horizontal $E_h$ components of the

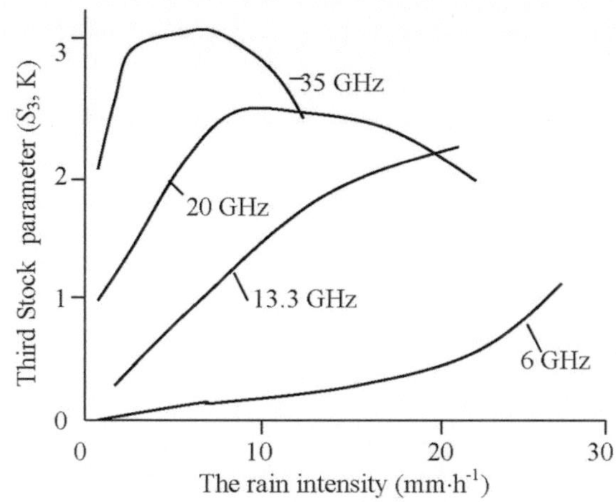


**Figure 7.8.** The dependence of the third Stock's parameter on rainfall intensity. The thickness of the rain layer is 3 km. Descending microwave emission with a 50° angle to the zenith is considered. The drop slope angle is 5° (Zagorin, 1999).

electric field intensity of radiothermal emission (Zagorin and Kutuza, 1998):

$$S = \begin{pmatrix} S_1 \\ S_2 \\ S_3 \\ S_4 \end{pmatrix} = \begin{pmatrix} T_V + T_h \\ T_V - T_h \\ T_{+45} - T_{-45} \\ T_l - T_r \end{pmatrix} = \frac{\lambda^2}{2k_g} \begin{pmatrix} |E_V|^2 + |E_h|^2 \\ |E_V|^2 - |E_h|^2 \\ 2\,\mathrm{Re}(E_V E_h^*) \\ 2\,\mathrm{Im}(E_V E_h^*) \end{pmatrix}.$$

Stock's parameters, measured directly in field experiments, together with other measurements and using cluster analysis allow clouds and rain types to be classified. Figures 7.8 and 7.9 give data about correlating Stock's parameters with the rain intensity calculated by Zagorin (1999). The maximum spectral dependences of the second, third, and fourth Stock's parameters on microwave Sun radiation dispersed as a result of rain intensity increasing are displaced to the low-frequency region (up to $\approx$6 GHz). Therefore, for a rain layer 4 km thick, when the Sun is observed by 20° regarding the horizon, the maximal values of $S_2$, $S_3$, and $S_4$ are achieved when rain intensity $r = 12.5\,\mathrm{mm \cdot h^{-1}}$ and frequency $\approx$13.5 GHz: $S_2 = 220$ K, $S_3 = 44.6$ K, and $S_4 = 5.95$ K. When $r = 50\,\mathrm{mm \cdot h^{-1}}$ and frequency $\approx$6 GHz these parameters become $S_2 = 483$ K, $S_3 = 101.5$ K, and $S_4 = 28.5$ K.

Polarization radiometry of precipitation is based on two effects: the non-spherical form of rain drops and the existence of a distinct direction in which the symmetrical axes of falling drops are oriented. This means that the emission and absorption of radio waves having vertical and horizontal polarizations that differ in directions from the orientation of symmetrical axes of rain drops can be distinguished. This is the reason the azimuth symmetry of the radiothermal field of rain emission is absent. This means that microwave emission from rain is partly polar-

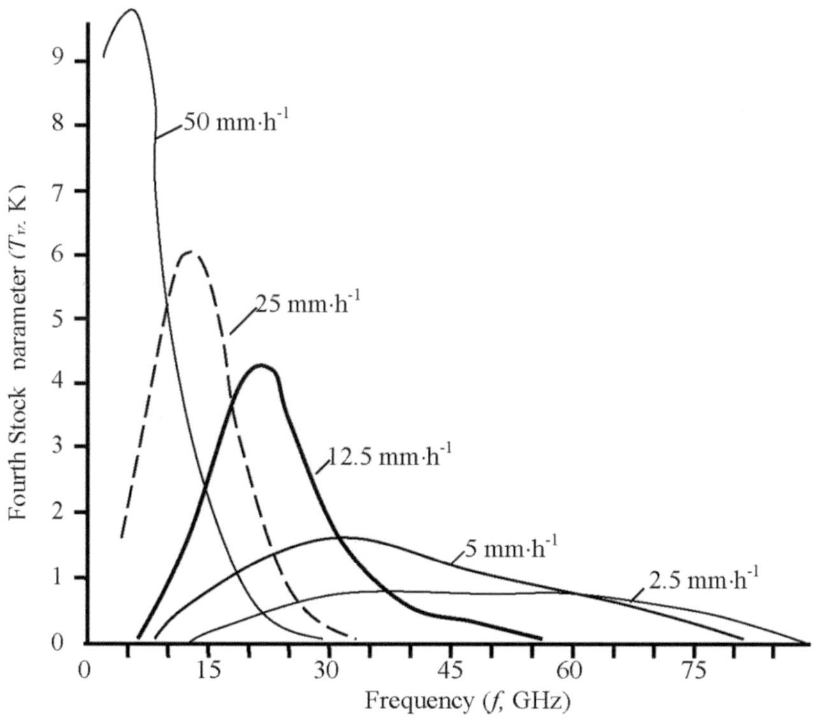


**Figure 7.9.** The spectral dependence of the fourth Stock's parameter on microwave Sun radiation dispersed by rain drops in the 4 km layer (Zagorin, 1999; Zagorin and Kutuza, 1998).

ized. Therefore, the task of precipitation diagnostics is reduced to finding the degree of dependence of emission polarization parameters (polarization degree, slope angle of polarization plane, polarization ellipse form, etc.) on characteristics of the state of the environment in which rain is dispersing.

The first polarization radiometric measurements of rain-descending microwave emission were taken by Kutuza (1977). They were then developed experimentally and theoretically in subsequent investigations. The results of measurements of $T_j$, in general, and different channels of microwave radiometer at $\lambda = 2.25$ cm (13.3 GHz) gave estimates of the degree of linear polarization of radio-emission from rain. When the intensity of a steady downpour was changed from 0.5 to $2.5\,\text{mm} \cdot \text{h}^{-1}$ it was equal to 5.5%. Convective rain with an intensity greater than $5\,\text{mm} \cdot \text{h}^{-1}$ had a degree of linear polarization that reached 8.5%. These estimates demonstrate the distinct delimitation of rain zones.

The methodology set up to estimate the parameters involved in the distribution of rain drops by size developed by Gasiewski and Kunkee (1994) is based on the measurements of two Stock's parameters, $S_1$ and $S_2$. The gist of the methodology consists in solving an inverse task under the preposition that non-spherical drops have an exponential distribution with two unknown parameters determined by data about $S_1$ and $S_2$. The part played by the third and fourth Stock's parameters in rain diagnostics is determined by expanding estimated parameters. Thus, $S_3$ is proportional to the differential relaxation, $\Delta_1$, of radio waves. The fourth Stock's parameter is proportional to the product of $\Delta_1$ and differential phase displacement, $\Delta_2$. The value $\Delta = \Delta_1 + i\Delta_2$ is called the rain anisotropy parameter and can be measured directly. Theoretical dependences of $\Delta$ on different rain parameters was studied by Gasiewski and Kunkee (1994). It was shown that the spread in distribution from the corner of the slope to the vertical projection of the symmetrical axis of rain drops in the plane orthogonal to the direction of the spread of waves exercises the largest influence on the precision of $\Delta$ estimation. Statistical models describing the orientation of drops allow the average value and dispersion to be connected with the Gaussian distribution law of drops according to the orientations of wind field parameters (Zagorin, 1999). It was discovered that when the orientation of the dispersion of drops is less than 30%, theoretical and experimental estimates of $\Delta$ are similar in value. This allows developed models of rain to be used in practice.

Atmospheric microwave control supposes the existence of a radiation source in the microwave range. Such a source can be the Sun, the radio-emission of which in its calm state (when flares are absent) within the frequency region $\lambda \in [2, 100]$ GHz can be characterized by $T_j \in [5 \cdot 10^3, 1.5 \cdot 10^4]$ K. Non-polarized microwave radiation from the Sun is relaxed and polarized by rain with an average relation $S_2/S_1 \approx 0.0528$. Radiometer dislocation and choice of ranges for such a microwave source is currently being studied by radiophysicists. This task is combined with formation of the $T_j$ level on the radiometer output as a function of the field composition on its antenna. This composition depends on the character of the litter surface, the repeated reflections of radio waves from which are to be taken into account to solve the inverse task arising in atmospheric radiometry. Under conditions when there is symmetry at the azimuth of the atmosphere–litter surface system, when emitting and scattering atmosphere particles are oriented chaotically, and when the litter surface is mirror reflecting or diffusely reflecting, then $S_1 \neq 0$, $S_2 \neq 0$, $S_3 = S_4 = 0$. When symmetry at the azimuth is absent, when atmospheric particles have an orientation that differs from the vertical, and when the spread of surface roughness is characterized by an asymmetrical distribution of slopes we receive: $S_1 \neq 0$, $S_2 \neq 0$, $S_3 \neq 0$, $S_4 \neq 0$. The fourth Stock's parameter $S_4$ which characterizes polarization ellipticity becomes unequal to zero when radiothermal emission undergoes no less than two-fold scattering into the hydrometeor volume or into surface roughness (Zagorin, 1999).

An important stage in precipitation monitoring is clearly optimizing registered radiation frequencies. This was partly done by Sasaki *et al.* (1989), who based on measurements showed that the frequencies near 1 GHz are the most informative. This result is corroborated by the curves in Figures 7.10 and 7.11.

The perspectivism of radiophysical methods used for the design of an atmospheric control system is accentuated by the many theoretical and empirical investigations. The basic problem consists in forming a knowledge base to maintain correlations between atmospheric characteristics (properties) and the behavior

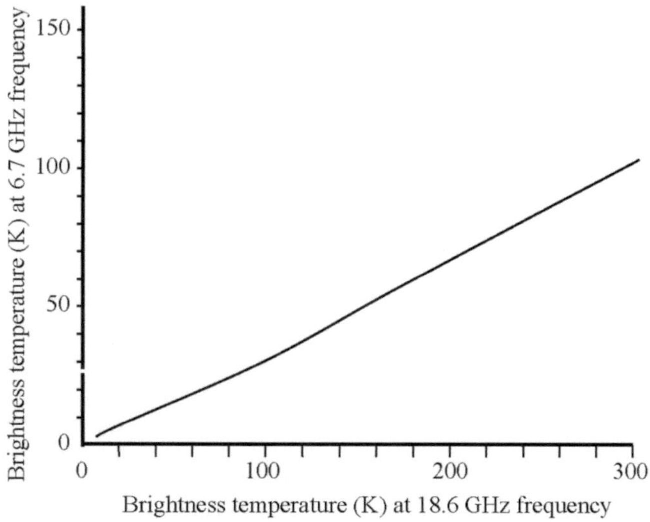


**Figure 7.10.** The state of the informational space of brightness temperatures at the most representative frequencies.

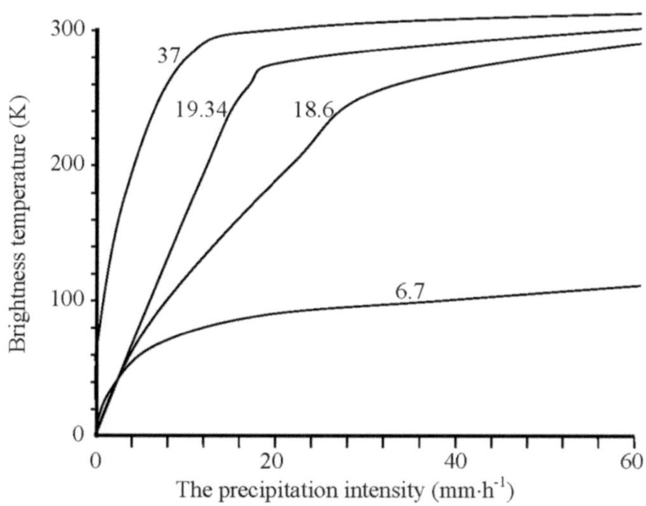


**Figure 7.11.** The comparative rainfall rate dependence of sky brightness temperatures at different frequencies (shown on the curves in GHz) (Sasaki *et al.*, 1989).

of electromagnetic waves inside the atmosphere. The basis on which atmospheric process models have been built reflecting the combinations of spatiotemporal scales is again an important element in the study of these correlations.

As a result of investigations by Yakovlev (2001) effective methodologies for designing an atmospheric monitoring system are connected with radio occultation studies of the atmosphere and ionosphere. The setup necessary to bring about such measurements is based on placing the transmitter on a satellite orbiting the Earth and placing the receiver on a geostationary satellite. As a result, such a structure secures a continuous vertical transection of the atmosphere with the objective of determining the distribution of parameters by altitude. The task is reduced to identifying the values of these parameters using data about signal absorption and fluctuation as a result of propagating along the changing tracks between satellites.

The influence of the atmosphere and ionosphere on the parameters of refraction, fluctuation, relaxation, absorption, decay, and change in signal frequency has been studied in detail. Depending on the orbit position of the satellite this influence makes it possible to receive data about the atmosphere and ionosphere in concrete regions of the planet. A total spatial image for part or all of the atmosphere can be had by data processing using GIMS technology.

It is possible to distinguish between the direct and inverse tasks of radio-translucence. The direct task arises when the atmospheric model is used to determine the abovementioned changes in electromagnetic waves for a given satellite trajectory. The inverse task consists in determining atmospheric and ionospheric parameters by means of experimental data describing signal variations.

The method of atmosphere radio-translucence can also be used when the microwave emission receiver is sited on the Earth's surface. Using the Doppler effect on wavelength 1.35 cm (22.235 GHz) in this case allows the vertical distribution of water vapor in the uppermost atmosphere to be received with acceptable precision.

Atmospheric microwave sensing can be expanded to consider the possibility of estimating a wider set of its parameters. To do so would mean solving a complex mathematical task connected with the prognosis of microwave relaxation according to sensing routes in the atmosphere. The millimeter range is especially interesting in this respect. According to the investigations by Meriakri (1992) and Strelkov (1995) many gases have discrete absorption spectra in waves of the millimeter and submillimeter ranges. This allows the average concentration of the gas component along the sensing route to be calculated based on estimating electromagnetic wave absorption. An example of such a calculation is given in Table 7.7.

A special feature of the millimeter range is the possibility to form narrowly directed radio-bundles because of the relatively small apertures of the emitters. The task of atmosphere parameter estimation can be transformed to registering variations in signal relaxation on the sensing route between the receiver and transmitter followed by solving a suitable integral equation that takes into account the refraction of electromagnetic waves in the troposphere, cross-diffraction of signal bundles, and possible variations in the vertical distribution of refraction indicators along the route. Monitoring systems of the low atmosphere also call for the influence of land covers on the propagation of millimeter waves to be taken into consideration.

Employing active and passive sensing in the millimeter and submillimeter ranges ensures reliable estimation of the atmospheric content of certain chlorine

**Table 7.7.** Gas concentrations based on measurements of signal relaxation at frequency $\nu_a$ along a 10 km atmospheric track (Meriakri, 1992).

| *Gas* | *Gas concentration* ($mg\,m^{-3}$) | $\nu_a$ (MHz) | *Atmospheric relaxation* ($db \cdot km^{-1}$) |
|---|---|---|---|
| CO | 1500<br>330 | 0.3<br>6.0 | 115.271<br>235.789 |
| $N_2O$ | 1600<br>240 | 0.2<br>6.0 | 100.492<br>351.666 |
| $SO_2$ | 2300<br>560 | 0.2<br>0.4 | 104.029<br>135.696 |
| $H_2S$ | 9 | 1.8 | 168.763 |

compounds, carbon oxides, nitrogen dioxide, ozone, and water vapor, in addition to determining wind speed and temperature (Strelkov, 1995, 1996). The prospect of these ranges bringing about bi-static radio-translucence of the atmosphere is down to the moderate price of field measurements and existing theoretical results concerning the methodologies used to reconstruct atmospheric spatial distribution parameters along observational routes.

Application of the microwave range to atmospheric diagnostics cannot be effective without creating measurement data-processing systems. Such systems are based on simulation models that make reconstructing atmospheric characteristics possible by making good use of multichannel sensing data. What is more, the obligatory presence of the optical and IR ranges between the channels of such systems results in an essential gain in monitoring system informativeness.

Within the framework of GIMS technology application, a simulation system for atmospheric pollution physics (SSAPP) has been designed as an experimental system to control and identify sources polluting the atmospheric basin (Bondur *et al.*, 2009; Krapivin *et al.*, 1996, 1997a, b). The SSAPP is intended to reconstruct multi-component non-stationary pollutant fields and identify local pollutant sources. The SSAPP uses the procedure of adaptive synthesis of simulation model to describe three components of the pollution field as a basic algorithm. This procedure consists in dynamic development of the simulation model as a result of estimating the divergence between model output and data registered by means of measuring devices in various ranges of the electromagnetic spectrum.

Putting current *a priori* information about the coordinates of measuring devices, pollutant sources, and parameters into a simulation model is brought about through coordination with a topographical map of the territory being controlled. Model optimization and control of the set of ranges are brought about by means of a database and available technical resources. The system adapts automatically to the spatial scales of the territory for which pollution is being simulated. The set of tasks solved by the SSAPP includes:

- calculation of the area of influence of each source in the system database;
- determination of the area of sensitivity of each measuring device;
- reconstruction of the spatial distribution of pollutants;
- identification of the causes behind dangerous pollution zones arising;
- search for unknown pollution sources.

The SSAPP is based on a set of models that describe the dynamics of the atmosphere and the spread of contaminants over restricted territories up to the whole planet. The meso and micro levels can be imitated by a set of Gaussian models, the choice of which depends on the parameters provided by the monitoring system's database. The global and macro levels of the spread of atmospheric contaminants can be described by Euler-type and Lagrange-type models.

## 7.8 MICROWAVE RADIOMETRY IN REMOTE MONITORING OF THE OCEAN

Reconstruction of radiothermal field emission of the ocean in the microwave range depends on water temperature and salinity, surface roughness, concentration of chlorophyll, and other components of the oceanic environment. Remote registration of this field calls for such factors as evaporation, precipitation, foam formation, cloudiness, and intensity variations of direct and scattered solar radiation to be taken into account as well. This is the reason the design of microwave remote-monitoring systems based on registering ocean emissions in the microwave range has taken into consideration the need to solve a set of complex mathematical tasks. Parameters reflecting the state of the atmosphere–ocean system have an important influence on how the spatial structure of the radiothermal field of this system is formed. Therefore, leaving zones subject to rain, snow, and hail to one side when total absorption by the atmosphere for wavelength $\lambda \geq 0.8\,\text{cm}$ is insignificant, the phenomenological theory of radiation transmission can be applied to determine the radiobrightness temperatures, $T_j$, of the atmosphere–ocean system. In zones subject to rain clouds, the relative contribution of scattering processes for emission on wavelength $\lambda \geq 0.8\,\text{cm}$ to the formation of field $\{T_j\}$ ranges from 5% to 50%. This means that the scattering effects in these zones need to be taken into account when processing microwave radiometry data received from the ocean surface.

When $T_j$ is registered by a satellite-based system from height $H$ under the observation angle $\theta$ with respect to the local vertical for wavelength $\lambda$ the following functional correlation of radiobrightness temperature with thermodynamic and physical characteristics of the atmosphere–ocean system is usually valid (Savorskij, 1992):

$$T_j(\lambda, \theta) = Q_1 + \cos^{-1}\theta \int_0^H Q_2(u)\,du + Q_3 \int_0^\infty Q_4(u)\,du, \tag{7.6}$$

where

$$Q_1 = \kappa_\lambda T_{SO} \exp[-\tau_\lambda(0,H)/\cos\theta],$$

$$Q_2 = T(u)\gamma_\lambda(u) \exp[-\tau_\lambda(u,H)\cos^{-1}\theta],$$

$$Q_3 = (1-\kappa_\lambda) \exp[-\tau_\lambda(0,H)\cos^{-1}\theta]\cos^{-1}\theta,$$

$$Q_4 = T(u)\gamma_\lambda(u) \exp[-\tau_\lambda(0,u)\cos^{-1}\theta];$$

$\gamma_\lambda(u)$ is the total relaxation coefficient of the atmosphere at altitude $u$; $T(u)$ is the thermodynamic temperature at altitude $u$; $\kappa_\lambda$ is surface emissive ability; $T_{SO}$ is ocean surface temperature; $\tau_\lambda(u_1,u_2)$ is the optical thickness of the atmospheric layer between levels $u_1$ and $u_2$ on wavelength $\lambda$:

$$\tau_\lambda(u_1,u_2) = \int_{u_1}^{u_2} \gamma_\lambda(z)\,dz.$$

Formula (7.6) is basic to calculation of ocean parameters using measurements $\{T_j\}$. The variety of modifications and simplifications of expression (7.4) is determined by the set of radiation models of the atmosphere–ocean system, where the dependences $\kappa_\lambda$ and $\gamma_\lambda$ on parameters of this system are theoretical or empirical. The coefficients of water surface emission on horizontal, $\kappa_\lambda^h$, and vertical, $\kappa_\lambda^V$, polarizations are calculated by the Fresnel formulas:

$$\kappa_\lambda^h = 1 - \left|\frac{\varepsilon_\lambda\cos\theta - (\varepsilon_\lambda - \sin^2\theta)^{1/2}}{\varepsilon_\lambda\cos\theta + (\varepsilon_\lambda - \sin^2\theta)^{1/2}}\right|^2, \tag{7.7}$$

$$\kappa_\lambda^V = 1 - \left|\frac{\cos\theta - (\varepsilon_\lambda - \sin^2\theta)^{1/2}}{\cos\theta + (\varepsilon_\lambda - \sin^2\theta)^{1/2}}\right|^2, \tag{7.8}$$

where $\varepsilon_\lambda$ is the dielectric permittivity of water at wavelength $\lambda$ and a function of salinity, temperature, and other parameters of the upper ocean layer:

$$\varepsilon_\lambda = \varepsilon'_\lambda - i\varepsilon''_\lambda(\varepsilon'_\lambda = 5.5 + a_1), \quad \varepsilon''_\lambda = \lambda_S\lambda^{-1}a_1, \quad a_1 = [\varepsilon_S - 5.5]/[1 + (\lambda_S/\lambda)^2],$$
$$\varepsilon_S = 0.0081t^2 - 0.40885t + 88.2, \quad t = T_{SO} - 273,$$
$$\lambda_S = 1.466\exp\{-0.0634t\} + 0.000136t^2 - 0.0272t + 1.8735.$$

Any correction to value $\varepsilon_\lambda$ as a result of water salinity variations $S$ can be taken into account by the approximate formula: $\varepsilon_S(S) = \varepsilon_S(0) - 17.2S$. From expressions (7.7) and (7.8) under vertical observation ($\theta = 0°$) we receive: $\kappa_\lambda^h = \kappa_\lambda^V = \kappa_\lambda^0 = P_\kappa$, where

$$P_\kappa = 4|\varepsilon_\lambda|^{1/2}\cos(\delta_\lambda/2)/|\varepsilon_\lambda| + 2|\varepsilon_\lambda|^{1/2}\cos(\delta_\lambda/2) + 1,$$

angle $\delta_\lambda$ is determined by the expression $\mathrm{tg}\,\delta_\lambda = \varepsilon''_\lambda/\varepsilon'_\lambda$.

Andreev *et al.* (1984) analyzed Shepovalov's invariant $(\kappa_\lambda^h)^2/(2\kappa_\lambda^h - \kappa_\lambda^V) = 1$ and proposed a constructive algorithm to calculate surface temperature $T_S$ using measurements of $T_{j,h}$ and $T_{j,V}$ from an observation angle of $\theta = 45°$:

$T_S = (T_{j,h})^2/(2T_{j,h} - T_{j,V})$. This algorithm has a precision of $\pm 3$ K. It can be used to calculate both $T_{SO}$ and the thermodynamic temperature of land. In practice, $T_j$ depends on $T_{SO}$ linearly with a slope from 0.35 to 0.5. An increase in salinity from 0 to 40‰ is provided by increasing $T_j$ under $\lambda < 3$ cm by values not exceeding 3 K and by decreasing $T_j$ under $\lambda \in [8, 10]$ cm by values up to 40 K. The dependence slope of this decrease is a function of $T_{SO}$.

Knowledge of the radiation dependences of water surface properties on the total spectrum of water surface parameters is the essential basis of a radiometric control system of the ocean surface.

For a water component with great conductivity (i.e., seawater) in the decimeter range the following correlation is valid: $\kappa \cong 0.5(f/\sigma)^{1/2}$, where $f$ is the radio wave frequency, and $\sigma$ is water conductivity. This expression allows water conductivity to be measured as long as water thermal emission is measured. If the substance dissolved in the water and its conductivity are known the concentration of the dissolved substance can be calculated by using thermal radio-emission. Hence, microwave radiometry can be used to monitor reservoir pollution.

The emissive and reflective abilities of a sea surface in the microwave region is closely connected with ocean environment parameters such as water temperature, salinity, dielectric constant, and roughness. Sasaki *et al.* (1988) showed the 6.7–18.6 GHz band to be highly informative when used to diagnose the atmosphere–ocean system. It was established that the correlation between the radiobrightness temperatures of the atmosphere and the sea surface is linear and that the reflective ability on horizontal polarization at the frequency 18.6 Ghz is higher than on vertical polarization. Moreover, it is observed that there exists an obvious dependence of radiobrightness temperature fluctuations of the sea surface on wind speed. For horizontal polarization these fluctuations equal $0.5\,\mathrm{K \cdot m^{-1}\,s}$ and $1.0\,\mathrm{K \cdot m^{-1}\,s}$ at 6.7 and 18.6 GHz, respectively. In the case of vertical polarization these values equal 0.3–0.8 $\mathrm{K \cdot m^{-1}\,s}$ and 0.6–1.5 $\mathrm{K \cdot m^{-1}\,s}$, respectively.

For P-type polarization, the sensitivity of the sea surface brightness temperature, $T_{BP}$ (K), to variations in wind speed, $V$ ($\mathrm{m \cdot s^{-1}}$). is determined by the value (Figures 7.12 and 7.13):

$$\omega = \left.\frac{\partial T_{BP}}{\partial V}\right|_{a_1,\ldots,a_k=\mathrm{const}},$$

where $a_i$ $(i = 1, \ldots, k)$ are the oceanic environmental parameters contributing to $T_{BP}$ change except for wind speed. The value of $\omega$ is almost linearly correlated with wind speed under constant sea surface temperatures and salinities, so $\partial^2 T_{BP}/\partial V^2 \approx 0$. Then $T_{BP} = T_{BOP}(\theta) + V\,\partial T_{BP}(\theta)/\partial V$, where $T_{BOP}$ is the brightness temperature of a calm sea surface (K), and $\theta$ is the zenith angle of incident radiation. According to Kirchoff's law we have the following wind speed sensitivity to reflectivity: $\partial\gamma_P(\theta)/\partial V = -\partial\varepsilon_P(\theta)/\partial V$. It gives the following expression for reflectivity at arbitrary wind speed: $\gamma_P(\theta) = \gamma_{OP}(\theta) - (V/T_{S_l}) \cdot \partial T_{BP}(\theta)/\partial V$, where $T_{SO}$ is sea surface temperature, and $\gamma_P$ and $\gamma_{OP}$ are the reflectivities of rough and calm sea surfaces, respectively.

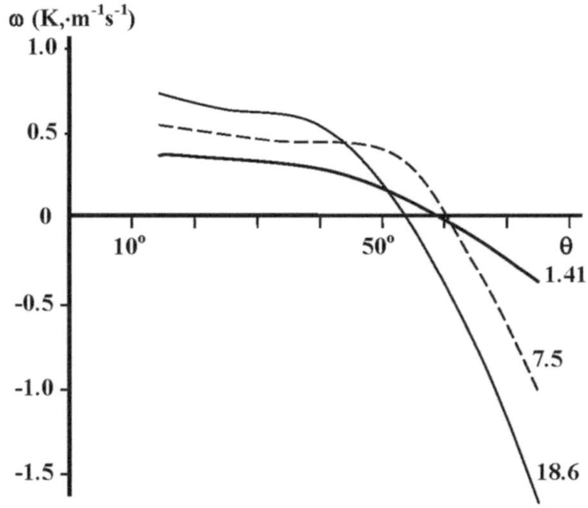


**Figure 7.12.** Empirical dependence of brightness temperatures of the ocean surface on wind speed at various wavelengths and observation angles (Sasaki *et al.*, 1988). Frequencies in gigahertz are given on the curves.

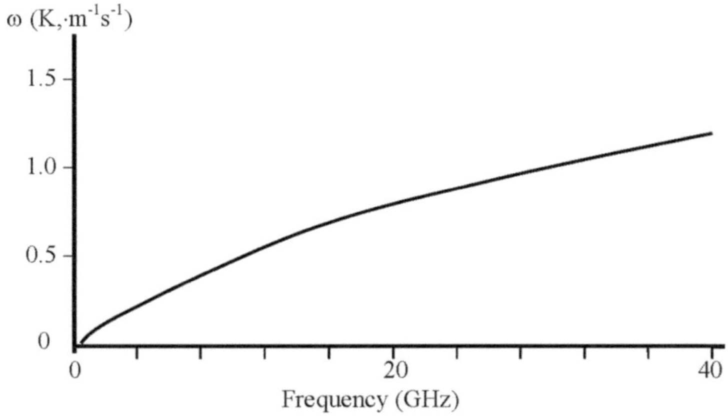


**Figure 7.13.** Empirical dependence of brightness temperature sensitivity on wind speed when observations are made at nadir (Sasaki *et al.*, 1988).

According to Sasaki *et al.* (1987a, b, 1988) the wind speed sensitivity of the sea surface brightness temperature at the satellite level is

$$\frac{\partial T_{BP}^{\text{sat}}(\theta)}{\partial V} = \frac{\partial T_{BP}(\theta)}{\partial V} \cdot \frac{T_{SO} - \{T_B^{\downarrow\text{sky}}(\theta) + T_B^{\text{space}}\}}{T_{SO}},$$

where $T_B^{\downarrow sky}$ is the downward sky brightness temperature at the sea surface (K); and $T_B^{space}$ is galactic noise (K).

For an arbitrary wavelength, sea surface brightness temperature can be represented in the form:

$$T_{BP}^{\lambda}(\theta) = \varepsilon_{1P}(\theta)T_{SO} + \gamma_P(\theta)T_B^{\lambda,\mathrm{sky}}(\theta_S, \phi_S),$$

where $\theta_S$ and $\phi_S$ are the zenithal and azimuthal angles,

$$\gamma_P(\theta) = \frac{1}{4\pi}\int_0^{2\pi}\int_0^{\pi/2} \gamma_P(\theta, \theta_S, \phi_S) \sin\theta_S \, d\theta_S \, d\phi_S.$$

Under conditions of wind-induced sea roughness, when sea waves are so big that radio waves of the centimeter and decimeter ranges cannot be used, scattering and emission characteristics can be estimated by means of Kirchoff's approximation.

In the centimeter wave range, where the electrophysical properties of water are comparatively poorly known, the expression (6.7) allows the water surface temperature to be determined when its radiothermal emission is measured.

The task of ocean surface diagnosis is normally the inverse of function $T_j = F(a_1, \ldots, a_k)$ in the way it relates to the set of geophysical parameters set $\{a_i\}$. The function $F$ reflects the type of aquatory radiation model. Linear models that allow the standard deviation of $T_j$ to be calculated are widely used: $\delta_i = k_i \delta a_i$, $k_i = \partial F/\partial a_i$, where $\delta a_i$ is the standard deviation of parameter $a_i$. Estimates of $k_i$ can be found in Nelepo *et al.* (1985).

Having synthesized a radiation model, $F$ can then be used to control ocean surface pollution and the atmosphere–ocean system is considered as a three-layer environment: atmosphere–pollutant–water. For calculation of Fresnel's coefficients to estimate the emissive property $\kappa_\lambda$ of this system, knowledge of the dielectric permeability of the pollutant layer is needed. Therefore, for oil spills in the microwave range there is weak dependence of the dielectric constant on $\lambda$ and the values of losses are typically small. For example, a crude oil spill is characterized by the value $\mathrm{Re}\,\varepsilon_\lambda \in [1.6, 3]$. Specifically, for $\lambda = 0.8$ cm we have $\varepsilon_\lambda = \delta_i = k_i \delta a_i$, $k_i = \partial F/\partial a_i$. Clearly, $\varepsilon_\lambda$ changes according to oil quality and the length of time oil has been in the water.

Any pollutant of the water environment forms an emulsion layer on the water surface which lasts a certain time. The dielectric permeability of this layer can be calculated by the mixture formula: $\varepsilon_\lambda = [(1 - W_1)(\varepsilon_{\lambda,1})^{1/3} + W_1(\varepsilon_{\lambda,2})^{1/3}]^3$, where $W_1$ is the amount of water in the emulsion volume, and $\varepsilon_{\lambda,1}$ and $\varepsilon_{\lambda,2}$ are the dielectric permeabilities of pollutants and water, respectively. Figure 7.14 demonstrates the frequency of $\varepsilon_\lambda$ for oil pollution.

The methodology behind radiation model synthesis for a three-layer environment is applied to microwave diagnostics of the ice cover. The dielectric permeability of the ice layer is given here as $\varepsilon_\lambda$ (Melentyev *et al.*, 1998).

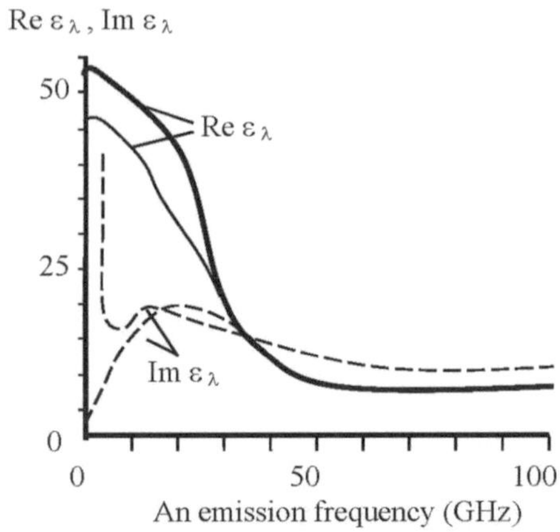


**Figure 7.14.** Dielectric properties of oil mixture with fresh (solid line) and salt (dashed line) water.

## 7.9 ALGORITHMS FOR REMOTE DATA PROCESSING

Spatial reconstruction of a radiometric image of vegetation, based on fragmentary data, calls for the development of new algorithms. There exist many approaches to this task. For example, Yoshioka (2004) discussed the problem of extraction of useful information from satellite imagery by means of algebraic manipulation. The concept of vegetation isolines underpins the three-layer soil–canopy–atmosphere system. The relationship between reflectances at two different wavelengths (650 and 940 nm) over this system which comprises atmosphere, canopy, and soil layers has been derived. Two algorithms are proposed below.

### 7.9.1 Data reconstruction using harmonic functions

The spread of heat in the plane of homogeneous media $G$ with constant thermophysical characteristics (density $\rho$, specific heat $C$, and conductivity coefficient $K$; $\rho, C, K = \text{const} > 0$) can be described by the equation:

$$\partial T/\partial t = a^2(\partial^2 T/\partial\varphi^2 + \partial^2 T/\partial\lambda^2), \tag{7.9}$$

where $T = T(\varphi, \lambda, t)$ is the temperature of the media at the point with spatial coordinates $(\varphi, \lambda) \in G$ at time $t$; $a^2 = K/C\rho$ is the temperature conductivity coefficient of $G$; and $\varphi$ and $\lambda$ are latitude and longitude, respectively.

If the thermal transport process is stationary then equation (7.9) transforms to the Laplace equation:

$$\text{div} \cdot \text{grad}\, T = \partial^2 T/\partial\varphi^2 + \partial^2 T/\partial\lambda^2. \tag{7.10}$$

The solution of equation (7.10) is the harmonic function of the spatial coordinates $\varphi$ and $\lambda$. In accordance with the Rayleigh–Jeans law (DeWitt and

Nutter, 1988) the brightness temperature as a result of remote microwave measurements is $T_a(\varphi,\lambda) = T_a(\varphi,\lambda,\mu,\theta)$, where $\mu$ is the wavelength, $\theta$ is the observation angle ($\mu,\theta = \text{const}$). It is proposed that for any small area $V_M$ at an arbitrary point $M \in G$ the brightness temperature is a linear function of the temperature of the media:

$$T_a(\varphi,\lambda) = A_M + B_M T(\varphi,\lambda); \quad (\varphi,\lambda) \in V_M, \tag{7.11}$$

where $A_M$ and $B_M$ are constants.

Formula (7.11) follows from theoretical and experimental estimates of $T_a$ (Burkov and Krapivin, 2009; Krapivin and Potapov, 2001). For media that are homogeneous in depth, the formula $T_a = \kappa T$ is valid. Here $\kappa = \kappa(\mu,\theta,\varepsilon)$ is the media radiation coefficient where $\varepsilon$ is dielectric permeability. According to the experimental estimates the radiance of fresh water measured by means of radiometers with wavelengths of 5–8 cm is a linear function of $T$. The slope of this dependence is 0.35–0.50 K/°C. An increase in water salinity $S$ from 0 to 16‰ is accompanied by a decrease in the sensitivity of the irradiation field to temperature variations. This effect is observed for wavelengths from 10 to 50 cm. The sensitivity of the irradiation field to variations in $T$ is a minimum when the following conditions are realized: $\mu S \cong 700$; $0 < T \leq 30°\text{C}$; $0 \leq S \leq 180‰$; $0 \leq \theta \leq 25°$.

From equation (7.11) it follows that $T_a$ at every point $M \in G$ satisfies the following condition:

$$T_a(\varphi,\lambda) = \frac{1}{2\pi}\int_0^{2\pi} T_a(\varphi + r\cdot\cos a, \lambda + r\cdot\sin a)\, da,$$

where the integral is over a circle of radius $r$ centered at $(\varphi,\lambda)$.

This condition is valid for any $r$ ($0 < r < r_M$) where $r_M$ is the radius of the area $V_M$. Therefore, $T_a$ is a harmonic function within $G$. A typical task here is finding the harmonic function $T_a(\varphi,\lambda)$ within $G$ when $T_a(\varphi,\lambda) = \tilde{T}_a(u)$ on the boundary $\Gamma$ of $G$ ($u \in \Gamma$, $u = \varphi + i\lambda$). Such a function is the real part of some analytic function $W(z)$ given by:

$$W(z) = \frac{1}{2\pi}\int_\Gamma \mu(\zeta)/(\zeta - z)\, d\zeta, \tag{7.12}$$

where $\mu(\zeta)$ is the real density; $\zeta \in \Gamma$; and $z = \varphi + i\lambda$ is an arbitrary internal point of $G$. We have $\text{Re}[W(u)]$ and $\text{Im}[d\zeta/(\zeta - u)] = -\cos(r,n)\, d\sigma/r$, where $r$ is the distance between $\zeta$ and $u$, $d\sigma$ is an element of $\Gamma$, and $n$ is an external normal to $\Gamma$. As $z \to u \in \Gamma$, $\mu(u)$ is approximated by the solution of the integral equation:

$$\mu(u) - \frac{1}{\pi}\int_\Gamma \mu(\zeta) r^{-1} \cos(r,n)\, d\sigma.$$

When $G$ is the circle $|z - z_0| < R$, the solution is the Poisson integral:

$$T_a(r,\psi) = \frac{1}{2\pi}\int_0^{2\pi} \tilde{T}_a(a)\frac{R^2 - r^2}{R^2 + r^2 - 2Rr\cos(\psi - a)}\, da,$$

where $\varphi + i\lambda = z_0 + re^{i\psi}$ ($r < R$, $0 \leq \psi \leq 2\pi$); and

$$\tilde{T}_a(a) = \tilde{T}_a(z_0 + R \cdot e^{ia}), \quad 0 \leq \alpha \leq 2\pi.$$

The combination of this procedure with other algorithms of spatiotemporal interpolation gives a full representation of environmental objects in the observation area by means of parametric estimates. For example, the TIM database was formed on the basis of measurements taken by the flying laboratory using a set of analogous harmonic algorithms. Table 7.8 illustrates the precision of this algorithm compared with measured values and the differential approximation method.

### 7.9.2 Method for parametric identification of environmental objects

Radiometers can be used to determine the brightness temperatures $Z_{ij}$ ($i = 1, \ldots, M$; $j = 1, \ldots, n$) given by $Z_{ij} = T_j + \xi_{ij}$, where $M$ is the number of measurements, $n$ is the number of radiometers, $T_j$ is the real value of the brightness temperature for wavelength $\mu_j$, and $\xi$ is noise with zero mean and dispersion $\sigma_j$. The problem is to determine the correlation function $T_j = f_j(X)$, where $X = \{x_1, \ldots, x_m\}$ are geophysical, ecological, biogeochemical, or other parameters. There are many algorithms that can be used to define function $f$. As a general rule, the mean-square criterion is used for this purpose. But such an approach has one defect: the impossibility of taking the dispersion properties of noise $E = \{\xi_{ij}\}$ into consideration.

Let function $f$ be linear. Then we have the following system of equations for parameters $A_{ij}$:

$$\|A_{ij}\|X = T + E. \tag{7.13}$$

It is necessary to solve equation (7.13) such that its solution has minimum dispersion. Such a solution is called the $\sigma$-solution.

The $i$th equation of system (7.13) is multiplied by the set of parameters $c_{1i}, \ldots, c_{mi}$. An additional condition is given:

$$\sum_{i=1}^{n} c_{ji} A_{il} = \delta_{jl}, \tag{7.14}$$

where

$$\delta_{jl} = \begin{cases} 1 & \text{for } j = l \\ 0 & \text{for } j \neq l \end{cases} \quad (l, j = 1, \ldots, m). \tag{7.15}$$

Under conditions (7.14) and (7.15) we have

$$x_1^0 = \sum_{i=1}^{n} c_{1i} T_i. \tag{7.16}$$

From (7.13) and (7.16) we obtain:

$$\tilde{x}_1 = \sum_{i=1}^{n} c_{1i} T_i + \sum_{i=1}^{n} c_{1i} \xi_i. \tag{7.17}$$

**Table 7.8.** Reconstruction of the hydrophysical field in the Nyuoc Ngot lagoon. Field measurements were carried out during March and April in 2001. Signs correspond to water inflow (+) and outflow (−). This lagoon is situated off central Vietnam in the South China Sea. Notation: $U$ is the measured value of water flow ($m^3\ s^{-1}$) at the lagoon–sea boundary; $M$ and $R$ are calculated values of this flow using the methods of differential approximation and harmonic functions, respectively.

| $U$ ($m^3\ s^{-1}$) | $M$ ($m^3\ s^{-1}$) | Error (%) | $R$ ($m^3\ s^{-1}$) | Error (%) |
|---|---|---|---|---|
| +10.3 | +11.3 | +10.1 | +11.6 | +12.9 |
| −4.0 | −3.6 | +9.5 | −4.4 | +9.8 |
| −8.7 | −15.3 | +12.4 | −9.9 | +13.3 |
| −14.2 | −12.3 | +13.1 | −16.0 | +12.6 |
| −24.1 | −22.7 | +5.7 | −22.2 | +7.7 |
| −22.4 | −24.4 | +8.9 | −20.4 | +9.0 |
| −16.3 | −17.9 | +9.6 | −18.2 | +11.5 |
| +2.3 | +2.6 | +11.5 | +2.0 | +11.4 |
| +15.0 | +13.2 | +12.3 | +12.6 | +15.8 |
| +35.1 | +30.9 | +12.2 | +30.1 | +14.2 |
| +39.2 | +35.1 | +10.4 | +43.0 | +9.7 |
| +38.2 | +42.6 | +11.6 | +33.2 | +13.1 |
| +34.8 | +39.6 | +13.8 | +30.0 | +13.7 |
| +26.5 | +25.2 | +4.8 | +28.5 | +7.6 |
| +19.2 | +21.1 | +9.9 | +21.2 | +10.3 |
| Average error (%) | 10.4 | | 11.5 | |
| Maximal error (%) | 13.8 | | 15.8 | |
| Minimal error (%) | 4.8 | | 7.6 | |

The dispersion of solution (7.17) is

$$D[\tilde{x}_1] = \sum_{i=1}^{n} c_{1i}^2 \sigma_i^2. \tag{7.18}$$

Dispersions of $\tilde{x}_i$ $(i = 2, \dots, m)$ are calculated by analogy with (7.18). To calculate min $D[\tilde{x}_1]$ the following additional equation is used

$$\psi(c_{11}, \dots, c_{1n}) = \sum_{i=1}^{n} c_{1i}^2 \sigma_i^2 + \tau_1 \left( \sum_{i=1}^{n} c_{1i} A_{i1} - 1 \right) + \sum_{j=2}^{m} \tau_j \sum_{i=1}^{n} c_{1i} A_{ij}.$$

The first derivatives of $\psi$ are equal to zero, giving the following set of equations:

$$2c_{1k}\sigma_k^2 + \sum_{j=1}^{m} \tau_j A_{kj} = 0 \quad (k = 1, \dots, n). \tag{7.19}$$

Conditions (7.14), (7.15), and (7.19) consist of a system of $(m + n)$ equations that need to be solved.

We have $D[x_j] = \tau_j/2$, where the set of $\tau_j$ are defined as solutions to the following equations:

$$\sum_{j=1}^{m} \mu_j \sum_{i=1}^{n} \frac{A_{ij}A_{i1}}{\sigma_i^2} = -2; \quad \sum_{j=1}^{m} \mu_j \sum_{i=1}^{n} \frac{A_{ij}A_{il}}{\sigma_i^2} = 0 \quad (l = 2, \dots, m).$$

These algorithms are used as sub-blocks of the Aral–Caspian Expert System (Bondur *et al.*, 2009). Forecasting the state of the ACS can be made by means of the TIM (see Chapter 5).

### 7.9.3 Method of differential approximation

The databases of environmental monitoring systems do not always correspond to the parametric fullness of the GIMS technology standard. Therefore, an algorithm that allows a database to be adapted to this standard should be considered. Let us suppose that $N$ characteristics, $x_i$ $(i = 1, \dots, N)$, of the environment are measured at times $t_s$ $(s = 1, \dots, M)$. The formal dependence between $x_i$ is represented by the system of differential equations with unknown coefficients $\{a_{ijk}, b_{ij}\}$:

$$\frac{d\xi_i}{dt} = \sum_{k,j=1}^{N} [a_{ijk}\xi_j(t)\xi_k(t) + b_{ij}\xi_j(t)]. \tag{7.20}$$

Putting the initial conditions as

$$\xi_i(0) \quad (i = 1, \dots, N) \tag{7.21}$$

the reconstruction task of $x_i(t)$ for the arbitrary time $t \in [0, T]$ comes down to the simple task of determining unknown coefficients based on the criterion:

$$E = \sum_{s=1}^{M} \left\{ \sum_{i=1}^{N} [\xi_i(t_s) - x_i(t_s)]^2 \right\} = \min_{\{a_{ijk}, b_{ij}, \xi_i(0)\}}.$$

There are many methods that can be used to solve this task. One is based on Bellman's dynamic programming method (Krapivin, 1969; Krapivin and Kondratyev, 2002).

### 7.9.4 Quasi-linearization method

The number of problems that crop up in ecoinformatics leads to the necessity of integrating generally non-linear integro-differential equations; but, in the majority of cases, these equations are not integrable by elementary or special functions. To solve them, it is generally necessary to make use of the latest achievements of calculating methods and technologies. In many problems the use of well-known numerical methods to solve initial value problems—even by means of modern high-speed electronic computers—does not come up with the desired results. The existing approximate methods of solving integro-differential equations are based as a rule on replacing derivatives by finite differences and represent a complicated multi-step process, which in practical problems cannot be solved on computers reasonably quickly. Therefore, to solve practical problems we have to search for other means of approximate solutions for integro-differential equations without using finite-difference methods.

In the method considered here, the integro-differential equation is substituted in each sub-interval of the independent variable by an easily integrable ordinary differential equation with constant coefficients; this method is not a new theoretical idea as it was known to Euler. However, error estimates are obtained here for the first time and methods applicable to various problems are developed in detail.

#### *7.9.4.1 Method of solution and estimation of error*

Let us consider the equation

$$L[y] - \lambda W[y] = f(x, y), \tag{7.22}$$

where $L[y]$ is the differential operator

$$L[y] = \sum_{i=0}^{n} P_i(x, y, y', \ldots, y^{(m_i)}) y^{(n-i)} \quad (m_i < n) \tag{7.23}$$

and $W[y]$ is the generalized Volterra operator

$$W[y] = \int_a^x \sum_{j=0}^{r} K_j(x, \xi) y^{(j)}(\xi) \, d\xi \quad (r < n), \tag{7.24}$$

where $\lambda$ is a real number; $P_i(x, y, y', \ldots, y^{(m_i)})$ and $f(x, y)$ are continuous functions with respect to their arguments in the finite interval $[a,b]$; $P_0 \neq 0$; and kernels $K_j(x,\xi)$, $j = 0,1,\ldots,r$ are continuous functions in the region $G\{a \le \xi \le x \le b\}$.

The initial conditions are

$$y^{(s)}(a) = y_0^{(s)}, \quad s = 0,1,\ldots,n-1. \tag{7.25}$$

Assuming equation (7.22) with initial conditions (7.25) has a unique continuous solution $y(x)$, let us construct an approximate solution $\tilde{y}(x)$ in $[a,b]$. Let us divide the interval $[a,b]$ by a sequence of points $x_0 = a$, $x_1,\ldots,x_m = b$, $h_k = x_{k+1} - x_k$. On each sub-interval $[x_k, x_{k+1}]$, $k = 0,1,\ldots,m-1$ let us replace equation (7.22) by the following linear differential equation of the $n$th order with constant coefficients

$$\tilde{L}_k[y] = \lambda\tilde{W}_k[y] + f(x_k,\tilde{y}_k) \tag{7.26}$$

with the initial conditions:

$$y^{(s)}(x_k) = \tilde{y}_k^{(s)}, \quad s = 0,1,\ldots,n-1, \tag{7.27}$$

where

$$\tilde{L}_k[y] = \sum_{i=0}^{n} P_i(x_k,\tilde{y}_k,\tilde{y}'_k,\ldots,\tilde{y}_k^{(m_i)}\tilde{y}^{(n-i)}, \tag{7.28}$$

$$\tilde{W}_k[y] = \sum_{j=0}^{r}(K_{j,k,0}\,\tilde{y}_0^{(j)}h_0 + K_{j,k,1}\,\tilde{y}_1^{(j)}h_1 + \cdots + K_{j,k,k}\,\tilde{y}_k^{(j)}h_k). \tag{7.29}$$

The general solution of equation (7.26) is known:

$$\tilde{y} = \tilde{y}(x, c_1^{(k)}, c_2^{(k)},\ldots,c_n^{(k)}), \tag{7.30}$$

where the constants are determined from the initial conditions at the beginning of each interval $[x_k, x_{k+1}]$. Calculations are carried out successively beginning with the interval $(k = 0)$.

Let us estimate the error in our solution of equation (7.22). Let $y(x)$ and $\tilde{y}(x)$ be the exact and approximate solutions, respectively. Let us denote

$$\tilde{P}_{ik} = P_i(x_k,\tilde{y}_k,\tilde{y}'_k,\ldots,\tilde{y}_k^{(m_i)}), \quad \tilde{f}_k = f(x_k,\tilde{y}_k), \quad \varepsilon_k = y(x_k) - \tilde{y}(x_k). \tag{7.31}$$

Let us integrate equations (7.22) and (7.26) from $x_k$ to $x$, and consider the final results for $x = x_{k+1}$. For the sake of convenience and brevity let us denote

$$\underbrace{\int_{x_k}^{x_{k+1}}\int_{x_k}^{x}\cdots\int_{x_k}^{x}}_{n}\varphi(x)\,dx\underbrace{\cdots}_{n}dx = \underbrace{\int_{x_k}^{x_{k+1}}}_{n}\varphi(x)\,dx,$$

giving us

$$y_{k+1} = y_k + y_k' h_k + h_k^2 \sum_{s=2}^{n-1} y_k^{(s)} \frac{h_k^{s-2}}{s!} - \sum_{i=1}^{n} \underbrace{\int_{x_k}^{x_{k+1}}}_{n} P_i\, y^{(n-i)}\, dx \underbrace{\int_{x_k}^{x_{k+1}}}_{n} f(x,y)\, dx$$

$$+ \lambda \underbrace{\int_{x_k}^{x_{k+1}}}_{n} W[y]\, dx \tag{7.32}$$

$$\tilde{y}_{k+1} = \tilde{y}_k + \tilde{y}_k' h_k + h_k^2 \sum_{s=2}^{n-1} \tilde{y}_k^{(s)} \frac{h_k^{s-2}}{s!} - \sum_{i=1}^{n} \underbrace{\int_{x_k}^{x_{k+1}}}_{n} \tilde{P}_{ik}\, \tilde{y}^{(n-i)}\, dx + \underbrace{\int_{x_k}^{x_{k+1}}}_{n} \tilde{f}_k\, dx$$

$$+ \lambda \underbrace{\int_{x_k}^{x_{k+1}}}_{n} \tilde{W}_k[y]\, dx. \tag{7.33}$$

From (7.32) and (7.33) we get:

$$\varepsilon_{k+1} = \varepsilon_k + \varepsilon_k' h_k + h_k^2 \sum_{s=2}^{n-1} \varepsilon_k^{(s)} \frac{h_k^{s-2}}{s!} - \sum_{i=1}^{n} \underbrace{\int_{x_k}^{x_{k+1}}}_{n} [P_i\, y^{(n-i)} - \tilde{P}_i\, \tilde{y}^{(n-i)}]\, dx$$

$$+ \underbrace{\int_{x_k}^{x_{k+1}}}_{n} [f - \tilde{f}_k]\, dx + \lambda \underbrace{\int_{x_k}^{x_{k+1}}}_{n} \{W_k[y] - \tilde{W}_k[y]\}\, dx.$$

We know that

$$\underbrace{\int_{x_k}^{x_{k+1}}}_{n} \tilde{f}_k\, dx = \frac{\tilde{f}_k h_k^n}{n!}, \qquad \underbrace{\int_{x_k}^{x_{k+1}}}_{n} \tilde{W}_k[y]\, dx = \frac{\tilde{W}_k[y]}{n!}.$$

Let us denote

$$E_k = \max_j |\varepsilon_k^{(j)}|, \qquad h_{\max} = \max_k h_k, \qquad p_i = \max_{[a,b]} |P_i|, \qquad M_{n-i} = \max_{[a,b]} |y^{(n-i)}|,$$

$$L_i = \max_{[a,b]} |\tilde{P}_{ik}|, \qquad N_{n-i} = \max_{[a,b]} |\tilde{y}^{(n-i)}|, \qquad F = \max_{[a,b]} |f|, \qquad G_0 = \max_{[a,b]} |\tilde{f}_k|,$$

$$T = \max_G |\tilde{W}_k[y]|, \qquad s = |b-a| \sum_{j=0}^{r} \max_G |K_j(x,\xi)| M_j \geq \max_G |W[y]|,$$

$$l = \frac{1}{n!}\left[\sum_{i=1}^{n} (p_i M_{n-i} + L_i N_{n-i}) + F + G_0 + |\lambda|(T+s)\right],$$

$$g = 1 + \sum_{s=2}^{n-1} \frac{h_{ax}^{s-1}}{s!},$$

and $M_{n-i} \approx N_{n-i}$. Then we get the following recurrent error estimation

$$E_{k+1} \leq (1 + gh_{\max})E_k + lh^n_{\max}. \tag{7.34}$$

Hence we get

$$E_k \leq (1 + gh_{\max})^k \varepsilon_0 + lh^{n-1}_{ax} g^{-1}[(1 + gh_{\max})^k - 1], \tag{7.35}$$

where $\varepsilon_0$ is the maximum error in the initial data. Obviously, if $\varepsilon_0 = 0$, then from (7.35) it follows that if $h_{\max} \to 0$ then $E_k \to 0$ (i.e., $\tilde{y}(x_k) \to y(x_k)$).

In case equation (7.22) has the form:

$$L[y] = f(x, y) + \int_a^x F(x, y, y', \ldots, y^{(m_l)}\, dx, \quad x \in [a, b],$$

then equation (7.35) will read

$$E_k \leq (1 + hp^{(0)})^k \varepsilon_0 + \frac{p^{(1)}h^n}{p^{(0)}}[(1 + hp^{(0)})^k - 1],$$

where

$$p^{(0)} = \sum_{s=1}^{n-1} \frac{h^{s-1}}{s!} + \frac{h^{n-1}}{n!}\left[\tilde{b} + \sum_{i=1}^{n}(p_i + (m_l + 1)\gamma_i\beta_{n-i}) + (b - a)c(m_l + 1)\right],$$

$$p^{(1)} = \frac{1}{(n+1)!}(\tilde{a} + \tilde{b}\beta_1 + 2B + \sum_{i=0}^{n} p_{0i}) + \frac{(b-a)q}{n!},$$

$$\tilde{a} = \max_{[a,b]}\left|\frac{\partial f}{\partial x}\right|, \qquad \tilde{b} = \max_{[a,b]}\left|\frac{\partial f}{\partial y}\right|, \qquad \gamma_i = \max_{s,[a,b]}\left|\frac{\partial P_i}{\partial y^{(s)}}\right|, \qquad \beta_i = \max_{[a,b]}\{|y^{(i)}|, |\tilde{y}^{(i)}|\},$$

$$c = \max_{s,[a,b]}\left|\frac{\partial F}{\partial y^{(s)}}\right|, \qquad q = 0.5\left(A + C\sum_{s=0}^{m_l} \beta_{s+1}\right),$$

$$p_{0i} = \beta_{n-i}\left(l_i + \gamma_i \sum_{s=0}^{m_l} \beta_{s+1}\right) + 2\alpha_i\beta_{n+1-i}, \qquad A = \max_{[a,b]}\left|\frac{\partial F}{\partial x}\right|, \qquad B = \max_{[a,b]}\{|F|, |\tilde{F}|\}.$$

### 7.9.4.2 *Solution of equation* $y^{(n)} = f(x, y, y', \ldots, y^{(n-1)})$

Let us apply the approximate method of the solution presented in Section 7.9.4.1 to integro-differential equations to solve the initial value problem:

$$y^{(n)} = f(x, y, y', \ldots, y^{(n-1)}), \quad (x, y) \in G, \tag{7.36}$$

$$y(x_0) = y_0, \qquad y^{(j)}(x_0) = y_0^{(j)}, \qquad j = 1, \ldots, n-1, \qquad (x_0, y_0) \in G, \tag{7.37}$$

where the function $f$ satisfies the Lipschitz condition

$$|f(x, y + \delta_0, \ldots, y^{(n-1)} + \delta_{n-1}) - f(x, y, y', \ldots, y^{(n-1)})| \leq K \sum_{i=0}^{n-1} |\delta_i|. \tag{7.38}$$

Let us divide the interval $[a, b]$ by a sequence of points $x_0 = a,\ x_1, \ldots, x_s = b$ into elementary intervals. Let $E = \{x_0, \ldots, x_s\}$. On each interval $\{x_\nu, x_{\nu+1}]$, let us solve the initial value problem:

$$y^{(n)} = f(x, \hat{y}_\nu, \hat{y}'_\nu, \ldots, \hat{y}_\nu^{(n-1)}), \qquad (x, \hat{y}_\nu) \in G, \qquad \nu = 0, 1, \ldots, s-1;$$
$$y(x_\nu) = \hat{y}_\nu, \qquad y^{(j)}(x_\nu) = \hat{y}_\nu^{(j)}, \qquad j = 1, \ldots, n-1, \qquad (x_\nu, \hat{y}_\nu) \in G.$$

Then, if function $f$ satisfies condition (7.38) and $\max_i [x_{i+1} - x_i] = h$, the solution of this problem is

$$\hat{y} = \{y_0, \hat{y}_1, \ldots, \hat{y}_s\}, \qquad \hat{y}^{(j)} = \{y_0^{(j)}, \hat{y}_1^{(j)}, \ldots, \hat{y}_s^{(j)}\}, \qquad j = 1, \ldots, n-1,$$

when $h \to 0$ tends to the solution of equations (7.36), (7.37), and an estimate for the rate of convergence is as follows:

$$\max_l |y_r^{(n-l)} - \hat{y}_r^{(n-l)}| \le \varepsilon_0 (1 + h\alpha_0)^r + \frac{h\alpha_1}{2\alpha_0}[(1 + h\alpha_0)^r - 1], \quad r = 1, 2, \ldots, s,$$

where

$$\alpha_0 = \sum_{i=1}^{n-1} \frac{h^{i-1}}{i!} + K\left(n + \frac{h}{2} \sum_{j=0}^{n-1} \sum_{s=0}^{n-j-2} \frac{h^s}{s!}\right),$$

$$\alpha_1 = K \sum_{j=0}^{n-1} \left[ M \frac{h^{n-j-1}}{(n-j-1)!} + \sum_{s=0}^{n-j-2} N_{s+j+1} \frac{h^s}{s!} \right],$$

$$M = \max_{[a,b]} |f(x, y, y', \ldots, y^{(n-1)})|, \qquad N_{s+j+1} = \max_r |\hat{y}_r^{(s+j+1)}|.$$

If the initial conditions are exactly given, the error estimate has the form:

$$\max_{1 \le l \le n-1} |y_r^{(n-l)} - \hat{y}_r^{(n-l)}| \le Dh^2, \quad r = 1, \ldots, s,$$

where

$$N = \max_{1 \le r \le s} |N_r|, \qquad D = K(M + nD) \frac{[1 + he^h(1 + 0.5Knh) + Kn]^r - 1}{4[1 + Kh(0.5h + e^{-h})]}.$$

### *7.9.4.3 Solution of a system of ordinary differential equations*

For the sake of simplicity let us confine ourselves to the important case of equations having the canonical form

$$y_i^{(m_i)}(t) = f_i(t, y_1, y'_1, \ldots, y_1^{(m_1 - 1)}, \ldots, y_n^{(m_n - 1)}), \quad i = 1, \ldots, n. \tag{7.39}$$

The system (7.39) can be replaced by an equivalent system of $m = m_1 + \cdots + m_n$ equations of first order, relative to the derivatives for all $m$ unknown functions. Then, standard software can be used to solve the last system.

Let the functions $f_i$, $i = 1, \ldots, n$ be continuous and differentiable with respect to all arguments. Let us suppose that the solution of system (7.39) with the initial conditions

$$y_i(t_0) = (y_i)_0, y_i'(t_0) = (y_i')_0, \ldots, y_i^{(m_i-1)}(t_0) = (y_i^{(m_i-1)})_0$$

exists and is unique in $t_0 \leq t \leq T$.

Divide the interval $[t_0, T]$ into elementary intervals $\Delta k = [t_k, t_{k+1}]$ by a sequence of points $t_0 < t_1 < \cdots < t_l = T$. On each such interval, let us search for the solution of system (7.39) in the form of a series:

$$\tilde{y}_i(t) = \tilde{y}_i(t_k) + \sum_{j=1}^{m_i-1} \frac{(t-t_k)^j}{j!} \tilde{y}^{(j)}(t_k) + \frac{(t-t_k)^{m_i}}{(m_i)!} (\tilde{f}_i)_k,$$

$$\tilde{y}_i^{(j)}(t_k) = \tilde{y}_i^{(j)}(t_{k-1}) + \sum_{s=1}^{m_i-j-1} \frac{(t_k - t_{k-1})^s}{s!} \tilde{y}_i^{(s+j)}(t_{k-1}) + \frac{(t_k - t_{k-1})^{m_i-j}}{(m_i - j)!} (\tilde{f}_i)_k$$

$$(j = 1, \ldots, m_i - 1).$$

The error of such a solution can be easily estimated by considering the exact expansion of functions $y_i(t)$ and $y_i^{(j)}(t)$ in a Taylor series:

$$|\varepsilon_i(t_{k+1})| \leq |\varepsilon_i(t_k)| + \sum_{j=1}^{m_i-1} \frac{h_k^j}{j!} |\varepsilon_i^{(j)}(t_k)| + M_i \frac{h_k^{m_i+1}}{(m_i+1)!}$$

$$+ M_i \frac{h_k^{m_i}}{(m_i)!} \sum_{j=1}^{n} \sum_{s=0}^{m_j-1} |\varepsilon_j^{(s)}(t_k)| \tag{7.40}$$

$$|\varepsilon_i^{(j)}(t_k) \leq |\varepsilon_i^{(j)}(t_{k-1})| + \sum_{s=1}^{m_i-j-1} \frac{h_{k-1}^s}{s!} |\varepsilon_i^{(s+j)}(t_{k-1})|$$

$$+ M_i \left( \frac{h_k^{m_i-j+1}}{(m_i-j+1)!} + \frac{h_k^{m_i-j}}{(m_i-j)!} \sum_{j=1}^{n} \sum_{s=0}^{m_j-1} |\varepsilon_j^{(s)}(t_{k-1})| \right), \tag{7.41}$$

where

$$M_i = \max_{[t_0,T]} \left\{ \left| \frac{\partial f_i}{\partial t} \right|, \left| \frac{\partial f_i}{\partial y_1} \right|, \ldots, \left| \frac{\partial f_i}{\partial y_n^{(m_n-1)}} \right| \right\}.$$

Formulas (7.40) and (7.41) give a recurrent estimate for error. It is possible to obtain an error estimate from them applicable to the entire interval $[t_0, T]$:

$$E_k \leq \varepsilon_0 (1 + hp_0)^k + \frac{hP_1}{p_0} [(1 + hp_0)^k - 1],$$

where the following notations are introduced

$$h = \max_k h_k, \qquad E_k = \max_{i,j} |\varepsilon_i^{(j)}(t_k)|, \qquad M = \max_i M_i, \qquad \nu = \min_s m_s,$$

$$P_1 = M \frac{h^{\nu-1}}{(\nu+1)!}, \qquad p_0 = \mu M \frac{h^{\nu-1}}{\nu!} + \sum_{s=1}^{\mu-1} \frac{h^{s-1}}{s!}, \qquad \mu = \max_s m_s.$$

#### 7.9.4.4 Solutions of equations with known moments on the right-hand side

Let us consider a particular case of equation (7.22):

$$L[y] = y^{(n)} + \sum_{i=1}^{n} p_i \, y^{(n-i)} = f(x), \quad x \geq 0, \tag{7.42}$$

where $p_i$ are constant coefficients; $f(x)$ is a single-valued and differentiable function; and $f(x) \to 0$ as $x \to \infty$ and its moments are known:

$$M_\nu f(x) = \int_0^\infty x^\nu f(x)\, dx < \infty, \quad \nu = 0, 1, \ldots, m. \tag{7.43}$$

It is necessary to solve equation (7.42) with the following initial conditions:

$$y^{(s)}(x_0) = y_0^{(s)} \quad (s = 0, 1, \ldots, n-1).$$

We shall approximate $f(x)$ in the following manner:

$$f(x) \approx \exp(-kx) \sum_{i=0}^{m} a_i x^i = P_m(x) \exp(-kx), \tag{7.44}$$

where $m > 0$ is an integer; and $k > 0$ and $a_i$ are constants to be determined. Then, from (7.43) and (7.44) we have

$$\tilde{M}_\nu f(x) = \int_0^\infty x^\nu \left[ \exp(-kx) \sum_{i=0}^{m} a_i x^i \right] dx - \sum_{i=0}^{m} a_i \frac{(\nu + i)!}{k^{\nu+i+1}}.$$

where $k$ is fixed according to conditions of best approximation by (7.44). Then equation (7.42) is replaced by the approximate equation:

$$L[\tilde{y}] = \exp(-kx) \sum_{i=0}^{m} a_i x^i, \tag{7.45}$$

which can easily be solved. For the error $\varepsilon(x) = y(x) - \tilde{y}(x)$, we obtain an equation from (7.42) and (7.45):

$$L[\varepsilon(x)] = f(x) - \exp(-kx) P_m(x) \equiv R_m(x). \tag{7.46}$$

Solving equation (7.46), we have

$$|\varepsilon(x)| \leq \frac{\eta}{n!} \sum_{k=0}^{\infty} \frac{(b-a)^{n+k}}{k!} M^k,$$

where

$$M = \max_{a \leq s \leq x \leq b} |K(x,s)|, \qquad \eta = \max_{[a,b]} |R_m(x)|, \qquad K(x,s) = \sum_{i=1}^{n} p_i \frac{(x-s)^{i-1}}{(i-1)!}.$$

#### *7.9.4.5 Refinements of approximate solutions of Volterra integral equations*

Let us consider Volterra integral equations of the first and second kind that crop up in remote monitoring problems:

$$\lambda \int_a^x G(x,y)\varphi(y)\,dy = g(x), \tag{7.47}$$

$$\varphi(x) - \lambda \int_a^x K(x,y)\varphi(y)\,dy = f(x), \tag{7.48}$$

where $x \in [a,b]$; the kernel $K(x,y)$ and its derivatives $K'_x(x,y)$ are continuous in the region $R\{a \le y \le x \le b\}$; $f(x)$ is a continuously differentiable function in $(a,b)$, the kernel $G(x,y)$ and $g(x)$ are twice continuously differentiable functions of $x$; and $G(x,x) \neq 0$. Then, as is well known, equations (7.47) and (7.48) have unique solutions $\varphi_1(x)$ and $\varphi_2(x)$, respectively, which are continuous and differentiable in $[a,b]$ for any value of $\lambda$. The case when $G(x,x) = 0$ for some point in the interval $[a,b]$ or for the entire interval needs special consideration. In our case, equation (7.47) is equivalent to the equation of the second kind

$$\varphi(x) + \int_a^x \frac{G'_x(x,y)}{G(x,x)}\varphi(y)\,dy = \frac{g'(x)}{\lambda G(x,x)}.$$

Therefore, the argument used to find an approximate solution of (7.48) is valid for equation (7.47) as well.

Equation (7.48) is solved traditionally by replacing the integral of the equation by a finite sum of some quadratic formula. Applying this approach we divide the interval $[a,b]$ by a sequence of points $x_0 = a < x_1 < x_2 < \cdots < x_m = b$ into elementary intervals $\Delta j = [x_j, x_{j+1}]$, and instead of (7.48) we write the equation:

$$\varphi(x_j) - \lambda \sum_{i=0}^{j-1} \int_{x_i}^{x_{i+1}} K(x_j,y)\varphi(y)\,dy = f(x_j) \quad (j = 0,1,\ldots,m). \tag{7.49}$$

Further, because of the assumptions made on $\varphi(x)$ and $K(x,y)$ we can write

$$\varphi(x) = \varphi(x_i) + (x - x_i)\varphi'(x_i) + \frac{(x-x_i)^2}{2!}\varphi''(\xi_i) \quad (x_i \le \xi_i \le x \le x_{i+1}) \tag{7.50}$$

and assuming the existence and differentiability of $K'_y(x,y)$ we have

$$K(x_j,y) = K(x_j,x_i) + (y - x_i)K'_y(x_j,x_i) + \frac{(y-x_i)^2}{2}K''_{yy}(x_j,\eta_i)$$
$$(x_i \le \eta_i \le y \le x_{i+1}). \tag{7.51}$$

Substituting (7.50) and (7.51) in (7.49) we get:

$$\varphi(x_j) - \lambda \sum_{i=0}^{j-1} \left\{ K(x_j, x_i)\varphi(x_i) + \frac{h_i}{2}[K(x_j, x_i)\varphi'(x_i) + K'_y(x_j, x_i)\varphi(x_i)] \right.$$
$$\left. + \frac{h_i^2}{3} K'_y(x_j, x_i)\varphi'(x_i) \right\} h_i + R_j = f(x_j),$$

where

$$R_j = -\lambda \sum_{i=0}^{j-1} \left\{ \frac{h_i^3}{2}\varphi''(\xi_i)\left[\tfrac{1}{3}K'_y(x_j, x_i) + \frac{h_i}{4}K'_y(x_j, x_i)\right] \right.$$
$$\left. + \frac{h_i^3}{2} K''_{yy}(x_j, \eta_i)\left[\tfrac{1}{3}\varphi(x_i) + \frac{h_i}{4}\varphi'(x_i)\right] + \frac{h_i^5}{20}\varphi''(\xi_i)K''_{yy}(x_j, \eta_i) \right\}.$$

Ignoring the small quantity $R_j$ in this expression, we get a recurrent formula for determining $\varphi(x_j)$ from the values of function $\varphi(x)$ at $x = x_0, x_1, \ldots, x_{j-1}$. By differentiating (7.48) we get a formula to calculate the values of the derivative $\varphi'(x)$:

$$\varphi'(x) = f'(x) + \lambda K(x, x)\varphi(x) + \lambda \int_a^x K'_x(x, y)\, dy.$$

From this equation, we have at $x = x_i$

$$\varphi'(x_i) = f'(x_i) + \lambda K(x_i, x_i)\varphi(x_i) + \lambda \sum_{s=0}^{i-1} \int_{x_s}^{x_{s+1}} K'_x(x_i, y)\varphi(y)\, dy, \tag{7.52}$$

where

$$\varphi(x_0) = f(a), \qquad \varphi'(x_0) = f'(a) + \lambda K(a, a)\varphi(a).$$

Ignoring the quantity

$$r_i = \lambda \sum_{s=0}^{i-1} \left\{ K''_{xy}(x_i, Q_s)\frac{h_s^2}{2}\left[\varphi(x_s) + \tfrac{2}{3}\varphi'(x_s)h_s + \frac{h_s^2}{4}\varphi''(\xi_s)\right] + K'_x(x_i, x_s)\frac{h_s^2}{6}\varphi''(\xi_s) \right\}$$
$$(x_s \le \xi_s, Q_s \le y \le x_{s+1}),$$

equation (7.52) can be written as:

$$\tilde{\varphi}'(x_i) = f'(x_i) + \lambda K(x_i, x_i)\tilde{\varphi}(x_i) + \lambda \sum_{s=0}^{i-1} h_s K'_x(x_i, x_s)\left[\tilde{\varphi}(x_s) + \frac{h_s}{2}\tilde{\varphi}'(x_s)\right].$$

We finally get an equation to determine $\tilde{\varphi}(x_j)$:

$$\tilde{\varphi}(x_j) - f(x_j) - \lambda \sum_{i=0}^{j-1} [\tilde{\varphi}(x_i)K(x_j, x_i)l_{ji} + h_i\gamma_{ji}\delta_i]h_i = 0, \tag{7.53}$$

where, for sake of brevity, the following notation is introduced:

$$l_{ji} = 1 + \frac{h_i}{2}\left[\lambda K(x_i, x_i) + \frac{K'_y(x_j, x_i)}{K(x_j, x_i)}[1 + \tfrac{2}{3}\lambda h_i K(x_i, x_i]\right],$$

$$\gamma_{ji} = \tfrac{1}{2}K(x_j, x_i)\left[1 + \tfrac{2}{3}h_i \frac{K'_y(x_j, x_i)}{K(x_j, x_i)}\right],$$

$$\delta_i = f'(x_i) + \lambda\sum_{s=0}^{i-1} h_s K'_x(x_i, x_s)\left[\tilde{\varphi}(x_s) + \frac{h_s}{2}\tilde{\varphi}'(x_s)\right].$$

Thus, starting our calculation of formula (7.53) from $j = 1$, we get values for the solution $\tilde{\varphi}_1, \tilde{\varphi}_2, \ldots, \tilde{\varphi}_m$ that have error $\varepsilon(x_i)$. Let us estimate the modulus of error $\varepsilon(x)$ from the above and introduce the following notation:

$$M = \max_{[a,b]}|\varphi(x)|, \qquad N = \max_{[a,b]}|\varphi'(x)|, \qquad L = \max_{[a,b]}|\varphi''(x)|,$$

$$Q = \max_R\{|K'_y(x,y)|, |K'_x(x,y)|\}, \qquad G = \max_R|K(x,y)|,$$

$$B = \max_R\{|K''_{yy}(x,y)|, |K''_{xy}(x,y)|\}, \qquad h = \max_s|h_s|,$$

$$l = \frac{m}{6}\left[B\left(M + \frac{3h}{4}N\right) + L(G + \tfrac{3}{4}Qh + \tfrac{3}{10}Bh^2)\right],$$

$$n = \frac{|\lambda|Bm}{2}\left(M + \frac{2h}{3}N + \frac{h^2}{4}L\right) + \frac{h}{6}BL.$$

Then we have

$$|\varepsilon_j| \le |\lambda|h(G + \tfrac{2}{3}Qh)\sum_{i=0}^{j-1}\left(|\varepsilon_i| + \frac{h}{2}|\varepsilon'_i|\right) + |\lambda|lh^3,$$

$$|\varepsilon'_i| \le |\lambda|G|\varepsilon_i| + |\lambda|Qh\sum_{s=0}^{i-1}\left(|\varepsilon_s| + \frac{h}{2}|\varepsilon'_s|\right) + nh^2,$$

where $\varepsilon_j = \varphi(x_j) - \tilde{\varphi}(x_j)$, $\varepsilon'_i = \varphi'(x_i) -{}^{\prime}(x_i)$. From these formulas we get roughly

$$|\varepsilon'_i| \le |\lambda|\left(G|\varepsilon_i| + hQ\sum_{s=0}^{i-1}|\varepsilon_s|\right) + (n + |\lambda|QNm)h^2,$$

$$|\varepsilon_j| \le |\lambda|h(G + \tfrac{2}{3}Qh)\sum_{i=0}^{j-1}\left[\left(1 + \frac{|\lambda|Gh}{2}\right)|\varepsilon_i| + \frac{|\lambda|Gh}{2}\sum_{s=0}^{i-1}|\varepsilon_s|\right] + |\lambda|th^3,$$

where

$$t = l + \frac{h}{2}m(G + \tfrac{2}{3}Qh)(n + |\lambda QNm|).$$

Let us denote

$$T = |\lambda|(G + 2Qh/3), \qquad b = 1 + |\lambda|Gh/2, \qquad r = |\lambda|(t + QMTm^2).$$

Then we get

$$|\varepsilon_j| \leq hTB \sum_{i=0}^{j-1} |\varepsilon_i| + rh^3$$

and, finally, we have

$$|\varphi_j - \tilde{\varphi}_j| \leq E_j = \varepsilon_0 Z_1^j + rh^3(1 - Thb_j)^{-1} \leq \varepsilon_0 + rh^3(1 - Thb_j)^{-1},$$

where $h \leq (Tb)^{-1}$; and $Z_1$ is the real root of the equation

$$Z^{j+1} - TbhZ^{j-1}/(Z - 1) = 0$$

between $Z = 1$ and $Z = hTb$.

## 7.10 GEOINFORMATION SYSTEM TO MONITOR AGRICULTURE

GIMS technology was the brainchild of the Institute of Radioengineering and Electronics of the Russian Academy of Sciences and was set up to solve the wide range of problems arising in agriculture (Chukhlantsev, 1986; Chukhlantsev and Shutko, 1988; Reutov and Shutko, 1986; Shutko, 1986; Shutko *et al.*, 2010). The properties of the interaction between electromagnetic energy and soil–plant formations have been of considerable interest to the scientists and engineers who designed the GeoInformation Monitoring System of Agricultural Functions (GIMSAF).

Microwave radiometric techniques to measure the characteristics of soil moisture have been developed and widely used in many regions (Shutko *et al.*, 2010). This experience was used in the GIMSAF to find a procedure to determine the water content and moisture profiles in soil layers a meter thick with and without vegetation cover by means of remote sensing. The procedure is based on the dependence of the microwave emissivity of soil on its moisture. The procedure takes into account the variations of temperature, density, and roughness of natural soils.

The GIMSAF carries out many tasks; for instance, it

- collects information pertaining to current parameters of the soil–vegetation system (moisture, shallow water table, soil salinity, biomass of the vegetation) over regions of various sizes;
- processes the above information within the framework of a model simulating the biogeocenotic, physical, and chemical processes in the soil–vegetation system with the aim of evaluating its current state and predicting its development and potential productivity;
- represents current and predicted information as schematic maps with spatiotemporal detailing within a controlled structure;
- makes technological decisions that are the best possible for maximizing the biological productivity of crops.

The GIMSAF structure includes units that

- collect information;
- process initial information and accumulate data to simulate crop productivity and water conditions;
- predict the geosystem state;
- evaluate the disparity between measured and predicted trends in geosystem processes;
- make decisions pertaining to measurement scheduling and geosystem control;
- offer service support to help users with input and output data.

The data collection unit is based on remotely sensed data and prior knowledge of the state of an object and meteorological parameters. Remote pickups switch on a microwave radiometric complex that runs computer programs to provide operational area information at the desired degree of detail concerning soil moisture content to a depth of 0 to 100 cm, subsoil water level to a depth of 0 to 3 m, salt concentration of the soil in the range 3 to $90\,\mathrm{g\cdot L^{-1}}$, vegetation biomass, and degree of mineralization and contamination in reservoirs. The operating range for moisture measurement is $0.0\text{–}0.6\,\mathrm{g\cdot cm^{-3}}$ with maximal absolute error equal to vegetation biomass of less or more than $200\,\mathrm{t\cdot ha^{-1}}$ of 0.05 and $0.05\text{–}0.08\,\mathrm{g\cdot cm^{-3}}$, respectively.

The operation of the information processing units is based on a computer program package providing spatiotemporal interpolation of remote trace measurements and point surface measurements in addition to information flow stabilization in the event of the information collection system failing. The program package also describes the dynamics of moisture and the biological, physical, chemical, and energetic processes involved in the interaction between the soil, vegetation, atmosphere, and energy sources; anthropogenic effects are also described.

The adaptation of information-processing units to actual conditions in the region studied is accomplished by introducing sets of coefficients and characteristic dependences.

The trend-correcting and decision-making units are based on evaluating the state of the object under consideration, predicting its development, and ascertaining the sensitivity of the system to potential production and technological measures put forward by experts.

Let us now list some of the features of the GIMSAF and some of its differences from similarly designated systems:

- remote radiophysical sensor pickups provide areal, operative, and quantitative information on the state of objects;
- ability to function at various levels of information support (remote, surface, and *a priori* datasets, partial datasets, or even in the absence of data);
- fully open to further development regarding technological and software support

and full compatibility with any sensor, information, consulting, and expert systems;
- algorithms that have the ability to adapt make the system self-regulating and, in particular, noise immune;
- compatible means of bringing about production and technological decisions.

Application results are expected to include

- getting reliable operative, areal, and quantitative information on the parameters of the object under consideration and predicting its general state;
- evaluating the expediency and efficiency of anthropogenic effects on agriculture and irrigation systems;
- making optimum production and technological decisions by means of a simulation experiment and providing an increase in object control efficiency.

## 7.11 AN ADAPTIVE TECHNOLOGY TO CLASSIFY AND INTERPRET REMOTE-SENSING DATA OF THE WATER SURFACE QUALITATIVELY

Collecting and processing information in a geoinformation monitoring system can only be done by effective monitoring of the object under consideration and involves using simulation modeling, information collecting, and information processing (Armand *et al.*, 1987; Burkov and Krapivin, 2009).

From the perspective of system analysis, the collection and processing of information in geoinformation monitoring means combining computers of various classes, databases, and advanced problem-oriented software. Such a system calls for the development of formalized descriptions of information flows and a special method to process them.

For geoinformation monitoring systems to develop, a decision needs to be made about the makeup of the set of problems that hinder the formation of data measurement flows. The classification of aquatories using remote-sensing measurements is an important one. Image recognition, statistical decision, and cluster analysis algorithms can be used to solve this problem.

At the present time, there are many image recognition methods, mainly because of the variety of statements about concrete tasks. Image recognition problems consist in dividing a group of objects into classes on the basis of certain requirements. Objects with general properties are related to one class. Initial data used to solve an image recognition problem include the results of observations or direct measurements, sometimes called initial attributes.

Taxonomy (clustering) is an important method of image recognition and classification.

Let us assume a set of $M$ is required to be divided into clusters and the elements included in these clusters should be close enough to each other from the point of view of the chosen criterion of nearness, and elements from different clusters should be far enough from each other. In one of many possible statements of this task two numbers, $a$ and $b$ ($0 < a < b$), are given. It is considered that two elements $x$ and $y$ are close enough to each other if $p(x, y) < a$, and far enough from each other if $p(x, y) > b$.

King's method is well known in taxonomy and gives good results when the quantity of available information in assumed clusters is moderate. According to this method the distance between groups of points in the space of attributes is defined as the distance between the centers of masses of these groups. Clustering in this case is based on the assumption that sites of increased density in the space of attributes correspond to similar situations.

A feature of remote measurements is information acquisition, when measurements acquired along the survey tracks of flying systems are sent as input to the processing system. As a result, a two-dimensional image of the object under investigation is registered. The statistical model of spottiness for the space being investigated can be used to acquire such an image.

Under real conditions, the study of spots, acquiring their statistical characteristics, and using them in a detection problem is in itself a complex enough problem. Criteria have to be developed to distinguish spots from other phenomena. For example, it is necessary to determine the threshold at which the spot indicator can be determined. It is further necessary to develop a model representation of spot detection processes.

The method of threshold determination is the most obvious and simple means of spot definition. In this case the part of space that belongs to the spot area, at which the environmental parameter within the chosen channel exceeds value ($l^+$) or, on the contrary, does not exceed value ($l^-$) is the threshold. Let $y = y(x_1, x_2)$ be a function of coordinates $(x_1, x_2)$ of points within the considered region. If "level surface" $y = \text{const}$ is outlined on the region surface, then the closed curves of level $y$ that bound the spots are projected onto it.

Algorithms to simulate spottiness are based on numerical solution of algebraic inequalities determining the coordinates of internal points of spots. It is generally impossible to write down the spot contour equation.

Therefore, the contours of spots are described by the system of simple algebraic equations interconnected by equation $\sum \varphi_i(x, y)$, where $\varphi_i(x, y)$ is the equation of an elementary curve. To simplify the way that software simulates an image of spottiness as equations $\varphi_i(x, y)$ the equation of a circle with varied coordinates of the center and radius is accepted. Complex forms of spots are formed by overlapping on a plane of the image several circles with different parameters that can be defined by a system of inequalities like:

$$\sum \{(x - a_i)^2 + (y - b_i)^2 - r_i\} \leq 0,$$

where $x, y$ are the cartesian coordinates of internal points of spots; $a_i$, $b_i$, and $r_i$ are coordinates of the center and radius of the $i$th circle, respectively; and $n$ is number of circles in the modeled image.

To simulate the randomness of the background distribution for spottiness, the

**Table 7.9.** Software modules of the simulation system to classify phenomena on land surfaces.

| *Software* | *Functional characteristics of software* |
|---|---|
| REICM | Reduction of experimental information in computer memory |
| DRHIM | Data reconstruction by means of the harmonics interpolation method |
| SIDSM | Spatial interpolation of data by means of a spline method |
| DRMOT | An optimal interpolation method |
| CASRS | Cluster analysis focused on the spatial signs of remote sensors |
| CALRS | Cluster analysis focused on local reading of sensors |
| RSC | Research into spottiness characteristics |
| RRAMDA | Realization of the recognition algorithm by discriminant analysis |
| CSAIRM | Sorting and accumulating *in situ* and remote measurements |
| CMS | Computer mapping of spots. |

spottiness model parameters $a_i$, $b_i$, and $r_i$ are set by means of random number generators. By changing the laws of distribution of random numbers and their statistical parameters, it is possible to receive statistically different images of spottiness.

The list of software modules of the simulation system to classify phenomena on land surfaces is given in Table 7.9. An important point about the system's algorithms and software is the possibility of spatial interpolation and data restoration using remote and *in situ* measurements.

A principal aspect of the practical importance of this system is qualitative interpretation and visualization of the results of remote measurements.

For primary processing of remote measurements it is useful to apply an average-connecting method of cluster analysis to detect specific informational zones. Such a method is effective when sampling volumes are small. Two variants of this approach can be brought about by the way in which algorithms and index spaces are organized.

In the system designed to classifiy phenomena on a terrestrial surface automatically, the two variants of this approach are program modules CASRS and CALRS, distinguished by the way in which attribute spaces and comparison algorithms are organized.

The CASRS module is focused on the attribute space of remote sensors. The CALRS module allocates areas of equal instability based on local variations in sensor data. To distinguish between the comparison algorithms of these items involves either taking into account or ignoring the interrelations between neighboring counts of sensors. The CASRS module forms clusters without taking the geographical generality of radiometer indications into account. The CALRS module forms continuous spatial clusters.

The modules automatically exclude calibration information from consideration. The CASRS module consists of six groups of operators. The first group organizes data files. Information averaging using the final set of radiometer readings is carried out by the second group of operators. The third group of operators finds minimal cluster distances and then allocates two points at which this distance is realized. The fourth group of operators investigates changes in minimal cluster distances at a given step of the procedure and, in the event of sharp changes, visualizes the cluster structure. The fifth group of operators combines the nearest clusters and recalculates their characteristics. The sixth group of operators develops a criterion for the algorithm to stop.

Analysis of the statistical characteristics of spottiness for three area types in the Atlantic and Pacific Oceans was conducted. These statistical characteristics were determined for the most informative thresholds. At the time, statistical characteristics of spottiness for the same areas were selected using criteria of the minimal value of the correlation coefficient for a joint sample of positive and negative spots. Analysis of these characteristics showed that the statistical characteristics of spottiness coincide in areas with temperate sea roughness and storm zones. The minimum correlation coefficient $\rho_{\min}$ breaks down at the most informative thresholds. However, for quiet areas the situation is different.

This means that statistical characteristics for the spottiness of brightness temperatures in microwaves can be used to detect and classify phenomena on the surface of the ocean caused by the degree of sea roughness.

Analysis of empirical histograms for spottiness of brightness temperatures in microwaves shows that in most cases $(l^+, l^-)$ are characteristics that can be coordinated with the exponential distribution while their amplitude can be coordinated with the normal distribution. Therefore, to detect and classify phenomena on ocean surfaces it is necessary to apply the best possible algorithms when programming/teaching the computer to make statistical decisions about these distributions (Mkrtchyan, 1982).

## 7.12 A DEVICE TO MEASURE GEOPHYSICAL AND HYDROPHYSICAL PARAMETERS

According to the algorithms of multi-channel microwave monitoring discussed in Section 6.4, it is possible to promptly measure wind speed, sea surface roughness, and water temperature as the base characteristics to calculate other parameters of the environment. From these measurements, mutual control of the accuracy of correction of estimates of the atmosphere–ocean gas exchange can be brought about. Solution of the correction equations generally requires the use of a discrepancy criterion between theoretical and empirical estimates of parameters or correlations. For this purpose, a criterion of mean square deviation is normally used. The reliability and efficiency of determination of solutions for the system of equations that relate theoretical and empirical estimates can be increased if their dispersive characteristics are taken as a discrepancy criterion of these solutions.

Let emission from the sea proper be measured simultaneously using $n$ radiometers, each operating at a fixed wavelength $\lambda_j$ ($j = 1, \dots, n$). As a result, at moment $t_i$ at the output of each radiometer the values $Z_{ij}$ ($i = 1, \dots, M$) are fixed, so that $Z_{ij} = T_j + \xi_{ij}$, where $T_j$ is the brightness temperature of a sea site at wavelength $\lambda_j$, and $\xi_{ij}$ is a random value (noise) with zero mean and dispersion $\sigma_j^2$.

Let the correlation between the brightness temperature and $m$ parameters of the sea be linear:

$$\begin{aligned} A_{11}x_1 + \cdots + A_{1m}x_m &= T_1 + \xi_1 \\ &\vdots \\ A_{n1}x_1 + \cdots + A_{nm}x_m &= T_n + \xi_n. \end{aligned} \tag{7.54}$$

where $A_{ij}$ are coefficients determined when monitoring the sea site with known parameters $\{x_i\}$. The $x_i^*$ parameter should be estimated for the $x_i^0$ parameters so that, first, the average $x_i^*$ coincides with the accurate solution of system (7.54) (i.e., $\bar{x}_i^* = x_i^0$ ($i = 1, \dots, m$)); second, a dispersion of the $x_i^*$ estimate is at a minimum (i.e., $D[x_i^*] \le D[\tilde{x}_i]$), where $\tilde{x}_i$ is the estimate of the $x_i^0$ parameter obtained by any other method; and, third, let the procedure used to estimate the $x_i^*$ parameters be as simple as possible so that the best discrete set of radiometric channels needed to realize these two conditions is chosen.

Let all the $x_i^*$ estimates meeting these conditions be $\sigma$-solutions of the system of linear equations whose right-hand parts have been disrupted. Multiply step by step the $i$th solution of the system (7.54) by $c_{1i}, \dots, c_{mi}$, and let

$$\sum_{i=1}^{n} c_{ji} A_{il} = \delta_{jl}, \tag{7.55}$$

where

$$\delta_{jl} = \begin{cases} 1 & \text{for } j = l \\ 0 & \text{for } j \neq l. \end{cases}$$

As a result, we have

$$x_1^0 = \sum_{i=1}^{n} c_{1i} T_i, \qquad \tilde{x}_1 = \sum_{i=1}^{n} c_{1i} T_i + \sum_{i=1}^{n} c_{1i} \xi_i, \qquad D[\tilde{x}_1] = \sum_{i=1}^{n} c_{1i}^2 \sigma_i^2.$$

For $\tilde{x}_i$ ($i \ge 2$) we write similar relationships. Derive an auxiliary expression:

$$\varphi(c_{11}, \dots, c_{1n}) = \sum_{i=1}^{n} c_{1i}^2 \sigma_i^2 + \mu_1 \left( \sum_{i=1}^{n} c_{1i} A_{i1} - 1 \right) + \sum_{j=2}^{m} \mu_j \sum_{i=1}^{n} c_{1i} A_{ij},$$

where $\mu_j$ are Lagrange multipliers that have not been estimated. Let us first put derivative functions $\varphi$ to zero. We then obtain:

$$2c_{1k}\sigma_k^2 + \sum_{j=1}^{m} \mu_j A_{kj} = 0 \quad (k = 1, \dots, n).$$

These equalities together with condition (7.55) form the system $(m+n)$ of equations whose solution enables us to find the desired optimal values of $c^*_{ij}$. As a result, $D[x_1] = -\mu_1/2$, and the remaining $\mu_j$ $(j \geq 2)$ values meet a system $m \times n$ of equations:

$$\sum_{j=1}^{m} \mu_j \sum_{i=1}^{n} A_{ij} A_{i1} \sigma_i^{-2} = -2, \quad \sum_{j=1}^{m} \mu_j \sum_{i=1}^{n} A_{ij} A_{i1} \sigma_i^{-2} = 0 \quad (l = 2, \ldots, m).$$

For $m = 2$ we have:

$$c^*_{1,k} = \left( A_{k1} \sum A_{i2}^2/\sigma_i^2 - A_{k2} \sum_{i=1}^{n} A_{i1} A_{i2}/\sigma_i^2 \right) \Delta^{-1} \sigma_k^{-2},$$

$$c^*_{2,k} = \left( A_{k2} \sum A_{i1}^2/\sigma_i^2 - A_{k1} \sum_{i=1}^{n} A_{i1} A_{i2}/\sigma_i^2 \right) \Delta^{-1} \sigma_k^{-2},$$

$$\Delta = \sum_{i=1}^{n} A_{i1}^2 \sigma_i^{-2} \sum_{i=1}^{n} A_{i2}^2 \sigma_i^{-2} - \left( \sum_{i=1}^{n} A_{i1} A_{i2} \sigma_i^{-2} \right)^2,$$

$$x_j^* = \sum_{i=1}^{n} c^*_{ji} Z_i \quad (j = 1, 2); \qquad D[x_1^*] = \Delta^{-1} \sum_{i=1}^{n} A_{i2}^2 \sigma_i^{-2}; \qquad D[x_2^*] = \Delta^{-1} \sum_{i=1}^{n} A_{i1}^2 \sigma_i^{-2}.$$

Real use of this algorithm in a radiometric monitoring system can be done in the following way. Put the $n$-channel commutation unit and $n$ units of dispersion calculation between the outputs of radiometers and respective inputs of the microprocessor in series. At the inputs of radiometers we plug in switches that are connected through a control unit to the first group of outputs of the microprocessor. The second group of outputs of the microprocessor are linked to the Resolver, whose second group of outputs is connected, respectively, to the outputs of radiometers. Each unit to calculate dispersion consists of in-series connected quantizers of the mean value calculator, divisor, summator, and second divisor, and the quantizer's output is connected to the second input of the summator. The Resolver consists of identical channels, each containing a summator, $n$ outputs of which are connected to $n$ multipliers. Some inputs of the multipliers form the first group of the Resolver's inputs, others the second group of its inputs—the number of channels being equal to the number of the parameters to be measured.

Figure 7.15 shows an arrangement to measure the parameters of geophysical and hydrophysical objects. The scheme in Figure 7.16 explains the structure of the unit to calculate dispersion. The Resolver is characterized in Figure 7.17 and operates as follows. When radiometers 1 are calibrated, signals $Z_{ij}$ from their outputs through the multi-channel switching unit 2 enter unit 3 of the dispersion calculator. The antennas on the radiometers are switched off. In unit 3 dispersion is calculated using the algorithm

$$\sigma_j^2 = \frac{1}{M-1} \sum_{i=1}^{M} (Z_{ij} - M_j), \qquad M_j - \frac{1}{M} \sum_{i=1}^{M} Z_{ij}.$$

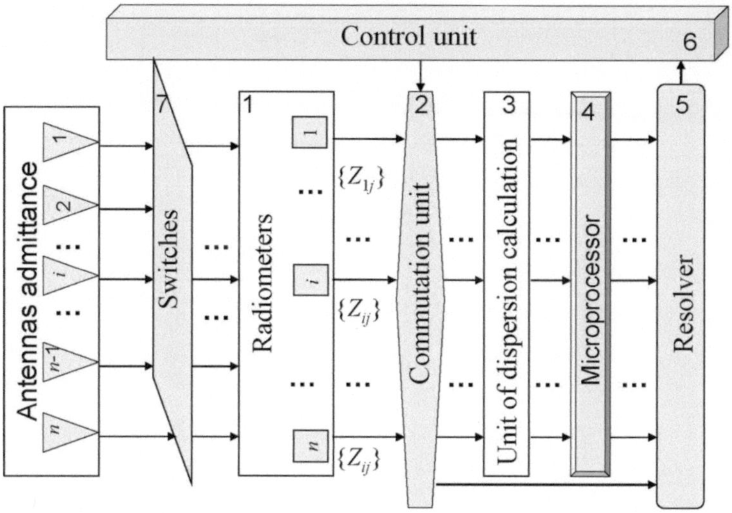


**Figure 7.15.** Recommended arrangement to measure the geophysical and hydrophysical parameters in microwave monitoring.

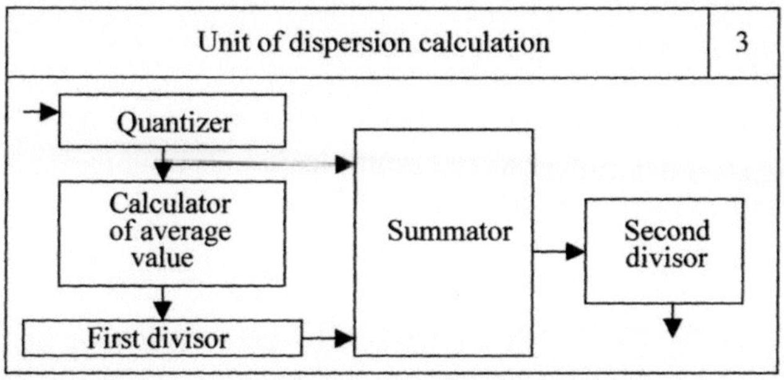


**Figure 7.16.** Scheme of operations in the unit of dispersion calculation.

The $M_j$ parameters are calculated in units of mean calculations. The summator calculates:

$$\sum_{i=1}^{M}(Z_{ij} - M_j).$$

Calculated values $\sigma_j^2$ enter the microprocessor 4 to calculate partial coefficients $c_{jl}$. From the first group of its outputs they enter the Resolver 5 where they are used to solve a system of equations, and from the second group of outputs they enter the control unit 6 which forms the signal that switches off radiometers 1, for which the partial coefficients are equal or close to zero (i.e., the information channels automatically get excluded). At the same time, the systems of equations to be solved get compressed. This step makes it possible to contract not only the volume

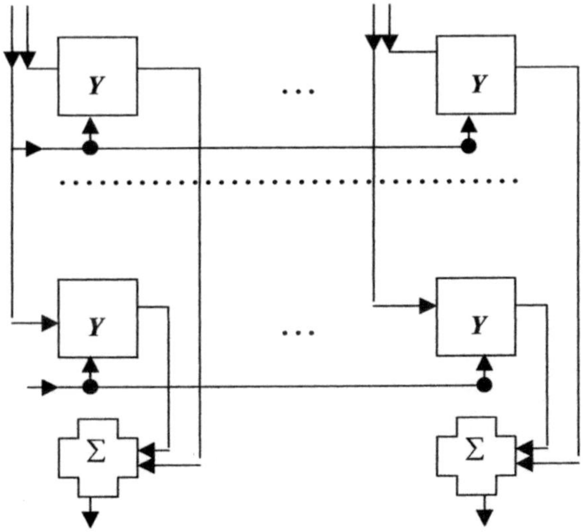


**Figure 7.17.** Resolver. Notation: $Y$, multiplier; $\Sigma$, accumulator.

of calculations but also to increase the accuracy of estimating the parameters of the environmental objects being sounded.

Once the dispersions have been determined, the commutator 2 blocks the input of data from non-information channels. The data from information channels enter the Resolver 5. The signal at the multiplier's output corresponds to the value of one of the parameters to be estimated.

## 7.13 DIRECT AND INVERSE PROBLEMS OF MICROWAVE MONITORING

An approach to the inverse problem of microwave monitoring is based on assuming a linear functional correlation between radiobrightness contrasts and the characteristics of the monitored object. The search for analytical relationships relating the radiobrightness temperature of the environment to its thermodynamic and electrophysical characteristics covers the more general classes of functional models. The most general form of such dependence is as follows:

$$T_{bi} = \kappa_i \int_0^{\infty} T(h)\gamma_i(h) \exp\left[-\int_0^h \gamma_i(x)\, dx\right] dh, \tag{7.56}$$

where $T(h)$ is the thermodynamic profile of the environment; $T_{bi}$ is the radiobrightness temperature at frequency $\nu_i$; $\gamma_i(h)$ is the profile of the absorption coefficient of the environment; and $\kappa_i$ is the emissivity of the environment.

Relationship (7.56) makes it possible using $T(h)$ and $\gamma_i(h)$ to find $T_{bi}$ (direct problem) and by means of $T_{bi}$ to calculate $T(h)$ and $\gamma_i(h)$ (inverse problem). The proposed method to solve these problems is based on the use of $T(h)$ and $\gamma_i(h)$ and their derivatives at point $h = 0$ with some additional limitations resulting from the transformations made below. Let us write relationship (7.56) in the form:

$$T_{b\lambda} = -\kappa_\lambda \int_0^\infty T(h)\, dF_\lambda(h), \tag{7.57}$$

where

$$F_\lambda(h) = \exp\left[-\int_0^h \gamma_\lambda(x)\, dx\right];$$

and $\lambda$ is wavelength.

Integrating (7.57) by parts, we obtain:

$$T_{b\lambda} = \kappa_\lambda\left[T_0 + \int_0^\infty T'(h) F_\lambda(h)\, dh\right] = \kappa_\lambda[T_0 + J_1]. \tag{7.58}$$

Let $J_1 = 0$, then $T_{b\lambda} = \kappa_\lambda T_0$. If $dT(h)/dh \equiv 0$, then $J_1 = 0$. Therefore, we search for conditions under which $J_1 = 0$. Let the following relationships be valid:

$$\gamma_\lambda(h) \equiv c_\lambda > 0, \qquad T(h) = \sum_{k=0}^{n} B_k h^k. \tag{7.59}$$

It follows from (7.59) that if coefficients $B_k$ meet the condition:

$$\sum_{k=1}^{n} k! \frac{B_k}{c_\lambda^k} = 0, \tag{7.60}$$

then $J_1 = 0$.

Integrating (7.58) by parts, we obtain:

$$T_{b\lambda} = \kappa_\lambda\left[T_0 + \frac{T_0'}{\gamma_\lambda(0)} + \int_0^\infty \frac{T''(h)\gamma_\lambda(h) - \gamma_\lambda'(h) T'(h)}{\gamma_\lambda^2(h)} F_\lambda(h)\, dh\right]. \tag{7.61}$$

Let us equate the integral in (7.61) to zero. This gives

$$T(h) = c \int \gamma_\lambda(h)\, dh,$$

where $c$ is an arbitrary constant. The integral in (9.20) will be equal to zero if

$$\sum_{k=2}^{n} B_k k(k-1) k! c_\lambda^{1-k} = 0. \tag{7.62}$$

Further integrating of (7.61) by parts gives

$$T_{b\lambda} = \kappa_\lambda\left[T_0 + \frac{T_0'}{\gamma_\lambda(0)} + \left(\frac{T'}{\gamma_\lambda(h)}\right)' \frac{1}{\gamma_\lambda(h)}\bigg|_{h=0} + J_3\right],$$

where

$$J_3 = \int_0^\infty \frac{T'''(h)\gamma_\lambda(h) - 3T''(h)\gamma_\lambda(h) - T'(h)\gamma_\lambda''(h) + 3T'(h)[\gamma_\lambda']^2}{\gamma_\lambda^4} F_\lambda(h)\, dh.$$

This integral will be equal to zero at

$$T(h) = c \int \gamma_\lambda(h) \int \gamma_\lambda(u)\, du\, dh.$$

If condition (7.59) is not met, then $J_3 = 0$ at

$$\sum_{k=3}^{n} B_k k(k-1)(k-2)k! c_\lambda^{2-k} = 0.$$

With the process of integration of (7.61) continued $N$ times, we obtain

$$T_{b\lambda} = \kappa_\lambda \left[ \sum_{k=1}^{N} P_k(0) + \int_0^\infty P'_N(h) F_\lambda(h)\, dh \right], \tag{7.63}$$

where

$$P_1(h) = T(h), P_2(h) = T'(h)/\gamma_\lambda(h), \ldots, P_k(h) = P'_{k-1}(h)/\gamma_\lambda(h).$$

But

$$J_N = \int_0^\infty P'_N(h) F_\lambda(h)\, dh = 0$$

for

$$\sum_{k=N}^{n} B_k k(k-1)\cdots(k-N+1)k! c_\lambda^{N-1-k} = 0 \quad (n > N).$$

Consider the case when

$$\gamma_\lambda(h) = \exp\{c_\lambda h\}.$$

From (7.63) we obtain:

$$T(h) = \sum_{k=0}^{N-1} B_k \exp[k c_\lambda h].$$

Let us now move on to the inverse problem. Let measurements be made at wavelengths $\lambda_1, \ldots, \lambda_N$. Solve the equation:

$$P'_{N-1}(h) = c\gamma_\lambda(h)$$

and find

$$P_{k-1}(h) = \int \gamma_\lambda(h) P_k(h)\, dh \quad (k = 1, \ldots, N).$$

Let $\gamma_\lambda(h) = \varphi(\lambda)\psi(h)$. In words, let the dependence of the absorption coefficient on the frequency and depth of the layer be divided by the product of the known function $\varphi(\lambda)$ and an unknown function $\psi(h)$. As a rule, the function $\varphi(\lambda)$ is set by an empirical table and its analytical approximation can be constructed. Then, excluding the second right-hand term (it is equal to zero) from (7.63), we write the relationship:

$$T_{b\lambda} = \kappa_\lambda \sum_{k=1}^{N} T^{(k-1)}(0) F_k[\varphi(\lambda), \psi^{(N-k-1)}(0), \ldots, \psi'(0), \psi(0)] \tag{7.64}$$

and the solution of the inverse problem is reduced to solving a system of algebraic equations with regard to functions:

$$T(0), T'(0), \ldots, T^{(N-1)}(0), \psi(0), \ldots, \psi^{(N-1)}(0).$$

Consider the case $\psi(h) = \text{const}$ (i.e., $\gamma_\lambda = \varphi(\lambda) > 0$). Equation (11) can be re-written as:

$$T_{b\lambda} = \kappa_\lambda \sum_{k=1}^{N} T^{(k-1)}(0)/\varphi^{(k-1)}(\lambda). \tag{7.65}$$

In the simplest version, if $T_{b\lambda}$ was measured at $N$ different wavelengths $\{\lambda_i; i = 1, \ldots, N\}$, then the determinant of system (7.65) will be equal to:

$$\Delta = \prod_{i=1}^{N} \kappa_i \prod_{i>k}^{N} [\varphi(\lambda_i) - \varphi(\lambda_k)]/[\varphi(\lambda_i)\varphi(\lambda_k)].$$

If $\varphi(\lambda_i) \neq \varphi(\lambda_k)$ with no combinations $i \neq k$, then $\Delta \neq 0$ and system (7.65) has a single solution. Now let

$$\psi(h) = h^m.$$

Then, it follows from (7.63) that

$$T(h) = \sum_{k=0}^{N-1} B_k h^{k(m+1)}$$

and we obtain a system of algebraic equations

$$P_k(0) = (m[k-1]! \underbrace{\ldots}_{(m+1)} !b_{k-1}/\varphi^{k-1}(\lambda)$$

with the following determinant:

$$\Delta = \prod_{i=1}^{N} \kappa_{\lambda_i} \prod_{p=1}^{N} [m(k-1)]^{N+1-p} \prod_{t>k}^{N} [\varphi(\lambda_t) - \varphi(\lambda_k)]/[\varphi(\lambda_t)\varphi(\lambda_k)].$$

When function $\psi$ is approximated by a polynomial, it is considered in a similar way.

# 8

# Microwave monitoring of vegetation using a GIMS-based method

## 8.1 INTRODUCTION

Vegetation covers are the most impacted by model reconstruction and forecasting. Keeping this in mind the objective of this chapter is threefold:

(1) to present a working methodology for the combined use of modeling technology and microwave remote-sensing measurements to assess how vegetation covers attenuate electromagnetic waves;
(2) to illustrate this methodology with computer calculations of the attenuation for various soil–plant formations;
(3) to give a perspective of the developed methodology applied to the study of global environmental change, including radiative forcing.

The majority of applied problems regarding agro-meteorology, forestry, animal husbandry, and other areas of human activity directed at protecting nature are difficult to solve simply because effective methods of control of the soil–plant formation (SPF) have not been developed sufficiently. In recent years, the global carbon cycle has acquired special significance because of the greenhouse effect. Knowledge of the state of the SPF allows getting a real picture of the spatial distribution of carbon sinks and sources on Earth's surface.

As is well known, of the types of remote-sensing techniques, microwave radiometry has proven the most effective at observing SPF environmental parameters. However, these observations are a function of different environmental conditions mainly depending on SPF type. This is the reason it is necessary to develop data-processing methods for microwave monitoring that allow SPF characteristics to be reconstructed by considering vegetation types and that provide the possibility of synthesizing their spatial distribution.

V.F. Krapivin and A.M. Shutko, *Information Technologies for Remote Monitoring of the Environment*, Springer Praxis Books, DOI 10.1007/978-3-642-20567-5_8, 

As noted by Chukhlantsev *et al.* (2003), microwave remote sensing of vegetation covers involves studying how electromagnetic waves (EMWs) within the vegetation layer are attenuated. The problems arising here can be resolved by combining experimental and theoretical studies. Vegetation covers are commonly characterized by varied geometry and additional parameters. Therefore, knowledge of the radiative characteristics of the SPF as functions of spatiotemporal coordinates can be acquired by means of a combination of on-site measurements and models. The general aspects of such an approach have been considered by many authors (Del Frate *et al.*, 2003; DeWitt and Nutter, 1989; Dong *et al.*, 2003; Engman and Chauhan, 1995; Ferrazzoli and Guerriero, 1996; Field, 2001; Font *et al.*, 2002; Friedi *et al.*, 2002; Goody and Jung, 1989; Karam *et al.*, 1992; Kirdiashev *et al.*, 1979; Sellers, 1985, 1987; Shutko, 1987, 1992, 1997; Shutko *et al.*, 1994, 1995; Vinogradova *et al.*, 2004). However, these investigations were mainly restricted to the investigation of models describing the dependence of the vegetation medium on environmental properties as well as the correlation between the morphological and biometrical properties of vegetation and its radiative characteristics.

GIMS technology is a prospective approach to solving the difficulties that arise here (Krapivin *et al.*, 2006). Combining an environmental acquisition system, a model of how a typical geo-ecosystem functions, a computer cartography system, and a means of artificial intelligence should facilitate creation of a geoinformation monitoring system for a typical natural element that is capable of solving many of the tasks that arise in microwave radiometry of the global vegetation cover. A GIMS-based approach to EMW attenuation by vegetation canopies allows synthesizing a knowledge base that establishes the relationships between experiments, algorithms, and models. The links between these areas are adaptive and, hence, provide the best strategy for experimental design and model structure. The goal of this chapter is to explain and assess application of the GIMS method to the tasks of reconstructing the spatiotemporal distribution of the characteristics of SHF radiation.

Experience gained from applying GIMS technology has shown that it opens up new prospects for microwave remote-sensing monitoring of vegetation covers making it possible to combine theoretical and empirical investigations of the role played by vegetation in microwave radiation of the land surface. The progress made in these results can be directly traced to the use of GIMS models. As a consequence of this progress, the energy balance of the biosphere can be evaluated with high precision. This is the reason the following tasks are currently subject to intense study:

- developing models to describe SPFs with as little demand on the parametrical environment as possible;
- analyzing the possibilities of different monitoring systems to assess the characteristics of the SPF;
- studying microwave radiation and its propagation in the atmosphere–vegetation–soil system both theoretically and empirically;

- synthesizing a GIMS database that includes theoretical and empirical assessments of model coefficients, radiation characteristics of vegetation, and spatial distribution of biomes.

## 8.2 THE ATTENUATION OF ELECTROMAGNETIC WAVES BY VEGETATION MEDIA

Many investigators (Basharinov *et al.*, 1979; Burkov and Krapivin, 2009; Hansen *et al.*, 2002; Zhan *et al.*, 2002) favor a technology for microwave monitoring of vegetation covers that is adaptive and can be constructed in a number of steps. However, the accuracy of such a technology has to be improved because various vegetation covers, especially canopies, are very dynamic and complicated for experimental study or model parameterization. There are two sides to this. One is connected with the creation of a highly productive technology to estimate the characteristics of the land cover. The other is connected with climate change as a consequence of the greenhouse effect (Kondratyev *et al.*, 2003c). In these cases GIMS technology recommends a balanced scheme to remove the difficulties that crop up in land cover monitoring. Armand *et al.* (1997) advocate combining experimental and theoretical studies into how vegetation covers attenuate microwave radiation. It is actually possible to synthesize the experimental dependence between attenuation and a restricted set of vegetation parameters. The attenuation of microwave radiation by vegetation cover in real time is only feasible by applying microwave models and interpolation algorithms.

From the point of view of microwave remote sensing, knowledge of the water content of vegetation is vital to synthesizing a model of the way in which vegetation attenuates microwave radiation. Such knowledge can be acquired by making use of the fact that water absorption features dominate the spectral reflectance of vegetation in the near-infrared (Sims and Gamon, 2003). Using such indexes as the normalized difference vegetation index (NDVI), plant water index (PWI), leaf area index (LAI), simple ratio vegetation index (SRVI), and the canopy structure index (CSI) it is possible to determine the canopy structure and photosynthetic tissue morphologies. The correlation between leaf water content and leaf level reflectance in the near-infrared has been successfully studied by many authors who linked leaf and canopy models to study the effects of leaf structure, dry matter content, LAI, and canopy geometry. As a result, it is possible to parameterize the forest structure including the canopy and crown closure, stem density, tree height, crown size, and other forest parameters (Tables 8.1 and 8.2).

Quantitative information about SPF properties may be obtained by means of different remote-sensing techniques based on passive microwave data. Such an approach is particularly effective when estimating the hydraulic parameters of soil–plant formations. Knowledge of these parameters allows the water balance of a territory occupied by vegetation of a given type to be parameterized and plant growth to be modeled. Under this approach comparison of remote data registered at different EMW ranges may provide very accurate estimates of the water content

**Table 8.1.** Some estimates of parameters for different types of branches (Karam *et al.*, 1992).

| *Branch group* | *Diameter range* (cm) | *Average diameter* (cm) | *Average length* (cm) | *Number density* ($m^{-3}$) |
|---|---|---|---|---|
| 1/Stems | 0.0–0.40 | 0.10 | 18 | 250.00 |
| 2 | 0.5–1.90 | 1.28 | 14 | 11.40 |
| 3 | 2.0–2.90 | 2.60 | 32 | 0.43 |
| 4 | 3.0–6.90 | 5.00 | 58 | 0.33 |
| 5/Trunk | 7.0–17.1 | 9.00 | 76 | 0.14 |

**Table 8.2.** Dielectric constant values (Karam *et al.*, 1992).

| *SPF element* | *Frequency band* | |
|---|---|---|
| | *L* | *X* |
| Leaves | 19.58–j5.54 | 14.9–j4.9 |
| Branches/Stems | 27.30–j8.4 | 20.0–j9.7 |
| Soil | 5.00–j0.7 | 5.0–j0.4 |

in various layers of the SPF (canopy, trunks, stems, and soil). Moreover, combining data analysis with remote sensing makes it possible to estimate the parameters of ground vegetation cover and the role vegetation cover plays in the fluxes of sensible and latent heat from the surface to the atmosphere.

Thus, a set of vegetation-related parameters need to be generated with specific spatiotemporal resolutions to calculate EMW attenuation by the vegetation layer. Available remote-monitoring data are to be used as input information for algorithms and models to synthesize the spatiotemporal distribution of attenuation effects. In this context, the most important parameter is the dielectric constant. Some estimates of this characteristic are given in Table 8.2. For example, the dielectric constant for stems has a representative value at the L-band of 27.3–j8.4. The dielectric value for leaves varies from 8.77–j2.88 to 19.58–j5.54. An average value of the dielectric constant for leaves equals 19.58–j5.54. The values of the dielectric constant at the X-band equal 14.9–j4.9 and 20.0–j9.7 for leaves and branches, respectively.

A detailed study of EMW attenuation by vegetation was undertaken in ISTC Project #2059 "Screening Effect of Vegetation in Active (SAR) and Passive (Radiometric) Remote Sensing of the Earth's Covers and in Radio-

Communication in UHF-, VHF-, and Microwave (X-, C-, S-, L-, and P-) bands" (Chukhlantsev, 2006). The goals of this project were the following:

- acquiring systematized data about electromagnetic wave attenuation by different types of vegetation covers, mainly in the meter wavelength band, but also in the decimeter and centimeter wavelength bands (UHF, VHF, P-, L-, S-, C-, and X-bands) using theoretical and experimental investigations;
- developing advanced, more sophisticated, and generalized models compared with the existing SAR and microwave radiometric models of different types of vegetation and paying especial attention to the electromagnetic properties of vegetative elements to satisfy the needs for reliable assessment of the vegetation impact (screening effect of vegetation) on soil condition sensing;
- developing methodologies to assess the vegetation impact in active (SAR) and passive (MR) remote sensing of the Earth's covers to reduce/eliminate/remove the vegetation impact on land surface observation;
- developing methodologies to assess the impact that vegetation has on radio communication.

The project carried out the following main tasks.

- building up a measuring complex to conduct investigations of how vegetation covers and plant elements (leaves, twigs, etc.) attenuate electromagnetic waves;
- systematizing and generalizing existing theoretical and experimental data about the attenuation of electromagnetic waves in vegetation;
- acquiring the experimental characteristics of electromagnetic wave attenuation by various types of vegetative covers under laboratory and field conditions in the meter, decimeter, and centimeter wavelength ranges;
- conducting model investigations of the attenuation of electromagnetic waves in vegetation covers in the meter, decimeter, and centimeter wavelength ranges;
- elaborating methodologies to assess the screening effect of vegetation by conducting single-wavelength measurements and by combining active and passive methods;
- elaborating methodologies to assess the screening effect by conducting multi-wavelength measurements and by combining active and passive remote-sensing systems;
- elaborating methodologies to assess the screening effect of vegetation in radio communication.

The technical approach and methodology used to carry out these tasks were founded on the following ideas. The theoretical part involved radio wave propagation methods for the two basic types of vegetation models:

(1) vegetation in the form of inhomogeneous randomized media (stochastically inhomogeneous or irregular), especially when discussing the vegetation model as a continuous medium with an effective value of dielectric permittivity;
(2) vegetation as a community of discrete separately located scatters and determining the attenuation cross-section of these elements.

In the continuous model, vegetative cover was described as a homogeneous medium with an effective value of dielectric permittivity. The effective values of permittivity were calculated using a quasi-static approach (the size of irregularities, disks, needles, strips is much less than when ascertained by a wavelength) by deriving different relationships based on the theory of dielectric mixtures. Corrections to the effective value of dielectric permittivity were made with due regard to the oscillating (wavy) nature of the process.

In the discrete model, the aggregated electromagnetic parameters of vegetation chosen to evaluate vegetation behavior were the attenuation cross-section and scattering cross-section of vegetation elements. Vegetation elements were modeled by disks, strips, and cylinders. The spectral dependence of the attenuation and scattering cross-sections were determined. Total attenuation in the vegetative media was studied by means of radiation transfer theory.

The experimental part of ISTC Project #2059 included laboratory and field measurements. A special laboratory complex was built to investigate attenuation by plant samples and plant elements at a wide range of frequencies. This complex included two waveguide transmission systems operating in the 0.2 to 8.0 GHz frequency range (Figures 8.1 and 8.2). Radio attenuation was studied by comparing data with and without vegetation samples in the waveguides. The basic electrical setup of the laboratory complex is shown in Figure 8.3. Attenuation was clearly found to be less than 5 dB in the 0.8 to 2.15 GHz range. The same dependence was observed in the 0.8 to 10.0 GHz range. Some experimental results are given in Figures 8.4–8.6.

Experiments under field conditions were conducted within forests using mobile measuring complexes (Figures 8.7 and 8.8). The territory consisted of grassy and forested areas with different plant types and densities. The generator was installed

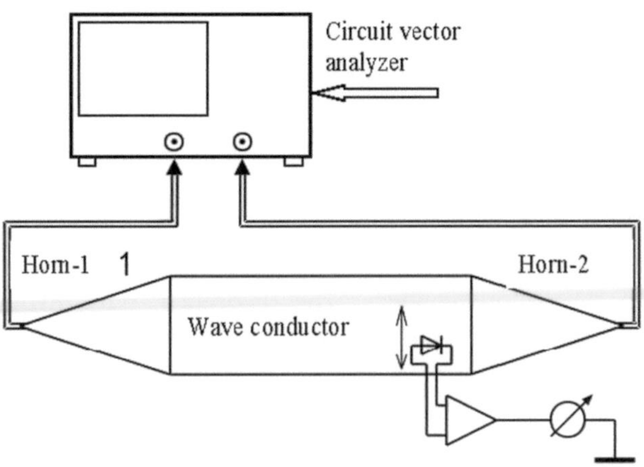


**Figure 8.1.** Block diagram of the super-wideband waveguide measuring system (Chukhlantsev, 2006).

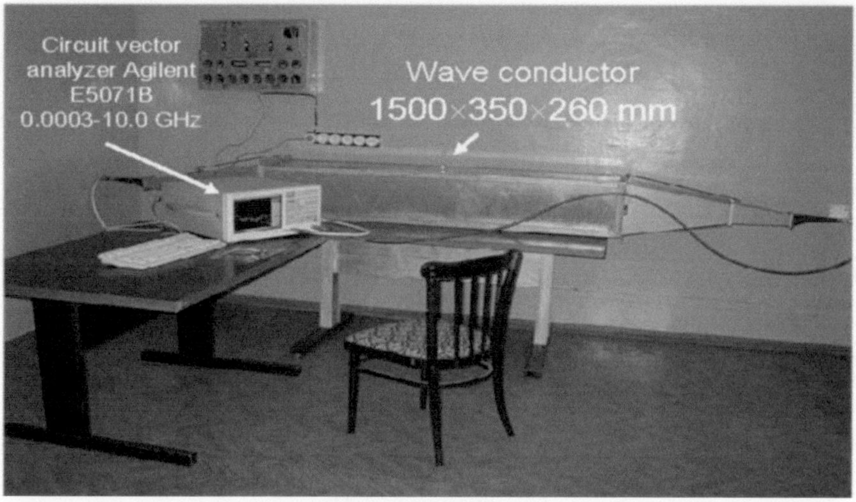


**Figure 8.2.** General view of the measuring system used to study how vegetation canopies attenuate electromagnetic waves in the 100–1,000 MHz frequency band (Chukhlantsev *et al.*, 2003).

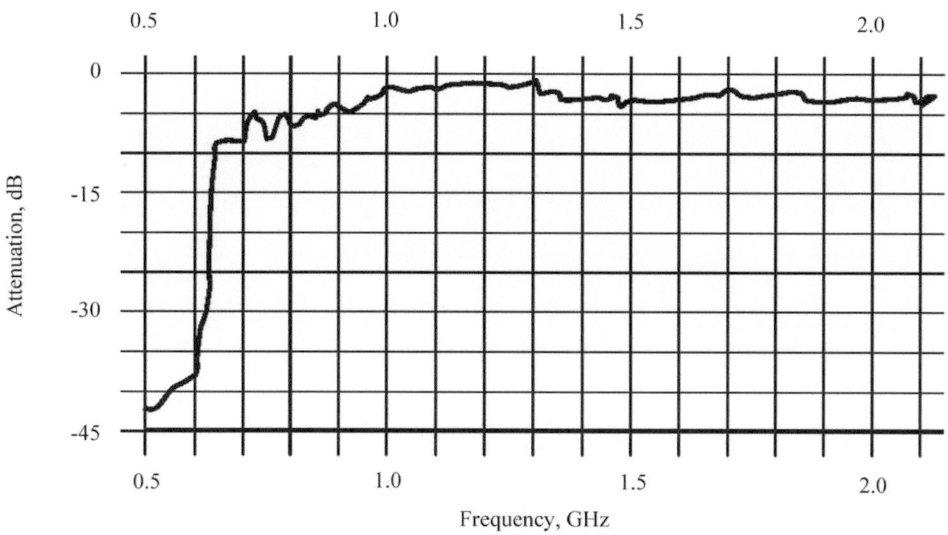


**Figure 8.3.** Frequency dependence of attenuation coefficient for the SHF channel of the measuring system.

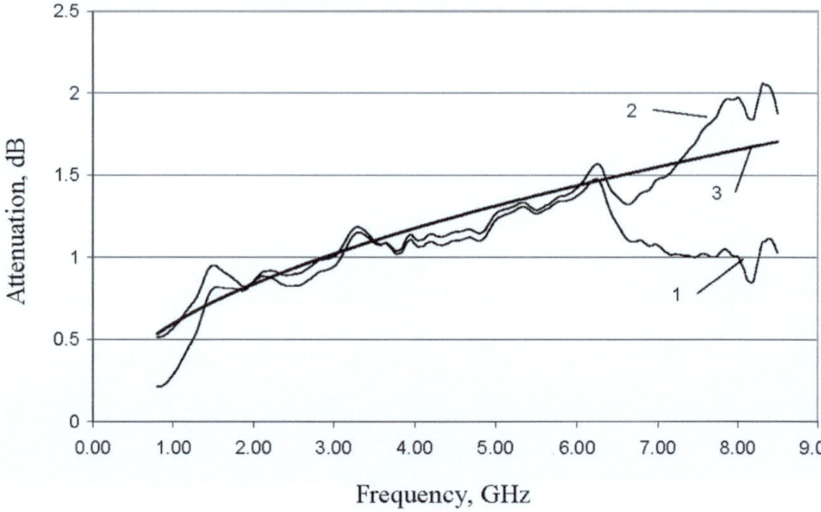


**Figure 8.4.** Frequency dependence of power attenuation in pine branches under 62.7% humidity. Notation: 1, attenuation values measured without correction; 2, corrected attenuation values; and 3, approximation of curve 2 by the power function. Measured attenuation values were corrected by neutralizing the power percolation effect from the waveguide (Chukhlantsev, 2006).

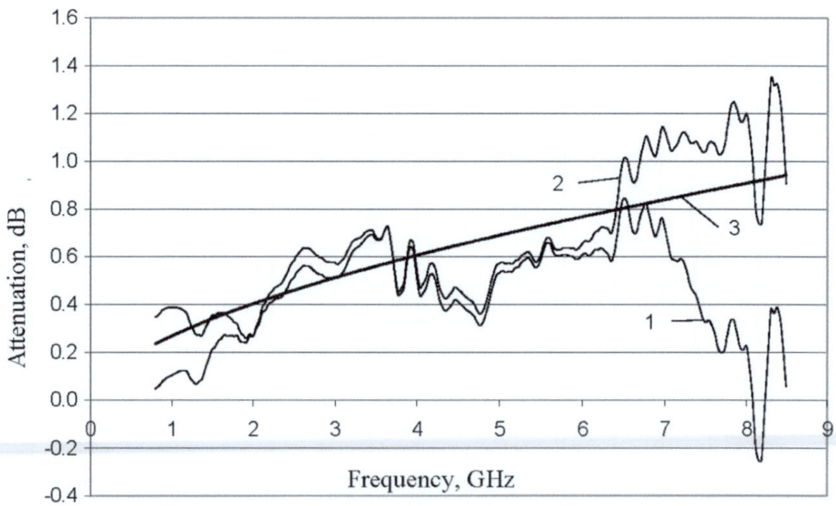


**Figure 8.5.** Frequency dependence of power attenuation in pine branches under 17.1% humidity. Notation: 1, attenuation values measured without correction; 2, corrected attenuation values; and 3, approximation of curve 2 by the power function. Measured attenuation values were corrected by neutralizing the power percolation effect from the waveguide (Chukhlantsev, 2006).

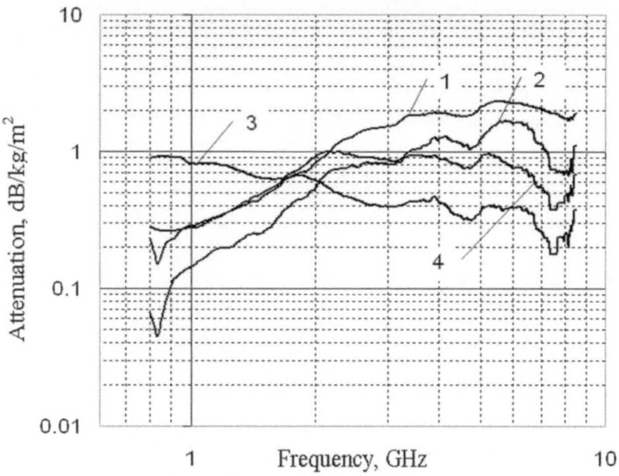


**Figure 8.6.** Frequency dependence of electromagnetic wave attenuation in aspen branches. Notation: 1, branches of 5 mm diameter with leaves; 2, branches of 5 mm diameter or less without leaves; 3, thick branches of 20 to 50 mm diameter; and 4, different branches of 5 to 20 mm diameter.

behind the trees. Receivers were located at different points and heights of the tree crowns. Assessments of the attenuation were obtained by comparing the intensity of signals at the different points of receivers in the tree. This led to a database of the EMW attenuation being synthesized (Golovachev *et al.*, 2004; Vinogradova *et al.*, 2004). Moreover, available microwave radiometric and SAR data are currently being processed and studied for inclusion in the database.

The methodological part of investigations also includes (Chukhlantsev, 2002, 2006; Chukhlantsev *et al.*, 2003):

- elaborating new technologies and approaches to assess the screening effect and its impact on SAR and microwave radiometric remote sensing, including synchronous active and passive observations and radio communication;
- determining the optimal frequencies or a set of frequencies for remote sensing of the Earth's covers taking account of the specifics of the areas of observation, terrain and vegetation type, climate, and current meteorological conditions;
- elaborating new methodologies for multi-channel measurements of the properties of both vegetation and soil by developing an advanced approach and adequate models to sense the soil–vegetation system in the absence of reliable prior knowledge-based information about the investigated object;
- elaborating an advanced approach to typical Earth environment observation from space by optimizing the instrument set and developing approaches to increase spatial resolution artificially by conducting multi-wavelength measurements.

**Figure 8.7.** Measuring system used to assess electromagnetic wave attenuation in *in situ* experiments (Chukhlantsev, 2006).

**Figure 8.8.** *In situ* measurement of electromagnetic wave attenuation by vegetation canopies in the 100–1,000 MHz frequency band.

## 8.3 LINKS BETWEEN EXPERIMENTS, ALGORITHMS, AND MODELS

Field observation is the initial stage in determining the important properties of the vegetation cover. Many problems exist in providing model descriptions of biological and physical processes that operate on different timescales. The tasks of describing EMW attenuation by the vegetation cover has timescales ranging from less than a second to hours or days and has various spatial scales too. Consequently, parametrical description of EMW attenuation by the vegetation cover calls for an enormous number of experimental observations, the planning of which depends on the model type and the vegetation component under investigation. By applying GIMS technology, effective synthesis of final results into given spatiotemporal scales can be brought about. This allows soil–vegetation–atmosphere models to be scaled and radiative transfer processes on a scale from individual plants or small plots to large-scale biomes to be described.

As mentioned above, satellite data can be used to calculate many parameters of the vegetation cover, including unstressed stomatal conductance, photosynthetic capacity, the proportion of photosynthetically active radiation absorbed by the green portion of the vegetation canopy, canopy reflectance, transpiration, and other important environmental characteristics. These data and the results of field experiments allow model experiments to be carried out.

The GIMS method uses *a priori* information to start the model experiment. Under this procedure, a set of algorithms operate over the whole dataset and determine the input parameters of the model used. The model structure and its coefficients are changed depending on differences between experimental and model estimations of the EMW attenuation effect. Usually, the central part of the model describes fundamental processes within the SPF, such as radiation balance, water circulation (evapotranspiration, water uptake by roots), photosynthesis, and mortality.

The combined use of algorithms, computer models, and remote-sensing technologies for environmental investigations is normally possible within a hybrid information-modeling system with the structure described by Soldatov (2011). This structure links experimental optical and microwave techniques to computer modeling, thus supporting correlations between different stages of a simulation experiment.

The model-based analysis and optimization of computational experiments have specific features that clearly call for new design points to be determined. Unfortunately, under many simulation experiment scenarios, the objective function (e.g., the response of the survivability indicator to impacts on the environmental system) cannot be assessed exactly. Thus, certainty can only be obtained by repeating the experiment and estimating the distribution of responses. GIMS technology based on evolutionary modeling and sequential analysis such as that shown in Figure 1.7 optimizes such an investigative process.

Complex engineering problems clearly cannot be resolved without using simulation experiments, which are usually brought about by numerical solution of a complex system of equations. In this case the researcher faces a conflict between wishing to enhance the accuracy of a simulation experiment and the limited capabilities of the algorithms used. Most difficulties arise when the researcher cannot use mathematical equations to parameterize the processes studied. These difficulties can be overcome as a result of the development of evolutionary computer technology that helps to design a new type of model (Bukatova *et al.*, 1991).

A GIMS-based approach normally reproduces a numerical simulation experiment derived from a given set of quantitative models and algorithms independently of the complexity of the problem under study. Of course, the researcher has to compile information about different parts of the simulation experiment including information about input and output data as well as about the models to use. All models used in the experiment must be identified, accessible, and fully described. Simulation models must be provided with all governing equations, parameter values, and necessary conditions such as initial state, boundary conditions, etc. The main step in a simulation experiment is defining the quality criteria needed to carry out the experiment. Many ways of resolving environmental problems using a GIMS-based approach are given in this book.

## 8.4 MICROWAVE MODEL OF VEGETATION COVER

### 8.4.1 Two-level model of vegetation cover

The atmosphere–vegetation–soil system (AVSS) is the commonly considered system of choice (Figure 8.9). Microwave emission from the AVSS is formed by absorption and diffraction processes combining within the media and by repeated reflection at their boundaries. Microwave emission models (MEMs) have been thoroughly studied in attempts to derive possible AVSS configurations by many authors (Ferrazzoli and Guerriero, 1996; Karam *et al.*, 1992; Krapivin *et al.*, 2006; Mitnik, 1977; Phillips *et al.*, 2009; Shutko, 1997; Shutko and Krapivin, 2011; Shutko *et al.*, 2007). AVSS emissivity is usually described as

$$e = 1 - |R|^2, \tag{8.1}$$

where

$$R = \frac{R_{21}\exp[-2i\chi h] + R_{32}}{R_{21}R_{32} + \exp[-2i\chi h]}, \qquad \chi = \chi_1 - i\chi_2 = \frac{2\pi}{\lambda}[\varepsilon_2 - \sin^2\theta]^{1/2},$$

where $\theta$ is the angle between the direction of electromagnetic wave propagation in the atmosphere and the atmosphere/vegetation interface; $R_{21}$ and $R_{32}$ are complex Fresnel coefficients for the atmosphere/vegetation and vegetation/soil interfaces, respectively; and $h$ is the vegetation cover height.

The Fresnel coefficients for the interface between the $i$th and $k$th medium are defined by the expression:

$$\hat{R}_{ik} = |R_{ik}|\exp(-iq_{ik}), \tag{8.2}$$

where $q_{ik}$ is phase shift.

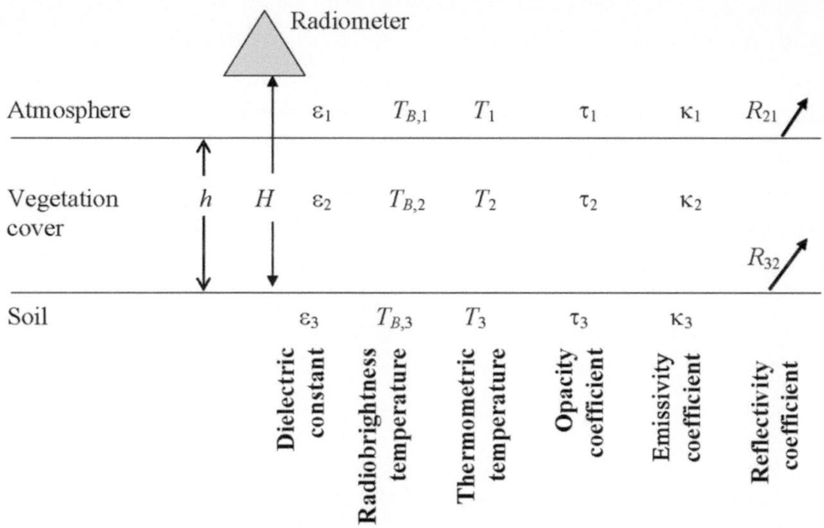


**Figure 8.9.** Model-based calculation of the shielding influence of vegetation cover in microwave monitoring (Krapivin *et al.*, 2006).

From (8.1) and (8.2) we get the following representation for emissivity:

$$e = \frac{r - \hat{R}_{32}^2 \exp(-4\chi_2 h) - \hat{R}_{21}^2 - 2\kappa_1 \sin q_{21} \sin \xi}{r + \kappa_1 \cos(q_{21} + \xi)},$$

where

$$\xi = q_{32} + 2\chi_1 h, \qquad \kappa_1 = 2\hat{R}_{21}\hat{R}_{32}\exp(-2\chi_2 h),$$

$$\hat{R}_{ik} = |R_{ik}|, \qquad r = 1 + \hat{R}_{21}^2 \hat{R}_{32}^2 \exp(-4\chi_2 h),$$

$$\hat{R}_{ik} = \left\{ \frac{(\gamma_i - \gamma_k)^2 + (\delta_i - \delta_k)^2}{(\gamma_i + \gamma_k)^2 + (\delta_i + \delta_k)^2} \right\}^{1/2}, \tag{8.3}$$

$$\operatorname{tg} q_{ik} = \frac{2(\delta_i \delta_k - \gamma_i \gamma_k)}{\gamma_i^2 - \gamma_k^2 + \delta_i^2 - \delta_k^2}, \tag{8.4}$$

$$\gamma_k = \begin{cases} \dfrac{Ak}{a_k} & \text{for horizontal polarization,} \\[2ex] \dfrac{A_k \varepsilon'_k + \beta_k \varepsilon''_k}{(\varepsilon'_k)^2 + (\varepsilon''_k)^2} & \text{for vertical polarization,} \end{cases}$$

$$\delta_k = \begin{cases} \dfrac{\beta_k}{a_k} & \text{for horizontal polarization,} \\[2ex] \dfrac{A_k \varepsilon''_k - \beta_k \varepsilon'_k}{(\varepsilon'_k)^2 + (\varepsilon''_k)^2} & \text{for vertical polarization,} \end{cases}$$

where

$$A_k = \sqrt{(a_k + b_k)/2}; \quad \beta_k = \sqrt{(a_k - b_k)/2}; \quad a_k = \sqrt{b_k^2 + (\varepsilon_k'')^2}; \quad b_k = \sqrt{(\varepsilon_k')^2 - \sin\theta};$$

$\varepsilon_2 = \varepsilon_2' - i\varepsilon_2''$ is the dielectric permittivity of the vegetation cover; $\varepsilon_3 = \varepsilon_3' - i\varepsilon_3''$ is the dielectric permittivity of the soil; and $\varepsilon_1 = \varepsilon_1' - i\varepsilon_1''$ is the dielectric permittivity of the air with $(\varepsilon_1' = 1, \varepsilon_1'' = 0)$.

Following Chukhlantsev and Shutko (1988), we determine Fresnel coefficients as functions of vegetation cover parameters. The dielectric constants of the vegetation and soil are input parameters to formulas (8.3) and (8.4). Considering the vegetation and soil as double-component mixtures of dry matter and water, the dielectric constants can be determined by the following expressions:

$$\sqrt{|\varepsilon_2|} = \rho_P\sqrt{\varepsilon_B} + (1 - \rho_P)\sqrt{\varepsilon_W}, \qquad \sqrt{|\varepsilon_3|} = \rho_B\sqrt{\varepsilon_B} + (1 - \rho_B)\sqrt{\varepsilon_W}, \quad (8.5)$$

where $\rho_P$ and $\rho_B$ are the relative volumetric concentrations of water in the plants and soil, respectively; and $\varepsilon_B$ and $\varepsilon_W$ are the dielectric constants of dry soil and water, respectively.

As shown empirically by Engman and Chauhan (1995), $\varepsilon_3' > \varepsilon_3''$ and both the real and imagery parts of the dielectric constant of soil are increasing functions of volumetric moisture content. These functions can be approximated by the following formulas:

$$\left.\begin{aligned} \varepsilon_3' &\approx c_1 + (c_2\rho_B + c_3\rho_B^2)\exp(c_4 - c_5/\lambda), \\ \varepsilon_3'' &\approx (d_1\rho_B + d_2\rho_B^2)\exp(d_3 - d_4/\lambda), \end{aligned}\right\} \quad (8.6)$$

where $\lambda$ is the wavelength (cm); and $c_i$ and $d_i$ are constants depending on the soil type. For example, soil that consists of 30.6% sand, 55.9% silt, and 13.5% clay is characterized by dependencies (8.6) with $c_1 = 2.35$, $c_2 = 52.4$, $c_3 = 31.1$, $c_4 = 0.057$, $c_5 = 1.22$, $d_1 = 7.1$, $d_2 = 46.9$, $d_3 = 0.0097$, and $d_4 = 1.84$.

Thus, taking dependencies (8.6) into account makes it possible to optimize the microwave range for passive remote sensing. Of course, a set of unsolved problems exists relating to the model corrections necessary to take surface roughness and other obstacles distorting brightness temperature $T_B$ into account, as well as a set of special features that arise when solving inverse tasks. A basis for future model refinements is the correlation between $T_B$ (the atmospheric transmissivity for a radiometer at height $H$ above the soil), smooth surface reflectivity $R$, and the thermometric temperatures of the SPF $T_{s-v}$ and atmosphere $T_a$. This correlation can be expressed by means of the Schmugge–Shutko formula (Schmugge, 1990; Shutko, 1986):

$$T_B = t(H)[RT_{\text{sky}} + (1 - R)T_{s-v}] + T_a, \quad (8.7)$$

where $T_{\text{sky}}$ is the contribution from the reflected sky brightness. Typical remote-sensing applications use microwave wavelengths $\lambda \geq 1$ cm and in this case atmospheric transmission approaches 99% and $T_a + T_{\text{sky}} \leq 5°$K (Engman and Chauhan, 1995).

A more precise correlation than (8.7) can be achieved by considering how surface roughness influences soil reflectivity: $R' = R\exp(-g \cdot \log^2 \omega)$, where $\omega$ is the angle of incidence, and $g$ is the roughness parameter ($g = 4\sigma^2 k^2$, where $\sigma$ is the root mean square height variation of the soil surface and $k = 2\pi/\lambda$).

Formulas (8.5) and (8.6) can be simplified by using the correlation between soil water content $\rho_B$ and vegetation water content $\rho_P$ (DeWitt and Nutter, 1989):

$$\rho_B = 78.9 - 78.4[1 - R' \exp(0.22\rho_P)].$$

For uniform media, expression (8.7) is often represented in the following way:

$$T_B = \frac{(1 - \hat{R}_{21}^2)\{[T_3(1 - \hat{R}_{32}^2)]e^{-\tau_2} + T_2(1 - e^{-\tau_2})(1 + \hat{R}_{32}^2 e^{-\tau_2})\}}{1 - \hat{R}_{21}^2 \hat{R}_{32}^2 \exp(-2\tau_2)},$$

where the opacity coefficient $\tau_2$ can be evaluated by means of the formula used by Kirdiashev *et al.* (1979):

$$\tau_2 = \frac{4\pi Q}{\nu \rho_1 \lambda I_m(\sqrt{\epsilon_2})},$$

where $\nu$ is vegetation volume density; and $\rho_1$ is the specific density of the wet biomass. Coefficient $\hat{R}_{21}$ is estimated by the following expression:

$$\hat{R}_{21} \approx \alpha_2[1 - \exp(-\tau_2)].$$

There are many other models of the AVSS—some simpler and others more complex. For example, the model describing the forest medium (as sketched in Figure 8.1) was developed by Ferrazzoli and Guerriero (1996) and is based on the Rayleigh–Jeans approximation of the electromagnetic properties of coniferous leaves (needles). The vegetation cover is represented as a two-layered structure. The first layer is the canopy composed of leaves and branches, represented by disks and cylinders, respectively. The second layer of the vegetation cover consists of trunks described by near-vertical cylinders. Such a model structure allows microwave forest emission to be simulated.

Layered vegetation models have been studied by many authors (Ferrazzoli and Guerriero, 1996; Karam *et al.*, 1992; Krapivin *et al.*, 2006). The main mathematical task involves solving the radiative transfer equation. Different models consider a wide frequency range for both deciduous and coniferous forests and account for the distributions of branch size and leaf orientation. These studies basically conclude that detailed knowledge of the forest structure is important for model adequacy. For example, lichens in taiga forests and forest tundra are characterized by lower photosynthesis and can influence the attenuation index. Resolving these tasks is the subject of future studies.

### 8.4.2 Analytical model of vegetation cover

The radiobrightness temperature registered by a radiometer at height $H$ above soil covered by plants depends on environmental parameters and parametrical descriptions of this dependence need to be considered. A simplified microwave model can be written as the following integral:

$$T_B(H) = e_s T_s \exp\left[-\int_0^H \gamma(x)\,dx\right] + \int_0^H T(y)\gamma(y)\exp\left[-\int_0^y \gamma(x)\,dx\right]dy + \zeta(H), \quad (8.8)$$

where $\gamma$ is the absorption coefficient; $e_s$ is the soil emissivity coefficient; $T_s$ is soil temperature; $T$ is atmospheric temperature; and $\zeta$ is model precision. There exist different simplifications of equation (8.8). For example, equation (8.8) is often used for various approximations of $T(y)$ and $\gamma(y)$:

$$T(y) = \begin{cases} T_2, & 0 \le y \le h, \\ T_1, & h < y \le H, \end{cases} \quad \text{or} \quad T(y) = T_0 + T_1 y + \cdots + T_n y^n$$

and

$$\gamma(y) = \gamma_0 + \gamma_1 y \quad \text{or} \quad \gamma(y) = \text{const.}$$

For satellite monitoring it can be shown that

$$T_B = e_s T_s p_{\text{sat}} + \sum_{l=1}^{n+1} D_l(0), \quad (8.9)$$

where $p_{\text{sat}}$ is atmospheric transmittance between the soil surface and the satellite,

$$D_1 = T(h), D_2 = \frac{1}{\gamma(h)} \cdot \frac{dT(h)}{dh}, \ldots, D_l = \frac{1}{\gamma(h)} \cdot \frac{dD_{l-1}(h)}{dh}, \ldots.$$

Taking different analytical representations of the vertical profile $T(h)$ into account, formula (8.9) can calculate brightness temperature $T_B$ with an error equal to $D_n(0)$.

The simplified microwave model (8.8) for brightness temperature of the canopy surface can be given as:

$$T_B(\lambda, \theta) = e_2 T_2 + \hat{R}_{21}(\theta) T_{B,\text{sky}}(\theta_1, \theta_2), \quad (8.10)$$

where $\theta$ is the angle of incidence; and $T_{B,\text{sky}}$ is the sky brightness temperature incident on the canopy from direction $(\theta_1, \theta_2)$; and $\theta_1$ and $\theta_2$ are the zenith and azimuth angles, respectively.

Model (8.10) simplifies analyzing the influence of canopy temperature and wind speed, for example, on variations in brightness temperature. This analysis is possible as long as canopy roughness as a function of wind speed is known.

## 8.5 VEGETATION DYNAMICS

### 8.5.1 General approach to modeling biocenology

Numerical biocenology has become an independent scientific direction as a result of authors combining the potentials of ecology and numerical modeling in their studies (Degermendzhy *et al.*, 2009; Krapivin and Kelley, 2009). This combination has resulted in many principles of traditional ecology being re-oriented, creating a kind of information interface between ecology and other sciences and elevating the role played by complex interdisciplinary studies. The sphere of interests of this new direction include the problems of complexity and generality in the development of ecological laws.

The latest models cover such problems as parameterizing vegetation cover succession, describing the production process as a function of environmental factors (Wirtz, 2000), modeling the competition for common resources (Alexeev *et al.*, 1992), modeling natural–anthropogenic systems (Polishchuk, 1992; Ugolnitsky, 1999), modeling the spatiotemporal dynamics of ecological systems, and many others.

Many authors model the succession of vegetation communities based on their personal views of succession processes. There are two fundamental interpretations of these processes based on the concepts of endo-ecogenesis and exo-ecogenesis. The dynamic phenomena in phytocenoses are deemed to involve a single process of vegetation development affected by both external and internal factors. The process of succession here covers such factors as soil development, layer-by-layer division of the vegetation community, change in the height of plants and their accumulated biomass, microclimate formation, change in species diversity, and relative stability of the community. Classifying the dynamics of the various forms that make up the vegetation community and the factors that affect it is important for further synthesis of the numerical model. Vasilevich (1983) introduced the following categories of successions: endogenic (autogenic), exogenic (allogenic), continuous, and post-disruptive (post-catastrophic). In the case of endogenic succession the source of changes is located in the vegetation proper and in the medium affecting it. There are no stages in continuous successions. In the course of post-disruptive succession, changes in the community take place irrespective of whether conditions are unstable or stable. Exo-ecogenesis is characterized by the external factors that prevail.

Population dynamics was a novel development in succession modeling. Hulst (1979) proposed three types of succession models. The first type was a model of succession with response but without competition (i.e., the reaction of the vegetation community to changes in environmental conditions). Succession was described by the usual differential equation:

$$dN_i/dt = r_i N_i (k_i - N_i)/k_i,$$

where $N_i$ is the number of individuals of the $i$th species; $r_i$ is the rate of exponential growth; and $k_i$ is the maximum population density.

The second type of succession model was based on suppression of one species by another in the process of intrusion. In this case the coefficient of competition was used:

$$\alpha_{ij} = \frac{\partial[dN_i/dt]}{\partial N_j} \Big/ \frac{\partial[dN_i/dt]}{\partial N_i},$$

and the model proper was written as:

$$dN_i/dt = r_i N_i \left(k_i - \sum \alpha_{ij} N_i\right)/k_i.$$

The third type of succession model was totally different. While population density was a function of the size of previous species of vegetation in the first two models, here the maximum population density $k_i$ depended on time.

Organic matter production by the vegetation community depends on the physiological processes in plant organisms and the availability of solar radiation, moisture, elements of mineral nutrition, temperature, and many other internal and external factors. Modeling the production process takes into account photosynthesis, respiration, transpiration, division of cells, etc. There are numerous correlations between these factors and plant production. Lieth's formulas are often the most used (Lieth, 1985). To estimate the annual production of plants $R_p$ (g/cm$^2$/yr) as a function of mean annual temperature $T$ (°C) and mean annual precipitation amount $r$ (mm), the following estimate can be used:

$$R_p = 3{,}000 \min\{(1 + \exp(1.315 - 0.119T))^{-1}, (1 - \exp(-0.000664r))\}.$$

Vegetation biomass density $P$ on a given territory is generally modeled by the balance equation:

$$dP/dt = \min\{\rho_P, R_{PP}\} - M_P - T_P,$$

where $\rho_P$ is the maximum $B/D$ coefficient of the type of vegetation considered; $R_{PP}$ is the vegetation productivity for $B/D$ coefficient values less than $\rho_P$ for present values of the $B/D$ coefficient; and $M_P$ and $T_P$ are parameters of decay and respiration expenditure per unit time, respectively.

Parametric descriptions of productivity in biogeocenotic models are an important addition to vegetation community monitoring. Unfortunately, the existing database and current knowledge do not provide synthesized information about a huge number of empirical correlations representing productivity dependences and other functional components of vegetation communities.

Model description of the competition between individual plants in numerical biogeocenology is especially important. On the whole, the mutual impact of plants can be described by so-called competition indexes (Bogatyrev, 1988). A freely growing plant is a plant whose functioning is independent of resource and is simply determined by species characteristics and external parameters. Interaction between plants starts with their influence on the same resource and with mutual intrusion on territories occupied by free growth. Biomes have more complicated and diverse interrelationships and, as a result, they differentiate niches and habitats. According to Mirkin (1986), who introduced five types of strategies for the way in which plants function in the environment ($K$, violents; $S$, ecotopic patients; $S_k$, phytocenotic patients; $R$, true regulator; and $R_k$, false or phytocenotic regulator). A systems approach is possible for model description of competition and survival. All these strategies to different degrees are inherent in individual species and, therefore, synthesizing a model of survival is determined by a concrete set of the plant's requirements to niche parameters. Usually, a plant's niche is an aggregation of the space it occupies, resources it uses, and additional conditions (e.g., the presence of pollinators, phytophagans, etc.).

Each plant or some totality of plants $N$ interacts with the environment $H$ (including other plants) exchanging some amount of resources $V_N$ for an obtained volume of resources $W_N$. The structure $|N|$ and the behavior $\underline{N}$ of plants or their communities are in equilibrium with the structure $|H|$ and behavior $\underline{H}$ of the environment, and this state can be described with a minimax criterion:

$$W_{N,\text{opt}} = \max_{\{|N|,\underline{N}\}} \min_{\{|H|,\underline{H}\}} W_N(V_N, |N|, \underline{N}, |H|, \underline{H})$$

$$W_{H,\text{opt}} = \max_{\{|H|,\underline{H}\}} \min_{\{|N|,\underline{N}\}} W_H(V_N, |N|, \underline{N}, |H|, \underline{H}).$$

The goals $N_S$ and $H_S$ of vegetation continuum $N$ and environment $H$, respectively, can be opposite, partially coinciding, or indifferent. According to the optimal strategy, plants of different kinds adjust to the levels of $(V, W)$ exchange that are possible on a given territory. Thus, there exist multi-level vegetation communities because of niche differentiation. For example, in the forest the factor of niche differentiation is many sided. It covers the special features of various kinds of plants such as the location of roots at different depths, different periods of blossoming, different response of plants to climate changes, different need of species for mineral nutrition, etc. In general, a plant's niche can vary in time both as a result of the dynamics of the plant proper and in connection with the drifting of adjacent niches. The diversity of possible situations has resulted in a series of models of vegetation community functioning.

A characteristic example of a vegetation community model is the discrete one proposed by Galitsky (1985). Let us consider it in more detail. The equation that describes a change in plant biomass $B$ can be written as

$$dB/dt = k(t)(\alpha B^k + (B/B_f)^\gamma B_f - \alpha B^k,$$

where $k(t) = \min\{A/A_f(t), 1\}$ is an indicator that the plant has a territory; $A$ is the size of the available territory for plant growth; $A_f$ is the size of the territory needed for free growth; $B_f$ is the biomass of a freely growing plant; $\alpha$ characterizes what the plant has spent toward the balance (for biomass maintenance); and $\gamma$ is the index of the deficit of territory.

The gist of the way in which Galitsky's model works is that the biomass of each plant located in a cell of its own in a common mosaic changes independently of other plants in the community until either this plant or any of its neighbors dies because it runs out of space. At the moment the plant dies, adjacent plants rush to occupy the vacant territory and, after changing parameter $A$, the equation of the model is valid again. As a result, as individual plants die, the original mosaic changes on a local scale, gradually covering larger areas. The criterion for death is simple: a plant is considered dead if its biomass drops below some threshold level $B_{\text{dead}}$ ($B_{\text{dead}}(t) = \varepsilon B_f(t), 0 < \varepsilon < 1$).

### 8.5.2 Biocenotic model

The parameters of vegetation cover change during the year depending on the weather situation (Table 8.3). The specific biomass $Q_i$ of the $i$th type of vegetation at time $t$ can be parameterized by means of the following equation:

$$\partial Q_i/\partial t = R_i - M_i - E_i, \tag{8.11}$$

where $R_i$ is biomass productivity; and $M_i$ and $E_i$ are biomass losses at the expense of withdrawal and transpiration, respectively.

The function $M_i(\varphi, \psi, t)$ reflects the set of natural $M_{Ni}$ and anthropogenic $M_{Ai}$ processes leading to vegetation biomass losses ($M_i = M_{Ni} + M_{Ai}$):

$$M_i(\varphi, \psi, t) = \mu_i(t) Q_i(\varphi, \psi, t),$$

where $\varphi, \psi$ are latitude and longitude, respectively.

**Table 8.3.** The dependence of annual production $R$ ($\mathrm{kg\,m^{-2}\,yr^{-1}}$) on average annual temperature $T_a$ and total annual rainfall $W$, $F(T_a, W)$.

| *Precipitation, W* (mm/yr) | *Atmospheric temperature $T_a$* (°C) | | | | | | | | | | | |
|---|---|---|---|---|---|---|---|---|---|---|---|---|
| | *−14* | *−10* | *−6* | *−2* | *2* | *6* | *10* | *14* | *18* | *22* | *26* | *30* |
| 3,125 | | | | | | | 3.4 | 3.5 | 3.7 | 3.8 | 3.9 | 4.0 |
| 2,875 | | | | | | | 3.2 | 3.3 | 3.5 | 3.6 | 3.7 | 3.8 |
| 2,625 | | | | | | | 3.0 | 3.2 | 3.3 | 3.4 | 3.5 | 3.6 |
| 2,375 | | | | | | | 2.8 | 2.9 | 3.0 | 3.1 | 3.2 | 3.3 |
| 2,125 | | | | | | | 2.5 | 2.6 | 2.7 | 2.9 | 2.9 | 3.0 |
| 1,875 | | | | | | 1.6 | 2.3 | 2.3 | 2.4 | 2.5 | 2.6 | 2.7 |
| 1,625 | | | | 0.4 | 0.6 | 1.3 | 2.0 | 2.1 | 2.1 | 2.2 | 2.3 | 2.4 |
| 1,375 | | 0.2 | 0.3 | 0.4 | 0.7 | 1.1 | 1.7 | 1.9 | 1.9 | 2.1 | 2.1 | 2.0 |
| 1,125 | 0.2 | 0.3 | 0.3 | 0.4 | 0.6 | 1.0 | 1.6 | 1.8 | 1.9 | 1.8 | 1.8 | 1.7 |
| 875 | 0.2 | 0.3 | 0.4 | 0.5 | 0.8 | 0.9 | 1.5 | 1.4 | 1.3 | 1.3 | 1.2 | 1.2 |
| 625 | 0.3 | 0.3 | 0.5 | 0.6 | 0.9 | 0.9 | 0.9 | 0.8 | 0.8 | 0.7 | 0.7 | 0.7 |
| 375 | 0.4 | 0.4 | 0.5 | 0.7 | 0.6 | 0.6 | 0.6 | 0.5 | 0.5 | 0.5 | 0.4 | 0.4 |
| 125 | 0.1 | 0.3 | 0.3 | 0.2 | 0.2 | 0.2 | 0.2 | 0.2 | 0.2 | 0.1 | 0.1 | 0.1 |

Flux $E_i$ is calculated by means of the formula:

$$E_i(\varphi, \psi, t) = \frac{\rho c_p[e^*(T_c) - e_a]}{\gamma_p(r_c + r_b)},$$

where $e^*(T_c)$ is saturated vapor pressure inside the canopy's foliage (in pascals, Pa); $e_a$ is the vapor pressure in the canopy's air space (Pa); $r_c$ is canopy resistance ($\mathrm{s\,m^{-1}}$); $r_b$ is the bulk leaf boundary layer resistance of the canopy ($\mathrm{s\,m^{-1}}$); $\rho$ is air density ($\mathrm{kg \cdot m^{-3}}$); $c_p$ is the specific heat of air ($\mathrm{J \cdot kg^{-1} \cdot K^{-1}}$); and $\gamma_p$ is the psychrometric constant ($\mathrm{Pa \cdot K^{-1}}$).

As shown in Figure 1.19, the GIMS makes use of 30 models for SPF simulation. The list of SPFs given in Table 8.4 was adapted to a spatial resolution of $4^\circ \times 5^\circ$ according to the classification of Bazilivich and Rodin (1967). In synthesizing these models, use was made of results obtained by Krapivin and Vilkova (1990) and Wirtz (2000). All models were based on equation (8.11) to ascertain the balance of biomass $Q_i(\varphi, \psi, t)$. Actual plant productivity was approximated as follows:

$$R_i = \delta_c^i(1 + \alpha_T^i \cdot \Delta T/100)\exp(-\beta_i/Q_i)\min\{\delta_e^i, \delta_Z^i, \delta_W^i, \delta_B^i\},$$

where $\alpha_T^i$ and $\beta_i$ are indexes of the dependence of production on temperature variation $\Delta T$ and biomass $Q_i$, respectively; $\delta_\zeta^i$ is the index of limitation of production by factor $\zeta$; $e$ is illumination; $Z$ is pollution; $W$ is soil moisture; $B$ is nutrient salts of the soil; and $c$ is atmospheric $CO_2$ concentration. The $\delta_\zeta^i$ functions actually used in real situations were calculated based on existing or preliminarily received data. Thus, the role played by atmospheric $CO_2$ concentration $C_A$ in photosynthesis was described by the relation $\delta_c^i = b_i C_A/(C_A + C_{0.5}^i)$, where $C_{0.5}^i$ is the $CO_2$ concentration for which $\delta_c^i = b_i/2$. The influence of solar radiation intensity $e(\varphi, \psi, t)$ on photosynthesis can be parameterized by the relation $\delta_e^i = \delta_i^* \exp(1 - \delta_i^*)$, where $\delta_i^* = e/e_i^*$ and $e_i^*$ is the optimal illumination for the $i$th type of plant. A more detailed description of the correlations between biocenotic processes is given by Kondratyev *et al.* (2002).

### 8.5.3 Classification of soil–plant formations

Figure 1.19 is a sample ecological subdivision of SPFs that takes many abiotic factors into account. From the standpoint of microwave monitoring, the set of principal vegetation features play a significant role in solving the electromagnetic wave attenuation problem. As is known, there are three priority factors for each type of vegetation cover: wet biomass $Q$, productivity $R$, and water content $\rho_P$. These factors are functions of climatic zones. Therefore, to assess the impact of climate change on the potential productivity of vegetation requires developing a model to describe the dynamics of vegetation. There exist many approaches to solving this. Some classifications reflect the principle variables describing the structure of the ground ecosystem: land cover, soil types, phenology, biomass, soil moisture, and temperature profiles. High-precision classifications take into account

**Table 8.4.** Identifiers (ID) adopted in the GIMS to designate soil–plant formations. Notation: $\sigma$ is the area of soil–plant formation ($10^6$ $km^2$) and $P$ is the productivity of soil–plant formations (kg $C \cdot m^{-2}$ $yr^{-1}$).

| *ID* | *Soil–plant formation* | $\sigma$ | $P$ |
|---|---|---|---|
| *A* | Arctic deserts and tundra | 2.55 | 0.17 |
| *B* | Alpine deserts | 1.15 | 0.47 |
| *C* | Tundra | 2.93 | 0.36 |
| *D* | Mid-taiga forests | 5.73 | 0.63 |
| *E* | Pampas and grass savannas | 1.66 | 1.11 |
| *F* | North taiga forests | 5.45 | 0.54 |
| *G* | South taiga forests | 6.60 | 0.65 |
| *H* | Subtropical deserts | 7.16 | 0.12 |
| *I* | Subtropical and tropical thickets of grass–tugai type | 0.90 | 1.96 |
| *J* | Tropical savannas | 17.10 | 1.35 |
| *K* | Solonchaks | 0.38 | 0.18 |
| *L* | Forest–tundra | 1.55 | 0.65 |
| *M* | Mountain tundra | 2.23 | 0.38 |
| *N* | Tropical xerophytic open woodlands | 9.18 | 1.42 |
| *P* | Subtropical broad-leaved forests and coniferous forests | 5.75 | 1.72 |
| *Q* | Alpine and subalpine meadows | 3.54 | 0.76 |
| *R* | Broad-leaved coniferous forests | 2.12 | 0.87 |
| *S* | Subboreal and saltwort deserts | 2.69 | 0.25 |
| *T* | Tropical deserts | 11.50 | 0.18 |
| *U* | Xerophytic open woodland and shrubs | 3.91 | 0.56 |
| *V* | Dry steppes | 2.66 | 0.38 |
| *W* | Moderately arid and arid steppes | 4.29 | 0.79 |
| *X* | Forest steppe (meadow steppe) | 3.72 | 0.74 |
| *Y* | Variable humid deciduous tropical forests | 7.81 | 2.46 |
| *Z* | Humid evergreen tropical forests | 10.40 | 3.17 |
| + | Broad-leaved forests | 7.21 | 1.25 |
| & | Subtropical semideserts | 2.08 | 0.45 |
| @ | Subboreal wormwood deserts | 2.77 | 0.61 |
| * | Absence of vegetation | 14.6 | 0.00 |

the vertical structure of ecosystems (layered ecosystems), their age structure, canopy chemistry, disturbance, decomposition, anaerobic state of the ecosystem, rate of succession, ecosystem stratification, ratio of $C_3$ to $C_4$ plants, etc. The classification of vegetation cover is clearly a complex problem of biocenology. The degree of correspondence between published classifications and real distributions of vegetation types influences final model estimates of EMW attenuation by vegetation cover. Therefore, GIMS technology stipulates that the spatial structure of the vegetation of the landscape studied be corrected using satellite observations. However, reliable observations are only available for relatively limited areas. Therefore, extrapolations are often used to correct images of the vegetation structure.

Ascertaining the effective global spatial structure of SPFs depends on the specifics of the problem to be solved. For instance, Bounoua *et al.* (2000) using the SiB2 model and spatial resolution of $7.2^\circ \times 9.0^\circ$ for 12 types of SPFs analyzed seasonal variations in NDVI as a result of climate change. Sud *et al.* (1996) used a horizontal resolution of $4^\circ \times 5^\circ$ for 13 types of SPFs to study the biogeophysical consequences of a tropical deforestation scenario being realized. The quality of results is of course a function of both soil–plant classification and spatial resolution. Unfortunately, at the present time there is no reliable global database with detailed parametrical description of SPFs. This represents an important problem in global monitoring that needs to be overcome.

### 8.5.4 Modeling forest ecosystems

The forest ecological system is an important constituent of the global continuum of soil–plant formations on Earth (see Table 8.4). Therefore, developing models to describe the dynamics of forests is a principal stage in synthesizing a global model. The importance of such models is all the more important because forest resources and their use affect the economic potentials of regions.

Many authors describe the dynamics of an individual plant's total biomass as a logistic equation:

$$dB/dt = k(B_{\text{lim}} - B)B,$$

where $B$ is biomass; $B_{\text{lim}}$ is the limit to the amount of biomass; and $k$ is the coefficient reflecting the effect of environmental factors and plant type on biomass growth.

To describe the dynamics of tree growth, the simplest functions are often used, such as Terazaki's equation $B = a\exp(-b/t)$, Koller's function $B = ax^b c^{-ct}$, Korsun and Bakman's approximation $B = a\exp(b\ln t + c\ln^2 t)$, etc. Bichele *et al.* (1980) were the first to propose a more complicated model of tree growth described by the system of ordinary differential equations:

$$dB_j = K_\Phi - k_R R_j - V_j + K_M,$$

where $K_\Phi = k_F \sum^4 \alpha_{ij}\Phi_i$; $K_M = M \sum^4 \beta_{ij}$; $B_j$ is the biomass of individual organs (i.e., self-contained parts with specific vital functions) of a plant ($j = 1, 2, 3, 4$); $M$

is the dry biomass of the whole plant; $\Phi_i$ is the amount of carbon dioxide assimilated in 24 hours by the $i$th organ of the plant ($k_F\Phi_I$ is equivalent dry biomass, $k_F$ is the coefficient of equivalence); $R_i$ is the amount of carbon dioxide emitted in 24 hours by the $i$th organ of the plant ($k_R R_i$ is equivalent dry biomass, $k_R$ is the coefficient of equivalence); $V_i$ is the dry biomass of the $i$th organ; $\alpha_{ij}$ is the share of "fresh" assimilates created in 24 hours in the $i$th organ of the plant moving during the same period of time into the $j$th organ of the plant; $\beta_{ij}$ is the exchange of "old" assimilates in 24 hours between the $i$th and $j$th organs per gram of dry biomass of the whole plant; $K_\Phi$ is biomass increment as a result of photosynthesis and re-distribution of fresh assimilates; $K_M$ is the increment connected with the re-distribution of old assimilates; and $k_R R_j$ is biomass expenditure on respiration.

The coefficients of this model are, of course, functions of environmental parameters. How the environment affects the dynamics of tree growth can generally be ascertained by applying Libikh's principle of limiting factors. Limiting factors of the environment include PAR, carbon dioxide, temperature, humidity, and mineral salts in the soil. By Libikh's principle of limiting factors, a plant grows until the resources of some factor run out. Using this principle can lead to uncertainty; therefore, in concrete situations it must be tested by means of experimental estimates.

The more developed models of tree growth, in contrast to a set of empirical parameters whose uncertainties are often a great barrier to practical application of models, include the functional dependences of tree growth elements on environmental factors. Among the most widespread and often applied is the Libikh criterion wherein the growth of a plant is confined to an element whose concentration is at a minimum and a plant or its organs grow until the resources of some element (e.g., photosynthetically active radiation or PAR, moisture, temperature, carbon dioxide, nutritive salts) are exhausted. There are other criteria that weight the influence of all environmental factors; these weights are characteristic indicators of plant type.

The process of tree growth can be presented as the interaction between three growth processes: change in leaf (needle) mass, growth of the trunk, and development of the root system. The tree growth process is schematically shown in Figures 8.10 and 8.11.

Let us denote the biomass of the canopy, trunk, and roots of the tree as $x_1$, $x_2$, and $x_3$, respectively. Then, the balance equations for a tree model can be written as

$$x_i(t+\Delta t) = x_i(t) + \varepsilon_i(t)\{R_x - T_x\}\Delta t \quad (i = 1, 2, 3),$$

where $\Delta t$ is the characteristic time step; $R_x$ is the gross productivity of a tree as a result of photosynthesis; $T_x$ is the biomass expenditure on respiration; $\varepsilon_i$ is the proportion of new biomass moving into the biomass of the $i$th organ of the tree.

The $\varepsilon_i$ parameters are functions of time just like other parameters of the environment. Following the principle of maximum survival, a supposition can be

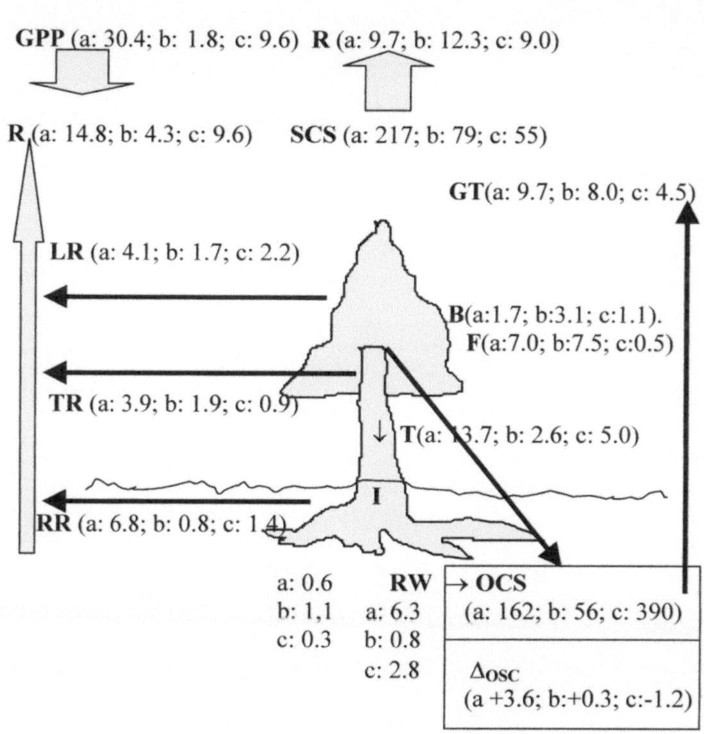


**Figure 8.10.** Annual supplies (t C · ha) and fluxes (t C · ha$^{-1}$ yr$^{-1}$) for tropical (a), temperate (b), and boreal (c) forests (Watson *et al.*, 2000). Notation: GPP, gross primary production; RA, autotrophic respiration; RH, heterotrophic respiration; RL, foliage respiration; RW, above-ground wood respiration; RR, root respiration; SOC, soil organic carbon stock; $\Delta_{SOC}$, net increment in soil organic carbon; E, below-ground detritus; K, below-ground net biomass carbon increment; B, above-ground net biomass carbon increment; O, above-ground detritus (leaf fall and mortality); and T, translocation from above to below ground.

made that the principle of maximum primary productivity is valid:

$$\tilde{Y}^*(t+\Delta t) = \max_{\{\varepsilon_i(t)\}} Y\{x_i(t) + \varepsilon_i(t)Y[x_i(t), E(t,\varphi,\lambda), C_A, T_A]\Delta t, E, C_A, T_A\}.$$

An important limiting factor to tree growth is water, the movement of which in the soil and the body of the tree determines the dynamics of tree biomass change. Following Kirilenko (1990), let us denote the soil as a homogeneous porous layer $[0, z_g]$, where $z_g$ is the depth of ground water. Let us introduce some notation: $W$ is relative soil moisture (kg $H_2O/m^3$/[kg soil/$m^3$]), $\rho$ is water density, and $\rho_s$ is volume density of soil. Let us also introduce a system of coordinates $(x, y, z)$, where the $z$-axis is directed downward from the soil surface ($z = 0$). Then the movement of water through the soil can be described by Darsi's 3-D equation:

$$V = -K[\nabla(\Phi_s - \rho g z)]/(\rho g), \tag{8.12}$$

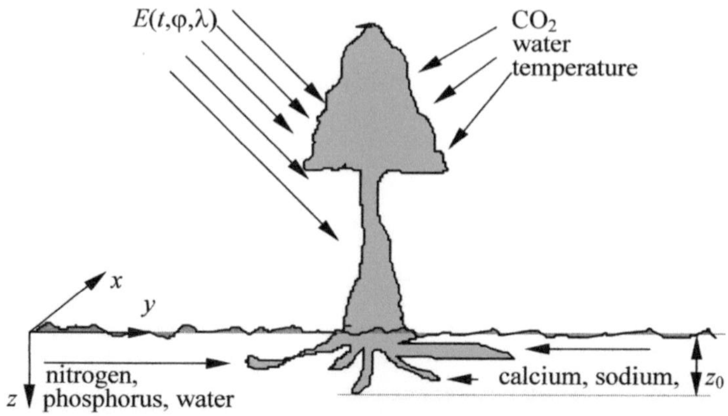


**Figure 8.11.** Growth of a tree as an element of forest biogeocenosis. Notation: $z_0$, thickness of the root system layer; and $E(t, \varphi, \lambda)$, solar radiation.

where $V(t, x, y, z)$ is the rate of water flow; $K(t, x, y, z)$ is the hydraulic conductivity of water; $g$ is gravity acceleration; and $\Phi(t, x, y, z)$ is the water potential of soil.

The equation for soil moisture dynamics is

$$\partial W/\partial t = -(\rho/\rho_s)(f + \text{div } V), \tag{8.13}$$

where $f(t, x, y, z)$ is the sink function.

The dependences of the water potential of soil and its hydraulic conductivity on soil moisture can be written as

$$K = K_s^* W^m, \qquad \Phi_s = \Phi_s^* W^{-n}$$

where

$$K_s^* = K_{FM} W_{FM}^{-m}; \qquad \Phi_s^* = \Phi_{MG} W_{MG}^n; \qquad m = 2n + 1;$$

$W_{FM}$ is the total water capacity of soil; $K_{FM}$ is soil conductivity corresponding to total water capacity; and $m$ and $n$ are constants.

The system of equations (8.12) and (8.13), using our revised notation, can be re-written as:

$$\partial\omega/\partial t = (-\rho/\rho_s)[(K_{FM}/W_{FM}) \text{ div}\{\mu\omega^n \nabla\omega + i\omega^n\} + f/W_{FM},$$
$$V = K_{FM}(\mu\omega^n \nabla\omega + i\omega^n),$$

where $\omega = n\Phi_{FM}/(\rho g)$; and $i$ is the unit vector directed along the $z$-axis.

In the case of homogeneous processes in the plane $(x, y)$ the problem can be simplified:

$$\partial\omega/\partial t = (-\rho/\rho_s)[(K_{FM}/W_{FM}) \, \partial/\partial z\{\mu\omega^n \, \partial\omega/\partial z + \omega^n\} + f(t, z)/W_{FM},$$
$$V(t, z) = K_{FM}(\mu\omega^n \, \partial\omega/\partial z + \omega^n).$$

Boundary and initial conditions can be given in the form:

$$\omega(t, z_g) = 1, \qquad V(t, 0) = P(t) - e(t), \qquad \omega(t_0, z_g) = \omega^0(z),$$

where $P(t)$ is precipitation intensity; $e(t)$ is the intensity of evaporation from the soil surface; and $\omega^0(z)$ is initial soil moisture.

The $f$ function describes the intensity of water absorption by a tree's root system:

$$f(t, z) = \xi_0[\Phi_s(t, z) - \psi_0(t)\} \, ds/dt,$$

where $\xi_0 = \xi_k \sigma_0$ is the conductivity of water entry to the root system; $\xi_k = r_k^{-1}$ is the specific conductivity of water entry to the root; $r_k$ is the specific resistance of water entry to the root; $\sigma_0$ is the general area of roots taking up water per unit soil surface area ($\sigma_0 = \beta_k m_k$); $m_k$ is the mass of roots beneath each unit of soil surface area; $\beta_k$ is an empirical constant; $\psi_0$ is the water potential within the root system; and $s(z)$ is a function of the vertical distribution of the root system ($s(0) = 0, s(z_0) = 1$).

To complete the synthesis of a tree's water regime, let us parameterize the processes of water flow in the plant and transpiration. Let us assume that water flows from the root system via the trunk to the canopy (branches, leaves) and then evaporates into the atmosphere (Lai *et al.*, 2000; Lal *et al.*, 1998). Let us divide the above-ground part of a tree of height $H$ into $n$ equal layers: $H = n\Delta z$. Across the boundaries of the $i$th layer the water moves as a result of the difference between water potentials $\psi_i$ and $\psi_{i-1}$ overcoming the resistance of xylem vessels:

$$r_{ks}^*(z) = (S_{ks}^*(z)\xi_{ks})^{-1},$$

where $\xi_{ks}$ is the specific conductivity of xylem vessels; and $S_{ks}^*$ is the cross-section of the tree at height $z$. As a result, supposing the area of the cross-section to be vertically constant, we obtain the following formula to describe the rate of water flow in the trunk across the $i$th layer:

$$v_{i-1,s}^* = -\xi_{ks} S_{ks}^* \Delta z^{-1} (\psi_i - \psi_{i-1} + \rho g \Delta z).$$

According to Kirilenko (1990), the intensity of transpiration from the $i$th layer can be described with the formula:

$$\gamma_i^* = \frac{d^i}{r_L^i} S_L^{*i},$$

where $d^i$ is any shortfall of absolute air humidity in the atmosphere; $S_L^{*i}$ is the leaf area of the $i$th layer; $r_L^i = r_{st}^i + r_a$; $r_{dt}^i$ is stomatal resistance; and $r_a$ is the boundary layer resistance of air.

To complete the description of Kirilenko's model, let us consider a standing wood of density $\rho_F$. The model equations can be written as

$$d^2\varphi/d\eta = Uv(\eta)\varphi(\eta), \quad \eta \in [0, \alpha]; \tag{8.14}$$

the boundary conditions are

$$\begin{cases} \varphi_0' = -[G(\tilde{\varphi}_s - \varphi_0) + F] \\ \varphi_\alpha' = -F \end{cases}, d\varphi/d\eta = -F \quad \text{for } \eta \in [\alpha, 1]. \tag{8.15}$$

At $\varphi_0 \leq 0$, the system of equations (8.14) and (8.15) becomes:

$$\begin{cases} \dfrac{d\varphi}{d\eta} = -F \\ \varphi_0 = \tilde{\varphi}_s, \end{cases}$$

where $\eta = z/H$; $\varphi(\eta) = 1 + \psi(\eta H)/\psi^*$; $\psi^*$ is the water potential of a leaf; $1/D_1$ is the minimum mouth resistance; $U = Hr_{ks}D_1 d\beta m_L/\psi^*$; $F = \rho gH/\psi^*$; $G = Hr_{ks}\xi_0$; $\alpha = a/H$; $\beta$ is the leaf index; and $h = -z$,

$$a = \begin{cases} H, & \psi(H) \geq \psi^*, \\ h, & \psi(h) = -\psi^*, \quad \psi(H) < -\psi^* < \psi(0), \\ 0, & \psi(0) \leq -\psi^*. \end{cases}$$

### 8.5.5 Modeling energy fluxes in the atmosphere–plant–soil system

Energy exchange in the atmosphere–plant–soil (APS) system plays an important role in the formation of climatic processes on every spatiotemporal scale and, hence, should be considered in models of the greenhouse effect to increase the reliability of estimates. There are a couple of problems that arise here: the development of methods to measure energy fluxes in the air–surface layer and modeling them. These problems are further aggravated because technologies need to be developed to estimate energy fluxes in the APS system on global scales so that these estimates can be taken into account in global climate systems. There are well-developed methods for remote-sensing a wide set of climatic parameters including energy fluxes (Krapivin and Kondratyev, 2002; Shutko *et al.*, 2010; Yakovlev, 2001). However, in the remote sounding of the APS system another complicated problem arises: assessment of the screening impact of vegetation on wavelength regions used in the monitoring system.

Ever since the 1980s, studies have been carried out on developing a model of photosynthesis in a bud. Farquhar *et al.* (1980) demonstrated how a biological model of $CO_2$ assimilation by leaves is connected with the fermentative kinetics and electronic properties of the chloroplast, which made it possible to plausibly describe the process of photosynthesis for plants of class *C*. Then, Ball (1988) and Collatz *et al.* (1990, 1991, 1992), using a model technology and observation data on the process of bud conductivity in photosynthesis, derived a semi-empirical model of how buds function that also considered changes in environmental temperature, humidity, $CO_2$ concentration in the atmosphere, and the intensity of assimilation of the plant species itself.

The consequences of the distribution of the photosynthetic potential of leaves and plant canopies as a result of illumination have also been considered.

A criterion has been developed to determine the common potential of the maximum photosynthetic abilities of a plant to assimilate $CO_2$.

Sellers (1985, 1987) developed methods to vertically integrate leaf canopies, models of light scattering and absorption, photosynthesis and bud conductivity, previously only known for a leaf. These studies revealed the interaction between the spectral characteristics of a plant species (spectral vegetation index, SVI) and important functional relationships that regulate the processes of photosynthesis of and evaporation from leaf canopies. Moreover, another analysis revealed a connection between the proportion of photosynthetically active radiance (PAR), absorbed by the canopy, and the respective characteristic coefficient of the plant species (simple ratio vegetation index, SR). We consider SR and relative difference (normalized difference vegetation index, NDVI) as SVI. The spectral characteristics of a plant species' SR and NDVI are based on the remote sounding of reflected emission in the visible and near-IR regions and determined as follows:

$$\mathrm{SR} = \frac{a_N}{a_V}, \qquad \mathrm{NDVI} = \frac{a_N - a_V}{a_N + a_V},$$

where $a_N$ and $a_V$ are hemispheric reflections and radiance in the visible and near-IR wavelength regions, respectively.

The parameters $a_N$ and $a_V$ are measured directly above a leaf canopy. Therefore, the effects of absorption and scattering of emission by the atmosphere are not considered here. For spaceborne remote sounding a method called FASIR (Fourier-adjusted, solar zenith angle corrected, interpolated and reconstructed data) has been developed to correct experimental data. A theoretical dependence (close to linear) was established between PAR and SR, confirmed later by experimental measurements in Kansas (Sellers *et al.*, 1995, 1996, 1997). Minimum and maximum FPAR are determined experimentally, and the final relationship becomes

$$\mathrm{PAR} = \frac{(\mathrm{SR} - \mathrm{SR}_{i,\min}) \cdot (\mathrm{PAR}_{\max} - \mathrm{PAR}_{\min})}{(\mathrm{SR}_{i,\max} - \mathrm{SR}_{i,\min})}, \tag{8.16}$$

where $\mathrm{PAR}_{\max} = 0.950$ and $\mathrm{PAR}_{\min} = 0.001$, but $\mathrm{PAR}_{\max}$ and $\mathrm{PAR}_{\min}$ generally depend on the type of vegetation; and $\mathrm{SR}_{i,\max}$ and $\mathrm{SR}_{i,\min}$ can be calculated from NDVI based on experimental data.

Sellers *et al.* (1996) use the $\Pi$ parameter to reflect the amount of cellulose taking part in photosynthesis. The $\Pi$ parameter changes from 0 (lack of vegetation) to $\approx 1.5$ for bushy species rich in photic cellulose and can be approximately estimated as

$$\Pi \approx \frac{\mathrm{PAR}}{\bar{k}},$$

where $\bar{k}$ is the coefficient of attenuation for PAR averaged over time.

Apart from the spectral characteristics of plants, Sellers *et al.* (1996) consider the photic characteristics (green leaf area index $L_g$) and the photic share (green part $N$) for a concrete plant species. Many authors have derived the dependences of $L_g$ on FPAR for species growing separately and in groups. The result can be presented in the form

$$L_g = (1 - F_{cl})L_{g,i,\max}\frac{\log(1 - \mathrm{PAR})}{\log(1 - \mathrm{PAR}_{\max})} + F_{cl}\frac{L_{g,i,\max}\mathrm{PAR}}{\mathrm{PAR}_{\max}},$$

where $F_{cl}$ is the share of the territory occupied by plant life determined for each vegetation class $L_s$.

The photic share $N$ of a plant species can be ascertained from the following considerations. The plant canopy consists both of leaf matter ($L_g$) participating in photosynthesis and of trunk cellulose $L_s$ (0.076 for forests, 0.05 for meadows and pastures), and a dead proportion $L_d$. Dead cellulose is supposed to remain in the leaf canopy for about a month before it is removed by leaf fall or another means of extraction. Thus, the photic share of cellulose is determined as

$$N = \frac{L_g}{L_T},$$

where $L_T = L_s + L_d + L_g$; $L_d$ is the dead proportion of cellulose $L_{g_{a-1}} - L_{g_{a1}}$, $L_d \geq 0$; $L_g$ is the photic proportion determined from PAR; $L_T$ is the total territory occupied by a plant species; and $L_s$ is the proportion of trunk cellulose.

### 8.5.6 A model of leaf canopy photosynthesis

In the SiB2 model of the biosphere derived by Sellers *et al.* (1996) there were equations to determine the temperature, humidity, and evaporation of plant canopies and three-layer soil covers. Moreover, a model of radiative transfer (soil–leaf canopy–atmosphere) and a model of photosynthesis were synthesized. The model of leaf canopy photosynthesis considered a single-layer plant species, surface layers, and root layers, as well as the re-charging zone. Equations of the SiB2 biospheric model were based on the concept of flux exchange in the soil–leaf canopy–atmosphere system. Notation was introduced for estimates of resistance $r$ and conductivity $g$ in inverse proportion ($r = 1/g$). Flux in this case was determined by analogy with the electrostatic form

$$\mathrm{Flux} = \frac{\text{Potential difference}}{\text{Resistance}}.$$

Sellers *et al.* (1996) considered the fluxes of sensible and latent heat transport involved in the evaporation of water vapor from plant canopies $\lambda E_c$ and surface soil layers, as well as $CO_2$ flux ($A_c$ and $R_{\mathrm{soil}}$ for soil). Potential differences connected with the fluxes were presented as differences in temperatures, pressures, and partial pressures of $CO_2$ for heat transport, latent heat transport, and $CO_2$, respectively. $CO_2$ fluxes ($A_c$ and $R_{\mathrm{soil}}$), flux in latent heat exchange $\lambda E_c$ of the leaf canopy, and its respiration constituent $\lambda E_{ct}$ are of interest to us in the model of photosynthesis.

The model of leaf canopy photosynthesis derived by Collatz *et al.* (1990) for vegetation species of classes $C_3$ and $C_4$ together with the bud model (Ball, 1988) were used to create the SiB2 biospheric model. Photosynthesis and conductivity are closely connected in this model (Sellers *et al.*, 1996).

According to Collatz *et al.* (1991), there are three factors limiting the intensity of plant photosynthesis. These are the efficiency of the photosynthetic fermentation system $w_c$ (the so-called rubisco limitation), the amount of PAR absorbed by cellulose chlorophyll $w_e$, and the ability of a plant species to assimilate and emit to the environment the products of photosynthesis $w_s$. As a result, the value of photosynthesis intensity $A$ is the minimum of these three limitations:

$$A \leq \min(w_c, w_e, w_s),$$

where $A$ is photosynthesis intensity (mol/m$^2$ s); $w_c$c is the rubisco (fermentation) limitation of photosynthesis intensity (mol/m$^2$ s); $w_e$ is the limitation of assimilation as a result of illumination (mol/m$^2$ s); and $w_s$ is the limitation of carbon available for a plant species (mol/m$^2$ s).

The physiological limitation of assimilation $w_c$ is a function of the supply of rubisco to a species—in other words, it is the ability of cellulose to realize biochemical reactions. The fermentation reserves of plants are described in the model by the $V_{\max}$ parameters which indicate the maximum catalytic supply of rubisco taking part in photosynthesis. The $w_c$ value also depends on leaf canopy temperature and soil moisture. Collatz *et al.* (1990) suggest the following expression for $w_c$ for vegetation classes $C_3$ and $C_4$:

$$w_c = \begin{cases} V_m \left[ \dfrac{c_i - \Gamma^*}{c_i + K_c(1 + O_2/K_o)} \right], & \text{for } C_3, \\ V_m, & \text{for } C_4, \end{cases}$$

where $V_m$ is a maximum catalytic supply of rubisco (mol/m$^2$/s); $c_i$ is the partial pressure of $CO_2$ in leaves (Pa); $O_2$ is the partial pressure of $O_2$ in leaves (Pa); $\Gamma^*$ is the point of $CO_2$ compensation (Pa) and equals $O_2/S$; $S$ is a special interaction between $CO_2$ and $O_2$; $K_c$ is the Michaelis–Menten constant for $CO_2$ (Pa); and $K_o$ is the constant of $O_2$ suppression (inhibition).

The $V_m$ parameter is expressed through $V_{\max}$, which is a function of the dependence on soil temperature and moisture. The $S, K_c, K_o$ parameters are functions of temperature.

The limitation of assimilation intensity $w_e$ is described as follows:

$$w_e = \begin{cases} (F_\pi \cdot n)\varepsilon_3(1 - \omega_\pi) \left[ \dfrac{c_i - \Gamma^*}{c_i + 2\Gamma^*} \right], & \text{for } C_3, \\ (F_\pi \cdot n)\varepsilon_4(1 - \omega_\pi), & \text{for } C_4, \end{cases}$$

where $F_\pi$ is a PAR vector (W/m$^2$); $n$ is the normal to the leaf surface; $\varepsilon_{3,4}$ is the quantum efficiency of $CO_2$ assimilation (mol); and $\omega_\pi$ is the dispersion coefficient for PAR.

The third limitation $w_s$ is written as:

$$w_s = \begin{cases} \dfrac{V_m}{2}, & \text{for } C_3, \\ \dfrac{V_m}{2}, & \text{for } C_4, \end{cases}$$

where $p$ is atmospheric pressure (Pa).

Photosynthesis intensity $A$ is a minimum of functions $w_c$, $w_e$, and $w_s$ and is estimated in the simplest way, but observations show that the transition from one limitation to another is not spasmodic. Therefore, Collatz *et al.* (1992) proposed a method of combining the contributions of all three limitations in the form of two equations, to be solved with the least values:

$$\beta_{ce} w_p^2 - w_p(w_c + w_e) + w_e w_c = 0,$$

$$\beta_{ps} A^2 - A(w_p + w_s) + w_p w_s = 0,$$

where $\beta_{ce}$ and $\beta_{ps}$ are correlation coefficients; and $w_p$ is a smoothed minimum between $w_c$ and $w_e$ (mol/m$^2$ s).

Theoretically, $\beta_{ce}$ and $\beta_{ps}$ can have values from unity (lack of relationships between limitations) to zero. Real estimates vary from 0.8 to 0.99.

Pure assimilation of $CO_2$ by a plant $A_n$ is equal to $A_n = A - R_d$, where $R_d$ is the loss of $CO_2$ resulting from respiration connected with the content of carboxylase in cellulose: $R_d = f_d V_m$, where $f_d = 0.015$ for $C_3$, $f_d = 0.025$ for $C_4$.

The connection between Collatz's model of photosynthesis and Ball's semi-empirical model of bud conductivity can be written as

$$g_s = m \frac{A_n}{c_s} h_s p + b,$$

where $g_s$ is bud conductivity (mol/m$^2$ s); $g_s$ (m/s) $= 0.0224 \dfrac{T}{T_f} \cdot \dfrac{p_0}{p} \cdot g_s$ (mol/m$^2$ s); $m$ is an empirical coefficient (9 for vegetation of class $C_3$, 4 for $C_4$, 6 for coniferous forests); $b$ is an empirical coefficient (mol/m$^2$ s or m/s, 0.01 for $C_3$, 0.04 for $C_4$); $h_s$ is relative moisture of leaf surfaces; $c_s$ is the $CO_2$ partial pressure in stems (Pa); $p$ is atmospheric pressure (Pa); $p_0$ is normal atmospheric pressure ($1.013 \times 10^5$ Pa); and $T_f$ is absolute zero temperature (273.16 K).

There is a relationship between the partial pressure of $CO_2$ ($c_i, c_s$) and relative humidity $h_s$. This dependence can be ascertained by analyzing processes taking place at the bud–atmosphere boundary, respective conductivities $g_i$ and $g_s$, pure $CO_2$ flux, and water vapor flux $E_{it}$ as a result of photosynthesis.

This relationship can be expressed as

$$E_{lt} = g_l(e_s - e_a)\frac{\rho c_p}{\lambda\gamma} = g_s(e_i - e_s)\frac{\rho c_p}{\lambda\gamma}, \qquad h_s = \left[\frac{e_s}{e_i}\right],$$

$$A_n = \frac{(c_a - c_s)}{p}\frac{g_l}{1.4} = \frac{(c_s - c_i)}{p}\frac{g_s}{1.6},$$

where $e_a$, $e_s$, and $e_i$ are partial pressures of water vapor in the open air, on the surface, and in the stem, respectively (Pa); $c_p$ is specific heat capacity of air (Joule/kg·K); $\gamma$ is a psychrometric constant (Pa/K); $g_l$ is water vapor conductivity on one side of the leaf (mol/m$^2$ s or m/s, $1/(2rl)$); $E_{lt}$ is respiration intensity (kg/m$^2$ s); $c_a$ and $c_s$ are partial pressures of carbon dioxide outside and inside the stem, respectively (Pa); and $e_i = e^*(T_c)$ (Pa).

The photosynthesis–conductivity model encompasses equations that describe the processes taking place in a bud, and the next step is to integrate the relationships between buds and the leaf canopy to obtain total values of $A_c$ and $g_c$. This was done in Sellers *et al.* (1996) by supposing the vertical distribution of fermentation in the canopy to be connected with the time-averaged distribution of PAR:

$$V_{\max_L} = V_{\max_0} e^{-\bar{k}L},$$

where $V_{\max_L}$, $V_{\max_0}$ are the $V_{\max}$ values at the LAI $= L$ level and at the top of the canopy (LAI $= 0$), respectively (mol/m$^2$ s); and $\bar{k}$ is a time-averaged value of the coefficient of PAR attenuation.

The amount of PAR assimilated by the leaf canopy is:

$$F_\pi \cdot n \approx F_{\pi_0} \left[\frac{G(\mu)}{\mu}\right] e^{-\bar{k}L},$$

where $F_{\pi_0}$ is the value of PAR flux caught by the leaf canopy (W/m$^2$); $G(\mu)$ is a projection of the orientation of canopy leaves towards incoming radiation flux ($\mu$). Sellers *et al.* (1996) established that the use of a time-averaged $\bar{k}$ value instead of the instant $k$ value leads to negligible uncertainties in calculations.

Now, since the $w_c$, $w_e$, and $w_s$ parameters are connected (depending on the vertical coordinate) with the LAI of the leaf canopy by means of an exponent, it is possible to use them to obtain an integral estimate of $A_c$. We have already discussed the difference between bushy and separately growing species ($V < 1$) and the notion of photic share ($N < 1$). Therefore, bearing in mind that photosynthesis intensity in dense thickets can be reduced to the lack of green rubisco-rich plant material, integration can be realized from 0 to $L_T/V$:

$$A_c = A_{n_0} \int_0^{L_T/V} VN\, e^{-\bar{k}L}\, dL = A_{n_0}\Pi, \tag{8.17}$$

where $A_{n_0} = A_n$ for leaves at the top of the canopy (mol/m$^2$ s):

$$A_{n_0} = f_c(V_{\max_0}, \ldots), f_e(F_{\pi_0}, \ldots), f_s(V_{\max_0}, \ldots);$$

$$\Pi = \frac{VN(1 - e^{-\bar{k}L_T/V})}{\bar{k}}; \qquad \Pi \approx \text{PAR}/\bar{k}.$$

The $A_c$ value can be used to determine the respective conductivity of the leaf canopy using the modified expression:

$$g_c = m\frac{A_c}{c_s}h_s p + bL_T,$$

where $h_S$ and $c_S$ are the volume analogs to $h_s$ and $c_s$ of the leaf canopy.

The $g_c$ parameter can be used to estimate respiration intensity $\lambda E_{ct}$:

$$\lambda E_{ct} = \left[\frac{e^*(T_c) - e_a}{1/g_c + 2r_b}\right]\frac{\rho c_p}{\gamma}(1 - W_c),$$

where $e^*(T_c)$ is saturated water vapor pressure at temperature $T_c$ (Pa); $e_a$ is water vapor pressure in an open atmosphere near the leaf canopy (Pa); $\rho$ and $c_p$ are air density and specific heat capacity, respectively (kg/m$^2$, Joule/kg/K); $\gamma$ is a psychrometric constant; and $W_c$ is the proportion of the leaf canopy that is wet.

To complete our description of the equations of the photosynthesis–conductivity model, the following expressions are given to estimate the rubisco (fermentation) supply of cellulose $V_m$:

$$V_m = V_{\max} f_T(T_c) f_W(W_2); \qquad K_c = 30 f_T(T_c); \qquad K_o = 30{,}000 f_T(T_c);$$

$$S = 2{,}600 f_T(T_c);$$

$$f_T(T_c) = \begin{cases} \dfrac{2Q_t}{\{1 + \exp[s_1(T_c - s_2)]\}}, & \text{for } C_3 \text{ and } V_m, \\ \dfrac{2Q_t}{\{1 + \exp[s_1(T_c - s_2)]\} \times \{1 + \exp[s_3(s_4 - T_c)]\}}, & \text{for } C_4 \text{ and } V_m, \\ \dfrac{2Q_t}{\{1 + \exp[s_5(T_c - s_6)]\}}, & \text{for } R_d \text{ and } V_m, \\ 2.1Q_t, & \text{for } K_c, \\ 1.2Q_t, & \text{for } K_o, \\ 0.57Q_t, & \text{for } S; \end{cases}$$

$$Q_t = \frac{(T_c - 298)}{10}; \qquad f_W(W_2) = \frac{1}{\{1 + \exp[0.02(\psi_c - \psi_r)]\}},$$

where $\psi_c$ is the critical water potential (m); $\psi_s$ is the saturation potential for moisture in the soil (m); $B$ is an empirical parameter; $s_l$ is the suppression of high and low temperatures; and $\psi_r = \psi_s W_2^{-B}$ is potential moisture in the root layer of the soil (m).

The physiological processes described by $A_c$ and $g_c$ do not respond instantly to changes in control derivatives. It takes about one minute to stabilize photosynthesis following changes in external conditions, but for conductivity the time step constitutes several minutes. Hence, for time steps longer than one minute photosynthesis intensity $A_c$ can be calculated using relationship (8.17), and the delay to changes in conductivity can be described reliably by the equation:

$$\frac{\partial g_c}{\partial t} = -k_g(g_c - g_{\text{inf}}), \tag{8.18}$$

where $k_g$ is a time constant for the response of cellulose ($0.00113c^{-1}$); $g_{c_{\text{inf}}} = g_c$ for $t \to \infty$; and $g_{c_{\text{inf}}} = m\frac{A_c}{c_s}h_s p + b$.

The solution (8.17) can be written as:

$$g_c = e^{-\bar{k}_g t} g_{c_0} + (1 - e^{-\bar{k}_g t}) g_{c_{\text{inf}}}, \qquad g_{c_0} = g_c \text{ for } t = 0,$$

and change in $g_c$ during time step $\Delta t$ will be:

$$\Delta g_c = g_c - g_{c_0} = (1 - e^{-\bar{k}_g \Delta t})(g_{c_{\text{inf}}} - g_{c_0}).$$

These correlations make it possible to calculate energy fluxes in the multi-layer structure of the atmosphere–plant–soil system. As a result, we can specify the elements used by climate models and assess from spaceborne monitoring data the impact of surface vegetation on the synoptic situation. This is especially important in arid and over-humid zones where the state of vegetation cover depends strongly on small changes in temperature and other meteorological parameters. Another important conclusion from these correlations can be drawn for the effect of vegetation cover on the processes of $CO_2$ assimilation from the atmosphere. Clearly, the use of such models increases the adequacy of global carbon cycle modeling.

### 8.5.7 Modeling production processes in coniferous forests

Coniferous forests are known to play an important role in the global carbon cycle. In this section we shall consider parameterizing the production process in the forest ecosystem using a stand of pine forest of one and the same age proposed and developed by Kirilenko (1990) for practical use in climate models. The model reflects correlations between mean annual and seasonal processes without considering the succession of vegetation communities. It describes the dynamics of carbon and water in the ecosystem using a system of five ordinary differential equations, one partial differential equation, and many algebraic equations reflecting biotic and abiotic processes in coniferous forests.

The model includes the following values: phase derivatives, input (external) and output (internal) derivatives, as well as parameters (constants, coefficients).

The phase derivatives include (kg/m$^2$ of soil):

- carbon held in the structural mass of young pine needles $m_1$ and old ones $m_2$, $m_L = m_1 + m_2$;
- carbon held in the above-ground parts of fully grown trees (branches, boughs, trunks) and growing roots $m_3$;
- carbon held in thin roots $m_4$;
- fund of mobile carbon assimilates $m_5$.

Input variables characterize these environmental conditions:

- incoming solar radiation intensity $E$;
- air temperature $T$ and humidity $H$;
- length of the daylight period $\tau$;
- daily amount of precipitation $P$;
- concentration of carbon dioxide in the atmosphere $C_o$.

Output variables are photosynthesis, average height and diameter of trees, transpiration, respiration, increase in the size of a tree's organs (i.e., self-contained parts with specific vital functions), and other characteristics of the coniferous forest ecosystem calculated in the model.

The biomass of phase variables in the model is measured in carbon units (the respective scheme of carbon fluxes is shown in Figure 8.12). The notation used for dynamic equations is:

- $s_3$ ($m_3$) is trunk area per $m^2$ of soil;
- $\theta_i$ ($\tau_B$) is a function of the distribution of assimilates (i.e., the share of the fund of assimilates spent on increasing the size of the $i$th organ);
- $\tau_B$ is biological time;
- $d_i$ is the share of dead biomass of the $i$th organ ($i = 1, \dots, 4$);
- $F$ is a photosynthesis value;
- $r_i$ is respiration intensity of the growth of the $i$th organ;
- $p_i$ is respiration intensity of the maintenance of the $i$th organ.

To parameterize the process of photosynthesis we use the Chartier formula:

$$F = \frac{\sigma(T_L)\Omega(\tilde{\omega}_S)}{2(r - r_c)} \frac{S_L}{2} \tau (C_0 + r\alpha E - \sqrt{(C_0 - r\alpha E)^2 + 4r_c \alpha E C_0}),$$

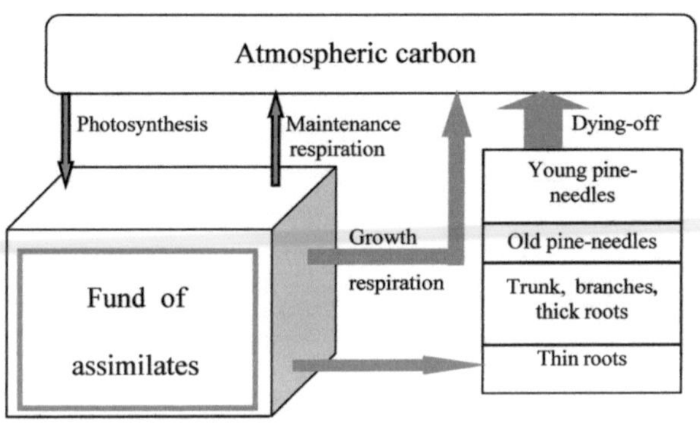


**Figure 8.12.** Conceptual scheme of the carbon cycle in a model of the production process in coniferous forests (Kirilenko, 1990).

where $\sigma(T_L)$ is a function of the dependence of photosynthesis intensity on leaf temperature $T_L$; $S_L = \beta m_L$ is the area of pine needles; $\tilde{\omega}_S$ is the weighted mean soil moisture; $\Omega(\tilde{\omega}_S)$ is a function of the impact of an over-wet root layer of soil on photosynthesis; $r = r_a^c + r_{st}^c + r_c$, $r_a^c$ and $r_{st}^c$ are diffuse resistances for $CO_2$ molecules in the leaf boundary layer and in the leaf stoma, respectively, $r_c$ is the efficient resistance of carboxylizing; $\tau$ is duration of the daylight period; and $\alpha$ is the initial angle of inclination of the light curve of photosynthesis:

$$\tilde{\omega}_S = \int_0^{A_0} \omega_S(z)\, ds(z),$$

where $s(z)$ is a function of the distribution of the root system along the vertical cross-section of the soil layer; and $A_0$ is the maximum capacity of the root layer of the soil.

For conifers $r_a^c \ll r_{st}^c$. Along the tree height the value of photosynthesis per unit mass of pine needles changes negligibly. Therefore, we shall consider some point, averaged over the tree height, for which the intensity of photosynthetically active radiation (PAR) is determined by the Beer–Lambert law:

$$E = \kappa_E E_e \exp(0.5 K_e h S_L H^{-1}),$$

where $E_e$ is the intensity of incoming solar radiation over the forest; $\kappa_E$ is the share of PAR in total solar radiation; and $K_e$ is the coefficient of extinction.

Assuming that the temperature of pine needles coincides with that of the atmosphere ($T_L = T$), we can use the following approximation for the $\sigma(T)$ function:

$$\sigma(T) = \max\left\{0, \frac{T - T_1}{T_0 - T_1} \exp\left[l^{-1} - l^{-1}\left(\frac{T - T_1}{T_0 - T_1}\right)^l\right]\right\},$$

where $l = 1.8$; $T_1$ is the minimum temperature at which photosynthesis still takes place; and $T_0$ is the optimal temperature for photosynthesis.

The $\Omega$ function characterizes the dependence of photosynthesis on soil moisture. During short-term soil over-wetness, photosynthesis does not decrease, and therefore $\Omega(\tilde{\omega}_S) = 1\ \forall \tilde{\omega}_S \in [0, 1]$. We normally use the dependence:

$$\Omega(\tilde{\omega}_S) = \begin{cases} 1, & \tilde{\omega}_S \in (0, \omega_{HB}); \\ (1 - \tilde{\omega}_S)/(1 - \omega_{HB}), & \tilde{\omega}_S \in [\omega_{HB}, 1]. \end{cases}$$

To describe the strategy used to ascertain how assimilates are distributed, let us introduce the function $\theta_i(\tau_B)$, where $\tau_B$ is biological time, and determine the ratio of the sum of efficient temperatures $T_l$ to the mean multi-year sum of efficient temperatures $T_\Sigma$ accumulated during the vegetation period:

$$\tau_B(j) = \sum_{l=1}^{j} T_l / T_\Sigma,$$

where $j$ is the day number.

The $\theta_i$ function can be determined from the expression:

$$\theta_i = a_i^\theta \exp[-b_i^\theta (c_i^\theta - \tau_B)^2] \quad (i = 1, \ldots, 4),$$

where $a_i^\theta = \max \theta_i$; $c_i^\theta$ is the biological time of maximum assimilate flux to the $i$th organ; and the $b_i^\theta$ coefficient determines the steepness of the $\theta_i$ function.

The flux of assimilates at time moment $\tau_B$ into the $i$th organ of the plant is equal to $m_5 \theta_i(\tau_B)$.

Let us assume that respiration used for the growth of the $i$th organ of a tree is proportional to the rate at which its size increases and is equal to $r_i\, dm_i/dt$, and that respiration used to maintain leaves and roots that uptake water is proportional to their mass and is equal to $p_i m_i$ $(i = 1, 2, 4)$. The $p_i$ coefficient depends on temperature:

$$p_i = a_i^p \exp[k_i^p (T - T^p)],$$

where $a_i^p$ is respiration used to maintain temperature $T^p$.

Respiration used to maintain the trunk and thick roots is proportional to the trunk area $s_3$ and is equal to $s_3 p_3 \sqrt{\beta_{st}}$, where $\beta_{st}$ is the share of the trunk mass in $m_3$. The rate at which organs die is considered to be proportional to their masses with coefficients $d_i$ $(i = 1, \ldots, 4)$.

The water cycle in a tree is described by a system of equations that parameterize water motion in the area occupied by the tree, in the soil, in the tree proper, and in its elements. Water is a limiting factor to tree growth by means of stomatal resistance, which affects the photosynthetic activity of plants. The water cycle in forests is the sum of two basic parts—abiotic and biotic—which describe the processes of water motion in the soil and its absorption by the root system, its motion in the plant, and transpiration. There are many models of these processes. The difference between them is determined by different suppositions about soil structure and distribution of the root system. Kirilenko (1990) proposed describing water motion in the soil by an equation that generalizes the Darsi classical model when the soil is considered a homogeneous porous layer:

$$\frac{\partial \omega}{\partial t} = -\frac{\rho}{\rho_S}\left[\frac{K_{\Pi B}}{\Phi_{\Pi B}}\frac{\partial}{\partial z}\left(\mu\omega^n \frac{\partial \omega}{\partial z} + \omega^n\right) + \frac{f(z,t)}{\Phi_{\Pi B}}\right],$$

$$V(z,t) = K_{\Pi B}\left(\mu\omega^n \frac{\partial \omega}{\partial z} + \omega^m\right)$$

with the boundary conditions:

$$\Phi(z_g, t) = 1, \qquad V(0,t) = W(t) - L(t)$$

and the initial condition:

$$\omega(z_g, t_0) = \omega^0(z),$$

where $W(t)$ is precipitation intensity; $L(t)$ is evaporation intensity from the soil surface; $\rho$ is water density; $\rho_S$ is the volume density of soil; $\Phi$ is relative soil moisture; $\phi_{\Pi B}$ is total water capacity of soil; $V$ is the rate of moisture motion in soil; $\mu = n\phi_{\Pi B}/(\rho g)$; $m = 1 + 2n$; $n$ is a constant; $g$ is gravity acceleration;

$\phi_{\Pi B} = \phi_S \omega^n$; $\phi_S$ is the water potential of soil; $K_{\Pi B}$ is conductivity corresponding to soil moisture $\Phi_{\Pi B}$; and $z$ is the vertical coordinate. The $f$ function describes the intensity of water absorption by the root system:

$$f(z,t) = [\xi_0 \bar{\phi} \varphi(z,t) + v(t)]\, dS/dz,$$

where $\varphi(z,t) = \varphi_S(z,t) - \tilde{\varphi}_S(t)$ is deviation of the dimensionless water potential from an average value for the root layer of the soil; $\xi_0 = \xi_k \sigma_0$ is the conductivity of water entry to the root system; $\xi_k = 1/r_k$ is the specific conductivity of water entry to roots; $r_k$ is the specific resistance of water entry to roots; $\sigma_0$ is the total area of tree roots taking up water per unit soil area

$$\varphi_S(z,t) = 1 - \left(\frac{\omega(z,t)}{\bar{\omega}}\right)^{-n}, \qquad \tilde{\varphi}_S(t) = 1 + \tilde{\phi}_S/\bar{\phi},$$

where $\tilde{\varphi}_S$ is the average dimensionless water potential for the root layer of the soil; and $v = \xi_0(\tilde{\phi}_S - \phi_0)$ is the rate of water absorption by roots from the soil.

Water from the root system moves upward along the trunk, through branches into leaves, and evaporates into the atmosphere. Let us assume that the amount of the moisture transpired is equal to that assimilated by the root system and that the rate of motion is equal to $v$. The total impact of coniferous forest on the water cycle in its territory is the sum of the effect of each tree. Practical application of this model needs information about the spatial heterogeneities and distributions of the stand of forest. Applying the model to deciduous and mixed forest systems would require additional units to describe two processes: the transformation of water potentials and the carbon cycle in leaves. Intensive studies are being carried out in this direction at the time of writing.

### 8.5.8 Modeling succession processes in the tundra–taiga system

Predicting the northern boundary of forests has become problematical as a consequence of worldwide anthropogenic processes and the way in which their effects are inexorably propagating towards northern latitudes. Connections between the observed propagation of this boundary and forest fires, reindeer grazing, man-induced change, and climate change have been widely discussed in the scientific literature. Various hypotheses have been put forward about correlations between the natural/technological processes near this boundary. For instance, there is one that forest fires can influence the northward propagation of birch forests and can lead to considerable changes in the nomadic way of life and reindeer migration. In the late 20th century the areas occupied by forest and those in which reindeer breed were found to have changed. For example, in Finland the traditional territory through which reindeer migrate moved 400–500 km south of the northern boundary of the forests (i.e., the boreal zone). Forest boundary change is followed by changes in the soil layer, and this makes predicting the northern boundary of forests very difficult as a result of shifts in biogeocenotic cycles. Moreover, any such prediction cannot be reliable, even when based on how this boundary has changed under natural conditions prehistorically. Therefore, the models and monitoring techniques of ecoinformatics constitute the only reliable means of estimating the dynamics of the northern boundary of forests. Nevertheless, many ecological problems need to be resolved:

- classifying the various elements that make up forest ecosystems at the northern boundary of forests and coming up with reasons for the presence or absence of certain types of trees;
- the role played by diseases in northern forests and their sensitivity to climate change and other external factors;
- the role played by fires, their location, and intensity in changing the structure and configuration of the transfer boundary between forests and tundra;
- the role played by insects (keeping in mind how their propagation and biomass depend on climate change) in the dynamics of the northern boundary of forests;
- the seed dispersal patterns of various plants and their effect on the dynamics of the northern boundary of forests;
- the impact of the dynamics of the northern boundary of forests on animal migration and habitats.

Expected climate warming is likely to result in some plant ecosystems replacing others (exogenic succession). This is especially likely to be so at high latitudes, where taiga transforms into tundra. Bogatyrev (1988) proposed a model to parameterize the transformation processes in the tundra–taiga system and make it possible to reveal some important regularities. He used phase variables for the biomass of moss ($X_1$), vegetation at the grass–bush level ($X_2$), conifers ($X_3$), forest cover ($X_4$), and the dead organic matter of soil ($X_5$). Phase variables were measured in tonnes of carbon per unit area (t/ha), and their spatial distribution was assumed to be uniform. Bogatyrev's carbon flux scheme is given in Figure 8.13.

The system of balance equations used to model transfer processes at the taiga–tundra border according to the scheme in Figure 8.13 is

$$dX_i/dt = R_{0i} - R_{i4} \quad (i = 1, 2, 3);$$

$$dX_4/dt = \sum_{i=1}^{3} R_{i4} - R_{40} - R_{45} - R_4;$$

$$dX_5/dt = R_{45} - R_{50} - R_5,$$

where $R_{ij}$ depend on many parameters (their determination is the principal stage of model formulation). Bogatyrev (1988) proposed the following formulas for these functions:

$$R_{0i} = F_i\varphi_i(T_M)G_i(X_3)f_i(X_5)H_i(W); \qquad R_{i4} = k_iX_i \quad (i = 1, 2);$$

$$R_{03} = F_3\varphi_3(T_M)G_3(X_3)f_3(X_5)H_3(W)\chi(X_3); \qquad R_4 = m_4X_4; \qquad R_5 = m_5X_5;$$

$$R_{34} = k_3X_3[1 + \kappa(T_M - T_A)]^{-1}; \qquad R_{40} = \sigma k_4X_4\varphi_4(T_M);$$

$$R_{50} = k_5X_5\varphi_5(T_M); \qquad R_{45} = (1 - \sigma)k_4X_4\varphi_4(T_M),$$

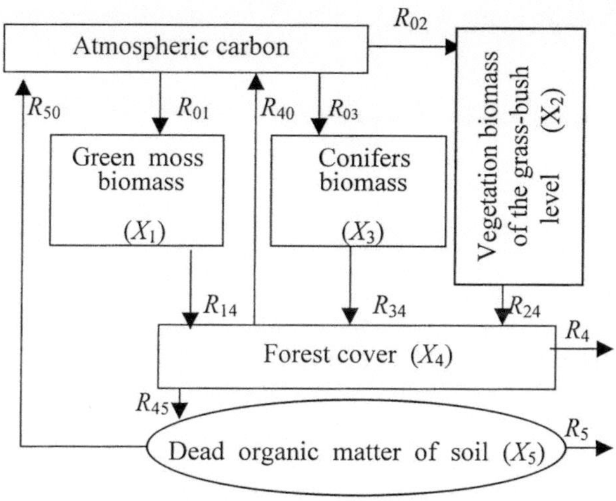


**Figure 8.13.** The carbon cycle in the tundra–taiga system according to the Bogatyrev model.

where $F_i$ are coefficients expressing the dependence of biomass increment on type of biomass (i.e., characteristic indicators of vegetation); $\varphi_i$ assign laws to changes in biomass increment as a function of atmospheric temperature beneath the canopy $T_M$; and $G_i$, $f_i$, and $H_i$ describe deviations in annual biomass increment from a maximum at a given temperature $T_M$ as a result of shortages in light, mineral resources in the soil, and moisture, respectively. The annual production of trees further depends on their biomass (this dependence is represented by $\chi$). The annual production of plants at the grass–bush level and moss is assumed independent of their biomass.

Determine the functions $\varphi_i$, $G_i$, $f_i$, $H_i$, and $\chi$:

$$\varphi_i(T_M) = 1 + (\delta_i/100)(T_M - T_Y) \quad (i = 1, 2); \qquad \varphi_3(T_M) = 1 + (\delta_3/100)(T_M - T_A);$$

$$G_3(X_3) = \exp(-\alpha_i X_3) \quad (i = 1, 2, 3); \qquad f_1(X_5) \equiv 1;$$

$$f_i(X_5) = \exp(-a_i/X_5) \quad (i = 2, 3);$$

$$H_i(W) = \exp\left\{-b_i\left(W - \frac{W_A - W_Y}{\sigma_i}\right)^2\right\} \quad (i = 1, 2, 3); \qquad \chi(X_3) = \left(\frac{X_3}{X_{3A}}\right)^\gamma,$$

where $T_Y$ and $T_A$ are the atmospheric temperatures of tundra and taiga, respectively; $\delta_i$ $(i = 1, 2, 3)$ are indicators of variability in annual biomass increment (in units of increasing temperature beneath the canopy of 1°C); $\alpha_i$, $a_i$, $b_i$, $\kappa$, and $\gamma \in (0, 1)$ are constant coefficients; $W_A$ and $W_Y$ are soil moisture in taiga and tundra, respectively; $k_i$ $(i = 1, 2, 3)$ are proportion coefficients; $\sigma \in (0, 1)$ is the proportion of forest cover transforming into the dead organic matter of soil as a result of decomposition.

To parameterize atmospheric temperature beneath the tree canopy, let us suppose that it cannot be below the environmental temperature $T$, increases with the mass of trees up to some level $X_3 = 0.5\beta_2/\beta_1$, and then decreases:

$$T_M(X_3, T) = T \cdot \begin{cases} f(X_3), & 1 \leq f(X_3) \leq T_A/T; \\ T_A/T, & f(X_3) > T_A/T; \\ 1, & f(X_3) < 1, \end{cases}$$

where

$$f(X_3) = 1 - \beta_1 X_3^2 + \beta_2 X_3;$$

and $\beta_1$ and $\beta_2$ are positive coefficients.

This model can be used as an independent unit of the global model, allows the spatial structure in Figure 1.19 to be changed, and specifies local elements of the water balance in the territory under study (as described in Krapivin and Kondratyev, 2002). To apply the model to real situations, its parameters should only be estimated by concrete attachment to the geographical latitude.

## 8.6 APPLICATIONS OF THE GIMS PERSPECTIVE

### 8.6.1 The greenhouse effect and forest ecosystems

A principal aspect of the anthropogenic impact on the environment is evaluation of the consequences of $CO_2$ emissions into the atmosphere. Published results estimating the greenhouse effect and excess $CO_2$ distribution in the biosphere vary widely, are sometimes contradictory, or too blandly stated. This is a natural consequence of all kinds of simplifications adopted in modeling the global $CO_2$ cycle. The GIMS makes it possible to create an effective monitoring system allowing the spatial distribution of carbon sinks and sources to be estimated in real time.

However, there are problems that first need to be solved before assessing the role played by anthropogenic use of the Earth's surface. In particular, there is the problem of formalized description of the processes involved in changing the structure of Earth's land covers, such as afforestation, forest reconstruction, deforestation, and associated carbon supplies. Understanding meteorological processes as functions of greenhouse gases is the key problem facing humankind in the first decade of the third millennium. Only when adequate knowledge of meteorological phenomena on various spatiotemporal scales under conditions of varying supplies of $CO_2$ and other greenhouse gases can correct and constructive decisions regarding global environmental protection be made.

The dynamics of surface ecosystems depend on interactions between those biogeochemical cycles that suffered significant anthropogenic modification in the last decade of the 20th century: especially, the cycles of carbon, nitrogen, and water. The surface ecosystems in which carbon remains in the living biomass, decomposing organic matter, and the soil play an important role in the global $CO_2$ cycle. Carbon exchange between these reservoirs and the atmosphere takes place through photosynthesis, respiration, decomposition, and burning. Human interference with this process takes place through changing the structure of vegetation covers, pollution of water basin surfaces and soil areas, as well as through direct emissions of $CO_2$ into the atmosphere.

The role played by various ecosystems in forming the carbon supply held in biospheric reservoirs determines the rate and direction of changes to regional meteorological situations and to the global climate. The accuracy with which the level of these changes is assessed depends on the reliability of data comprising the inventory of surface ecosystems.

Existing environmental data show knowledge of the rates and trends of carbon accumulation in surface ecosystems to be rather uncertain. However, it is clear that surface ecosystems are important assimilators of excess $CO_2$. Understanding the details of such assimilation is only possible by modeling the process of plant growth (i.e., considering the effect that the soil's nutrient elements and other biophysical factors have on plant photosynthesis). Therefore, forest ecosystems and associated processes of natural afforestation, forest reconstruction, and deforestation should be studied in detail. In forested areas the volume of the reservoir of $CO_2$ in the atmosphere is a function of the density of the forest canopy, and in a given period of time any change in this volume is determined by the level and character of the dynamic processes involved in the transition of one type of forest into another. The causes of this transition can be natural, anthropogenic, or mixed. Biocenology has tried to create a universal theory of such transitions, but so far has only succeeded in describing observed transitions qualitatively.

### 8.6.2 A new type of environmental monitoring

A GIMS-based method opens up the prospect of developing a new type of environmental monitoring wherein global databases are formed on demand by the GSM. A departure from established global monitoring techniques, based on new information technology, makes it possible to create a global monitoring system in which the GSM provides part of the support needed for the system. The structure of such a system is presented in Figure 1.7. Application of evolutionary computer technology allows the whole system to be categorized as a class of subsystems with variable structures and makes the system adaptable to changes in the natural processes or entities under observation. Furthermore, it becomes possible to detail the natural system under study in the space of the phase variables heterogeneously, and to select a non-uniform geographical grid when sampling the planetary surface (i.e., the arbitrary insertion of significant regularities at the regional level becomes possible).

This system, which processes global information automatically, is aimed at acquiring combined models that reflect climatic and anthropogenic changes in the biosphere in real time and those known historically (or, rather, their simulation). The system relies on a set of models of biospheric processes that allows prompt assessment of the current and future environmental state in this scenario, using software from other units and a scenario of anthropogenic behavior formulated at input. A system that processes global information automatically also has the advantage of formulating the entry and boundary conditions of particular built-in models for the study of regional systems and virtually substitutes them for field measurements of conditions.

As shown by Kondratyev *et al.* (2003b) GIMS technology marks a new approach in biocenology. Figure 8.13 demonstrates this approach, which manifests itself biocenologically as numerous models of concrete biocenoses aimed at finding ways of predicting their dynamics and technologies to control them. In this respect simple models are growing in popularity, as are the numerous modifications of them that introduce into the model factors limiting some functions of biocenosis by means of environmental parameters.

### 8.6.3 A preliminary simulation experiment

Knowledge of the global attenuation of electromagnetic waves by vegetation is important for many different applications. The concept of attenuation is complex. It can be estimated as $\tau_2$ or as the $T_{B,3}(h)/T_{B,3}(0)$ ratio (see Figure 8.9). Detailed calculations call for all sets of electromagnetic wave transformations within the vegetation layer to be considered. Let us now consider a simple and rough scenario of the vegetation layer.

A GIMS-based approach allows EMW attenuation by vegetation of different types to be reconstructed based on fragmentary measurements. One example of this approach is given below. Several vegetation types from Table 8.4 were selected for this purpose. The input to the microwave emission model (MEM) consists of data computed by the GSM (Kondratyev *et al.*, 2004a, b). The microwave surface roughness parameter $g$ was set to zero. Dielectric constants $\varepsilon_i$ were estimated using average data from Basharinov *et al.* (1979), Chukhlantsev and Shutko (1988), Chukhlantsev *et al.* (2003), Karam *et al.* (1992), and Shutko (1986). The gravimetric moisture content of tree canopies and tree trunks were taken as 0.45 and 0.55, respectively. The average height of the canopy was set equal to $0.6h$, where the height, $h$, of the vegetation cover was modified by about $h_0 = 25\,\text{m}$ as a result of seasonal variations according to the biocenotic model (8.11).

Following the plant classification of Table 8.4 and taking into account the quantitative characteristics of forest ecosystems, attenuation can be estimated as given in Table 8.5. The values obtained have a dispersion of about 5.5%. This example certainly shows the possibility of getting operative global information about the spatial distribution of the electromagnetic properties of the vegetation cover. More detailed calculations taking various microwave ranges with $H$ and $V$ polarizations into consideration may be useful in view of the many potential applications.

## 8.7 CLOSING REMARKS

As shown by many authors (Armand *et al.*, 1987, 1997; Kahn *et al.*, 1976; Kondratyev *et al.*, 2002, 2003b, c, 2004a, b; Krapivin and Kondratyev, 2002;

**Table 8.5.** Seasonal distribution of averaged estimations of the crown attenuation (dB) at the wavelength $\lambda = 27$ cm.

| *Vegetation cover (ID from Table 8.4)* | *Season* | | | |
|---|---|---|---|---|
| | *Summer* | *Fall* | *Winter* | *Spring* |
| *F* | 1.4 | 1.3 | 1.2 | 1.3 |
| *D* | 2.6 | 2.4 | 2.3 | 2.4 |
| *G* | 2.8 | 2.6 | 2.4 | 2.6 |
| *R* | 2.9 | 2.7 | 2.6 | 2.8 |
| + | 3.4 | 2.3 | 1.9 | 3.1 |
| *P* | 3.2 | 3.1 | 3.1 | 3.2 |
| *Z* | 3.6 | 3.5 | 3.5 | 3.6 |
| *Y* | 3.5 | 3.4 | 3.4 | 3.5 |
| *N* | 1.4 | 1.4 | 1.4 | 1.5 |
| *I* | 3.4 | 3.3 | 3.2 | 3.3 |

Krapivin and Phillips, 2001; Krapivin and Shutko, 1989, 2002), balanced criteria exist when choosing information covering the hierarchy of causal investigative constraints in the biosphere: the coordination of measurement tolerances, the depth of spatial quantization when describing land covers, the degree of detailing of the biome, etc. As confirmed by researchers evaluating the results of computing experiments, at the empirical level these criteria facilitate selection of the informational structure of a geoinformational monitoring system by representing the hierarchical subordination of models at various levels (Figure 8.14).

It has become clear that a GIMS-based approach marks a new milestone in microwave remote-sensing monitoring by making it possible to combine theoretical and field investigations of the role played by vegetation in attenuating microwave electromagnetic waves. The GIMS database has to encompass a wide range of items—from estimates of model coefficients to spatial distributions of biomes—for practical results to be achieved. The scope of the GIMS will be determined by a set of natural phenomena such as the energy balance in the biosphere directly related to vegetation covers, especially that of forests. This is the reason the latest work of authors in this area foresees the formulation and solution of the following problems:

- development of models to describe the dynamics of SPFs;
- analysis of the possibilities of different monitoring systems to see how good they are at estimating the parameters of vegetation covers;

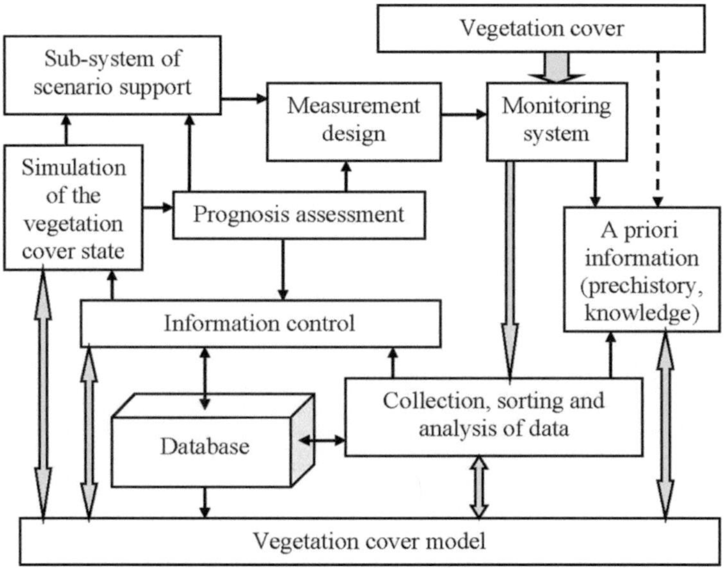


**Figure 8.14.** The adaptive regime of geoinformation monitoring combining a vegetation cover model and experimental measurements.

- theoretical studies of the dependence of microwave radiation propagation in the AVSS on characteristics of the vegetation media;
- synthesis of the GIMS database including the theoretical and experimental estimates of model coefficients, vegetation radiometric characteristics, and the spatial distributions of biomes.

# References


Abrahamson D.E. (1989). *Challenge of Global Warming*. Island Press, Washington, D.C., 376 pp.

Aitkenhead M.J., McDonald A.J.S., Dawson J.J., Couper G., Smart R.P., Billett M., Hope D., and Palmer S. (2003). A novel method for training neural networks for time-series prediction in environmental systems. *Ecological Modelling*, **162**, 87–95.

Alcamo J., Kreileman G.J.J., Krol M.S., and Zuidema G. (1994). Modeling the global society–biosphere–climate system: Description and testing. *Water, Air & Soil Pollution*, **76**(1/2), 1–35.

Alexandrov G. and Oikawa T. (2002). TsuBiMo: A biosphere model of the $CO_2$–fertilization effect. *Climate Research*, **19**, 265–270.

Alexeev V.V., Kryshev I.I., and Sazykina T.G. (1992). *Physical and Numerical Modeling of Ecosystems*. Gidrometeoizdat, St. Petersburg, Russia, 367 pp. [in Russian].

Alfultis M.A. and Martin S. (1987). Satellite passive microwave studies of the Sea of Okhotsk ice cover and its relation to oceanic processes, 1978–1982. *Journal of Geophysical Research*, **92**(C12), 13013–13028.

Aloyan A.E. (2004). Numerical modelling of minor gas constituents and aerosols in the atmosphere. *Ecological Modelling*, **179**, 163–175.

Anderson R.P., Lew D., and Peterson A.T. (2003). Evaluating predictive models of species' distributions: Criteria for selecting optimal models. *Ecological Modelling*, **162**(3), 211–232.

Andreev G.A., Borodin L.F., Zaentsev L.V., and Pubtsov S.N. (1984). On the remote determination of land cover temperature using the polarizational invariant of microwave radiothermal radiation. *Research of the Earth from Space*, **4**, 120–123 [in Russian].

Aota M., Shirasawa K., and Krapivin V.F. (1991a). Simulation model for study of the Okhotsk Sea. *Proceedings of the International Symposium "Engineering Ecology '91", October 22–24, 1991, Zvenigorod, Russia*. IREE RAS Publishing, Moscow, pp. 110–113.

Aota M., Shirasawa K., Krapivin V.F., and Mkrtchyan F.A. (1991b). The system for data processing in Okhotsk Sea monitoring. *Proceedings of the Sixth International Symposium on Okhotsk Sea and Sea Ice, February 3, 1991, Mombetsu, Japan*. Okhotsk Sea & Cold Ocean Research Association, Mombetsu, Japan, pp. 317–318.



V.F. Krapivin and A.M. Shutko, *Information Technologies for Remote Monitoring of the Environment*, 467
Springer Praxis Books, DOI 10.1007/978-3-642-20567-5, © Springer-Verlag Berlin Heidelberg 2012

Aota M., Shirasawa K., Krapivin V.F., and Mkrtchyan F.A. (1992). Simulation model of the Okhotsk Sea geoecosystem. *Proceedings of the Seventh International Symposium on Okhotsk Sea and Sea Ice, February 2–5, 1992, Mombetsu, Japan*. Okhotsk Sea & Cold Ocean Res. Assoc., Mombetsu, Japan, pp. 311–313.

Aota M., Shirasawa K., Krapivin V.F., and Mkrtchyan F.A. (1993). A project of the Okhotsk Sea GIMS. *Proceedings of the Eighth International Symposium on Okhotsk Sea and Sea Ice and ISY/Polar Ice Extent Workshop, February 1–5, 1993, Mombetsu, Japan*. Okhotsk Sea & Cold Ocean Res. Assoc., Mombetsu, Japan, pp. 498–500.

Archer F., Shutko A., Coleman T., Haldin A., Sidorov I., and Novichikhin E. (2006). Microwave remote sensing of land surface from mobile platform: The Alabama 2003–2005 experiment. *Abstracts of the Seventh International IEEE-2006 Geoscience and Remote Sensing Symposium (IGARSS '06), Denver, CO, July 31–August 4, 2006*, pp. 117–121.

Armand N.A., Krapivin V.F., and Mkrtchyan F.A. (1987). *Methods for Data Processing in the Radiophysical Investigation of the Environment*. Science Publishing House, Moscow, 270 pp. [in Russian].

Armand N.A., Krapivin V.F., and Shutko A.M. (1997). GIMS-technology as a new approach to the informational service of environment study. *Problems of the Environment and Natural Resources*, **3**, 31–50 [in Russian].

Arutiunian V.O., Aloyan A.E., Chi J., and Kuznetsov Yu.A. (2004). Numerical modeling of regional transport of gas components taking into consideration of photochemical transformation. *Proceedings of RAS: Physics of the Atmosphere and Ocean*, **49**(4), 501–513 [in Russian].

Asrar G. and Dozier J. (1994). *EOS: Science Strategy fot the Earth Observing System*. AIP Press, Woodbury, NY, 119 pp.

Bacastow R. (1981). Numerical evaluation of the evasion factor. In: B. Bolin (Ed.), *Carbon Cycle Modelling: SCOPE 16*, Wiley, Chichester, U.K., pp. 95–101.

Ball J.T. (1988). An analysis of stomatal conductance. Ph.D. thesis, Stanford University, 89 pp.

Bard S. (1999) Global transport of anthropogenic contaminants and the consequences for the Arctic marine ecosystems. *Marine Pollution Bulletin*, **38**(5), 356–379.

Barenbaum A.S. (2002). *Galaxy, Solar System, The Earth: Subordinate Processes and Evolution*. Geos, Moscow, 393 pp. [in Russian].

Barenbaum A.S. (2004). Mechanism for the formation of gas and oil accumulation. *Annals of Acad. Sci.*, **399**(6), 1–4 [in Russian].

Bartsev S.I., Degermendzhi A.G., and Erokhin D.V. (2003). Global generalized models of carbon dioxide dynamics. *Problems of the Environment and Natural Resources*, **12**, 11–28 [in Russian].

Basharinov A.E., Zotova E.N., Naumov M.I., and Chukhlantsev A.A. (1979). Radiation characteristics of plant covers in microwave range. *Radiotechnics*, **34**(5), 16–20 [in Russian].

Bazilevich N.I. and Rodin L.E. (1967). Diagrammatic map of productivity and biological cycle of basic terrestrial plant types. *Proceedings of All-Union Geographical Soc.*, **99**(3), 190–194 [in Russian].

Belkin I.M. and Carnillon P.C. (2004). Surface thermal fronts of the Okhotsk Sea. *Pacific Oceanography*, **2**(1/2), 6–19.

Benduhn F. and Renard P. (2004). A dynamic model of the Aral Sea water and salt balance. *Journal of Marine Systems*, **47**, 35–50.

Berdnikov S.V., Dombrovsky Yu.A., Ostrovskaya A.G., Prichodko M.V., Titova L.I., and Tjutjunov Yu.V. (1989). Simulation model of basic components of Okhotsk Sea ecosystem. *Marine Hydrophysical Journal*, **3**, 52–57 [in Russian].

Bichele I., Moldau X., and Ross Yu. (1980). *Sub-model for the Assimilation, Distribution and Vegetation Growth under Condition of Water Deficit*. Tartu Astrophysical Observatory Report A-5, Tartu, 22 pp. [in Russian].

Bjorkstrom A. (1979). A model of $CO_2$ in interaction between atmosphere, oceans and land biota. In: B. Bolin (Ed.), *Global Carbon Cycle: SCOPE 13*, Willey, Chichester, U.K., pp. 403–457.

Bobylev L.P., Kondratyev K.Ya., and Johannessen O.M. (Eds.) (2003). *Arctic Environment Variability in the Context of Global Change*, Springer/Praxis, Heidelberg, Germany/Chichester, U.K., 471 pp.

Bodenbender J., Wassmann R., Papen H., and Rennenberg H. (1999). Temporal and spatial variation of sulfur-gas-transfer between coastal marine sediments and the atmosphere. *Atmospheric Environment*, **33**(21), 3487–3502.

Bogatyrev B.G. (1988). Modelling the transition processes on the border of vegetation zones in conditions of anthropogenic climate changes. Ph.D. thesis, Moscow State University, Moscow, 114 pp. [in Russian].

Bondur V.G., Krapivin V.F., and Savinikh V.P. (2009). *Monitoring and Prediction of Natural Disasters*. Scientific World, Moscow, 691 pp. [in Russian].

Bondur V.G., Krapivin V.F., and Potapov I.I. (2010) Methods to reconstruct the aerospace diagnostics of forest ecosystems. *Ecological Systems and Devices*, **3**, 17–27 [in Russian].

Borisov P. M. (2005). *Can Man Change Climate?* Science Publishing House, Moscow, 270 pp. [in Russian].

Borodin L.F. and Krapivin V.F. (1998). Remote measurements of the land cover characteristics. *Problems of the Environment and Natural Resources*, **7**, 38–54 [in Russian].

Borodin L.F., Krapivin V.F., and Kulikov Yu.N. (1987). Air monitoring system in regional exploration of the nature–economic situation. *Problems of Desert Development*, **1**, 80–88 [in Russian].

Borodin L.F., Krapivin V.F., and Long B.T. (1996). Application of GIMS technology to Aral–Caspian aquageosystem monitoring. *Problems of the Environment and Natural Resources*, **10**, 46–61 [in Russian].

Bortnik V.N. and Chistyakova S.P. (Eds.) (1990). *Hydrometeorology and Hydrochemistry of the USSR Seas, Vol. 7: Aral Sea*. Gidrometeoizdat, Leningrad, U.S.S.R., 196 pp. [in Russian].

Bortnik V.N., Lopatina S.A., and Krapivin V.F. (1994). Simulation system to study hydrophysical fields in the Aral Sea. *Meteorology and Hydrology (Moscow)*, **9**, 102–108 [in Russian].

Boschetti L., Brivio P.A., and Gregoire J.M. (2003). The use of Meteosat and GMS imagery to detect burned areas in tropical environments. *Remote Sensing of Environment*, **85**, 78–91.

Bounoua L., Collatz G.J., Los S.O., Sellers P.J., Dazlich D.A., Tucker C.J., and Randall D.A. (2000). Sensitivity of climate to changes in NDVI. *Journal of Climate*, **13**, 2277–2292.

Boysen M. (Ed.) (2000). *Biennial Report 1998 & 1999*. Potsdam Institute for Climate Impact Research, Potsdam, Germany, 130 pp.

Brandhorst W. (1966). Las condiciones oceanográficas y la pesca de anchoa en la Zona Norte de Chile. *En Memoria Primer Seminario Latino Americano sobre el Ocêano Pacífico Oriental*. Universidad Nacional Mayor de San Marcos, Lima, Peru, pp. 17–28.

Bras R.L. (1990). *Hydrology*. Addison-Wesley, New York, 643 pp.

Brebbia C.A. (2004). *Risk Analysis IV*. WIT Press, Southampton, U.K., 400 pp.

Bréon F.-M. (2004). *POLDER Level-2 Products: Data Format and User Munual.* CNES NASDA, Tokyo, 37 pp.

Brucan S. (1984). The global crisis. *International Studies Quarterly*, **28**(1), 97–109.

Bui T. L. (1999). Data processing automatization in systems of geophysical monitoring in Vietnam. Doctoral dissertation, Institute of Radioengineering and Electronics, Russian Academy of Sciences, Moscow, 354 pp.

Bukatova I.L. (1992). *Evolutionary Neurocomputer Technology*. IREE Press, Moscow, 34 pp. [in Russian].

Bukatova I.L. and Makrusev V.V. (2003). *Modern Information Technologies for Management*. RCA Publishing, Moscow, 252 pp. [in Russian].

Bukatova I.L., Kipyatkov V.A., and Sharov A.M. (1991). Simulation evolutionary technology of multi-channel processing of signals. *Problems of Radio Electronics*, **3**, 5–26 [in Russian].

Burkov V.D. and Krapivin V.F. (2009). *Ecoinformatics: Algorithms, Methods, and Technologies*. Moscow State University of Forest Publishing, Moscow, 430 pp. [in Russian].

Canadel I.G., Dickinson R., Hibbard K., Raupach M., and Young O. (Eds.) (2003). *Global Carbon Project: The Science Framework and Implementation* (Report No. 1), Earth System Science Partnership, Canberra, 69 pp.

Cao Van Phuong, Nguyen Boi Khue, Dang Manh Cuong, Nguyen Ngoc Tan, Krapivin V.F., and Mkrtchyan F.A. (2009a). Thunderstorms as interactive component of global ecodynamics. *Proceedings of the A.S. Popov Russian Society for Radiotechnics, Electronics, and Communication. Series: "Engineering Ecology"*, **5**, 9–14.

Cao Van Phuong, Nguyen Boi Khue, Dang Manh Cuong, Nguyen Ngoc Tan, Krapivin V.F., Mkrtchyan F.A., and Soldatov V.Yu. (2009b). Synthesis of geoecological information-modeling systems. *Proceedings of the Popov A.S. Russian Society for Radiotechnics, Electronics, and Communication. Serie: "Engineering Ecology"*, **5**, 18–22.

Charvat K., Alberts M., and Harakova S. (Eds.) (2010) *INSPIRE, GMES and GEOSS Activities, Methods and Tools towards a Single Information Space in Europe for the Environment*. Wireless Information, London, 170 pp.

Chen W., Chen J., and Cihlar J. (2000). An integrated terrestrial ecosystem carbon-budget model based on changes in disturbance, climate, and atmospheric chemistry. *Ecological Modelling*, **135**(1), 55–79

Chernenko I.M. (1987). More on the Aral Sea problem. *Problems of Desert Development (Ashkhabad)*, **4**, 53–57 [in Russian].

Christensen J.H. (1997). The Danish Eulerian hemispheric model: A three-dimensional air pollution model used for the Arctic. *Atmospheric Environment*, **31**(24), 4169–4191.

Chukhlantsev A.A. (1986). Scattering and absorption of microwave radiation by elements of plants. *Radio Engineering and Electronic Physics*, **31**, 1095–1104 [in Russian].

Chukhlantsev A.A. (2002). Microwave attenuation by coniferous branches. *Forest News*, **1**(21), 110–112 [in Russian].

Chukhlantsev A.A. (2006). *Microwave Radiometry of Vegetation Canopies*. Springer-Verlag, Berlin, 287 pp.

Chukhlantsev A.A. and Shutko A.M. (1988). Microwave radiometry of the Earth's surface: Effect of vegetation. *Research of the Earth from Space*, **2**, 67–72 [in Russian].

Chukhlantsev A.A., Shutko A.M., and Golovachev S.P. (2003). The attenuation of electromagnetic waves by vegetation covers. *Journal of Radio Electronics*, **48**(11), 1285–1311 [in Russian].

CIESIN (2011). *Migration and Global Environmental Change: Future Challenges and Opportunities*. Center for International Earth Science Information Network, Ottawa, 129 pp.

Cocknell C., Koeberl C., and Gilmour I. (Eds.) (2006). *Biological Processes Associated with Impact Events*. Springer-Verlag, Berlin, 376 pp.

Collatz G.J., Berry J.A., Farquhar J.A., and Pierce J. (1990). The relationship between the rubisco reaction mechanism and models of leaf photosynthesis. *Plant Cell Environment*, **13**, 219–225.

Collatz G.J., Ball G.J., Grivet J.T., and Berry J.A. (1991). Physiological and environmental regulation of stomatal conductance, photosynthesis and transpiration: A model that includes a laminar boundary layer. *Agric. For. Meteorol.*, **54**, 107–136.

Collatz G.J., Ribas-Carbo M., and Berry J.A. (1992). Coupled photosynthesis stomatal conductance model for leaves of $C_4$ plants. *Australian Journal of Plant Physiology*, **19**, 519–538.

Collatz G.J., Bounoua L., Los S.O., Randall D.A., Fung I.Y., and Sellers P.J. (2000). A mechanism for the influence of vegetation on the response of the diurnal temperature range to changing climate. *Geophysical Research Letters*, **27**(20), 3381–3384.

Corcoran P.P. (Ed.) (2005). *The Earth Charter in Action: Toward a Sustainable World*. KIT Publishing, Amsterdam, 192 pp.

Cracknell A.P., Krapivin V.F., and Varotsos C.A. (eds.) (2009). *Global Climatology and Ecodynamics: Anthropogenic Changes to Planet Earth*. Springer/Praxis, Heidelberg, Germany/Chichester, U.K. 518 pp.

Dalton L.A. and Dougherty E.R. (2011). Bayesian minimum mean-square error estimation for classification error, Part II: Linear classification of Gaussian models. *IEEE Transactions on Signal Processiong*, **59**(1), 130–144.

Degermendzhy A.G., Bartsev S.I., Gubanov V.G., Erokhin D.V., and Shevirnogov A.P. (2009). Forecast of biosphere dynamics using small-scale models. In: A.P. Cracknell, V.F. Krapivin, and C.A. Varotsos (eds.), *Global Climatology and Ecodynamics*. Springer/Praxis, Heidelberg, Germany/Chichester, U.K., pp. 241–300.

Del Frate F., Ferrazzoli P., and Schiavon G. (2003). Retrieving soil moisture and agricultural variables by microwave radiometry using neural networks. *Remote Sensing of Environment*, **84**(2), 174–183.

Demers S., Legendre L., Therriault J.C., and Ingram R.G. (1986). Biological production at the ice–water ergocline. In: J.C.J. Nihoul (ed.), *Marine Interfaces Ecohydrodynamics*. Elsevier, Amsterdam, pp. 31–55.

Demirchian K.S. and Kondratyev K.Ya. (2004). Global carbon cycle and climate. *Proc. of the Russian Geographical Society*, **136**(1), 16–25 [in Russian].

Demirchian K.S., Demirchian K.K., Danilevich Ya.B., and Kondratyev K.Ya. (2002). Global climate warming, energetics and geopolicy. *Annals of RAS: Energetics*, **3**, 221–235 [in Russian].

DeWitt D.P. and Nutter G.D. (eds.) (1989). *Theory and Practice of Radiation Thermometry*. Wiley, New York, 1,137 pp.

DOE (2004). *Our Changing Planet: The U.S. Climate Change Science Program for Fiscal Years 2004 and 2005*. DOE, Washington, D.C., 159 pp.

DOE (2009). *Our Changing Planet: The U.S. Climate Change Science Program for Fiscal Year 2010*. DOE, Washington, D.C., 244 pp.

Dong J., Kaufmann R.K., Myneni R.B., Tucker C.J., Kauppi P.E., Liski J., Buermann W., Alexeyev V., and Hughes M.K. (2003). Remote sensing estimates of boreal and temperate forest woody biomass: Carbon pools, sources, and sinks. *Remote Sensing of Environment*, **84**(3), 393–410.

Dore, S. E., Likas, R., Sadler, D. W., and Karl, D. M. (2003). Climate-driven changes to the atmospheric $CO_2$ sink in the subtropical North Pacific Ocean. *Nature (London)*, **424**(6950), 754–757.

Doronin Yu.P. and Heisin D.E. (1975). *Sea Ice*. Gidrometeoizdat, Leningrad, U.S.S.R., 318 pp. [in Russian].

Edwards P.N. (2010). *A Vast Machine: Computer Models, Climate Data, and the Politics of Global Warming*. MIT Press, Cambridge, MA, 552 pp.

Eliasson B., Riemer P., and Wokaun A. (1999). *Greenhouse Gas Control Technologies*. Elsevier, Amsterdam, 1,095 pp.

Elinson M.I. (1983). On the microchip building-block base for one class of cybernetic systems. *Microelectronics*, **12**(5), 403–412 [in Russian].

Engman E.T. and Chauhan N. (1995). Status of microwave soil moisture measurements with remote sensing. *Remote Sensing of Environment*, **51**(1), 189–198.

EPA (2001). *Non-$CO_2$ Greenhouse Gas Emissions from Developed Countries: 1990–2010* (EPA-430-R-01-007). U.S. Environmental Protection Agency, Washington, D.C., 79 pp.

EPA (2005). *Proceedings of EPA Science Forum 2005: Collaborative Science for Environmental Solutions, May 16–18, 2005, Washington*. U.S. Environmental Protection Agency, Washington, D.C., 144 pp.

Essam J.W. (1980). Percolation theory. Reports on Progress in Physics, 43, 833-912.

Fang H. and Liang S. (2003). Retrieving leaf area index with a neural network method: Simulation and validation. *IEEE Transactions on Geoscience and Remote Sensing*, **41**(9), 2052–2062.

FAO (2010). *The State of Food Insecurity in the World 2010*. Food & Agriculture Organization of the United Nations, Rome, 58 pp.

FAO (2011a). *The State of Food and Agriculture*. Food & Agriculture Organization of the United Nations, Rome, 148 pp.

FAO (2011b). *The State of Food Insecurity in the World 2011*. Food & Agriculture Organization of the United Nations, Rome, 5 pp.

Farquhar G.D., von Caemmerer S., and Berry J.A. (1980). A biochemical model of photosynthetic $CO_2$ fixation in leaves of $C_3$ species. *Planta*, **149**, 78–90.

Ferm M. and Hultberg H. (1999). Dry deposition and internal circulation of nitrogen, sulfur and base cations to a coniferous forest. *Atmospheric Environment*, **33**(27), 4421–4430.

Ferrazzoli P. and Guerriero L. (1996). Passive microwave remote sensing of forests: A model investigation. *IEEE Trans. on Geoscience and Remote Sensing*, **34**(2), 433–443.

Field C.B. (2001). Sharing the garden. *Science*, **294**(5551), 2490–2491.

Field C.B and Raupach M.R. (Eds.) (2004). *Global Carbon Cycle: Integrating Humans, Climate, and the Natural World*. Island Press, Washington, D.C., 584 pp.

Field J.G., Hempel G., and Summerhayer C.P. (Eds.) (2002). *Oceans 2020: Science Trends and the Challenge of Sustainability*. Island Press, Washington, D.C., 296 pp.

Fleishman B.S. (1965). Regular method for solution of games with step-by-step gain functions. *Technical Cybernetics*, **3**, 17–23 [in Russian].

Fleishman B.S. (1970). The state of theory of potential efficiency of complex systems. *Cybernetica*, **13**(4), 199–211.

Fogel L.J., Owens A.J., and Walsh M.J. (1966). *Artificial Intelligence through Simulated Evolution*. Wiley, New York, 230 pp.

Font J., Lagerloef G.S.E., Vine D.L., and Camps A. (2002). Open issues for the soil moisture and ocean salinity (SMOS) satellite mission salinity retrieval. *Proceedings of the First*

*Results Workshop on EuroSTARRS, WISE, LOSAC Compaigns, November 4–6, 2002, Toulouse, France*. ESA, Noordwijk, The Netherlands, pp. 7–16.

Forest C.E., Stone P.H., Sokolov A.P., Allen M.R., and Webster M.D. (2002). Quantifying uncertainties in climate system properties with the use of recent climate observations. *Science*, **295**(5552), 113.

Forrester J.W. (1971). *World Dynamics*. Wright-Allen Press, Cambridge, MA, 189 pp.

Friedi M.A., McIver D.K., Hodges J.C.F., Zhang X.Y., Muchoney D., Strahler A.H., Woodcock C.E., Gopal S., Schneider A., Cooper A. *et al.* (2002). Global land cover mapping from MODIS: Algorithms and early results. *Remote Sensing of Environment*, **83**(1/2), 287–302.

Galitsky V.V. (1985). Horizontal structure and dynamics of an aged vegetation community: Mathematical modeling. In: Yu.M. Svirezhev (Ed.), *Mathematical Modeling of Biogeocenotic Processes*, Science Publishing House, Moscow, pp. 59–69 [in Russian].

Gardner J.S. (2002). Natural hazards risk in the Kullu District, Himachal Pradesh, India. *Geographical Review*, **92**, 172–177.

Garsey E. and McGlade J. (Eds.) (2006). *Complexity and Co-Evolution: Continuity and Change in Socio-economic Systems*. Edward Elgar, Cheltenham, U.K., 219 pp.

Gasiewski A.J. and Kunkee D.B. (1994) Polarized microwave emission from water waves. *Radio Science*, **29**(6), 1449–1466.

George R.E. (2002). *Socioeconomic Democracy: An Advanced Socioeconomic System* (Praeger Studies on the 21st Century). Praeger, Santa Barbara, CA, 328 pp.

Germeyer Yu.B. (1974). *Games with Non-antagonistic Interests*. Science Publishing House, Moscow, 289 pp. [in Russian].

GCOM (2008). *Global Change Observation Mission*. Earth Observation Research Center, Japan Aerospace Exploration Agency, Tokyo, 10 pp.

Givant S. and Halmos P. (2009). *Introduction to Boolean Algebras* (Undergraduate Texts in Mathematics, XIV). Springer-Verlag, Berlin, 574 pp.

Golitsin G.S. (1995). Caspian Sea level rise is the task of diagnosis and forecasting of regional changes in climate. *Atmosphere and Ocean Physics*, **31**(3), 385–391 [in Russian].

Golovachev S.P., Krapivin V.F., Lalaian M.V. Shutko A.M., and Chukhlantsev A.A. (2004). Global mapping of microwave attenuation by the vegetation canopy. *Problems of the Environment and Natural Resources*, **1**, 59–69 [in Russian].

Goody R.M. and Jung Y.L. (1989). *Atmospheric Radiation*. Oxford University Press, New York, 519 pp.

Gorshkov V.G., Gorshkov V.V., and Makarieva A.M. (2000). *Biotic Regulation of the Environment*. Springer/Praxis, Heidelberg, Germany/Chichester, U.K. 364 pp.

Gorshkov V., Makarieva A., Mackey B., and Gorshkov V. (2002). Biological theory and global change science. *Global Change Newsletter*, **48**, 11–14.

Grankov A.G. (1994). Passive radiometric diagnosis of vertical turbulent heat fluxes through the ocean surface from satellites. In: M.A. Donelan (Ed.), *The Air–Sea Interface*. University of Toronto Press, Toronto, Canada, pp. 55–58.

Grankov A.G. and Mil′shin A.A. (1994). The correlation between moisture/water content and temperature of the atmosphere in the near-water layer. *Research of the Earth from Space*, **10**, 78–81 [in Russian].

Grankov A.G. and Mil′shin A.A. (2004) *Interconnection between Radio Emission by the Ocean–Atmosphere System and Heat and Dynamic Processes at the Interface*. Phizmatlit, Moscow, 166 pp. [in Russian].

Grigoryev Al.A. and Kondratyev K.Ya. (2001). *Ecodynamics and Geopolitics, Vol. II: Ecological Catastrophes*. Research Centre for Ecological Safety Press, St. Petersburg, Russia, 688 pp. [in Russian].

Guha A., Jacobs J.M., Jackson T.J., Cosh M.N., Hsu E.-C., and Judge J. (2003). Soil moisture mapping using ESTAR under dry conditions from the Southern Great Plains Experiment (SGP99). *IEEE Transactions on Geoscience and Remote Sensing*, **41**(10), 2392–2397.

Gulyaev Yu.V., Krapivin V.F. and Bukatova I.L. (1987). On the road to evolutionary informatics. *Bulletin of the U.S.S.R. Acad. Sci.*, **11**, 53–61 [in Russian].

Gulyaev Yu.V., Bukatova I.L., Golubeva L.N., and Krapivin V.F. (1989). *Evolution Informatics and Intellectual Special Processors* (Preprint 6(507)). IREE Press, Moscow, 38 pp.

Gulyaev Yu.V., Krapivin V.F., Cherepenin V.A., Chukhlantsev A.A., Shutko A.M., and Potapov I.I. (2009) Radioelectronics methods in environmental investigations. *Problems of the Environment and Natural Resources*, **1**, 3–150 [in Russian].

Haarbrink R., Krapivin V.F., Krisilov A., Krisilov V., Novichikhin E.P., Shutko A., and Sidorov I. (2011). Intelligent data processing in global monitoring for environment and security. ITHEA, Sofia-Kiev, the Ukraine, pp. 189–217.

Hales B., Takahashi T., and Bandstra L. (2005). Atmospheric $CO_2$ uptake by a coastal upwelling system. *Global Biogeochemical Cycles*, **19**(GB1009), doi: 10.1029/2004GB002295, 1–11.

Hansen M.C., DeFries R.S., Townshend J.R.G., Sohlberg R., Dimiceli C., and Carroll M. (2002). Towards an operational MODIS continuous field of percent tree cover algorithm: Examples using AVHRR and MODIS data. *Remote Sensing of Environment*, **83**(1/2), 303–319.

Harvey L.D.D. and Huang Z. (2001). A quasi-one-dimensional coupled climate-change cycle model, 1: Description and behavior of the climate component. *Journal of Geophysical Research*, **106**(C10), 22339–22353.

Hasegawa Y. and Kasagi N. (2001). The effect of Schmidt number on air–water interface mass transfer. *Proceedings of the Fourth International Conference on Multiphase Flow, New Orleans, May 27–June 1, 2001*. University of Nottingham, New Orleans, LA, pp. 296–292.

Hasegawa Y. and Kasagi N. (2005). Turbulent mass transfer mechanism across a contaminated air–water interface. *Proceedings Fourth International Symposium on Turbulence and Shear Flow Phenomena (TSFP-4), Williamsburg, Virginia, June 27–29, 2005*, pp. 971–976.

Holdren J.P. (2003). Environmental change and human condition. *Bulletin American Academy of Arts and Sciences, New York*, **57**(1), 25–31.

Holmberg M., Rankinen K., Johansson M., Forsius M., Kleemola S., Ahonen J., and Syri S. (2000). Sensitivity of soil acidification model to deposition and forest growth. *Ecological Modelling*, **135**(2/3), 311–325

Houghton J.T., Ding Y., Griggs D.J., Noguer M., Van der Linden P.J., Dai X., Maskell K., and Johnson C.A. (2001). *Climate Change 2001: The Scientific Basis* (contribution of Working Group I to the Third Assessment Report of the Intergovernmental Panel Group on Climate Change). Cambridge University Press, Cambridge, U.K., 881 pp.

Hublaryan M.G. (1995). The Caspian Sea phenomenon. *Russian Academy of Sciences Bulletin*, **65**(7), 616–630 [in Russian].

Huesemann M.N., Skillman A.D., and Crecelius E.A. (2002). The inhibition of marine nitrification by ocean disposal of carbon dioxide. *Marine Pollution Bulletin*, **44**(2), 142–148.

Hulst R. van (1979). The dynamics of vegetation: Succession in model communities. *Vegetatio*, **39**(2), 85–96.

Hushon J.M. (Ed.) (1990). *Expert Systems for Environmental Applications*. American Chemical Society, Washington, D.C., 232 pp.

IPCC (2005). Meeting Report, *IPCC Workshop on New Emission Scenarios, June 29–July 1, 2005, Laxenburg, Austria*. WMO/UNEP, Laxenburg, Austria, 47 pp.

IPCC (2007). *Climate Change 2007: The Physical Science Basis*. WMO/UNEP, Geneva, Switzerland, 18 pp.

Islam M.M., Sado K., and Soon C.E. (2005). Analysis of satellite derived sea surface temperature data for South China Sea and Java Sea. *Asian Journal of Geoinformatics*, **5**(3), 12–23.

Itoh M., Ohshima K. I., and Wakatsuchi M. (2003). Distribution and formation of Okhotsk Sea intermediate water: An analysis of isopycnal climatological data. *Journal of Geophysical Research*, **108**(C8), 3258–3276.

Ivanov B.V. and A.P. Makshtas (1990). *Quasi-stationary Zero-measure Model of Arctic Ices* (Proceedings of Arctic and Antarctic Institute No. 420). Gidrometeopress, Leningrad, U.S.S.R., pp. 18–31 [in Russian].

Ivanov-Rostovtsev A.G., Kolotilo L.G., Tarasiuk Yu.F., and Sherstiankin P.P. (2001). *Self-Organization and Self-Regulation of Natural Systems*. Russian Geographical Society, St. Petersburg, Russia, 216 pp. [in Russian].

Justice C.O., Bailey G.B., Maiden M.E., Rassol S.I., Strebel D.E., and Tarpley J.D. (1995). Recent data and information system initiatives for remotely sensed measurements of the land surface. *Remote Sensing of Environment*, **51**(1), 235–244.

Kadukin A.I. and Klige R.K. (1991). The water balance of the Caspian Sea and Aral Sea. In: G. Schiller, R. Lemmelä, and M. Spreafica (Eds.), *Hydrology of Natural and Manmade Lakes*. International Association of Hydrological Sciences, Oxford, U.K., Vol. 206, pp. 55–60.

Kahl J.D., Harris J.M., Herbert G.A., and Olson M.P. (1988). Intercomparison of long-range trajectory models applied to Arctic haze. In: H. van Dop (Ed.), *Air Pollution Modeling and Its Applications* (VII, 13). Plenum Press, New York, pp. 175–185.

Kahn H., Brown W., and Martel L. (1976). *The Next 200 years: A Scenario for America and the World*. Morrow. New York, 241 pp.

Karam M.A., Fung A.K., Lang R.H., and Chauhan N.S. (1992). A microwave scattering model for layered vegetation. *IEEE Transactions on Geoscience and Remote Sensing*, **30**(4), 767–784.

Karimova G.U. and Chukanin K.I. (1988). A scheme for pollutant transport in the Arctic troposphere. In: A.I. Voskresensky (Ed.), *Monitoring the Arctic Climate*. Gidrometeoizdat, Leningrad, U.S.S.R., pp. 168–180.

Karpychev Yu.A. (2007). Variations in the sedimentation in Kara Bogaz Gol Bay related to sea level fluctuations during the novocaspian time. *Oceanology*, **47**(6), 857–864.

Kawasaki Y. and Kono T. (1993). Water exchange between the Okhotsk Sea and Pacific Ocean through the middle of Kuril islands. *Proceedings of Eighth International Symposium on Okhotsk Sea and Sea Ice and ISY/Polar Ice Extent Workshop, February 1–5, 1993, Mombetsu, Japan*. Okhotsk Sea & Cold Ocean Research Association, Mombetsu, Japan, pp. 60–63.

Keeling C.D. and Bacastow R.B. (1977). Impact of industrial gases on climate. In: R. Revelle and W. Munk (Eds.), *Energy and Climate*. National Academy of Sciences, Washington, D.C., pp. 72–95.

Kelley J.J. and Gosink T. (1992). *The Arctic Environment–Air–Sea–Land Exchange of Trace Gases* (Report CP 92-7). University of Alaska Fairbanks, Fairbanks, AK, 29 pp.

Kelley J.J., Krapivin V.F., and Vilkova L.P. (1992a). A model concept to estimate the role played by Arctic ecosystems in the global carbon budget. *Proceedings of International Symposium "Problems of Ecoinformatics", December 12–18, 1992, Zvenigorod, Russia.* IREE Press, Moscow, pp. 19–20 [in Russian].

Kelley J.J., Rochon G.L., Novoselova O.A., Krapivin V.F., and Mkrtchyan F.A. (1992b). Global geo-eco-information monitoring. *Proceedings of the International Symposium "Problems of Ecoinformatics", December 12–18, 1992, Zvenigorod, Russia.* IREE Press, Moscow, pp. 3–7 [in Russian].

Kelley J.J., Krapivin V.F., and Popovich P.R. (1999a). Problems in Arctic environment monitoring. *Problems of the Environment and Natural Resources*, **6**, 32–40 [in Russian].

Kelley J.J., Krapivin V.F., and Popovich P.R. (1999b). Problems in Arctic monitoring. *Problems of the Environment and Natural Resources*, **1**, 2–6 [in Russian].

Kiefer D.A. and Mitchel B.G. (1983). A simple, steady state description of phytoplankton growth based on absorption cross section and quantum efficiency. *Limnol. Oceanogr.*, **28**, 770–776.

Kiehl J., Hack J., Gent P., Large W., and Blackman M. (2003). *Community Climate System Model Strategic Business Plan.* NCAR, Washington, D.C., 28 pp.

Kiehl J.T. and Gent P.R. (2004). The community climate system model: Version Two. *J. Climate*, **17**, 3666–3682.

Kirdiashev K.P., Chukhlantsev A.A., and Shutko A.M. (1979). Microwave radiation of the Earth's surface in the presence of vegetation cover. *Radio Eng. Electron. Phys. (Engl. Transl.)*, **24**, 256–264.

Kirilenko A. P. (1990). Numerical modeling of the production process and of the water cycle of forest ecosystems. Ph.D. thesis, Computing Center of RAS, Moscow, 151 pp. [in Russian].

Klein E. (2008). *GIGAS: GEOS, INSPIRE, and GMES an Action in Support*. Institute for Computer Graphics Research, Darmstadt, Germany, 13 pp.

Klimov V.V., Kovalev V.I., Krapivin V.F., Mkrtchyan F.A., and Nitu C. (1999). An expert system to diagnose pollutant spills on the water surface. *Proceedings of International Conference on Control Systems and Computer Science, Bucharest, May 3–5, 1999.* Politechnik University of Bucharest, Bucharest, pp. 277–283.

Klyuev V.V. (Ed.) (2000). *Ecological Diagnostics*. Znanie, Moscow, 495 pp. [in Russian].

Kochergin V.P. (1978). *Theory and Methods for Calculation of Oceanic Currents.* Science Publishing House, Moscow, 126 pp.

Kochetov S.V. (1995). *Estimation of Heat Exchange between the Atmosphere and Ocean across the Fine Ice of the Arctic Basin* (Proceedings of Arctic and Antarctic Institute No. 389), Leningrad, Russia, pp. 11–15 [in Russian].

Kohler M.A. and Richards M.A. (1962). Multicapacity basin accounting for predicting runoff from storm precipitation. *Journal of Geophysical Research*, **67**(13), 5187–5197.

Kondratyev K.Ya. (1990). The biosphere and climate. *Atmósfera*, **3**, 165–182.

Kondratyev K. Ya. (1998a). Ecological risk: Real and hypothetical. *Bulletin of the Russian Geographical Society*, **130**(3), 13–24 [in Russian].

Kondratyev K.Ya. (1998b). *Multidimensional Global Change.* Wiley/Praxis, Chichester U.K. 771 pp.

Kondratyev K.Ya. (1999). *Ecodynamics and Geopolicy, Vol. 1: Global Problems.* St. Petersburg Sci. Center of RAS, St. Petersburg, 1,036 pp. [in Russian].

Kondratyev K.Ya. (2000a). Global changes on the border of two millennia. *Herald of Russian Academy of Sciences*, **70**(9), 788–796 [in Russian].

Kondratyev K.Ya. (2000b). Research of the Earth from space: Scientific plane EOS. *Research of the Earth from Space*, **3**, 82–91 [in Russian].

Kondratyev K.Ya. and Fedchenko P.P. (1992). *Precise Structure of the Sun's Spectrum and Its Role in Biospheric Evolution*. Propo Press, St. Petersburg, Russia, 40 pp. [in Russian].

Kondratyev K.Ya. and Galindo I. (2001). *Global Change Situations: Today and Tomorrow*. Universidad de Colima, Colima, Mexico, 164 pp.

Kondratyev K.Ya. and Johannessen O.M. (1993). *Arctic and Climate*. Propo Press, St. Petersburg, Russia, 139 pp. [in Russian].

Kondratyev K.Ya. and Krapivin V.F. (2001a). An expert system to control the greenhouse effect. *Problems of the Environment and Natural Resources*, **6**, 23–39 [in Russian].

Kondratyev K.Ya. and Krapivin V.F. (2001b). Biocomplexity and global ecological monitoring. *Research of the Earth from Space*, **1**, 1–8 [in Russian].

Kondratyev K.Ya. and Krapivin V.F. (2001c). Land ecosystems and their global dynamics. *Problems of the Environment and Natural Resources*, **5**, 24–38 [in Russian].

Kondratyev K.Ya. and Krapivin V.F. (2001d). Global dynamics of basic land ecosystems. *Research of the Earth from Space*, **4**, 3–12 [in Russian].

Kondratyev K.Ya. and Krapivin V.F. (2004). Monitoring and prediction of natural disasters. *Il Nuovo Cimento*, **27**(6), 657–672.

Kondratyev K.Ya. and Losev K.S. (2002). Present problems facing global civilization development and possible outlooks. *Earth Research from Space*, **2**, 3–23 [in Russian].

Kondratyev K.Ya. and Varotsos C.A. (2000). *Atmospheric Ozone Variability: Implications for Climate Change, Human Health, and Ecosystems*. Springer/Praxis, Heidelberg, Germany/Chichester, U.K., 758 pp.

Kondratyev K.Ya., Fedchenko P.P., Kozoderov V.V., and Topchiev A.G. (1990). *Biosphere: Methods and Results of Remote Sensing*. Science Publishing House, Moscow, 223 pp. [in Russian].

Kondratyev K.Ya., Krapivin V.F., and Phillips G.W. (2002). *Global Environmental Change: Modelling and Monitoring*. Springer-Verlag, Berlin, 316 pp.

Kondratyev K.Ya., Krapivin V.F., and Phillips G.W. (2003a). Arctic Basin pollution dynamics. In: L.P. Bobylev, K.Ya. Kondratyev, and O.M. Johannessen (Eds.), *Arctic Environment Variability in the Context of Global Change*. Springer/Praxis, Heidelberg, Germany/Chichester, U.K., pp. 309–362.

Kondratyev K.Ya., Krapivin V.F., and Savinykh V.P. (2003b). *Outlook for Civilization Development: Multidimensional Analysis*. Logos, Moscow, 574 pp. [in Russian].

Kondratyev K.Ya., Krapivin V.F., and Varotsos C.A. (2003c). *Global Carbon Cycle and Climate Change*. Springer/Praxis, Heidelberg, Germany/Chichester, U.K., 372 pp.

Kondratyev K.Ya., Losev K.S., Ananicheva M.D., and Chesnokova I.V. (2003d). *Scientific–Natural Bases for Life Sustainability*. Institute of Geography, Moscow, 340 pp. [in Russian].

Kondratyev K.Ya., Krapivin V.F., Savinykh V.P., and Varotsos C.A. (2004a). *Global Ecodynamics: A Multidimensional Analysis*. Springer/Praxis, Heidelberg, Germany/Chichester, U.K., 658 pp.

Kondratyev K.Ya., Losev K.S., Ananicheva M.D., and Chesnokova I.V. (2004b). *Stability of Life on Earth*. Springer/Praxis, Heidelberg, Germany/Chichester, U.K., 162 pp.

Kondratyev K.Ya., Krapivin V.F., and Varotsos C.A. (2006). *Natural Disasters as Interactive Components of Global Ecodynamics*. Springer/Praxis, Heidelberg, Germany/Chichester, U.K., 574 pp.

Korgenevsky A.V., Krapivin V.F., and Cherepenin V.A. (1989). Modelling the global processes of magnetosphere. In: E.P. Novichikhin (Ed.), *Methods of Informatics in Radiophysical Investigations of Environment*. Science Publishing House, Moscow, pp. 25–43 [in Russian].

Kram P. (1999). Application of the forest–soil–water model. *Ecological Modelling*, **120**(1), 9–30.

Kramer H.J. (1995). *Observation of the Earth and Its Environment*. Springer-Verlag, Berlin, 832 pp.

Krapivin V.F. (1969). On approximate solutions of initial value problems for integro-differential equations with quasilinear differential operator and generalized Volterra operator. *Casopis pro pestovani matematiky (Praha)*, **94**, 21–33.

Krapivin V.F. (1972). *Game Theory Methods for the Synthesis of Complex Systems in Conflict States*. Soviet Radio Press, Moscow, 192 pp. [in Russian].

Krapivin V.F. (1978). *Theory of Complex System Survivability*. Science Publishing House, Moscow, 248 pp.[in Russian].

Krapivin V.F. (1993). Mathematical model for global ecological investigations. *Ecological Modelling*, **67**, 103–127.

Krapivin V.F. (1995). Simulation model for the investigation of pollution dynamics in the Arctic basin. *Oceanology (Moscow)*, **35**(3), 366–375 [in Russian].

Krapivin V.F. (1996). Estimation of the Peruvian current ecosystem by a mathematical model of the biosphere. *Ecological Modelling*, **91**, 1–14.

Krapivin V. F. (2000a). The greenhouse effect and the global biogeochemical cycle of carbon dioxide. *Problems of the Environment and Natural Resources*, **2**, 2–18 [in Russian].

Krapivin V.F. (2000b). Radiowave ecological monitoring. In: V.V. Klyuev (Ed.), *Ecological Diagnostics*. Znanie, Moscow, pp. 295–311 [in Russian].

Krapivin V.F. and Kelley J.J. (2009) Model based method for the assessment of global change in the nature–society system. In: A.P. Cracknell, V.F. Krapivin, and C.A. Varotsos (Eds.), *Global Climatology and Ecodynamics*. Springer/Praxis, Heidelberg, Germany/Chichester, U.K., pp. 133–184.

Krapivin V.F. and Klimov V.V. (1995). Valuation of convergence of "physical mixture" strategies in matrix games. *Theory and Control Systems*, **6**, 209–217 [in Russian].

Krapivin V.F. and Klimov V.V. (1997). Stable strategies in games with gain functions $M(x-y)$. *Methods of Cybernetics and Informational Technologies*, **2**, 36–45 [in Russian].

Krapivin V.F. and Kondratyev K.Ya. (2002). *Global Changes of the Environment*. St. Petersburg University, St. Petersburg, Russia, 724 pp. [in Russian].

Krapivin V.F. and Mkrtchyan F.A. (1991). Applications in study of environment. *Proceedings of the Eighth International Conference on Control Systems and Computer Science, May 22–25, 1991, Bucharest*. Polytechnical Institute, Bucharest, pp. 49–56.

Krapivin V.F. and Mkrtchyan F.A. (2009). Remote sensing radiometry technology for the Okhotsk Sea ecosystem biocomplexity assessment. *North Pacific Marine Science Organization Scientific Report*, **36**, 219–223.

Krapivin V.F. and Nazaryan N.A. (1995). *Game Theory Approach to Biosphere Survivability*. IREE Press, Moscow, 28 pp. [in Russian].

Krapivin V.F. and Nazaryan N.A. (1997). A numerical model to study the global sulfur cycle. *Numerical Modeling*, **9**(8), 36–50 [in Russian].

Krapivin V.F. and Phillips G.W. (2001). A remote sensing-based expert system to study the Aral–Caspian aquageosystem water regime. *Remote Sensing of Environment*, **75**, 201–215.

Krapivin V.F. and Potapov I.I. (2001). Algorithms for the reconstruction of two-dimensional image fragments in monitoring problems. *Problems of the Environment and Natural Resources*, **6**, 16–23 [in Russian].

Krapivin V.F. and Potapov I.I. (2002). *Methods of Ecoinformatics*. ARISTI, Moscow, 496 pp. [in Russian].

Krapivin V.F. and Shutko A.M. (1989). Observation and prognosis of the state of environmental resources, ecological and meteorological situations by geoinformational monitoring system. *Proceedings of the Fourth International Symposium on Okhotsk Sea and Sea Ice, February 5–7, 1989, Mombetsu, Japan*. Okhotsk Sea & Cold Ocean Research Association, Mombetsu, Japan, pp. 1–5.

Krapivin V.F. and Shutko A.M. (2002). Investigations in microwave monitoring of land covers. *Problems of the Environment and Natural Resources*, **4**, 44–53 [in Russian].

Krapivin V.F. and Varotsos C.A. (2007). *Globalization and Sustainable Development: Environmental Agendas*. Springer/Praxis, Heidelberg, Germany/Chichester, U.K., 304 pp.

Krapivin V.F. and Varotsos C.A. (2008). *Biogeochemical Cycles in Globalization and Sustainable Development*. Springer/Praxis, Heidelberg, Germany/Chichester, U.K., 562 pp.

Krapivin V.F. and Vilkova L.P. (1990). Model estimation of excess $CO_2$ distribution in biosphere structure. *Ecological Modelling*, **50**, 57–78.

Krapivin V.F., Svirezhev Yu.M., and Tarko A.M. (1982). *Mathematical Modeling of Global Biosphere Processes*. Science Publishing House, Moscow, 272 pp. [in Russian].

Krapivin V.F., Shutko A.M., Strelkov G.M., and Loskutov V.S. (1990). The modeling of ecological situation in upper sea water layer for climate conditions in northern regions. *Proceedings of the Fifth International Symposium on Okhotsk Sea and Sea Ice, February 4–6, 1990, Mombetsu, Japan*. Okhotsk Sea & Cold Ocean Research Association, Mombetsu, Japan, pp. 243–247.

Krapivin V.F., Mkrtchyan F.A., Nitu C., Petrache Gh., Tertisco M., and Graur A. (1991). Multidirectional data processing automatization for remote study of environment. *Proceedings of the Eighth International Conference on Control Systems and Computer Science, May, 22–25, 1991, Bucharest*. Polytechnical Institute, Bucharest, Vol. 2, pp. 57–62.

Krapivin V.F., Bui T.L., Rochon G.L., and Hicks D.R. (1996). A global simulation model as a method for estimation of the role of regional area in global change. *Proceedings of Second Hô Chi Minh City Conference on Mechanics, September 24–25, 1996*. Institute of Applied Mechanics, Hô Chi Minh City, pp. 68–69.

Krapivin V.F., Cherepenin V.A., Nazaryan N.A., Phillips G.W., and Tsang F.Y. (1997a). Simulation model for radionuclides transport in the Angara–Yenisey river system. P*Problems of the Environment and Natural Resources*, **2**, 41–58.

Krapivin V.F., Bui T.L., Dean C., Nguyen M.N., Rochon G.L., and Hicks D.R. (1997b). System of survey and simulation for air pollution over large industrial regions. *Proceedings of the IASTED International Conference "Modelling, Simulation and Optimization" (MSO '97), August 11–13, 1997, Singapore*. IASTED/Acta Press, Anaheim, CA, pp. 307–311.

Krapivin V.F., Cherepenin V.A., Phillips G.W., August R.A., Pautkin A.Yu., Harper M.J., and Tsang F.Y. (1998a) An application of modeling technology to the study of radio-nuclear pollutants and heavy metals dynamics in the Angara–Yenisey river system. *Ecological Modelling*, **111**, 121–134.

Krapivin V.F., Michalev M., Mkrtchyan F.A., Nitu C., and Shutko A.M. (1998b). A project of intelligent information technology and its application to agricultural system monitoring. *Proceedings of the Third International Symposium on Ecoinformatics Problems, December 8–9, 1998*. Institute of Radioengineering and Electronics, Moscow, pp. 6–12. [in Russian].

Krapivin V.F., Shutko A.M., Chukhlantsev A.A., Golovachev S.P., and Phillips G.W. (2006). GIMS-based method for vegetation microwave monitoring. *Environmental Modelling and Software*, **21**, 330–345.

Krapivin V.F., Nitu C., and Soldatov V.Yu. (2007) An adaptive geoinformation technology based approach to the monitoring and prediction of natural disasters. *Proceedings of the 16th International Conference on Control Systems and Computer Science, May 22–25, 2007, Bucharest*. Printech, Bucharest, pp. 450–455.

Krapivin V.F., Nitu C., and Soldatov V.Yu. (2010). Synthesis of geoecological information-modeling systems. *Scientific Bulletin of Electrical Engineering Faculty, Valahia University of Targoviste, Romania*, **10**(1), 101–105.

Kuksa V.I. (1994). *Southern Seas under Anthropogenic Stress*. Hydrometeoizdat, St. Petersburg, Russia, 369 pp. [in Russian].

Kump L.R., Kasting J.F., and Crane R.G. (2003). *Earth System*. Prentice Hall, Englewood Cliffs, NJ, 432 pp.

Kutuza B.G. (1977). Measurement of atmospheric radio-brightness polarization during rainfall at the 2.25 cm wavelength. In: K.Ya. Kondratyev (Ed.), *Radiophysical Investigations of the Atmosphere*. Hydrometeoizdat, Leningrad, U.S.S.R., pp. 201–204 [in Russian].

Kutuza B.G., Zagorin G., Hornbostel A., and Schroth A. (1998). Physical modeling of passive polarimetric microwave observations of the atmosphere with respect to the third Stokes parameter. *Radio Science*, **33**(3), 677–695.

Kutuza B., Shutko A., Plushchev V., Ramsey E., Logan B., DeLoach S., Haldin A., Novichikhin E., Sidorov I., Manakov V. *et al.* (2000). Advantages of synchronous multi-spectral SAR and microwave radiometric observations of land covers from aircraft platforms. *Proceedings of EUSAR '2000, Third European Conference on Synthetic Aperture Radar, May 23–25, 2000, Munich, Germany*, pp. 663–666.

Kuzmin P.O. (1957). Hydrological investigations of land waters. *Int. Assoc. Sci. Hydrol.*, **3**, 468–478.

Lai Ch.T., Katul G., Oren R., Ellsworth D., and Schäfer K. (2000). Modeling $CO_2$ and water vapor turbulent flux distributions within a forest canopy. *Journal of Geophysical Research*, **105**(D21), 26333–26351.

Lai T.L. (2001). Sequential analysis: Some classical problems and new challenges. *Statistica Sinica*, **11**, 303–408.

Lal R. (2008). Sequestration of atmospheric $CO_2$ in global carbon pools. *Energy and Environmental Science*, **1**, 86–100.

Lal R. and Stewart B.A. (Eds.) (1994). *Soil Process and Water Quality*. CRC Press, Boca Raton, FL, 285 pp.

Lal R., Follett R.F., and Stewart B.A. (Eds.) (1998). *Soil Processes and the Carbon Cycle*. Springer/CRC Press, Berlin, 609 pp.

Lebedev S.A. and Kostianoy A.G. (2005). *Satellite Altimetry of the Caspian Sea*. Sea Publishing, Moscow, 366 pp. [in Russian].

Legendre L. and Demers S. (1985). Auxilary energy, ergoclines and aquatic biological production. *Naturaliste canadien*, **112**, 5–14.

Legendre L. and Krapivin V.F. (1992). Model for vertical structure of phytoplankton community in Arctic regions. *Seventh International Symposium on Okhotsk Sea and Sea*

*Ice, February 2, 1992, Mombetsu, Japan* (Abstracts). Okhotsk Sea & Cold Ocean Research Association, Mombetsu, Japan, pp. 314–316.

Legendre P. and Legendre L. (1998). *Numerical Ecology*. Elsevier, Amsterdam, 853 pp.

Leimbach M. and Toth F.L. (2003). Economic development and emission control over the long term: The ICLIPS aggregated economic model. *Climate Change*, **56**(1/2), 139–165.

Lieth H. (1985). A dynamic model of the global carbon flux through the biosphere and its relations to climatic and soil parameters. *International Journal of Biometeorology*, **29**(2), 17–31.

Likens G.E., Bormann F.H., and Johnson N.M. (1981). Interactions between major biogeochemical cycles in terrestrial ecosystems. In: G.E. Likens (Ed.), *Some Perspectives of the Major Biogeochemical Cycles*. Wiley, New York, pp. 93–112.

Lin D.-L., Sakoda A., Shibasaki R., Goto N., and Suzuki M. (2000). Modelling a global biogeochemical nitrogen cycle in terrestrial ecosystems. *Ecological Modelling*, **135**(1), 89–110.

Liou Y.A., Pavelyev A.G., Matyugov S.S., Yakovlev O.I., and Wickert J. (2010). *Radio Occultation Method for Remote Sensing of the Atmosphere and Ionosphere*. InTech, Rijeka, Croatia, 169 pp.

Logofet D.O. (2002). Matrix population models: Construction, analysis and interpretation. *Ecological Modelling*, **148**(3), 307–310.

Lomborg B. (2001). *The Sceptical Environmentalist: Measuring the Real State of the World.* Cambridge University Press, Cambridge, U.K., 496 pp.

Lomborg B. (ed.) (2004). *Global Crisis, Global Solutions*. Cambridge University Press, Cambridge, U.K., 670 pp.

Lovett R.A. (2002). Rain might be leading carbon sink factor. *Science*, **296**(5574), 1787.

Luecken D.J., Berkowitz C.M., and Easter R.C. (1991). Use of a three-dimensional cloud–chemistry model to study the transatlantic transport of soluble sulfur species. *Journal of Geophysical Research D*, **96**(12), 22477–22490

Lyapunov A.A. (1971). Constructing a mathematical model of the balance relationships in ecosystems of tropical ocean waters. In: M.E. Vinogradov (Ed.), *Functioning of Pelagic Communities of Tropical Areas of the Ocean*. Science Publishing House, Moscow, pp. 13–24 [in Russian].

Maguire D.J., Goodchild M.F., and Rhind D.W. (Eds.) (1991). *Geographical Information Systems*, Vols. 1 and 2. Longman Scientific & Technical, New York, 649 and 447 pp.

Maki M., Ishihara M., and Tamura M. (2004). Estimation of leaf water status to monitor the risk of forest fires by using remotely sensed data. *Remote Sensing of Environment*, **90**, 440–450.

Marchuk G.I. and Aloyan A.E. (2009). Dynamics and kinetics of gas admixtures and aerosols in the atmosphere and their significance for the biosphere. *Biosphere*, **1**(1), 48–57 [in Russian].

Marchuk G.I. and Kondratyev K.Ya. (1992). *Priorities of Global Ecology*. Science Publishing House, Moscow, 262 pp. [in Russian].

Marshall T.P. (2001). Birth of the Fujita Scale. *Storm Track*, **24**(3), 6–10.

McCarl B.A. and Schneider U.A. (2001). Greenhouse gas mitigation in US agriculture and forestry. *Science*, **294**(5551), 2481–2482.

McCauley L.L. and Meier M.F. (Eds.) (1991). *Arctic System Science: Land/Atmosphere/Ice Interactions*. ARCUS, Fairbanks, AK, 48 pp.

McCloy K.R. and Lucht W. (2004). Comparative evaluation of seasonal patterns in long time series of satellite image data and simulations of a global vegatation model. *IEEE Transactions on Geoscience and Remote Sensing*, **42**(1), 140–153.

Mcintyre A.D. (1999). The environment and the oil companies. *Marine Pollution Bulletin*, **3**(3), 155–156

Meaden G.T. (1985) *A Study of Tornadoes in Britain* TORRO HQ, Trowbridge, U.K. 131 pp.

Meadows D.H., Meadows D.L., Ranger J., and Behrens W.W. (1972). *The Limits to Growth*. N.Y. University Books, New York, 208 pp.

MEAP (2007). *Millenium Ecosystem Assessment*. Island Press, Washington, D.C., 25 pp.

Melentyev V.V., Jochannessen O.M., Kondratyev K.Ya., Bobilev L.P., and Tichomirov A.I. (1998). Satellite-based radiolocation diagnostics of ice lake cover: Ecology and history. *Research of the Earth from Space*, **2**, 91–101 [in Russian].

Meriakri V.V. (1992). Spectroscopy of the millimeter and sub-millimeter range. *Bulletin of the Moscow State University: Physics and Astronomy*, **33**(3), 86–94 [in Russian].

Mertens E.W. and Gould J.R. (1979). The effects of oil on marine life. *Erdöl und Köhle Erdgas-Petroleum*, **32**, 162–166.

Metternicht G.I. (1998). Fuzzy classification of JERS-1 SAR data: An evaluation of its performance for soil salinity mapping. *Ecological Modelling*, **111**(1), 61–74.

Milford C., Hargreaves K.J., Sutton M.A., Loubet B., and Cellier P. (2001). Fluxes of $CO_2$ over upland moorland in the vicinity of agricultural land. *Journal of Geophysical Research*, **106**(D20), 24169–24181.

Milne A. (2004). *Doomsday: The Science of Catastrophic Events*. Praeger, Westport, CT, 194 pp.

Mintzer I.M. (1987). *A Matter of Degrees: The Potential for Controlling the Greenhouse Effect* (Research Report No. 15). World Resources Institute, Washington, D.C., 70 pp.

Mirkin B.M. (1986). *Vegetation Communities: What Are They?*. Science Publishing House, Moscow, 160 pp. [in Russian].

Mitnik L.M. (1977). *Physical Foundations of Environment Remote Sensing*. Politechnical Institute Press, Leningrad, U.S.S.R., 56 pp. [in Russian].

Mkrtchyan F.A. (1982). *Optimal Distinction of Signals and Monitoring Problems*. Science Publishing House, Moscow, 186 pp. [in Russian].

Mkrtchyan F.A., Krapivin V.F., Kovalev V.I., Klimov V.V., Rukovishnikov A.I., and Golovachev S.P. (2004). An adaptive spectroellipsometric technology for the ecological monitoring of the aquatic environment. *Proceedings of 25th ACRS, Chiang-Mai, Thailand*. AARS, Bangkok, pp. 13–15.

Moir J.W.B. (Ed.) (2011). *Nitrogen Cycling in Bacteria: Molecular Analysis*. Caister Academic Press, Norwich, U.K., 270 pp.

Monin A.S. and Krasitskyi V.P. (1985). *Phenomena on the Ocean Surface*. Hydrometeoizdat, Leningrad, U.S.S.R., 375 pp. [in Russian].

Morgan J. and Codispoti L. (Eds.) (1995). *Department of Defense Arctic Nuclear Waste Assessment Program: FYs 1993–1994* (ONR 322-95-5). Office of Naval Research, Arlington, VA, pp. 15–30.

Moroshkin K.V. (1964). A new scheme of Okhotsk Sea surface currents. *Oceanology (Moscow)*, **4**(4), 641–643 [in Russian].

Muller R. and Peter T. (1992). The numerical modeling of the sedimentation of polar stratospheric cloud particles. *Ber. Bunsenges phys. Chem.*, **96**, 353–361.

Myerson R.B. (1997). *Game Theory: Analysis of Conflicts*. Harvard University Press, Cambridge, MA, 600 pp.

Nazaretyan A.P. (2004). Anthropogenic crises: Hypothesis of technological–humanitarian balance. *Herald of Russian Academy of Sciences*, **74**(4), 319–330 [in Russian].

Nefedova E.I. (1994). Mathematical modeling of the global carbon cycle in the atmosphere–ocean system. Ph.D. thesis, Computing Center RAS, Moscow, 112 pp. [in Russian].

Nefedova E.I. and Tarko A.M. (1993). Study of the global carbon cycle using the zonal model in the atmosphere–ocean system. *Proc. of Russian Acad. Sci.*, **333**(5), 645–647 [in Russian].

Nelepo B.A., Korotaev G.K., Suetin V.S., and Terechin Yu.V. (1985). *Investigations of Ocean from Space*. Naukova Dumka, Kiev, the Ukraine, 185 pp. [in Russian].

Nguyen Xuan Man, Krapivin V.F., Mkrtchyan F.A., and Potapov I.I. (2011a). GIMS technology in remote monitoring of forest ecosystems. *Problems of the Environment and Natural Resources*, **3**, 3–13 [in Russian].

Nguyen Xuan Man, Krapivin V.F., Mkrtchyan F.A., Klimov V.V., Kovalev V.I., Rukovishnikov A.I., and Soldatov V.Yu. (2011b). Measurements and identification of water solutions based on the spectroellipsometer. *Problems of the Environment and Natural Resources*, **2**, 3–22 [in Russian].

Nishida K., Nemani R.R., Glassy J.M., and Running S.W. (2003). Development of an evapotranspiration index from Aqua/MODIS for monitoring surface moisture status. *IEEE Transactions on Geoscience and Remote Sensing*, **41**(2), 493–501.

Nitu C., Krapivin V.F., and Bruno A. (2000a). *Intelligent Techniques in Ecology*. Printech, Bucharest, 150 pp.

Nitu C., Krapivin V.F., and Bruno A. (2000b). *System Modelling in Ecology*. Printech, Bucharest, 260 pp.

Nitu C., Krapivin V.F., and Pruteanu E. (2004). *Ecoinformatics: Intelligent Systems in Ecology*. Magic Print, Onesti, Bucharest, 411 pp.

Njoki E.G. and O'Neil P.E. (1982). Multifrequency microwave radiometer measurements of soil moisture. *IEEE Transactions on Geoscience and Remote Sensing*, **GE-20**(4), 468–475.

Norris K. and McCulloch N. (2003). Demographic models and the management of endangered species: A case study of the critically endangered Seychelles magpie robin. *Journal of Applied Ecology*, **40**(5), 890–899.

Osterberg C.L. (1985). *Nuclear War and the Ocean*. Wiley, New York, 467 pp.

Papakyriakou T.N. and McCaughey J.H. (1991). An evaluation of evapotranspiration for a mixed forest. *Canadian Journal of Forestry Research*, **21**(11), 1622–1631

Park S.U., In H.J., and Lee Y.H. (1999). Parametrization of wet deposition of sulfate by precipitation rate. *Atmospheric Environment*, **33**(27), 4469–4475

Payne J.R., McNabb G.D., and Clayton J.R. (1991). Oil-weathering behavior in Arctic environments. *Polar Research*, **10**(2), 631–662.

Peng C. (2000). From static biogeographical model to dynamic global vegetation model: A global perspective on modelling vegetation dynamics. *Ecological Modelling*, **135**(1), 33–54

Peng C., Jiang H., Apps M.J., and Zhang Y. (2002). Effects of harvesting regimes on carbon and nitrogen dynamics of boreal forests in central Canada: A process model simulation. *Ecological Modelling*, **155**(2/3), 177–189.

Penman H.L. (1948). Natural evaporation from open water, bare soil and grass. *Proceedings of the Royal Society (London)*, **193**, 120–145.

Perrie W., Zhang W., Ren X., and Long Z. (2004). The role of midlatitude storms on air–sea exchange of $CO_2$. *Geophysical Research Letters*, **31**(L09306), doi: 10.1029/2003GL019212, 1–4.

Pervaniuk V.S. (2001). A spatial numerical model of the global biogeochemical cycles of carbon and nitrogen in the atmosphere–ocean system. Ph.D. thesis, Computing Center RAS, Moscow, 118 pp. [in Russsian].

Petty G.W. (1995) The status of satellite-based rainfall estimation over land. *Remote Sensing of Environment*, **51**(1), 125–137.

Phillips G.W., August R.A., Cherepenin V.A., Harper M.J., King S.E., Krapivin V.F., Pautkin A.Yu., and Tsang F.Y. (1997). Radionuclear pollutants in the Angara and Yenisey rivers of Siberia. *Radioprotection-Colloques*, **32**, 299–304.

Phillips G.W., Nitu C., Chukhlantsev A.A., Golovachev S.P., Krapivin V.F., and Shutko A.M. (2009). A new technology of vegetation microwave monitoring. *Scientific Bulletin of Electrical Engineering Faculty, Valahia University of Targoviste, Romania*, **2**(11), 95–98.

Pielke R.A., Sr. (2001a). Carbon sequestration: The need for an integrated climate system approach. *Bulletin of the American Meteorological Society*, **82**(11), 20–21.

Pielke R.A., Sr. (2001b). Earth system modeling: An integrated assessment tool for environmental studies. In: T. Matsuno and H. Kida (Eds.), *Present and Future of Modeling Global Environmental Change: Toward Integrated Modeling*. Terra Science Publishing, Tokyo, pp. 311–337.

Pielke R.A., Sr. (2002). Overlooked issues in the U.S. national climate and IPCC assessments. *Climate Change*, **52**(1/2), 1–11.

Plotnikov V.V. (1996). Long-term prognosis of Okhotsk Sea ice conditions with large-scale atmospheric processes taken into account. *Meteorology and Hydrology*, **12**, 93–100 [in Russian].

Podlazov A.V. (2001). Self-organizing criticism and risk analysis. *Proceedings of High Schools: Applied Nonlinear Dynamics*, **9**(1), 49–88 [in Russian].

Polishchuk Yu.M. (1992). *Simulation Modelling of Systems with Natural Components*. Science Publishing House, Novosibirsk, Russia, 228 pp. [in Russian].

Pollak R.A. (1990). Two-sex demographic models. *Journal of Political Economy*, **98**(2), 399–420.

Porte A. and Bartelink H.H. (2002). Modelling mixed forest growth: A review of models for forest management. *Ecological Modelling*, **150**(1/2), 141–188.

Power H.C. (2000). Estimating atmospheric turbidity from climate data. *Ecological Modelling*, **135**(1), 125–134

Preller R.H. and Cheng A. (1999). Modeling the transport of radioactive contaminants in the Arctic. *Marine Pollution Bulletin*, **38**(2), 71–91.

Prieur L. and Legendre L. (1988). Oceanographic criteria for new phytoplankton production. In: B.J. Rothschild (Ed.), *Toward a Theory on Biological–Physical Interactions in the World Ocean*. Kluwer Academic, Quebec, Canada, pp. 71–112.

Purkis S.J. and Klemas V.V. (2011). *Remote Sensing and Global Environmental Change*. Wiley/Blackwell, Chichester, U.K., 367 pp.

Qi Y., Xu M., and Wu J. (2002). Temperature sensitivity of soil respiration and its effects on ecosystem carbon budget: Nonlinearity begets surprises. *Ecological Modelling*, **153**(1/2), 131–142.

Radakov D.V. (1972). *Shoaling of Fish as an Ecological Phenomenon*. Science Publishing House, Moscow, 174 pp. [in Russian].

Randall D.A., Wood R.A., Bony S., Colman R., Fichefet T., Fyfe J., Kattsov V., Pitman A., Shukla J., Srinivasan J. *et al.* (2007). Cilmate models and their evaluation. In: S. Solomon, D. Qin, M. Manning, Z. Chen, M. Marquis, K.B. Averyt, M.Tignor, and H.L. Miller (Eds.), *Climate Change 2007: The Physical Science Basis* (contribution of Working Group I to the Fourth Assessment Report of the Intergovernmental Panel on Climate Change). Cambridge University Press, Cambridge, U.K., 1,008 pp.

Rastetter E.B.; Aber J.D.; Peters D.P.C.; Ojima D.S.; and Burke I.C. (2003). Using mechanistic models to scale ecological processes across space and time. *BioScience*, **53**(1), 1–9.

Ratkovich D.Ya. (1997). Water balance of the Caspian Sea and its level regime. *Power Technology and Engineering*, **31**(6), 355–363.

Reutov E.A. and Shutko A.M. (1986). Prior-knowlege-based soil-moisture determination by microwave radiometry. *Soviet Journal of Remote Sensing*, **5**(1), 100–125 [in Russian].

Riedlinger S.H. and Preller R.H. (1991). The development of a coupled ice–ocean model for forecasting ice conditions in the Arctic. *Journal of Geophysical Research*, **96**, 16955–16977.

Riedo M., Gyalistras D., and Fuhrer J. (2000). Net primary production and carbon stocks in differently managed grasslands: Simulation of site-specific sensitivity to climate change. *Ecological Modelling*, **134**(2/3), 207–227.

Ries J.E. and Hibler W.D. (1989). Interannual characteristics of an 80 km resolution diagnostic Arctic ice–ocean model. *Ann. Glaciol.*, **15**, 155–162.

Rochon G.L., Krapivin V.F., Watson M., Fauria S., Tsang F.Y., and Fernandez M. (1996). Remote characterization of the land–sea interface: A case study of the Vietnam/South China Sea coastal zone. *Proceedings of Second Hô Chi Minh City Conference on Mechanics, September 24–25, 1996*. Institute of Applied Mechanics, Hô Chi Minh City, pp. 72–74.

Rovinsky F.Ya. Chernogaeva G.M., and Paramonov S.G. (1995). The role played by river flow and atmospheric transport in the pollution of Russian northern seas. *Meteorology and Hydrology (Moscow)*, **9**, 22–29 [in Russian].

Safai P.D., Rao P.S.P., Momin G.A., All K., Chate D.M., and Praveen P.S. (2004). Chemical composition of precipitation during 1984–2002 at Pune, India. *Atmospheric Environment*, **38**(12), 1705–1714.

Salokhiddinnov A.T. and Khakimov Z.M. (2004). Ways the Aral Sea behaves. *Journal of Marine Systems*, **47**, 127–136.

Sarkisyan A.S. (2000). Synthesis of observational data and modeling results as a perspective direction for the investigation of oceans, seas, and lakes. *Physics of Atmosphere and Ocean (Moscow)*, **2**, 202–210 [in Russian].

Sasaki Y., Asanuma I., Muneyama K., Naito G., and Suzuki T. (1987a). A simplified microwave model and its application to the determination of some oceanic environmental parametric values, using multichannel microwave radiometric observations. *IEEE Transactions on Geoscience and Remote Sensing*, **GE-25**(3), 384–392.

Sasaki Y., Asanuma I., Muneyama K., Naito G., and Suzuki T. (1987b). The dependence of sea-surface microwave emission on wind speed, frequency, incidence angle, and polarization over the frequency range from 1 to 40 GHz. *IEEE Transactions on Geoscience and Remote Sensing*, **GE-25**(2), 138–146.

Sasaki Y., Asanuma I., Muneyama K., Naito G., and Suzuki T. (1988). Microwave emission and reflection from the wind-roughened sea surface at 6.7 GHz. *IEEE Transactions on Geoscience and Remote Sensing*, **26**(6), 860–868.

Sasaki Y., Naito G., and Wilheit T.T. (1989). Rainfall rate dependence of brightness temperature of microwave emission from rain. *Proceedings of Annual Spring Meeting of Oceanographical Society of Japan, Tokyo, April 5–9, 1989*. Oceonographic Society of Japan, Tokyo, pp. 65–66.

Savin A.I. and Bondur V.G. (2000). Scientific principles for the creation and diversification of global aerospace systems. *Optics of the Atmosphere and Ocean (Moscow)*, **1**, 46–63 [in Russian].

Savinikh V.P., Krapivin V.F., and Potapov I.I. (2007). *Information Technologies in Ecological Monitoring Systems*. Geodezkartizdat, Moscow, 388 pp. [in Russian].

Savorskij V.P. (1992). Spatiotemporal structure of the radiothermal field in the atmosphere–ocean system. PhD thesis, Institute of Radioengineering and Electronics, Russian Academy of Sciences, Moscow, 199 pp. [in Russian].

Schaum J.L., Segna J.J., Yong J.S., Benes C.M., and Muir W.R. (1990). Computerized system for performing risk assessment for chemical constituents of hazardous waste. In: J.M. Huston (Ed.), *Expert Systems for Environmental Applications*. American Chemical Society, Washington, D.C., pp. 176–193.

Schimel D.S. (1995). Terrestrial biogeochemical cycles: Global estimates with remote sensing. *Remote Sensing of Environment*, **51**(1), 49–56.

Schmugge T. (1990). Measurements of surface soil moisture and temperature. In: R.J. Hobbs and H.A. Mooney (Eds.), *Remote Sensing of Biosphere Functioning*. Springer-Verlag, New York, pp. 31–62.

Sellers D. (1983). Quasi-three-dimensional climatic model. *J. Clim. and Appl. Meteorol.*, **22**, 1557–1574.

Sellers P.J. (1985). Canopy reflectance, photosynthesis and transpiration. *International Journal of Remote Sensing*, **6**, 1335–1372.

Sellers P.J. (1987). Canopy reflectance, photosynthesis and transpiration, II: The role of biophysics in the linearity of their interdependence. *Remote Sensing of Environment*, **21**, 143–183.

Sellers P.J., Meeson B.W., Hall F.G., Asrar G., Murphy R.E., Schiffer R.A., Bretherton F.P., Dickinson R.E., Ellingson R.G., Field C.B. *et al.* (1995). Remote sensing of the land surface for studies of global change: Models—algorithms—experiments. *Remote Sensing of Environment*, **51**(1), 3–26.

Sellers P.J., Bounoua L., Collatz G.L., Randall D.A., Dazlich D.A., Los S.O., Berry J.A., Fung I., Tucker C.J., Field C.B. *et al.* (1996). Comparison of radiative and physiological effects of doubled atmospheric $CO_2$ on climate. *Science*, **271**, 1402–1406.

Sellers P.J., Dickinson R.E., Randall D.A., Betts A.K., Hall F.G., Berry J.A., Collatz G.J., Denning A.S., Mooney H.A., Nobre C.A. *et al.* (1997). Modeling the exchanges of energy, water, and carbon between continents and the atmosphere. *Science*, **275**, 502–509.

Sergin S.Ya. (Ed.) (1974). *Study of the Glaciers–Ocean–Atmosphere System*. U.S.S.R. Far East Sci. Centre, Vladivostok, Russia, 138 pp. [in Russian].

Shahramantyan M.A. (2003). *New Information Technologies in Tasks of the Provision of National Safety of Russia (Natural–Technological Aspects)*. Russian Ministry for Excessive Situations, Moscow, 358 pp. [in Russian].

Shapovalov S.M. (Ed.) (2010). *Physical, Geological, and Biological Researches of Oceans and Seas*. Scientific World, Moscow, 628 pp. [in Russian].

Shinohara Y. and Shikama N. (1988) (Eds.). *Marine Climatological Atlas of the Sea of Okhotsk* (Technical Report No 23). Meteorological Research Institute, Hokkaido, Japan, 57 pp.

Shultze E.-D., Heimann M., Harrison S., Holland E., Lloyd J., Prentice I.C., and Schimel D.S. (2001). *Global Biogeochemical Cycles in the Climate System*. Academic Press, New York, 350 pp.

Shutko A.M. (1986). *Microwave Radiometry of Water Surfaces and Ground Surfaces*. Science Publishing House, Moscow, 204 pp. [in Russian].

Shutko A.M. (1987). Remote sensing of waters and lands via microwave radiometry (the principles of method, problems feasible for solving, economic use). *Pontificiae Academiae Scientiarum Scripta Varia*, **68**, 413–441.

Shutko A.M. (1992). Soil–vegetation characteristics at microwave wavelengths. In: P. Mather (Ed.), *Understanding the Terrestrial Environment: The Role of the Earth Observations from Space*. Taylor & Francis, London, pp. 53–66.

Shutko A.M. (1997). Remote sensing of soil moisture and moisture related parameters by means of microwave radiometry. In: S. Sorooshian, H. Gupta, and J. Rodda (Eds.), *Global Environmental Change and Land Surface Processes in Hydrology: The Trials and Tribulations of Modeling and Measuring* (NATO ASI Series, Serie I: Global Environmental Changes, vol. 46). NATO, pp. 263–273.

Shutko A.M. and Krapivin V.F (Eds.) (2011). *Operational Diagnostics, Estimation of the Scale of Damage and Aftermath Reduction of Stressful Natural Processes*. Professor Marin Drinov Academic Publishing House, Sofia, 287 pp. [in Russian].

Shutko A.M., Krapivin V.F., Mkrtchyan F.A., Reutov E.A., Novichkhin E.P., Leonidov V.A., Mishanin V.G., and Tsankov N.S. (1994). Econo-ecological estimates of the effectiveness of utilizing remotely sensed data and GIS information for soil moisture and moisture related parameters determination (Geoinformation Monitoring System Approach—GIMS). *Proceedings of ICID, May 7–9, 1994, Varna, Russia* , pp. 185–189.

Shutko A.M., Reutov E.A., and Golovachev S.P. (1995). Estimation of soil moisture profiles and root zone moisture content by means of microwave radiometry and *a priori* information. In: B. Choudhury, Y. Kerr, E. Njoku, and P. Pampaloni (Eds.), *Proceedings of the ESA/NASA International Workshop on Passive Microwave Remote Sensing of Land–Atmosphere Interaction, January 11–15, 1993, Saint-Larry, France*. Utrecht Publishing, The Netherlands, pp. 461–474.

Shutko A.M., Haldin A., Krapivin V., Novichikhin E., Sidorov I., Tishchenko Yu., Haarbrink R., Georgiev G., Kancheva R., Nikolov H. *et al.* (2007). Microwave radiometry in monitoring and emergency mapping of water seepage and dangerously high groundwaters. *Journal of Telecommunications and Information Technology*, **1**, 76–82.

Shutko A.M., Krapivin V.F., Haarbrink R.B., Sidorov I.A., Novichikhin E.P., Archer F., and Krisilov A.D. (2010). *Practical Microwave Radiometric Risk Assessment*. Professor Marin Drinov Academic Publishing House, Sofia, 88 pp.

Sims D.A. and Gamon J.A. (2003). Estimation of vegetation water content and photosynthetic tissue area from spectral reflectance: A comparison of indices based on liquid water and chrolophyll absorption features. *Remote Sensing of Environment*, **84**(4), 526–537.

Singer S.F. (2008). *Nature, Not Human Activity, Rules the Climate* (report of the Nongovernmental International Panel on Climate Change). Heartland Institute, Chicago, IL, 30 pp.

Smirnova I.P., Subbotin V.V., and Letunovsky V.A. (1988). Monitoring of the Arctic thermal regime. In: A.I. Voskresensky (Ed.), *Monitoring of the Arctic Climate*. Gidrometeoizdat, Leningrad, U.S.S.R., pp. 24–32.

Soldatov V.Yu. (2006). Microwave diagnostics of the ocean–atmosphere system. *Problems of the Environment and Natural Resources*, **12**, 28–35 [in Russian].

Soldatov V.Yu. (2007). Diagnostics of the ocean–atmosphere system by means of percolation model. *Problems of the Environment and Natural Resources*, **5**, 52–64 [in Russian].

Soldatov V.Yu. (2009a). An information technology for the operative diagnostics of the ocean–atmosphere system. *Proceedings of the International Symposium on Mathematical Modelling of Process of Tropical Hurricane Beginning and Searching for Its Indicator Precursors, December 16, 2009, Hô Chi Minh City, Vietnam*. Institute of Applied Mechanics, Hô Chi Minh City, pp. 65–85.

Soldatov V.Yu. (2009b). Possibilities of monitoring systems to detect tropical hurricanes. *Proceedings of the International Symposium on Mathematical Modelling of Process of Tropical Hurricane Beginning and Searching for Its Indicator Precursors, December 16, 2009, Hô Chi Minh City, Vietnam*, pp. 41–45.

Soldatov V.Yu. (2010). Multi-functional information-modeling system for hydrophysical experiments. *Proceedings of the Ninth International Symposium on Ecoinformatics Problems, December 9–11, 2010, Moscow*, pp. 96–100 [in Russian].

Soldatov V.Yu. (2011). Multi-functional information-modeling system for hydrophysical experiments. PhD thesis, IREE, Moscow, 145 pp. [in Russian].

Soldatov V.Yu., Krapivin V.F., Borodin L.F., and Startsev A.A. (2008). Direct and inverse problems of the microwave monitoring of the environment. *Proceedings of the Eighth International Symposium "Ecoinformatics Problems", December 16–17, 2008, Moscow*. The Moscow Sciences Engineering A.S. Popov Society for Radio, Electronics and Communication, Moscow, pp. 19–25.

Soldatov V.Yu., Nitu C., and Krapivin V.F. (2010). Diagnosis of transition processes in the ocean–atmosphere system. *Proceedings of the International Conference DAS (Data Application System), May 27–29, 2010, Suceava, Romania*. The Stefan cel Mare University of Suceava, Romania, pp. 36–39.

Somes N.L.G. (1999) Numerical simulation of wet land hydrodynamics. *Environment International*, **25**(6/7), 773–779.

Sorokin Yu.I. (1977). Characteristics of the Peruvian upwelling ecosystem. *Proc. of the Russian Academy of Sci.*, **236**(2), 497–500 [in Russian].

Sorokin Yu.I. and Mikheev V.N. (1979). On characteristics of the Peruvian upwelling ecosystem. *Hydrobiologia*, **62**, 165–189.

Spilsbury L. and Spilsbury R. (2008). *Plant Classification (The Life of Plants)*. Heinemann-Raintree, Oxford, U.K., 48 pp.

Sportisse B. (2000). Box models versus Eulerian models in air pollution modeling. *Atmospheric Environment*, **35**(1), 173–178.

Starke L. (Ed.) (2004). *State of the World—2004: Progress towards a Sustainable Society*. Earthscan, London, 246 pp.

Sternberg L. and De Angells D. (2002). Carbon isotope composition of ambient $CO_2$ and recycling: A matrix model. *Ecological Modelling*, **154**(1/2), 179–192.

Stoddard J.L., Kahl J.S., Deviney F.A., DeWalle D.R., Driscoll C.T., Herlihy A.T., Kellogg J.H., Mudoch P.S., Webb J.R., and Webster K.E. (2003). *Response of Surface Water Chemistry to the Clean Air Act Amendments of 1990* (EPA/620/R-03/001). U.S. Environmental Protection Agency, Corvallis, OR, 92 pp.

Straub C.P. (Ed.) (1989). *Practical Handbook of Environmental Control*. CRC Press, Boca Raton, FL, 537 pp.

Streets D.G., Jiang K., Hu X., Sinton J.E., Zhang Z.-Q., Xu D., Jacobson M.Z., and Hansen H.E. (2001). Recent reductions in China's greenhouse gas emissions. *Science*, **294**(5548), 1835–1837.

Strelkov G.M. (1995). Possibility of determining the content of admixed gases CO and $N_2O$ using an active sounding method of the atmosphere in the millimeter range. *Research of the Earth fron Space*, **4**, 3–7 [in Russian].

Strelkov G.M. (1996). Active sounding of water vapor within the mesopause with the use of millimeter and sub-millimeter waves. *Research of the Earth from Space*, **14**, 169–177 [in Russian].

Strelkov G.M. and Soldatova I.V. (1989). Radio-impulse propagation within the line of oxygen absorption. *Radiotechnics (Moscow)*, **3**, 16–17 [in Russian].

Sud Y.C., Walker G.K., Kim J.-H., Liston G.E., Sellers P.J., and Lau W.K.-M. (1996). Biogeophysical consequences of a tropical deforestation scenario: A GCM simulation study. *Journal of Climate*, **9**(12), 3225–3247.

Sukov A.I. and Soldatov V.Yu. (2007) Multi-channel sequential procedure of statistical decision making in the detection task of tropical hurricanes. *Proceedings of the A.S. Popov Russian Society for Radiotechnics, Electronics, and Communication. Series: "Engineering Ecology"*, **4**, 35–41.

Sukov A.I., Soldatov V.Yu., Krapivin V.F., Cracknell A.P., and Varotsos C.A. (2008). A sequential analysis method for the prediction of tropical hurricanes. *International Journal of Remote Sensing*, **29**(9), 2787–2798.

Suzuki A. (1992). Results of collecting fishes: Trophic to temperate migrant fishes off the Okhotsk Sea coast during 1988 to 1991 in northern Hokkaido, Japan. *Proceedings of the Seventh International Symposium on Okhotsk Sea and Sea Ice, February 2–5, 1992, Mombetsu, Japan*. Okhotsk Sea & Cold Ocean Research Association, Mombetsu, Japan, pp. 225–231.

Svirezhev Yu.M. (1987). *Ecological and Demographic Consequences of a Nuclear War*. Akademie-Verlag, Berlin, 112 pp.

Svirezhev Yu.M. (2002). Simple spatially distributed model of the global carbon cycle and its dynamic properties. *Ecological Modelling*, **155**(1), 53–69.

Tait E.J. (1987). Research policy and review, 14: Environmental issues and the social sciences. *Environment and Planning*, **19**(4), 437–445.

Tanaka N. (2010). *World Energy Outlook 2010*. International Energy Agency, Beijing, 36 pp.

Tarko A.M. (2001). Investigation of global biosphere processes with the aid of a global spatial carbon dioxide cycle model. *Proceedings of the Sixth International Carbon Dioxide Conference, October 1–2, 2001* (Extended Abstracts). Tohoku University, Sendai, Japan, Vol. 2., pp. 899–902.

Tarko A.M. (2003). *Analysis of Global and Regional Changes in Biogeochemical Carbon Cycle: A Spatially Distributed Model* (Interim Report, IR-03-041). International Institute for Applied Systems Analysis, Laxenburg, Austria, 28 pp.

Tarko A.M. (2005). *Mathematical Modeling of Anthropogenic Changes in Global Biospheric Processes*. Physics-Mathematics Publishing, Moscow, 278 pp. [in Russian].

Terziev F.S. (Ed.) (1992). *Hydrometeorology and Hydrochemistry of U.S.S.R. Seas, Vol. 1: Barents Sea*. Gidrometeoizdat, St. Petersburg, Russia, 182 pp. [in Russian].

Terziev F.S., Zatuchnoy B.M., and Gershanovitch D.E. (1993). *The Okhotsk Sea*. Gidrometeopress, St. Petersburg, Russia, 167 pp. [in Russian].

Tianhong L., Yanxin S., and An X. (2003). Integration of large scale fertilizing models with GIS using minimum unit. *Environmental Modelling*, **18**(3), 221–229.

Timoshevskii A., Yeremin V., and Kalkuta S. (2003). New method for ecological minitoring based on the method of self-organising mathematical models. *Ecological Modelling*, **162**(2/3), 1–13.

Trites A.W., Livingston P.A., Mackinson S., Vasconcellos M.C., Springer A.M., and Pauly D. (1999). *Ecosystem Change and the Decline of Marine Mammals in the Eastern Bering Sea*. Fisheries Center Res. Rep., Fairbanks, AK, 100 pp.

Trudinger P.A. (Ed.) (1980). Biogeochemistry of ancient and modern environments. *Proceedings of the Fourth International Symposium, September 26–August 4, 1979, Canberra*. Springer-Verlag, Berlin, 723 pp.

Truskov P.A., Astafiev V.N., and Surkov G.A. (1992). Problems of choice of sea ice cover parameters design criteria. *Proceedings of the Seventh International Symposium on*

*Okhotsk Sea and Sea Ice, February 2–5, 1992, Mombetsu, Japan*. Okhotsk Sea & Cold Ocean Research Association, Mombetsu, Japan, pp. 21–25.

Tsegaye T.D., Metzl R., Wang X., Schamschula M., Tadesse W., Clendenon D., Golson K., Coleman T.L., Archer F., and Schaefer G. (2005). Near real time variability of soil moisture and temperature under different land use and cover: The Alabama Mesonet. *Journal of Environmental Hydrology*, **13**, 1–13.

Turekian K.K. (1996). *Global Environmental Change: Past, Present, and Future*. Prentice Hall, New Haven, CT, 200 pp.

Ugolnitsky G.A. (1999). *Management of Ecologo-economic Systems*. High School Book Publishing, Moscow, 132 pp. [in Russian].

USEPA (1983). *Methods for the Chemical Analysis of Water and Wastes* (USEPA-600/4-79-020). U.S. Environmental Protection Agency, Washington, D.C.

Valette-Silver N.J., Hameedi M.J., Efurd D.W., and Robertson A. (1999). Status of the contamination in sediments and biota from the western Beaufort Sea (Alaska). *Marine Pollution Bulletin*, **38**(8), 702–722.

Vasilevich V.I. (1983). *Essays on Theoretical Phytocenology*. Science Publishing House, Leningrad, U.S.S.R., 226 pp. [in Russian].

Vernadsky V.I. (1944). A few words about the noosphere. *Successes of Modern Biology*, **18**(2), 49–93 [in Russian].

Vinogradov M.E., Menshutkin V.V., and Shushkina E.A. (1972). On mathematical simulation of a pelagic ecosystem in tropical waters of the ocean. *Marine Biology*, **16**(4), 261–268.

Vinogradov M.E., Krapivin V.F., Fleishman B.S., and Shushkina E.A. (1975). Using a mathematical model to analyze the behavior of ecosystems in the oceanic pelagic zone. *Oceanology (Moscow)*, **10**(2), 313–320 [in Russian].

Vinogradov M.E., Shushkina E.A., and Kukina I.N. (1977). Structural and functional analysis of pelagic communities in equatorial upwelling. *Pol. arch. hydrobiol.*, **24**, 503–524.

Vinogradova T.O., Golovachev S.P., Grankov A.G., Mil′shin A.A. Priadko A.V., Chukhlantsev A.A., Shelobanova N.K., and Shutko A.M. (2004). Spectral measurements of microwave attenuation by crowns of a fir and chestnut. *Problems of the Environment and Natural Resources*, **1**, 69–76 [in Russian].

Wakatsuchi M. and Martin S. (1991). Water circulation in the Kuril Basin of the Okhotsk Sea and its relation to eddy formation. *Journal of the Oceanographical Society of Japan*, **47**, 152–168.

Walker G. (2003). *Snowball Earth: The Story of the Great Global Catastrophe that Spawned Life as We Know It*. Crown Publishers, New York, 269 pp.

Wang L.K., Shammas N.K., and Hung Y.-T. (Eds.) (2011). *Advanced Biological Treatment Processes*, Vol. 9. Humana Press, New York, 737 pp.

Wange G. and Archer D.J. (2003). Evaporation of groundwater from arid playas measured by C-band SAR. *IEEE Transactions on Geoscience and Remote Sensing*, **41**(7), 1641–1650.

Wania F., Hoff J.T., Jai C.Q. and Mackay D. (1998) The effects of snow and ice on the environmental behavior of hydrophobic organic chemicals. *Environmental Pollution*, **102**(1), 25–41.

Ward R.C., Loftis J.C., and McBride G.B. (1990). *Design of Water Quality Monitoring System*. Van Nostrand Reinhold, New York, 231 pp.

Watson R.T., Noble I.R., Bolin B., Ravindranath N.H., Verardo D.J., and Dokken D.J. (Eds.) (2000). *Land Use, Land-Use Change, and Forestry*. Cambridge University Press, Cambridge, U.K., 377 pp.

WB (2009). *World Development Report 2010: Development and Climate Change*. World Bank, New York, 424 pp.

Webster P.J., Holland G.J., Curry J.A., and Chang H.-R. (2005). Changes in tropical cyclone number, duration, and intensity in a warming environment. *Science*, **309**, 1844–1846.

Wielgolaski F.E. (Ed.) (1997). *Polar and Alpine Tundra.* Elsevier, New York, 930 pp.

Wilson M.F. and A. Henderson-Sellers (1985). A global archive of land cover and soils data for use in general circulation climate models. *Journal of Climate*, **5**, 119–143.

Wirtz K.W. (2000). Second order up-scaling: Theory and an exercise with a complex photosynthesis model. *Ecological Modelling*, **126**(1), 59–71.

Yakimov S.P. (1996). Algorithms to assess the risk of forest fire by remote-sensing data. PhD thesis, State Technical University, Krasnoyarsk, Russia, 155 pp. [in Russian].

Yakovlev O.I. (2001). *Space Radio Science.* Taylor & Francis, London, 320 pp.

Yakovlev O.I., Paveliev A.G., and Matyugov S.S. (2009). *Satellite Monitoring of the Earth.* Book House Publishing, Moscow, 206 pp. [in Russian].

Yan Q. and Blum R.S. (2001). Distributed signal detection under the Neyman–Pearson criterion. *IEEE Transactions on Information Theory*, **47**(4), 1368–1377.

Yang X., Wang M., Huang Y., and Wang Y. (2002). A one-compartment model to study soil carbon decomposition rate at equilibrium situation. *Ecological Modelling*, **151**(1), 63–73.

Yasanov N.A. (2003). Climate of the Phanerozoic and the greenhouse effect. *Herald of the Moscow State University, Ser. 4: Geology*, **6**, 3–11 [in Russian].

Yemshanov D. and Perera A.H. (2002). A spatially explicit stochastic model to simulate boreal forest cover transitions: General structure and properties. *Ecological Modelling*, **150**(1/2), 189–209.

Yokozawa M. (1998). Effects of competition mode on spatial pattern dynamics in plant communities. *Ecological Modelling*, **106**(1), 1–16

Yoshioka H. (2004). Vegetation isoline equations for an atmosphere–canopy–soil system. *IEEE Transactions on Geoscience and Remote Sensing*, **42**(1), 166–175.

Yude P., Birdsey R., Fang J., Houghton R., Kauppi P., Kurz W.A., Phillips O.L., Shvidenko A., Lewis S.L., Canadell J.G. *et al.* (2011). A large and persistent carbon sink in the world's forests. *Science*, **333**, 988.

Zagorin G.K. (1999). Polarizational characteristics (Stock's parameters) of natural and scattered microwave radiation in rain. PhD thesis, Institute of Radioengineering and Electronics, Russian Academy of Sciences, Moscow, 178 pp. [in Russian].

Zagorin G.K. and Kutuza B.G. (1998). Special features of the transportation of polarizational heat microwave radiation in clouds and precipitation. *Radiotechnics (Moscow)*, **10**, 566–575 [in Russian].

Zenkevich L.A. (1963). *Biology of the USSR Seas.* U.S.S.R. Acad. Sci. Press, Moscow, 739 pp. [in Russian].

Zhan X., Sohlberg R.A., Townshend J.R.G., DiMiceli C., Carroll M.L., Eastman J.C., Hansen M.C., and DeFries R.S. (2002). Detection of land cover changes using MODIS 250 m data. *Remote Sensing of Environment*, **83**(1/2), 336–350.

Zhou Xiuji, Zou Chengzhi, and Yang Peicai (1986). Global annual climate with cloud, vapor, and cause–effect connections of $CO_2$ taken into account. *Advances in Atmospheric Science*, **3**(3), 314–329.

Zonneveld C. (1998). A cell-based model of the chlorophyll a to carbon ratio in phytoplankton. *Ecological Modelling*, **113**(1/3), 55–70.

Zuev V.V. (2000). Destruction of the ozone layer: An ecological catastrophe or the greatest mystery of the century? *Proceedings of the Second Symposium on the Ecology of Siberian Rivers and the Arctic, Tomsk, November 24–26, 2000.* Tomsk University Press, Tomsk, Russia, pp. 36–40 [in Russian].

# Index



V.F. Krapivin and A.M. Shutko, *Information Technologies for Remote Monitoring of the Environment*, Springer Praxis Books, DOI 10.1007/978-3-642-20567-5, 

MIX
Papier aus verantwortungsvollen Quellen
Paper from responsible sources
FSC® C105338
FSC
www.fsc.org

If you have any concerns about our products,
you can contact us on
**ProductSafety@springernature.com**

In case Publisher is established outside the EU,
the EU authorized representative is:
**Springer Nature Customer Service Center GmbH**
**Europaplatz 3, 69115 Heidelberg, Germany**

Printed by Libri Plureos GmbH
in Hamburg, Germany